Advanced mathematical methods

Advanced mathematical methods

ADAM OSTASZEWSKI

Senior Lecturer in Mathematics,
London School of Economics and Political Science

CAMBRIDGE
UNIVERSITY PRESS

CAMBRIDGE UNIVERSITY PRESS
Cambridge, New York, Melbourne, Madrid, Cape Town, Singapore, São Paulo, Delhi

Cambridge University Press
The Edinburgh Building, Cambridge CB2 8RU, UK

Published in the United States of America by Cambridge University Press, New York

www.cambridge.org
Information on this title: www.cambridge.org/9780521247887

© Cambridge University Press 1990

First published 1990
Reprinted 1999

A catalogue record for this publication is available from the British Library

ISBN 978-0-521-24788-7 hardback
ISBN 978-0-521-28964-1 paperback

Transferred to digital printing 2009

Dedicated to my Parents
and to a fond memory

Contents

Preface   ix

Some notes and conventions   xii

I   Linear algebra   **1**
1   Vector spaces (revision)   1
2   Geometry in $\mathbb{R}^n$ 15
3   Matrices   30
4   Projections   48
5   Spectral theory   59
6   Reduction to upper triangular form   80
7   Reduction to tridiagonal form   92
8   Inverses   101
9   Convexity   113
10   The separating hyperplane theorem   122
11   Linear inequalities   135
12   Linear programming and game theory   143
13   The simplex method   162
14   Partial derivatives (revision)   178
15   Convex functions   184
16   Non-linear programming   195

II   Advanced calculus   **207**
17   The integration process   207
18   Manipulation of integrals   257
19   Multiple integrals   300
20   Differential and difference equations (revision)   347
21   Laplace transforms   380
22   Series solution of linear ordinary differential equations   406
23   Calculus of variations   423

III   Solutions to selected exercises   **443**
         Miscellany   522
         Appendix A   529
         Appendix B   533

         Index   541

Preface

This is a two-semester text covering techniques in linear algebra and multivariable calculus which every practitioner of mathematical sciences ought to know. It is intended as an advanced course for degrees that require a strong mathematical background. In keeping with its subject, the book adopts a how to do it approach. However, a broad explanatory perspective is offered to underpin this approach and so to skirt the proverbial pitfall of the cook-book method – a student all stuffed up with recipes and not knowing where to go when faced with a nonstandard problem.

So the aim here is two-fold: to equip students with the routine skills necessary in most modern applications, and to help develop a firm understanding of the underlying mathematical principles. Emphasis has therefore been placed equally on heart-of-the-matter, down-to-earth explanations, and on the presentation of an abundant stock of worked examples and exercises.

As regards the explanatory material, the guiding principle has been to keep the argument informal but nonetheless careful and accurate. This does justice to the mathematics without becoming obsessive over the kind of detail that a fully rigorous approach would require. The explanations are always concrete and geometric in nature. In fact, the geometric approach naturally leads to the adoption of linear algebra, the language of vectors, as a unifying frame for both parts of the course. A topic like non-linear programming then comes across as a fine exemplar of the interplay of the two areas.

Though unified, the two parts can each stand alone and serve as one-semester courses in, respectively, advanced linear algebra and advanced calculus.

A textbook's teaching strength ultimately lies in the exercise it poses for the reader's mind; here at worst, that is served in cornucopian helpings. All the problems are either hand-picked from the received literature or hand-crafted, and much mileage is expected of them. Some are plainly routine, others are challenging and rewarding; quite a few are accompanied by hints. The special attention accorded to the exercises is reflected in the inclusion of a very large number of complete solutions. These have often allowed the author's lectures to diverge from the set piece, responding to the needs of the moment and of the particular audience – like a performer's improvised coda.

The material, essentially in its entirety, forms a staple course for second and third year students at LSE reading for the mathematically based degrees: mathematics, statistics, actuarial science, computer science, operational research, management science, mathematical economics, econometrics. Topics required for such wide-ranging mathematical modelling of human activity are therefore extensive and include almost all the familiar techniques of classical applied mathematics. Notable exceptions to such syllabuses, and consequently not covered in the book, are: contour integrals (with the associated complex analysis), divergence and curl. On the other hand the material contained here steps outside the classical tradition by considering convexity, the separating hyperplane theorem, some game theory (all beloved of the Economist), and Riemann–Stieltjes integration (to allow Probabilists a simultaneous treatment of discrete and continuous random variables). However, tools and techniques are at the forefront; any applications to the social sciences are rare and indeed interspersed discreetly, so that traditional applied mathematicians and engineers will feel at home with this book.

The text is practically self-contained, but, not unnaturally for a second course, the topics included by way of revision are dealt with rapidly. Even so, such passages are garnished with some novelties and surprises to keep up the reader's attention and appetite. For example, the linear algebra preliminaries hint at infinite-dimensional spaces (as a preamble to the calculus of variations). In the theory of ordinary differential equation a novel approach is offered via a simple algebraic sum-theorem; this theorem allows difference equations and differential equations the 'exact same' treatment and is seen to be responsible for the analogies between the two topics.

The book's other claims to attention and originality (any humour apart) include: the development of an approach for 'mere mortals' to the Riemann–Stieltjes integral (usually a topic banished to theory books only) – appropriately kitted out with 'calculus-like' exercises; a very extensive discussion of multiple integrals (a topic more feared than revered by students); a geometric approach to the simplex method, based on oblique co-ordinate systems; a marshalling of exercises on convex sets and convex functions (usually a Cinderella affair in textbooks at this level) and likewise on non-linear programming.

However, in most other respects this text book, as any, owes much to its many nameless predecessors, which have all played a rôle in setting the received tradition; but there is a particular debt to the books suggested to LSE students for further reading, which often inspired drafts of lecture notes at various stages. These are:

Applied Linear Algebra, B. Noble, Prentice-Hall, 1969;
Applied Linear Algebra, R.A. Usmani, Dekker, 1987;
Mathematical Analysis, T.M. Apostol, Addison-Wesley, 1965;
Differential and Integral Calculus, S. Banach, PWN, 1957;
Advanced Calculus, M.R. Spiegel, McGraw-Hill, 1974;
Laplace Transforms, M.R. Spiegel, McGraw-Hill, 1965;
Differential Equations and the Calculus of Variations, L. Elsgolts, Mir, 1970;
Differential and Integral Calculus, G.M. Fichtenholz, Fizmatgiz, 1958.

Above all, the book owes a special debt to Ken Binmore, from whom the author inherited not only the pleasant duty of teaching the original methods course at LSE but also some notes – a valuable graft of teaching method. There are thus some exercises and at least one passage (cf. pages 326–327) shamelessly quoted verbatim from Binmore's *Calculus*, the antecendent text in the LSE series. That is, in part, testimony to series continuity and, in all truth, as well an intentional personal tribute.

Warmest thanks are due also to my depertmental colleagues for reading, or proof reading, parts of this book, especially to Mr. Adonis Demos, my teaching assistant, and to Dr. Graham Brightwell and Dr. Elizabeth Boardman. My last, but not least, thanks go to David Tranah of CUP for much patience and help in effecting the transition from lecture notes to book.

Adam Ostaszewski

Some notation and conventions

We follow standard notation and conventions such as those of K.G. Binmore's *Calculus*.

$\approx$ denotes an approximate equality.
$\bar{\lambda}$ denotes the complex conjugate of λ.
$\mathbb{R}$ denotes the set of real numbers.

For a, b real numbers $(a, b]$ denotes the interval of numbers x with $a < x \leqslant b$; similarly $[a, b)$ denotes the numbers x with $a \leqslant x < b$. (a, b) may ambiguously denote the point with co-ordinates a and b or the interval of numbers x with $a < x < b$.

A number b is an upper bound for a set $S \subseteq \mathbb{R}$ if for each $s \in S$ we have $s \leqslant b$. The number d is a lower bound for S if for each $s \in S$ we have $d \leqslant s$.

A set in the plane (in the space $\mathbb{R}^3$, etc) is 'closed' if the boundary of the set belongs to the set. On the other hand a set in the plane is 'open' if for each point of the set the whole of some disc centred at the point is contained entirely in the set (thus the set is open to small movements in all directions away from any point). Notice that if a set is closed it is *not* open since arbitrarily small movements in certain directions away from a point on the boundary lead out of the set.

(a_{ij}) denotes a matrix whose general entry in row i and column j is a_{ij}.

If $A = (a_{ij})$ then A^* denotes the matrix $(\bar{a}_{ji})$, its complex conjugate transpose.

A_{ij} denotes the appropriate minor of A – it is the determinant of the matrix obtained from A by deleting the i-th row and j-th column.
$[\mathbf{a}, |\mathbf{b}]$, $[\mathbf{a} : \mathbf{b}]$, $(\mathbf{a}|\mathbf{b})$, $[\mathbf{a}, \mathbf{b}]$ all denote a matrix with columns a and b. The notation

is supposed to draw the readers attention to the fact that $\mathbf{a}, \mathbf{b}, \ldots$ etc are columns in the displayed matrix.

$(A|\mathbf{b})$ denotes the matrix obtained by adding the column $\mathbf{b}$ to the matrix A. $T : A \to B$ means that T is a transformation with A as its domain of definition and taking values in B. The set of values which T assumes on A is denoted by $T(A)$. This is called the range or image of A under T.

It is also denoted by $R(T)$.

The transformation T is one-to-one (or, injective) if $Ta_1 = Ta_2$ implies that $a_1 = a_2$.

If S is a set of vectors lying in a vector space V, $\mathrm{Lin}\,(S)$ is the smallest vector subspace of V containing the vectors of S; it is called the linear span of S.

$\#(S)$ denotes the cardinality of the set S (i.e. how many elements it has).

If $x(t)$ is a function of t, $\dot{x}$ denotes the derived function.

$\Rightarrow$ means 'implies'

$\Leftrightarrow$ means 'implies and is implied by', i.e. 'if and only if'

$\exists x$ means 'there is an x such that'

n always denotes an integer

$f^{(n)}(x)$ denotes the n-th derivative of $f(x)$.

$f_x(x, y)$ denotes $(\partial f / \partial x)\,(x, y)$.

Whenever convenient we shall write

$$\int_c^d dy \int_a^b f(x, y)dx$$

to denote

$$\int_c^d \left\{ \int_a^b f(x, y)dx \right\} dy.$$

$E[X]$ denotes the expected value of the random variable X which is defined to be

$$\int_{-\infty}^{\infty} xf(x)dx,$$

where $f(x)$ is the probability density function of the random variable X. A 'function is integrable' will always mean that the appropriate integral of the function (as per context) exists. The Reader should be wary that $[\cdots]$ may denote just square bracketting or the staircase function (see p. 236).

I LINEAR ALGEBRA

1

Vector spaces (revision)

In both Part I and Part II we shall be working in the n-dimensional space $\mathbb{R}^n$. The current chapter concentrates on linear transformations and how to represent them most conveniently by matrices. The next chapter will consider the geometric properties of $\mathbb{R}^n$.

$\mathbb{R}^n$ is the mathematician's first generalisation of familiar Euclidean three space $\mathbb{R}^3$ whose points are described by the three co-ordinates (x_1, x_2, x_3). Realising that the handling of large amounts of data, that the description of the behaviour of a complicated machine (and more generally of a dynamical system) via, say, the readings on several of many dials (Fig. 1.1), all have in common the requirement of storing parameters quite like x_1, x_2, x_3 though rather more numerous than three, the mathematician introduces the collection of column n-vectors

$$\begin{bmatrix} x_1 \\ x_2 \\ \vdots \\ x_n \end{bmatrix} \qquad (1)$$

(alternatively written using the symbol for transposition as $\mathbf{x} = (x_1, \ldots, x_n)^t$).

Not content with this he introduces a more general notion, that of a *vector space* (of which $\mathbb{R}^n$ is in a sense the prime and canonical example). He then strives to formulate analogues of familiar notions of three-dimensional geometry which help him to understand the more formidable examples of vector spaces. In practice he tries to show that 'the vector space' under consideration either is $\mathbb{R}^n$ in disguise or that $\mathbb{R}^n$ approximates to it fairly well.

It will be a relief to know that although our main concern is with $\mathbb{R}^n$, here and there we shall point to naturally occurring 'vector spaces' of a different sort.

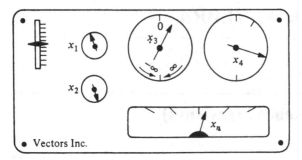

Fig. 1.1

1.1　Vector spaces

In order for a collection of objects X to be regarded as a *vector space* we have to be able to do two things with these objects: (1) *add* any two objects x and y to get an object $\mathbf{x} + \mathbf{y}$ also in X; (2) *scale* up any $\mathbf{x}$ in X by any factor α to give $\alpha\mathbf{x}$, another object in X. The scaling factors, like α, known as *scalars*, are either from $\mathbb{R}$ or $\mathbb{C}$ ($=$ complex numbers). (It is possible to consider more general scalars, but we don't.) How we choose to add or scale depends on the context or application but there is a proviso nevertheless that certain 'laws of addition etc.' are satisfied. We leave aside such formalities, taking them as read.

1.2　Examples

1. $\mathbb{R}^n$: $(x_1, x_2, \ldots, x_n)^t + (y_1, y_2, \ldots, y_n)^t = (x_1 + y_1, x_2 + y_2, \ldots, x_n + y_n)^t$

   $\alpha(x_1, x_2, \ldots, x_n)^t = (\alpha x_1, \alpha x_2, \ldots, \alpha x_n)^t.$

2. *Polynomials of degree* $\leqslant n$ with real coefficients, i.e. of the form:

   $$a(x) = a_n x^n + a_{n-1} x^{n-1} + \cdots + a_1 x + a_0 \quad \text{with all} \quad a_i \in \mathbb{R}.$$

The natural way to add and scale is:

$$\{a_n x^n + a_{n-1} x^{n-1} + \cdots + a_1 x + a_0\}$$
$$+ \{b_n x^n + b_{n-1} x^{n-1} + \cdots + b_1 x + b_0\}$$
$$= (a_n + b_n)x^n + (a_{n-1} + b_{n-1})x^{n-1} + \cdots + (a_1 + b_1)x$$
$$+ (a_0 + b_0),$$
$$\alpha\{a_n x^n + a_{n-1} x^{n-1} + \cdots + a_1 x + a_0\}$$
$$= (\alpha a_n)x^n + (\alpha a_{n-1})x^{n-1} + \cdots + (\alpha a_0).$$

Although the objects under consideration here are distinct from those in $\mathbb{R}^{n+1}$ nevertheless we can regard them as forming $\mathbb{R}^{n+1}$ in masquerade. The coefficients in $a(x)$ can be arranged as an $(n+1)$tuple thus:

$$(a_n, a_{n-1}, \ldots, a_1, a_0)^t.$$

Now note that $\alpha a(x)$ gives rise to the $(n+1)$tuple
$$(\alpha a_n, \alpha a_{n-1}, \ldots, \alpha a_1, \alpha a_0)^t,$$

whilst $a(x) + b(x)$ as above, gives

$$(a_n + b_n, a_{n-1} + b_{n-1}, \ldots, a_1 + b_1, a_0 + b_0)^t.$$

So these $(n+1)$tuples are added and scaled by the same rules as in Example 1, i.e. the polynomials of degree $\leqslant n$ differ from $\mathbb{R}^{n+1}$ in notation only. (That is, if we choose to ignore other properties of polynomials such as the fact that the variable in $a(x)$ can be substituted for.)

3. *All polynomials with real coefficients.* These can be added and scaled in the obvious way, but now that the restriction on degree has been removed, this vector space 'is' not $\mathbb{R}^n$ for any n.

4. *All real-valued functions.* The set of functions whose domain of definition is S taking real values, is denoted by $\mathbb{R}^S$. If f, g are in $\mathbb{R}^S$ we can define a function h, so that $h = f + g$, thus (see Figure 1.2)

$$h(s) = f(s) + g(s),$$

i.e.

$$(f + g)(s) = f(s) + g(s)$$

and

$$(\alpha f)(s) = \alpha f(s).$$

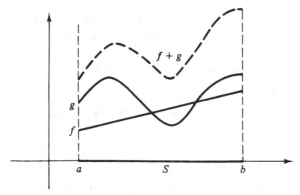

Fig. 1.2

Again this space is not in general $\mathbb{R}^n$ particularly if S is the interval $[a, b]$. Nevertheless the function defined by $f(x) \equiv e^x$ on $[0, 1]$ can be approximated reasonably by the polynomial function

$$1 + x + \frac{x^2}{2} + \frac{x^3}{6} + \frac{x^4}{24} + \frac{x^5}{120} + \frac{x^6}{720}.$$

Observe that if $S = \{1, 2, 3\}$ then each function $f \in \mathbb{R}^{\{1,2,3\}}$ is identifiable with the element of $\mathbb{R}^3$ $(f(1), f(2), f(3))^t$. Furthermore, just as in Example 2, the rules for addition and scaling are identical; thus $\mathbb{R}^{\{1,2,3\}}$ 'is' $\mathbb{R}^3$. $\mathbb{R}^S$ is particularly useful in the mathematical study of dynamical systems, when S is the time domain.

In brief: Vector space theory is the study of systems of objects that can be added and scaled.

1.3   Check your vocabulary

In this paragraph we refresh rusty old memories and remind ourselves of a stock of words used in studying $\mathbb{R}^n$. We have to remember that these concepts are applicable to *all* vector spaces.

Linear combinations
If $x_1, \ldots, x_k$ are vectors and $\alpha_1, \ldots, \alpha_k$ are scalars and x satisfies

$$x = \alpha_1 x_1 + \alpha_2 x_2 + \cdots + \alpha_k x_k,$$

then we say that x is a *linear combination* of $x_1, \ldots, x_k$.
If for some scalars $\alpha_1, \ldots, \alpha_k$ not all zero we have

$$0 = \alpha_1 x_1 + \alpha_2 x_2 + \cdots + \alpha_k x_k,$$

then we say that the vectors $x_1, \ldots, x_k$ are *linearly dependent*. Indeed it follows in this case that one of these vectors (at least) is a linear combination of the others.
If, for given vectors $x_1, \ldots, x_k$, it is true that the equation

$$0 = \alpha_1 x_1 + \cdots + \alpha_k x_k$$

can hold only when $\alpha_1 = \alpha_2 = \cdots = \alpha_k = 0$, then we say that $x_1, \ldots, x_k$ are *linearly independent*.

Examples
(i) $(1, 3, 2)^t = (1, 1, 0)^t + 2(0, 1, 1)^t$ so $\{(1, 3, 2)^t, (1, 1, 0)^t, (0, 1, 1)^t\}$ are linearly dependent vectors.

(ii) $(1,0,0)^t$, $(0,1,0)^t$, $(0,0,1)^t$ are linearly independent because the equation

$$(0,0,0)^t = \alpha_1(1,0,0)^t + \alpha_2(0,1,0)^t + \alpha_3(0,0,1)^t$$
$$= (\alpha_1, \alpha_2, \alpha_3)^t.$$

implies $\alpha_1 = \alpha_2 = \alpha_3 = 0$.

Basis and dimension

We say that the set of vectors $\{x_1, \ldots, x_k\}$ *spans* the vector space X if each vector in X is a linear combination of the vectors $x_1, \ldots, x_k$.

The trivial calculation of the last paragraph shows that any $(\alpha_1, \alpha_2, \alpha_3)^t$ is a linear combination of $(1,0,0)^t$, $(0,1,0)^t$ and $(0,0,1)^t$ so these three vectors span $\mathbb{R}^3$.

Example

The solutions (see Part II, Chapter 20) of the equation

$$\frac{d^2 y}{dx^2} + y = 0$$

are known to take the form $y = A \cos x + B \sin x$ where A and B are constants. This says that the two vectors in $\mathbb{R}^\mathbb{R}$ $f_1(x) = \sin x$ and $f_2(x) = \cos x$ span the space of solutions of the differential equation (see also Section 1.7).

A *finite dimensional* space is one that is spanned by a finite set of vectors $\{x_1, \ldots, x_k\}$.

A set $\{x_1, \ldots, x_k\}$ is a *basis* if each vector can be *uniquely* expressed as a linear combination of $x_1, \ldots x_k$. This is equivalent to saying that $\{x_1, \ldots, x_k\}$ spans the space *and* $x_1, \ldots, x_k$ are linearly independent. To see why this is, suppose

$$x = \alpha_1 x_1 + \cdots + \alpha_k x_k$$

and

$$x = \beta_1 x_1 + \cdots + \beta_k x_k,$$

subtracting

$$0 = (\alpha_1 - \beta_1)x_1 + \cdots + (\alpha_k - \beta_k)x_k.$$

So if $x_1, \ldots, x_k$ are linearly independent $(\alpha_1 - \beta_1) = 0 = (\alpha_2 - \beta_2) = \cdots$ and hence we have uniqueness. We leave the converse as an exercise.

In any given finite dimensional vector space it is true that *any two bases contain the same number of vectors.* (This is non-trivial.) The number involved is called the *dimension*. If $\{\mathbf{x}_1, \ldots, \mathbf{x}_k\}$ is a basis and

$$\mathbf{x} = \alpha_1 \mathbf{x}_1 + \cdots + \alpha_k \mathbf{x}_k,$$

then the uniquely determined scalars $\alpha_1, \ldots, \alpha_k$ are called the *co-ordinates* of $\mathbf{x}$ *with respect to the basis* $\{\mathbf{x}_1, \ldots, \mathbf{x}_k\}$.

1.4   Subspaces

A subspace of a vector space X is a subset Y of X which is rather special in that addition and scaling applied to vectors from Y produces vectors in Y. Thus Y can be regarded as a vector space in its own rights. Compare the example in Figure 1.3.

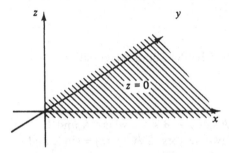

Fig. 1.3. *The plane $z = 0$ is a vector space in its own right. It is of course $\mathbb{R}^2$ in disguise*

Symbolically, Y is a subspace if both conditions

(i) $\mathbf{y}_1 + \mathbf{y}_2 \in Y$ for $\mathbf{y}_1, \mathbf{y}_2 \in Y$,
(ii) $\alpha \mathbf{y} \in Y$ for $\mathbf{y} \in Y$ and all α,

are satisfied or, the single (more compact) condition holds:

$\alpha \mathbf{y}_1 + \beta \mathbf{y}_2 \in Y$ for all $\mathbf{y}_1, \mathbf{y}_2 \in Y$ and all scalars α, β.

Example 1
$Y = \{(\lambda, \mu, \lambda + \mu)^t : \lambda, \mu \in \mathbb{R}\}$ is a subspace of $\mathbb{R}^3$.
We can see this intuitively by noting that Y is the plane with equation

$$z = x + y.$$

(See Figure 1.4.) Formally, for $(\lambda, \mu, \lambda + \mu)^t$, $(\Lambda, M, \Lambda + M)^t$ in Y

$$\alpha(\lambda, \mu, \lambda + \mu)^t + \beta(\Lambda, M, \Lambda + M)^t$$
$$= (\alpha\lambda + \beta\Lambda, \alpha\mu + \beta M, \alpha\lambda + \alpha\mu + \beta\Lambda + \beta M)^t$$
$$= ((\alpha\lambda + \beta\Lambda), (\alpha\mu + \beta M), (\alpha\lambda + \beta\Lambda) + (\alpha\mu + \beta M))^t \in Y.$$

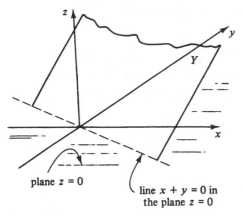

plane $z = 0$

line $x + y = 0$ in
the plane $z = 0$

Fig. 1.4

Example 2

In the space $X = \mathbb{R}^{(0,1)}$ let $D(0, 1)$ be the set of differentiable functions. Clearly,

$$\frac{d}{dx}(\alpha f_1(x) + \beta f_2(x)) = \alpha f_1'(x) + \beta f_2'(x).$$

The function $\alpha_1 f_1 + \alpha_2 f_2$ is also differentiable, i.e. is in $D(0, 1)$. Thus $D(0, 1)$ is a subspace. See Part II, Chapter 20 for applications of this fact.

1.5 Linear equations and rank

The linear equations

$$\left.\begin{array}{c} a_{11}x_1 + \cdots + a_{1n}x_n = b_1 \\ a_{21}x_1 + \cdots + a_{2n}x_n = b_2 \\ \vdots \\ a_{m1}x_1 + \cdots + a_{mn}x_n = b_m \end{array}\right\}$$

are a bore to write out in full. The wily way to save time is to rewrite this as follows

$$\begin{bmatrix} a_{11} & \cdots & a_{1n} \\ a_{21} & & a_{2n} \\ \vdots & & \\ a_{m1} & & a_{mn} \end{bmatrix} \begin{bmatrix} x_1 \\ x_2 \\ \vdots \\ x_n \end{bmatrix} = \begin{bmatrix} b_1 \\ b_2 \\ \vdots \\ b_m \end{bmatrix},$$

whence arise *matrices*. To save more time in theoretical deliberations the abbreviated form of the above equation is used:

$$Ax = b.$$

It is useful to place the problem of solving the equations 'in perspective' and to focus interest on the transformation that takes the column vector in $\mathbb{R}^n$

$$\mathbf{x} = \begin{bmatrix} x_1 \\ x_2 \\ \vdots \\ x_n \end{bmatrix}$$

into the vector $\mathbf{y} = (y_1, \ldots, y_m)^t$ in $\mathbb{R}^m$ where

$$\mathbf{y} = A\mathbf{x}.$$

The set

$$N(A) = \{\mathbf{x} : A\mathbf{x} = 0\}$$

is known as the *kernel* or *null space* of the transformation, and the set

$$R(A) = \{\mathbf{y} : \text{for some } \mathbf{x}, \mathbf{y} = A\mathbf{x}\}$$

is known as the *range* of the transformation. They are of particular interest; both are subspaces (check!). Note that $A\mathbf{x} = \mathbf{b}$ is soluble if and only if $\mathbf{b} \in R(A)$. (See Figure 1.5.) That is, of course, a mere restatement. Recall that

rank$(A) =$ dimension of the range of A,

nullity $(A) =$ dimension of the kernel of A.

Observe that

$$A\mathbf{x} = \begin{bmatrix} a_{11}x_1 + a_{12}x_2 + \cdots + a_{1n}x_n \\ a_{21}x_1 + a_{22}x_2 + \cdots + a_{2n}x_n \\ \vdots \\ a_{m1}x_1 + a_{m2}x_2 + \cdots + a_{mn}x_n \end{bmatrix} = x_1 \begin{bmatrix} a_{11} \\ a_{21} \\ \vdots \\ a_{m1} \end{bmatrix} + x_2 \begin{bmatrix} a_{12} \\ a_{22} \\ \vdots \\ a_{m2} \end{bmatrix} + \cdots$$

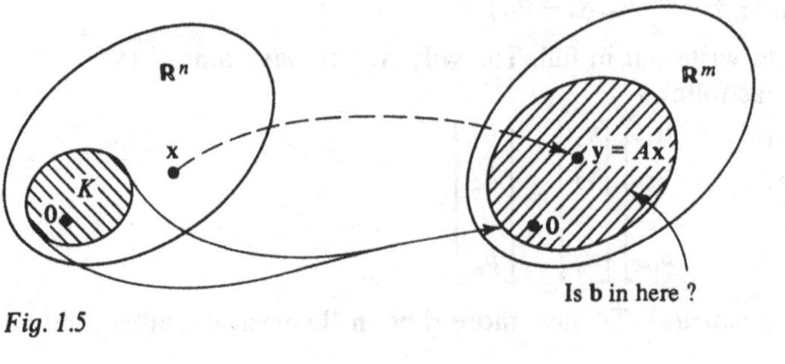

Is b in here ?

Fig. 1.5

so the range of A is the subspace of $\mathbb{R}^m$ spanned by the columns of A. Hence

ran(A) = maximum number of independent columns = column-rank
and in fact, also
rank(A) = maximum number of independent rows = row-rank.

1.6   Calculation of rank

In general this may be done by reducing a matrix to echelon form and counting the number of non-zero rows. Thus, for example:

$$
\begin{bmatrix} 1 & 1 & 2 & 3 \\ 3 & 1 & 1 & 2 \\ -3 & 1 & 4 & 5 \end{bmatrix}
\xrightarrow[\substack{\mathbf{r}_2 + \mathbf{r}_3 \\ -3\mathbf{r}_1 + \mathbf{r}_2}]{}
\begin{bmatrix} 1 & 1 & 2 & 3 \\ 0 & -2 & -5 & -7 \\ 0 & 2 & 5 & 7 \end{bmatrix}
\xrightarrow[\substack{\mathbf{r}_2 + \mathbf{r}_3 \\ -\mathbf{r}_2}]{}
\begin{bmatrix} 1 & 1 & 2 & 3 \\ 0 & 2 & 5 & 7 \\ 0 & 0 & 0 & 0 \end{bmatrix}
$$

so the rank here is 2. (We have shown the appropriate row manipulations below the arrow.) We note one more step in the reduction

$$
\begin{bmatrix} 2 & 0 & -1 & -1 \\ 0 & 2 & 5 & 7 \\ 0 & 0 & 0 & 0 \end{bmatrix} [2\mathbf{r}_1 - \mathbf{r}_2].
$$

Notice that the 'row operations' used here correspond exactly to manipulation of equations. In reality it is the equations that are being cast into an echelon form while redundant equations are deleted. Clearly this does not alter row rank. The fact that the column rank is at the same time unchanged will be explained later (see Chapter 3, Section 3.6).

If we are dealing with a *square* matrix, we can use determinants to check whether a matrix is of full rank:

$$
\begin{bmatrix} 1 & 1 & 2 \\ 2 & 0 & 1 \\ 1 & 2 & 0 \end{bmatrix} = 1 \cdot (-2) - (-1) + 2 \cdot (4) = 7 \neq 0.
$$

The matrix is non-singular so the rank is 3. This is also a convenient method for testing whether given vectors are linearly independent. The first example above tells us that the row vectors $(1, 1, 2, 3)$, $(3, 1, 1, 2)$, $(-3, 1, 4, 5)$ are not linearly independent. The second, on the other hand, tells us that $(1, 1, 2)$, $(2, 0, 1)$, $(1, 2, 0)$ are linearly independent.

Nullity may be calculated from rank by using the relationship

rank + nullity = dimension.

Note that the term 'dimension' above refers to the dimension of the domain space ('dim-dom' would be the appropriate mnemonic); so if the matrix A which we are testing is of size $m \times n$ and the transformation we have in mind is $\mathbf{y} = A\mathbf{x}$ the dimension required by the equation is n.

Thus, for example,

$$\text{nullity} \begin{bmatrix} 1 & 1 & 2 & 3 \\ 3 & 1 & 1 & 2 \\ -3 & 1 & 4 & 5 \end{bmatrix} = 4 - 2 = 2.$$

Notice that, however,

$$\text{nullity} \begin{bmatrix} 1 & 3 & -3 \\ 1 & 1 & 1 \\ 2 & 1 & 4 \\ 3 & 2 & 5 \end{bmatrix} = 3 - 2 = 1.$$

To determine the kernel of the matrix A above, we need to solve the equations

$$\left. \begin{aligned} x_1 + x_2 + 2x_3 + 3x_4 &= 0 \\ 3x_1 + x_2 + x_3 + 2x_4 &= 0 \\ -3x_1 + x_2 + 4x_3 + 5x_4 &= 0 \end{aligned} \right\},$$

or, equivalently,

$$\left. \begin{aligned} 2x_1 \quad - \ x_3 - \ x_4 &= 0 \\ 2x_2 + 5x_3 + 7x_4 &= 0 \end{aligned} \right\},$$

Thus $\mathbf{x}$ belongs to the kernel of A if

$$\begin{aligned} (x_1, x_2, x_3, x_4)^t &= (\tfrac{1}{2}\{x_3 + x_4\}, -\tfrac{1}{2}\{5x_3 + 7x_4\}, x_3, x_4)^t \\ &= x_3(\tfrac{1}{2}, -\tfrac{5}{2}, 1, 0)^t + x_4(\tfrac{1}{2}, -\tfrac{7}{2}, 0, 1)^t. \end{aligned}$$

Hence $\ker(A)$ is the span of $(1, -5, 2, 0)^t$, $(1, -7, 0, 2)^t$. (We have scaled up by a factor of 2 for convenience.)

1.7   The solution set

Consider the equation $A\mathbf{x} = \mathbf{b}$. If $k = \text{nullity } A = \dim N(A)$, there is a basis $\{x_1, \dots, x_k\}$ for the subspace $N(A)$; i.e. any vector $\mathbf{x}$ satisfying $A\mathbf{x} = 0$ can be written in the form

$$\mathbf{x} = \alpha_1 \mathbf{x}_1 + \cdots + \alpha_k \mathbf{x}_k.$$

Now suppose that $\mathbf{w}$ is one *particular solution* of the equation $A\mathbf{x} = \mathbf{b}$

while $\mathbf{z}$ is any other solution. Thus

$$A\mathbf{w} = \mathbf{b} \quad \text{and} \quad A\mathbf{z} = \mathbf{b}.$$

Subtracting we obtain $A(\mathbf{z} - \mathbf{w}) = 0$, so $\mathbf{z} - \mathbf{w}$ belongs to $N(A)$. Thus for some $\alpha_1, \ldots, \alpha_k$ we have

$$\mathbf{z} - \mathbf{w} = \alpha_1 \mathbf{x}_1 + \cdots + \alpha_k \mathbf{x}_k,$$

or

$$\mathbf{z} = \mathbf{w} + \{\alpha_1 \mathbf{x}_1 + \cdots + \alpha_k \mathbf{x}_k\}.$$

This expression is significant as we shall see in coming chapters (Part II, Chapter 20, Section 20.8.3). Here we note that the set

$$\mathbf{w} + N(A)$$
$$= \{\mathbf{w} + \mathbf{x} : \mathbf{x} \in N(A)\}$$

is obtained by translating all vectors in $N(A)$ by $\mathbf{w}$. Thus the solution set of $A\mathbf{x} = \mathbf{b}$ is the translate of the subspace $N(A)$. (See Figure 1.6.) Of course if $N(A) = \{0\}$ the solution set is just $\{\mathbf{w}\}$, a set consisting of a unique point.

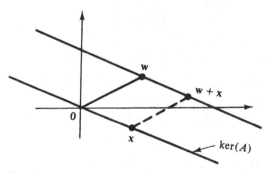

Fig. 1.6

1.8 The Wronskian

Suppose given some functions $f_1, f_2, \ldots, f_n$ in $\mathbb{R}^{(a,b)}$, say for example: $\cos x$, $\sin x$, 1 ($=$ the function which is identically 1). Are they linearly independent of each other?

If, as in the example, $f_1, f_2, \ldots, f_n$ can be differentiated a number of times ($n - 1$ times, to be precise), we can apply some familiar vector space methods.

We have to consider the equation

$$\alpha_1 f_1 + \cdots + \alpha_n f_n = 0.$$

Thus for all x it is being asserted that

$$\alpha_1 f_1(x) + \cdots + \alpha_n f_n(x) = 0.$$

The function on the left-hand side of the equation is identically zero. Hence its derivative is identically zero. Thus:

$$\alpha_1 f'_1(x) + \cdots + \alpha_n f'_n(x) = 0.$$

Repeating this argument, we have

$$\alpha_1 f''_1(x) \quad + \cdots + \alpha_n f''_n(x) \quad = 0,$$
$$\vdots$$
$$\alpha_1 f_1^{(n-1)}(x) + \cdots + \alpha_n f_n^{(n-1)}(x) = 0.$$

We now have that for a fixed value of x, the n equations above are satisfied by some numbers $\alpha_1, \alpha_2, \ldots, \alpha_n$. Let us treat these numbers as the unknowns in a system of n simultaneous equations where the (known) coefficients are the numbers $f_j^{(i)}(x)$. Suppose that for some value of x, say x_0, where $a < x_0 < b$, the determinant of coefficients

$$W(x_0) = \begin{vmatrix} f_1(x_0) & f_2(x_0) \cdots & f_n(x_0) \\ f'_1(x_0) & f'_2(x_0) & f'_n(x_0) \\ \vdots & & \\ f_1^{(n-1)}(x_0) & f_2^{(n-1)}(x_0) & f_n^{(n-1)}(x_0) \end{vmatrix}$$

is non-zero. Then the equation

$$\begin{bmatrix} f_1(x_0) & f_2(x_0) \cdots & f_n(x_0) \\ f'_1(x_0) & f'_2(x_0) & f'_n(x_0) \\ \vdots & & \\ f_1^{(n-1)}(x_0) & f_2^{(n-1)}(x_0) & f_n^{(n-1)}(x_0) \end{bmatrix} \begin{bmatrix} \alpha_1 \\ \alpha_2 \\ \vdots \\ \alpha_n \end{bmatrix} = 0$$

has precisely one solution, viz. $\alpha_1 = \alpha_2 = \cdots = \alpha_n = 0$. Consequently we have:

> The functions $f_1, f_2, \ldots, f_n$ in $\mathbb{R}^{(a,b)}$ are linearly independent if for some value of x in (a, b), the Wronskian determinant $W(x)$ is non-zero.

Example

For the functions $\cos x, \sin x, 1$ we compute that

$$\begin{vmatrix} 1 & \cos x & \sin x \\ 0 & -\sin x & \cos x \\ 0 & -\cos x & -\sin x \end{vmatrix} = \sin^2 x + \cos^2 x = 1 \neq 0,$$

so, the functions $\cos x$, $\sin x$, 1 are linearly independent in *any* interval (a, b). (Note that in this example the choice of x_0 is almost irrelevant since the Wronskian is non-zero at *all* points; this is evidently a stronger property than that required.)

We note that though the Wronskian condition is sufficient for linear independence it is by no means a necessary condition i.e. $W(x) \equiv 0$ does not imply linear dependence. (It implies 'functional' dependence.)

1.9 Exercises

1. Determine whether the following sets of vectors in $\mathbb{R}^4$ are linearly independent
   (i) $(2, 1, 1, 1)^t, (1, 2, -1, 1)^t, (3, 1, 0, 2)^t, (1, 2, 3, 4)^t$
   (ii) $(1, 2, 0, 1)^t, (2, 1, 0, 1)^t$
   (iii) $(2, 1, 1, 1)^t, (1, 2, -1, 1)^t, (3, 1, 0, 2)^t, (3, -3, 0, 2)^t$.
2. In each case in question 1 find a basis for the subspace spanned by the given set of vectors.
3. Calculate the rank of the following matrix and find a basis for its kernel.

$$\begin{bmatrix} 1 & 2 & 3 & 1 \\ 2 & 3 & 1 & 2 \\ -1 & 0 & 7 & -1 \\ 3 & 4 & -1 & 3 \\ 1 & 3 & 8 & 1 \end{bmatrix}$$

4. Prove that the matrix A given below is non-singular.

$$A = \begin{bmatrix} 1 & 2 & 3 \\ 3 & 1 & 2 \\ 2 & 3 & 1 \end{bmatrix}.$$

Calculate its inverse and hence solve $Ax = (1, 2, 3)^t$. Is this the easiest way to solve the equation?

5. Find all the solutions to the equation

$$\begin{bmatrix} 1 & 2 & 1 \\ 2 & 4 & 3 \\ 1 & 2 & 2 \end{bmatrix} x = b$$

when (i) $b = (1, 2, 1)^t$ and (ii) $b = (1, 2, 2)^t$. In case (i) write down a basis for the subspace of which the solution set is a translate.

6. Justify the following formula for van der Monde's determinant:

$$\begin{bmatrix} 1 & 1 & \cdots & 1 \\ x_0 & x_1 & & x_n \\ x_0^2 & x_1^2 & & x_n^2 \\ \vdots & & & \\ x_0^n & x_1^n & & x_n^n \end{bmatrix} = \prod_{i>j}(x_j - x_i).$$

7. Calculate the Wronskian for the functions $\exp(\alpha_1 x), \exp(\alpha_2 x), \ldots, \exp(\alpha_n x)$ and hence show that the set is linearly independent provided that $\alpha_1, \alpha_2, \ldots, \alpha_n$ are all distinct.

8. Prove that $e^x, xe^x, x^2 e^x$ are linearly independent.

9. Write down a basis for the space of polynomials of degree n or less.

2

Geometry in $\mathbb{R}^n$

2.1 Affine sets

The general equation of a line ℓ *in the plane* is, as we already know, given by the single linear equation

$$a_1 x_1 + a_2 x_2 = b.$$

At a geometric level there is little to distinguish this line from the one

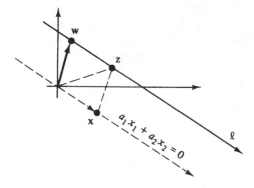

Fig. 2.1

parallel to it passing through the origin (Fig. 2.1). The latter has equation

$$a_1 x_1 + a_2 x_2 = 0.$$

However this particular line constitutes a *one-dimensional vector subspace.* Let us call it V. Thus

$$V = \{(x_1, x_2)^t : a_1 x_1 + a_2 x_2 = 0\}.$$

If **w** is some fixed point on the line ℓ the general point **z** of ℓ satisfies $\mathbf{z} - \mathbf{w} \in V$. Writing $\mathbf{x} = \mathbf{z} - \mathbf{w}$ we see that $\mathbf{z} = \mathbf{w} + \mathbf{x}$, i.e. **z** is obtained by *shifting* a general point **x** of V through the fixed vector **w**.

   Now note that in *three-dimensional space* the equation $a_1 x_1 + a_2 x_2 = b$ no longer describes a line, but a plane. Actually, of course, the general equation of a plane is now

$$a_1 x_1 + a_2 x_2 + a_3 x_3 = b.$$

A line may be obtained as the intersection of two planes in 3-space; it takes, for that same reason, $n - 1$ equations to characterise a *one-dimensional* subspace of $\mathbb{R}^n$. The general result is embodied in the next theorem.

Theorem

   A subset V of $\mathbb{R}^n$ (or $\mathbb{C}^n$) is a vector subspace of dimension k if and only if it is the solution set of an equation $\mathbf{A}\mathbf{x} = 0$ where $\mathbf{A}$ is of rank $n - k$.

Proof. If A has rank $n - k$ then its nullity is k as required. Conversely, let V have dimension k and suppose $\mathbf{v}_1, \mathbf{v}_2, \ldots, \mathbf{v}_k$ is a basis for V. Extend this to a basis $\mathbf{v}_1, \mathbf{v}_2, \ldots, \mathbf{v}_k, \mathbf{v}_{k+1}, \mathbf{v}_{k+2}, \ldots, \mathbf{v}_n$ of the whole of $\mathbb{R}^n$. Any vector **x** may thus be represented uniquely as

$$\mathbf{x} = \alpha_1 \mathbf{v}_1 + \cdots + \alpha_k \mathbf{v}_k + \alpha_{k+1} \mathbf{v}_{k+1} + \cdots + \alpha_n \mathbf{v}_n,$$

or, equivalently, in matrix form as

$$\mathbf{x} = \begin{bmatrix} | & | & & | & | & & | \\ \mathbf{v}_1 & \mathbf{v}_2 & \cdots & \mathbf{v}_k & \mathbf{v}_{k+1} & \cdots & \mathbf{v}_n \\ | & | & & | & | & & | \end{bmatrix} \begin{bmatrix} \alpha_1 \\ \alpha_2 \\ \vdots \\ \alpha_n \end{bmatrix}.$$

We shall abbreviate this to

$$\mathbf{x} = M\alpha.$$

   Now the matrix M is of size $n \times n$ and has rank n (since its n columns are linearly independent). Thus an inverse M^{-1} exists and we have

$$\alpha = M^{-1}\mathbf{x}.$$

Let us write this out in full:

$$\begin{bmatrix} \alpha_1 \\ \vdots \\ \alpha_k \\ \alpha_{k+1} \\ \vdots \\ \alpha_n \end{bmatrix} = \begin{bmatrix} \rule{1cm}{0.4pt} & m_1 & \rule{1cm}{0.4pt} \\ & \vdots & \\ \rule{1cm}{0.4pt} & m_k & \rule{1cm}{0.4pt} \\ \rule{1cm}{0.4pt} & m_{k+1} & \rule{1cm}{0.4pt} \\ & \vdots & \\ \rule{1cm}{0.4pt} & m_n & \rule{1cm}{0.4pt} \end{bmatrix} \begin{bmatrix} x_1 \\ \vdots \\ x_k \\ x_{k+1} \\ \vdots \\ x_n \end{bmatrix}.$$

From it we may extract the equation

$$\begin{bmatrix} \alpha_{k+1} \\ \vdots \\ \alpha_n \end{bmatrix} = \begin{bmatrix} \rule{1cm}{0.4pt} & m_{k+1} & \rule{1cm}{0.4pt} \\ & \vdots & \\ \rule{1cm}{0.4pt} & m_n & \rule{1cm}{0.4pt} \end{bmatrix} \begin{bmatrix} x_1 \\ \vdots \\ x_n \end{bmatrix}.$$

Denoting the right-hand side by $A\mathbf{x}$ we have

$$\mathbf{x} \in V \Leftrightarrow \alpha_{k+1} = \cdots = \alpha_n = 0 \Leftrightarrow A\mathbf{x} = \mathbf{0}.$$

Since the rows of M^{-1} are linearly independent, the $n - k$ rows of A (viz. $m_{k+1}, m_{k+2}, \dots, m_n$) are also linearly independent; hence the rank of A is $n - k$.

Definition
*An **affine set** is any set of the form*

$$\{\mathbf{x} : A\mathbf{x} = \mathbf{b}\}. \tag{*}$$

Motivated by the last theorem we say that the affine set (∗) has dimension k when A has rank $n - k$. Let us consider why this definition depends on the set itself rather than the choice of matrix A.

We have already seen Chapter 1, Section 1.7 that any solution of the equation $A\mathbf{x} = \mathbf{b}$ takes the form $\mathbf{w} + \mathbf{z}$ where $\mathbf{w}$ is any particular solution of the equation and $\mathbf{z}$ satisfies $A\mathbf{x} = 0$. (See Figure 2.2.) The affine set (1) is thus a translate of a vector subspace $N(A)$ (of the dimension k) and here it is clear that k depends on the subspace

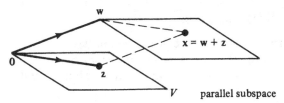

Fig. 2.2

rather than on its representation via A. Of course, any vector subspace may be characterised as an affine set that passes through the origin.

Example
The line ℓ illustrated in Figure 2.3 is the solution set of the equation

$$\begin{bmatrix} 1 & 2 & 1 \\ 2 & 1 & 1 \\ 1 & -1 & 0 \end{bmatrix} \begin{bmatrix} x \\ y \\ z \end{bmatrix} = \begin{bmatrix} 1 \\ 2 \\ 1 \end{bmatrix}.$$

(Evidently the matrix has rank 2.)

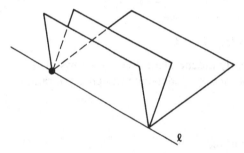

Fig. 2.3

In $\mathbb{R}^n$ the affine set with equation

$$a_1 x_1 + a_2 x_2 + \cdots + a_n x_n = b$$

(where at least one coefficient is non-zero) is called a *hyperplane*. It is of dimension $n-1$ (i.e. one less than the space) since the corresponding row matrix $A = (a_1, a_2, \ldots, a_n)$ is of rank 1. Observe that this notion generalises the concept of a plane in $\mathbb{R}^3$.

2.2   Equations of lines and planes

A line ℓ is a one-dimensional affine set, i.e. it is the translate of a one-dimensional vector subspace S. Let **s** be any one non-zero vector in S. Thus an arbitrary vector **y** in S is just of the form α**s**. Thus a point **x** on the line ℓ may be written as

$$\mathbf{x} = \mathbf{a} + \alpha\mathbf{s}, \tag{1}$$

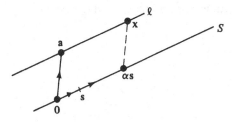

Fig. 2.4

where **a** is a fixed point of ℓ. This we call the *parametrisation* of a 'line through **a** in direction **s**' the parameter being α. See Figure 2.4.

If we want to obtain a parametrisation for the line ℓ 'through **a** and **b**' we note that the line is necessarily in direction $\mathbf{b} - \mathbf{a}$ and so its general point **x** is

$$\mathbf{x} = \mathbf{a} + \alpha(\mathbf{b} - \mathbf{a})$$
$$= (1 - \alpha)\mathbf{a} + \alpha\mathbf{b}. \tag{2}$$

Note that if **x** lies between **a** and **b** the scaling factor α satisfies $0 \leqslant \alpha \leqslant 1$ and in fact is equal to the ratio $AX : AB$. See Figure 2.5.

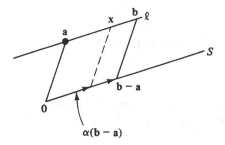

Fig. 2.5

Writing $\mathbf{s} = (s_1, s_2, \ldots, s_n)^t$, $\mathbf{a} = (a_1, a_2, \ldots, a_n)^t$ and $\mathbf{x} = (x_1, x_2, \ldots, x_n)^t$ the last two equations become

$$\alpha = \frac{x_1 - a_1}{s_1} = \frac{x_2 - a_2}{s_2} = \cdots = \frac{x_n - a_n}{s_n}, \tag{3}$$

$$\alpha = \frac{x_1 - a_1}{b_1 - a_1} = \frac{x_2 - a_2}{b_2 - a_2} = \cdots = \frac{x_n - a_n}{b_n - a_n}. \tag{4}$$

Observe that if α is given, the co-ordinates of the point are determined.

Suppressing any reference to the value α gives the following system of equations in the case of (3).

$$\left.\begin{array}{c} \dfrac{x_1 - a_1}{s_1} = \dfrac{x_2 - a_2}{s_2} \\[2ex] \dfrac{x_1 - a_1}{s_1} = \dfrac{x_3 - a_3}{s_3} \\[2ex] \vdots \\[1ex] \dfrac{x_1 - a_1}{s_1} = \dfrac{x_n - a_n}{s_n} \end{array}\right\}.$$

In the matrix reformulation of the above system of equations, the matrix of coefficients will clearly have rank $n - 1$.

Remarks

The restriction $\alpha \geqslant 0$ in (1) gives the *ray* with vertex **a** in direction **s**.

The restriction $0 \leqslant \alpha \leqslant 1$ in (2) gives the parametrisation of a line segment joining **a** to **b**.

2.3   The notion of length

The distance of the point $(x_1, x_2, x_3)^t$ from the origin of $\mathbb{R}^3$ is, by Pythagoras' theorem

$$\sqrt{\{x_1^2 + x_2^2 + x_3^2\}}.$$

Since a similar formula holds in two dimensions, we generalise this notion of length (Fig. 2.6) by defining the *norm* or *length* of the vector $\mathbf{x} = (x_1, x_2, \ldots, x_n)^t$ in $\mathbb{R}^n$ to be the number

$$\|\mathbf{x}\| = \sqrt{\{x_1^2 + x_2^2 + \cdots + x_n^2\}}.$$

Thus, for example, $\|(1, 1, 2)^t\| = \sqrt{\{1 + 1 + 4\}} = \sqrt{6}$.

Fig. 2.6

Remark

Recall that a vector f in $\mathbb{R}^{[0,1]}$ may be regarded as having $f(t)$ as its t co-ordinate, for $0 \leqslant t \leqslant 1$. One may go on to define its length by means of an analogous formula

$$\|f\| = \sqrt{\int_0^1 |f(t)|^2\, dt}.$$

Evidently, this formula is valid for continuous functions. The formula here is not purely an analogy. It provides the basis for analysing dynamical (i.e. time dependent) systems by vector space methods.

2.4 Inner product and the notion of angle

We define the *inner product* or *scalar product* of two vectors $\mathbf{x} = (x_1, x_2, \ldots, x_n)^t$ and $\mathbf{y} = (y_1, y_2, \ldots, y_n)^t$ in $\mathbb{R}^n$, written as $\langle \mathbf{x}, \mathbf{y} \rangle$ to be the number

$$\langle \mathbf{x}, \mathbf{y} \rangle = x_1 y_1 + x_2 y_2 + \cdots + x_n y_n.$$

It is obvious that

1. $\langle \mathbf{x}, \mathbf{x} \rangle = \|\mathbf{x}\|^2$,
2. $\langle \mathbf{x}, \mathbf{y} \rangle = \langle \mathbf{y}, \mathbf{x} \rangle$,
3. $\langle \alpha \mathbf{x} + \beta \mathbf{y}, \mathbf{z} \rangle = \alpha \langle \mathbf{x}, \mathbf{z} \rangle + \beta \langle \mathbf{y}, \mathbf{z} \rangle$.

Using these facts, we may therefore compute that

$$\begin{aligned}
\|\mathbf{x} - \mathbf{y}\|^2 &= \langle \mathbf{x} - \mathbf{y}, \mathbf{x} - \mathbf{y} \rangle = \langle \mathbf{x}, \mathbf{x} - \mathbf{y} \rangle - \langle \mathbf{y}, \mathbf{x} - \mathbf{y} \rangle \\
&= \langle \mathbf{x}, \mathbf{x} \rangle - \langle \mathbf{x}, \mathbf{y} \rangle - \langle \mathbf{y}, \mathbf{x} \rangle + \langle \mathbf{y}, \mathbf{y} \rangle \\
&= \|\mathbf{x}\|^2 - 2\langle \mathbf{x}, \mathbf{y} \rangle + \|\mathbf{y}\|^2.
\end{aligned} \tag{5}$$

Now consider the two dimensional plane through the three points $\mathbf{0}, \mathbf{x}, \mathbf{y}$ (Fig. 2.7). Since our general notion of length agrees with that in two and

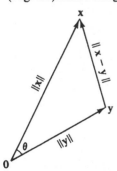

Fig. 2.7

three dimensions we may quote the Cosine Rule as saying

$$\| \mathbf{x} - \mathbf{y} \|^2 = \| \mathbf{x} \|^2 + \| \mathbf{y} \|^2 - 2 \| \mathbf{x} \| \cdot \| \mathbf{y} \| \cos \theta.$$

Comparing this equation with (5) we see that

$$\langle \mathbf{x}, \mathbf{y} \rangle = \| \mathbf{x} \| \cdot \| \mathbf{y} \| \cos \theta.$$

We are thus able to introduce the notion of angle between two vectors in $\mathbb{R}^n$ by reference only to length and inner product.

2.5   Hyperplanes

We have already defined hyperplanes in $\mathbb{R}^n$ as being given by an equation of the form

$$a_1 x_1 + a_2 x_2 + \cdots + a_n a_n = p.$$

Let us rewrite this in inner product notation:

$$\langle (a_1, \ldots, a_n)^t, (x_1, \ldots, x_n)^t \rangle = p. \tag{6}$$

Suppose for the moment that the vector $\mathbf{a} = (a_1, a_2, \ldots, a_n)^t$ has *unit* length. Thus the vector $\mathbf{p} = p \cdot (a_1, a_2, \ldots, a_n)^t$ satisfies the equation (6). Now if $\mathbf{x}$ is any point such that the (two-dimensional) triangle OxP is right-angled at P we shall have (see Figure 2.8)

$$\begin{aligned} \langle \mathbf{a}, \mathbf{x} \rangle &= \| \mathbf{x} \| \cdot \| \mathbf{a} \| \cos \theta \\ &= \| \mathbf{x} \| \cos \theta \\ &= OP = p. \end{aligned}$$

Conversely, if $\langle \mathbf{a}, \mathbf{x} \rangle = p$, then the triangle OxP is right-angled at P. We thus see that if $\| \mathbf{a} \| = 1$ the equation (6) describes a hyperplane perpendicular to a and at a distance p from the origin. Clearly, if $p > 0$

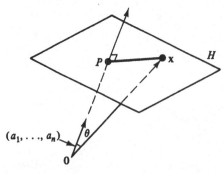

Fig. 2.8

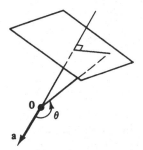

Fig. 2.9. $p < 0$

then a points towards the half-space separated from $\mathbf{0}$ by the hyperplane. Directions are reversed if $p < 0$ as in Figure 2.9. (Evidently the hyperplane passes through $\mathbf{0}$ if $p = 0$.)

Example
Find the equation of the plane through $(1, 2, 3)^t, (3, 1, 2)^t, (2, 3, 1)^t$. What is the distance of the hyperplane from the origin?
Let the equation be $ax + by + cz = p$, where $(a, b, c)^t$ is a unit vector. We then have

$$a + 2b + 3c = p,$$
$$3a + b + 2c = p,$$
$$2a + 3b + c = p.$$

Thus

$$\begin{bmatrix} a \\ b \\ c \end{bmatrix} = \begin{bmatrix} 1 & 2 & 3 \\ 3 & 1 & 2 \\ 2 & 3 & 1 \end{bmatrix}^{-1} \begin{bmatrix} p \\ p \\ p \end{bmatrix} = \frac{p}{18} \begin{bmatrix} -5 & 7 & 1 \\ 1 & -5 & 7 \\ 7 & 1 & -5 \end{bmatrix} \begin{bmatrix} 1 \\ 1 \\ 1 \end{bmatrix}$$

$$= \frac{p}{18} \begin{bmatrix} 3 \\ 3 \\ 3 \end{bmatrix} = \frac{p}{6} \begin{bmatrix} 1 \\ 1 \\ 1 \end{bmatrix}.$$

Hence

$$\| (a, b, c)^t \|^2 = \frac{p^2}{36}(1 + 1 + 1) = \frac{p^2}{12} = 1.$$

Thus the distance from the origin is $\sqrt{12}$. The equation of the plane is however more elegantly written in the scaled up form

$$x + y + z = 6.$$

2.6   The inner product in $\mathbb{C}^n$

In the space $\mathbb{C}^n$ we define the inner product by the formula:

$$\langle \mathbf{x}, \mathbf{y} \rangle = x_1 \bar{y}_1 + x_2 \bar{y}_2 + \cdots + x_n \bar{y}_n.$$

This ensures that

$$\langle \mathbf{x}, \mathbf{x} \rangle = |x_1|^2 + |x_2|^2 + \cdots + |x_n|^2,$$

so that $\langle \mathbf{x}, \mathbf{x} \rangle$ is real and non-negative and takes the same form as the Pythagorean formula for the square of the distance from $\mathbf{0}$ to $(x_1, \ldots, x_n)^t$. Let us define the norm in $\mathbb{C}^n$ by the formula

$$\| \mathbf{x} \| = \sqrt{\{|x_1|^2 + |x_2|^2 + \cdots + |x_n|^2\}}.$$

The following properties for the inner product resemble those for $\mathbb{R}^n$

(i) $\langle \mathbf{x}, \mathbf{x} \rangle = \| \mathbf{x} \|^2$

(ii) $\langle \mathbf{x}, \mathbf{y} \rangle = \overline{\langle \mathbf{y}, \mathbf{x} \rangle}$

(iii) $\langle \alpha \mathbf{x} + \beta \mathbf{y}, \mathbf{z} \rangle = \alpha \langle \mathbf{x}, \mathbf{z} \rangle + \beta \langle \mathbf{y}, \mathbf{z} \rangle$

Notice that by (ii) we have

(iv) $\langle \mathbf{x}, \alpha \mathbf{y} + \beta \mathbf{z} \rangle = \bar{\alpha} \langle \mathbf{x}, \mathbf{y} \rangle + \bar{\beta} \langle \mathbf{x}, \mathbf{z} \rangle$

Remark

In the space of functions $\mathbb{C}^{[0,1]}$ the appropriate inner product is

$$\langle f, g \rangle = \int_0^1 f(t)\overline{g(t)}dt.$$

We can use the properties noted above to derive the very important **Cauchy–Schwartz Inequality**

$$|\langle \mathbf{x}, \mathbf{y} \rangle| \leqslant \| \mathbf{x} \| \cdot \| \mathbf{y} \|.$$

Proof. Let $\alpha = re^{i\theta}$ be an arbitrary complex number. Then

$$0 \leqslant \| \mathbf{x} + \alpha \mathbf{y} \|^2 = \langle \mathbf{x} + \alpha \mathbf{y}, \mathbf{x} + \alpha \mathbf{y} \rangle = \langle \mathbf{x}, \mathbf{x} + \alpha \mathbf{y} \rangle + \alpha \langle \mathbf{y}, \mathbf{x} + \alpha \mathbf{y} \rangle$$

$$= \langle \mathbf{x}, \mathbf{x} \rangle + \bar{\alpha} \langle \mathbf{x}, \mathbf{y} \rangle + \alpha \langle \mathbf{y}, \mathbf{x} \rangle + \alpha \bar{\alpha} \langle \mathbf{y}, \mathbf{y} \rangle$$

$$= \| \mathbf{x} \|^2 + 2Re(\alpha \overline{\langle \mathbf{x}, \mathbf{y} \rangle}) + |\alpha|^2 \| \mathbf{y} \|^2.$$

Write $\langle \mathbf{x}, \mathbf{y} \rangle$ in the form $Re^{i\phi}$ so that $\langle \mathbf{y}, \mathbf{x} \rangle = Re^{-i\phi}$. We now choose the complex number α so that its argument, θ, satisfies $\theta = \phi$. Then $\alpha \langle \mathbf{y}, \mathbf{x} \rangle = re^{i\theta} Re^{-i\phi} = rR$ which is a real number. Hence we now have

$$0 \leqslant \| \mathbf{x} \|^2 + 2rR + r^2 \| \mathbf{y} \|^2 = \| \mathbf{x} \|^2 + 2r|\langle \mathbf{x}, \mathbf{y} \rangle| + r^2 \| \mathbf{y} \|^2.$$

Since this quadratic expression in r is non-negative for all values of r, its discriminant must be non-positive. (The quadratic then has at most one real root.) Thus

$$4|\langle \mathbf{x}, \mathbf{y} \rangle|^2 - 4 \cdot \|\mathbf{x}\|^2 \cdot \|\mathbf{y}\|^2 \leqslant 0,$$

which is the desired result.

We may deduce the corollary known as the **Triangle Inequality**

$$\|\mathbf{x} + \mathbf{y}\| \leqslant \|\mathbf{x}\| + \|\mathbf{y}\|.$$

Proof. Taking $\alpha = 1$ in our earlier computation we have:

$$\|\mathbf{x} + \mathbf{y}\|^2 = \|\mathbf{x}\|^2 + 2Re(\langle \mathbf{x}, \mathbf{y} \rangle) + \|\mathbf{y}\|^2$$
$$\leqslant \|\mathbf{x}\|^2 + 2|\langle \mathbf{x}, \mathbf{y} \rangle| + \|\mathbf{y}\|^2.$$

But by the Cauchy–Schwartz inequality:

$$\|\mathbf{x}\|^2 + 2|\langle \mathbf{x}, \mathbf{y} \rangle| + \|\mathbf{y}\|^2 \leqslant \|\mathbf{x}\|^2 + 2\|\mathbf{x}\| \cdot \|\mathbf{y}\| + \|\mathbf{y}\|^2$$
$$= (\|\mathbf{x} + \mathbf{y}\|)^2.$$

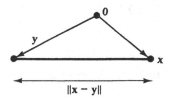

Fig. 2.10

The justification for the name of the inequality is given in Figure 2.10. The triangle here has vertices at $\mathbf{0}, \mathbf{x}$ and $\mathbf{y}$. We have

$$\|\mathbf{x} - \mathbf{y}\| \leqslant \|\mathbf{x}\| + \| - \mathbf{y}\|$$
$$= \|\mathbf{x}\| + \|\mathbf{y}\|.$$

2.7 Orthogonality

We say that $\mathbf{x}$ and $\mathbf{y}$ are *orthogonal* (and write $\mathbf{x} \perp \mathbf{y}$) if

$$\langle \mathbf{x}, \mathbf{y} \rangle = 0.$$

Evidently, this says that the angle between the two vectors is a right angle. We have chosen to write the definition in terms of the inner product notation; this makes it applicable to any vector space in which there exists an inner product formula satisfying the 'three laws'.

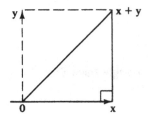

Fig. 2.11

Note how Pythagoras' theorem can now be stated (cf. Figure 2.11):

$$\text{if } \quad \mathbf{x} \perp \mathbf{y} \quad \text{then} \quad \|\mathbf{x}\|^2 + \|\mathbf{y}\|^2 = \|\mathbf{x} + \mathbf{y}\|^2.$$

The proof in $\mathbb{R}^n$ (or $\mathbb{C}^n$) is as follows:

$$\begin{aligned}
\|\mathbf{x} + \mathbf{y}\|^2 &= \langle \mathbf{x} + \mathbf{y}, \mathbf{x} + \mathbf{y} \rangle \\
&= \langle \mathbf{x}, \mathbf{x} \rangle + \langle \mathbf{x}, \mathbf{y} \rangle + \langle \mathbf{y}, \mathbf{x} \rangle + \langle \mathbf{y}, \mathbf{y} \rangle \\
&= \|\mathbf{x}\|^2 + 0 + 0 + \|\mathbf{y}\|^2.
\end{aligned}$$

Definition

*A set of vectors $\{\mathbf{e}_1, \mathbf{e}_2, \ldots, \mathbf{e}_n\}$ is said to be **orthonormal**, if*

$$\left. \begin{aligned}
\langle \mathbf{e}_i, \mathbf{e}_j \rangle &= 0 \quad \text{if } \quad i \neq j, \\
\langle \mathbf{e}_i, \mathbf{e}_i \rangle &= 1,
\end{aligned} \right\}$$

*i.e. the vectors are of **unit** length and **mutually orthogonal**.*

Observe that if $\{\mathbf{e}_1, \mathbf{e}_2, \ldots, \mathbf{e}_n\}$ is an orthonormal set lying in an n dimensional space then it is necessarily a basis. To see this, it is sufficient to prove that the vectors are linearly independent. With this in mind suppose that

$$\alpha_1 \mathbf{e}_1 + \cdots + \alpha_n \mathbf{e}_n = 0.$$

But, for each i,

$$\begin{aligned}
0 = \langle \alpha_1 \mathbf{e}_1 + \cdots + \alpha_n \mathbf{e}_n, \mathbf{e}_i \rangle &= \alpha_1 \langle \mathbf{e}_1, \mathbf{e}_i \rangle + \cdots + \alpha_n \langle \mathbf{e}_n, \mathbf{e}_i \rangle \\
&= \alpha_i.
\end{aligned}$$

Note that generally, if

$$\mathbf{x} = \alpha_1 \mathbf{e}_1 + \cdots + \alpha_n \mathbf{e}_n,$$

then

$$\langle \mathbf{x}, \mathbf{e}_i \rangle = \alpha_i,$$

and so

$$\|\mathbf{x}\|^2 = \alpha_1^2 + \alpha_2^2 + \cdots + \alpha_n^2.$$

2.8   The Gram–Schmidt process

Suppose given a basis $\{\mathbf{v}_1, \mathbf{v}_2, \ldots, \mathbf{v}_n\}$ for a vector space. We can construct from it a new orthonormal basis $\{\mathbf{e}_1, \mathbf{e}_2, \ldots, \mathbf{e}_n\}$ by choosing appropriate combinations of the $\mathbf{v}$'s. The Gram–Schmidt process does this by considering successively the subspaces spanned by $\mathbf{v}_1$, $\{\mathbf{v}_1, \mathbf{v}_2\}$,

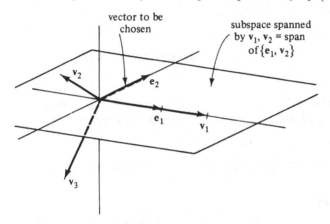

Fig. 2.12

$\{\mathbf{v}_1, \mathbf{v}_2, \mathbf{v}_3\}, \ldots$, etc. (see Figure 2.12). Then we choose $\mathbf{e}_1, \mathbf{e}_2, \ldots, \mathbf{e}_n$ in turn so that

$$\mathbf{e}_1 = \alpha_{11}\mathbf{v}_1,$$
$$\mathbf{e}_2 = \alpha_{21}\mathbf{v}_1 + \alpha_{22}\mathbf{v}_2,$$
$$\mathbf{e}_3 = \alpha_{31}\mathbf{v}_1 + \alpha_{32}\mathbf{v}_2 + \alpha_{33}\mathbf{v}_3,$$
$$\cdot \quad \cdot \quad \cdot \quad \cdot \quad \cdot \quad \cdot \quad \cdot$$
$$\mathbf{e}_n = \alpha_{n1}\mathbf{v}_1 + \cdots + \alpha_{nn}\mathbf{v}_n,$$

where the scalars α_{ij} are selected so that

$$\|\mathbf{e}_i\|^2 = 1 \quad \text{for all } i \quad \text{and} \quad \langle \mathbf{e}_i, \mathbf{e}_j \rangle = 0 \quad \text{if} \quad i \neq j.$$

It is not necessary to solve for the α_{ij} as though they were unknowns. A more convenient trick is this. To obtain $\mathbf{e}_1$ rescale $\mathbf{v}_1$ to unit length, thus

$$\mathbf{e}_1 = \frac{\mathbf{v}_1}{\|\mathbf{v}_1\|}.$$

Now suppose e_1 and e_2 span the same subspace as e_1 and v_2. Since v_2 is a combination of e_1 and e_2 we know, by Section 2.7, that

$$v_2 = \langle v_2, e_1 \rangle e_1 + \langle v_2, e_2 \rangle e_2.$$

Hence e_2 is parallel to the vector

$$v_2 - \langle v_2, e_1 \rangle e_1,$$

so rescaling the latter vector will give us e_2 (or its negative). The procedure can be repeated. Thus e_3 may be selected as a rescaled version of

$$v_3 - \langle v_3, e_1 \rangle e_1 - \langle v_3, e_2 \rangle e_2.$$

Example

Find an orthonormal basis for the subspace spanned by $(1, 2, 2)^t, (2, 1, 2)^t$.

Solution. We take $e_1 = (1, 2, 2)^t / \sqrt{\{1 + 4 + 4\}} = 1/3(1, 2, 2)^t$. Now $\langle v_2, e_1 \rangle = \langle (2, 1, 2)^t, 1/3(1, 2, 2)^t \rangle = 8/3$. The vector e_2 is parallel to

$$(2, 1, 2)^t - \tfrac{8}{9}(1, 2, 2)^t = \tfrac{1}{9}(10, -7, 2)^t.$$

So we take $e_2 = (10, -7, 2)^t / \sqrt{\{100 + 49 + 4\}} = 1/\sqrt{153}(10, -7, 2)^t$.

2.9   Exercises

1. Determine the equation of the hyperplane in $\mathbb{R}^3$ which passes through the points $(1, 2, 3)^t, (1, 0, 1)^t, (1, 2, 1)^t$.

2. Write down the equation of a line which passes through the points $(1, 2, 3)^t, (1, 0, 1)^t$.

3. Explain why $(1, 2/3, 5/3)^t$ lies on the line segment joining $(1, 2, 3)^t$ and $(1, 0, 1)^t$.

4. Prove that

$$\| x + y \|^2 + \| x - y \|^2 = 2 \cdot \| x \|^2 + 2 \cdot \| y \|^2.$$

Interpret this result geometrically. [Hint: parallelograms?]

5. If $\langle x, y \rangle = 0$ for each y in $\mathbb{R}^n$, prove that $x = 0$.

6. (i)   If x and y are in $\mathbb{R}^n$ prove that

$$\langle x, y \rangle = \tfrac{1}{4} \| x + y \|^2 - \tfrac{1}{4} \| x - y \|^2.$$

(ii)   If x and y are in $\mathbb{C}^n$ prove that

$$\langle x, y \rangle = \tfrac{1}{4} \| x + y \|^2 - \tfrac{1}{4} \| x - y \|^2 + \frac{i}{4} \| x + iy \|^2 - \frac{i}{4} \| x - iy \|^2.$$

7. A line in $\mathbb{R}^3$ is defined by the equations

$$\frac{x_1 - 3}{2} = \frac{x_2 - 1}{1} = \frac{x_3 - 2}{1}.$$

Find a unit vector which is parallel to this line.

8. Find the equation of a plane in $\mathbb{R}^3$ which passes through the point $(1, 2, 1)^t$ and is orthogonal to $(2, 1, 2)^t$. What is the distance of this plane (i) to the origin, (ii) to the point $(1, 2, 3)^t$.

9. Find the equation of a plane in $\mathbb{R}^3$ which passes through the point $(3, 1, 2)^t$ and is parallel to both vectors $(1, 0, -1)^t$ and $(1, 2, 3)^t$.

10. Find the equation of a plane in $\mathbb{R}^3$ which passes through the points $(3, 1, 2)^t$ and $(2, 1, 3)^t$ and is parallel to $(1, 2, 3)^t$.

11. Use the Gram–Schmidt process to find an orthonormal basis for the span of $(2, 1, 2)^t$ and $(2, 0, 1)^t$.

3

Matrices

3.1 Linear transformations

We begin with a rapid revision of a number of facts concerning matrices and rank; a more leisurely pace is taken up from Section 3.7 onwards.

Let V and W be vector spaces with respective dimensions n and m. We say that the transformation $T:V \to W$ is *linear* if, for all vectors $\mathbf{v}_1, \mathbf{v}_2$ in V and scalars α_1, α_2,

$$T(\alpha_1 \mathbf{v}_1 + \alpha_2 \mathbf{v}_2) = \alpha_1 T\mathbf{v}_1 + \alpha_2 T\mathbf{v}_2.$$

As usual we denote the range of T by

$$R(T) = \{T\mathbf{v} : \mathbf{v} \in V\}$$

and its kernel or null space by

$$N(T) = \{\mathbf{v} \in V : T\mathbf{v} = 0\},$$

as illustrated in Figure 3.1. We begin with the important relation

$$\text{rank}\,(T) + \text{nullity}\,(T) = \dim V,$$

where

$$\text{rank}\,(T) = \dim R(T) \quad \text{and} \quad \text{nullity}\,(T) = \dim N(T).$$

Let $\mathbf{v}_1, \ldots, \mathbf{v}_k$ be a basis for $N(T)$. We may extend this to a basis $\mathbf{v}_1, \ldots, \mathbf{v}_k, \mathbf{v}_{k+1}, \ldots, \mathbf{v}_n$ of V. Let $r = n - k$. We put

$$\mathbf{u}_1 = T\mathbf{v}_{k+1}, \quad \mathbf{u}_2 = T\mathbf{v}_{k+2}, \ldots, \mathbf{u}_r = T\mathbf{v}_{k+r}.$$

Consider a vector $T\mathbf{v}$ in $R(T)$. Let us write

$$\mathbf{v} = \alpha_1 \mathbf{v}_1 + \cdots + \alpha_n \mathbf{v}_n,$$

then

$$T\mathbf{v} = \alpha_1 T\mathbf{v}_1 + \cdots + \alpha_k T\mathbf{v}_k + \alpha_{k+1} T\mathbf{v}_{k+1} + \cdots + \alpha_n T\mathbf{v}_n$$
$$= 0 + \alpha_{k+1} T\mathbf{v}_{k+1} + \cdots + \alpha_n T\mathbf{v}_n,$$

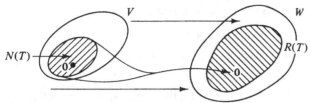

Fig. 3.1

since $Tv_1 = Tv_2 = \cdots = Tv_k = 0$. Thus $\{u_1, \ldots, u_r\}$ spans $R(T)$. To see that these vectors are independent suppose

$$0 = \beta_1 u_1 + \cdots + \beta_r u_r.$$

Thus

$$0 = \beta_1 Tv_{k+1} + \cdots + \beta_r Tv_{k+r},$$

or

$$0 = T(\beta_1 v_{k+1} + \cdots + \beta_r v_{k+r}).$$

Hence we are in $N(T)$ so for some constants $\gamma_1, \ldots, \gamma_k$

$$\beta_1 v_{k+1} + \cdots + \beta_r v_{k+r} = \gamma_1 v_1 + \cdots + \gamma_k v_k,$$

or

$$\gamma_1 v_1 + \cdots + \gamma_k v_k - \beta_1 v_{k+1} - \cdots - \beta_r v_n = 0.$$

But $v_1, \ldots, v_n$ are linearly independent vectors, so every coefficient β_i and γ_i must be zero. Thus $r = \dim R(T)$ as required.

3.2   Representation by a matrix

As in the last section let T be a linear transformation of an n dimensional vector space V into an m dimensional vector space W. Let $v_1, \ldots, v_n$ be a basis for V. Thus we may write any vector v in V in the form

$$v = x_1 v_1 + \cdots + x_n v_n$$

and so we may regard the column vector $(x_1, \ldots, x_n)^t$ in $\mathbb{R}^n$ as a representation of the vector v in V. Now let $w_1, \ldots, w_m$ be a basis for W. Since $Tv_1, \ldots, Tv_n$ are vectors in W we may write

$$Tv_1 = a_{11} w_1 + a_{21} w_2 + \cdots + a_{m1} w_m$$
$$Tv_2 = a_{12} w_1 + a_{22} w_2 + \cdots + a_{m2} w_m$$
$$\vdots$$
$$Tv_n = a_{1n} w_1 + a_{2n} w_2 + \cdots + a_{mn} w_m$$

We can now find $T\mathbf{v}$ from the equations above:

$$
\begin{aligned}
T\mathbf{v} = {} & x_1(a_{11}\mathbf{w}_1 + a_{21}\mathbf{w}_2 + \cdots + a_{m1}\mathbf{w}_m) \\
& + x_2(a_{12}\mathbf{w}_1 + a_{22}\mathbf{w}_2 + \cdots + a_{m2}\mathbf{w}_m) \\
& + \cdots + x_n(a_{1n}\mathbf{w}_1 + a_{2n}\mathbf{w}_2 + \cdots + a_{mn}\mathbf{w}_m) \\
= {} & (a_{11}x_1 + a_{12}x_2 + \cdots + a_{1n}x_n)\mathbf{w}_1 + (a_{21}x_1 + \cdots + a_{2n}x_n)\mathbf{w}_2 \\
& + \cdots + (a_{m1}x_1 + \cdots + a_{mn}x_n)\mathbf{w}_m.
\end{aligned}
$$

Hence if we write

$$
T\mathbf{v} = y_1\mathbf{w}_1 + y_2\mathbf{w}_2 + \cdots + y_m\mathbf{w}_m
$$

and compare the two expressions for $T\mathbf{v}$ we obtain

$$
\begin{bmatrix} y_1 \\ y_2 \\ \vdots \\ y_m \end{bmatrix}
=
\begin{bmatrix}
a_{11} & a_{12} & \cdots & a_{1n} \\
a_{21} & a_{22} & & a_{2n} \\
\vdots & & & \\
a_{m1} & a_{m2} & & a_{mn}
\end{bmatrix}
\begin{bmatrix} x_1 \\ x_2 \\ \vdots \\ x_n \end{bmatrix},
$$

or $\mathbf{y} = A\mathbf{x}$. Thus the matrix A represents the transformation T relative to the basis $\mathbf{v}_1, \ldots, \mathbf{v}_n$ for V and the basis $\mathbf{w}_1, \ldots, \mathbf{w}_m$ for W. Of course, given an $m \times n$ matrix A we can use A to define a transformation of $\mathbb{R}^n$ into $\mathbb{R}^m$ by sending the column vector $\mathbf{x}$ to the column vector $\mathbf{y}$ where $\mathbf{y} = A\mathbf{x}$. However, this is not the only way to obtain a transformation from a matrix A. Given a basis $\mathbf{v}_1, \ldots, \mathbf{v}_n$ for V and a basis $\mathbf{w}_1, \ldots, \mathbf{w}_m$ for W we may send $\dot{\mathbf{v}} = x_1\mathbf{v}_1 + \cdots + x_n\mathbf{v}_n$ to $y_1\mathbf{w}_1 + y_2\mathbf{w}_2 + \cdots + y_m\mathbf{w}_m$ where $\mathbf{y} = A\mathbf{x}$. For this reason it is usual to confuse a matrix A with the transformation which it defines, but only provided the context makes clear what spaces and bases are implied.

3.3   Invertible transformations

A transformation $T : V \rightarrow W$ is said to be invertible if it is one-to-one and its range is all of W, these being the necessary and sufficient conditions to find for any vector $\mathbf{w}$ in W precisely one vector $\mathbf{v}$ in V with $T\mathbf{v} = \mathbf{w}$. We will write $T^{-1}\mathbf{w}$ for the unique $\mathbf{v}$ such that $T\mathbf{v} = \mathbf{w}$, provided T is invertible.

The condition that T be one-to-one is equivalent to $N(T) = \{\mathbf{0}\}$. For if $\mathbf{0} \neq \mathbf{v} \in N(T)$ then $T\mathbf{v} = \mathbf{0} = T\mathbf{0}$, so that two vectors are mapped to $\mathbf{0}$. On the other hand, if $T\mathbf{v}_1 = T\mathbf{v}_2$, so that $T(\mathbf{v}_1 - \mathbf{v}_2) = \mathbf{0}$, then $\mathbf{v}_1 - \mathbf{v}_2 \in N(T)$; so if two distinct vectors are mapped to the same image, the null space will contain a vector different from zero.

Evidently if T is invertible dim $W = $ dim $R(T) = $ rank $T = $ dim V; this is because rank $T = $ dim $V - $ dim $N(T) = $ dim $V - 0$. Conversely, if $R(T) = W$ and dim $W = $ dim V, then once again nullity $T = $ dim $V - $ rank $T = 0$ and so T is invertible.

Note that T^{-1} is also a linear transformation (assuming T is invertible). For if $\mathbf{w}_1 = T\mathbf{v}_1$ and $\mathbf{w}_2 = T\mathbf{v}_2$ then for any scalars α, β

$$T(\alpha\mathbf{v}_1 + \beta\mathbf{v}_2) = \alpha\mathbf{w}_1 + \beta\mathbf{w}_2.$$

Hence

$$T^{-1}(\alpha\mathbf{w}_1 + \beta\mathbf{w}_2) = \alpha\mathbf{v}_1 + \beta\mathbf{v}_2 = \alpha T^{-1}\mathbf{w}_1 + \beta T^{-1}\mathbf{w}_2.$$

So $T^{-1}: W \to V$ is also linear. This rather simple observation enables us to derive a non-trivial fact. Let the matrices A and B represent T and T^{-1} relative to preselected bases in V and W. Both matrices are square and of the same size (why?) and $AB = BA = I$. Indeed, for any vectors $\mathbf{v}$ in V and $\mathbf{w}$ in W, $T(T^{-1}\mathbf{w}) = \mathbf{w}$ and $T^{-1}(T\mathbf{v}) = \mathbf{v}$.

Notice that if for some matrices C and D, we have $CA = I = AD$, then

$$C = CI = CAD = ID = D.$$

Thus the matrix B with $AB = BA = I$ is unique. *Exercise*: if $AB = I$ and A is square show that $BA = I$.

3.4 Change of basis

Let $\{\mathbf{v}_1, \mathbf{v}_2, \ldots, \mathbf{v}_n\}$ be a basis of V. If $\mathbf{v}$ is in V then the column vector $\mathbf{x} = (x_1, \ldots, x_n)^t$ with

$$\mathbf{v} = x_1\mathbf{v}_1 + \cdots + x_n\mathbf{v}_n$$

gives the co-ordinates of $\mathbf{v}$ with respect to the basis $\{\mathbf{v}_1, \mathbf{v}_2, \ldots, \mathbf{v}_n\}$ and thus represents $\mathbf{v}$. But if we switch to another basis $\{\mathbf{u}_1, \mathbf{u}_2, \ldots, \mathbf{u}_n\}$ for V we will obtain a new co-ordinate vector $\mathbf{X} = (X_1, \ldots, X_n)^t$ for $\mathbf{v}$ where this time

$$\mathbf{v} = X_1\mathbf{u}_1 + \cdots + X_n\mathbf{u}_n.$$

We examine the transformation $(x_1, \ldots, x_n)^t \to (X_1, \ldots, X_n)^t$. Exactly as in Section 2 we may write

$$\left.\begin{aligned}
\mathbf{u}_1 &= p_{11}\mathbf{v}_1 + \cdots + p_{n1}\mathbf{v}_n \\
\mathbf{u}_2 &= p_{12}\mathbf{v}_1 + \cdots + p_{n2}\mathbf{v}_n \\
&\vdots \\
\mathbf{u}_n &= p_{1n}\mathbf{v}_1 + \cdots + p_{nn}\mathbf{v}_n
\end{aligned}\right\}.$$

Thus

$$X_1\mathbf{u}_1 + \cdots + X_n\mathbf{u}_n = (p_{11}X_1 + p_{12}X_2 + \cdots + p_{1n}X_n)\mathbf{v}_1$$
$$+ \cdots + (p_{n1}X_1 + \cdots + p_{nn}X_n)\mathbf{v}_n$$
$$= x_1\mathbf{v}_1 + \cdots + x_n\mathbf{v}_n.$$

Thus

$$\begin{bmatrix} x_1 \\ x_2 \\ \vdots \\ x_n \end{bmatrix} = \begin{bmatrix} p_{11} & p_{12} & \cdots & p_{1n} \\ p_{21} & p_{22} & & p_{2n} \\ \vdots & \vdots & & \\ p_{n1} & p_{n2} & & p_{nn} \end{bmatrix} \begin{bmatrix} X_1 \\ X_2 \\ \vdots \\ X_n \end{bmatrix}.$$

This should not really surprise us since $P = (p_{ij})$ represents the identity transformation $\mathbf{v} \to \mathbf{v}$ of V into V when $\{\mathbf{v}_1, \mathbf{v}_2, \ldots, \mathbf{v}_n\}$ acts as a basis in V regarded as the domain of the transformation, while $\{\mathbf{u}_1, \mathbf{u}_2, \ldots, \mathbf{u}_n\}$ acts as a basis for V regarded as the range space of the transformation.

We note that the identity transformation is very obviously invertible, so the matrix P has an inverse and we are justified in writing the equation

$$\mathbf{X} = P^{-1}\mathbf{x}.$$

Examples
(i) In $\mathbb{R}^2$ the basis change is made from the natural basis $\mathbf{e}_1 = (1, 0)^t, \mathbf{e}_2 = (0, 1)^t$ to the basis $(1, 1)^t, (1, 2)^t$. Find new co-ordinates of an arbitrary vector in terms of its original co-ordinates.

We have

$$\begin{bmatrix} 1 \\ 1 \end{bmatrix} = p_{11}\begin{bmatrix} 1 \\ 0 \end{bmatrix} + p_{21}\begin{bmatrix} 0 \\ 1 \end{bmatrix}$$

and

$$\begin{bmatrix} 1 \\ 2 \end{bmatrix} = p_{12}\begin{bmatrix} 1 \\ 2 \end{bmatrix} + p_{22}\begin{bmatrix} 0 \\ 1 \end{bmatrix},$$

thus

$$P = \begin{bmatrix} 1 & 1 \\ 1 & 2 \end{bmatrix}.$$

Hence

$$P^{-1} = \begin{bmatrix} 2 & -1 \\ -1 & 1 \end{bmatrix}.$$

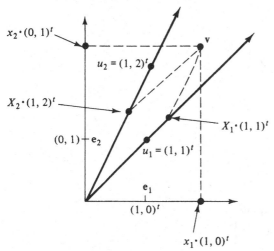

Fig. 3.2. $\mathbf{v}$ represented by $(x_1, x_2)^t$ and by $(X_1, X_2)^t$ as $x_1\mathbf{e}_1 + x_2\mathbf{e}_2 = X_1\mathbf{u}_1 + X_2\mathbf{u}_2$

So

$$\begin{bmatrix} X_1 \\ X_2 \end{bmatrix} = \begin{bmatrix} 2 & -1 \\ -1 & 1 \end{bmatrix} \begin{bmatrix} x_1 \\ x_2 \end{bmatrix}.$$

See Figure 3.2.

Remark

We remark here that if the original basis $\{\mathbf{v}_1, \mathbf{v}_2, \ldots, \mathbf{v}_n\}$ is the natural basis then, for example,

$$\mathbf{u}_1 = p_{11}\mathbf{e}_1 + \cdots + p_{n1}\mathbf{e}_n = \begin{bmatrix} p_{11} \\ p_{21} \\ \vdots \\ p_{n1} \end{bmatrix}$$

and this is the first column of P. We have in this case the interesting and rather simple relationship

$$P = [\mathbf{u}_1, \mathbf{u}_2, \ldots, \mathbf{u}_n],$$

i.e. the columns of P consist of the new basis vectors.

(ii) In $\mathbb{R}^2$ a basis change is made from the natural basis to $(\cos\theta, \sin\theta)^t$, $(-\sin\theta, \cos\theta)^t$ (Figure 3.3). By the remark above

$$P = \begin{bmatrix} \cos\theta & -\sin\theta \\ \sin\theta & \cos\theta \end{bmatrix},$$

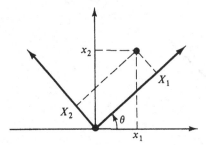

Fig. 3.3

so
$$P^{-1} = \begin{bmatrix} \cos\theta & \sin\theta \\ -\sin\theta & \cos\theta \end{bmatrix}.$$

Remark

Observe that if a change of basis is made from the natural basis to an orthonormal basis, then

$$P^t P = \begin{bmatrix} \mathbf{u}_1^t \\ \mathbf{u}_2^t \\ \vdots \\ \mathbf{u}_n^t \end{bmatrix} [\mathbf{u}_1, \mathbf{u}_2, \ldots, \mathbf{u}_n] = I,$$

so that
$$P^{-1} = P^t.$$

3.5   Equivalent matrices

Let $T: V \to W$ be represented by A relative to the basis $\{\mathbf{v}_1, \mathbf{v}_2, \ldots, \mathbf{v}_n\}$ in V and $\{\mathbf{w}_1, \mathbf{w}_2, \ldots, \mathbf{w}_n\}$ in W. If we make a change of basis in both V and W we have for some matrices P, Q

$$\mathbf{x} = P\mathbf{X}, \quad \mathbf{y} = Q\mathbf{Y},$$

where $\mathbf{X}, \mathbf{Y}$ are new co-ordinate vectors in V and W respectively. Hence T is now represented by $Q^{-1}AP$, as a simple substitution for $\mathbf{Y}$ and $\mathbf{X}$ in the equation $\mathbf{y} = A\mathbf{x}$ will prove. We say that the two matrices A, B are *equivalent* if for some non-singular matrices P, Q we have

$$B = Q^{-1}AP.$$

Thus two matrices are equivalent if they represent the same transformation relative to different bases.

Let A be $m \times n$ and consider the transformation $T: \mathbf{x} \to \mathbf{y} = A\mathbf{x}$ of $\mathbb{R}^n$ into $\mathbb{R}^m$. Let $r = \operatorname{rank} A = \dim R(T)$. Of course $R(T)$ is the column space of A because if $A = [\mathbf{a}_1, \mathbf{a}_2, \ldots, \mathbf{a}_n]$ where $\mathbf{a}_1, \mathbf{a}_2, \ldots, \mathbf{a}_n$ are the columns of A then

$$A\mathbf{x} = x_1\mathbf{a}_1 + x_2\mathbf{a}_2 + \cdots + x_n\mathbf{a}_n.$$

Thus r is the maximum number of linearly independent columns. Let $\{\mathbf{w}_1, \mathbf{w}_2, \ldots, \mathbf{w}_r\}$ be a basis of $R(T)$; for example, take any r linearly independent columns of A. Pick $\mathbf{v}_1, \mathbf{v}_2, \ldots, \mathbf{v}_r$ so that

$$\mathbf{w}_1 = A\mathbf{v}_1, \quad \mathbf{w}_2 = A\mathbf{v}_2, \ldots, \mathbf{w}_r = A\mathbf{v}_r.$$

For example, if $\mathbf{a}_1, \mathbf{a}_2, \ldots, \mathbf{a}_r$ are linearly independent take $\mathbf{v}_1 = \mathbf{e}_1$, $\mathbf{v}_2 = \mathbf{e}_2, \ldots, \mathbf{v}_r = \mathbf{e}_r$.

Since nullity $A = n - r$ we may pick $\mathbf{v}_{r+1}, \mathbf{v}_{r+2}, \ldots, \mathbf{v}_n$ to be a basis for $N(A)$. Then $\mathbf{v}_1, \ldots, \mathbf{v}_r, \mathbf{v}_{r+1}, \ldots, \mathbf{v}_n$ is a basis for $\mathbb{R}^n$ (why?). Now extend $\{\mathbf{w}_1, \mathbf{w}_2, \ldots, \mathbf{w}_r\}$ to a basis $\{\mathbf{w}_1, \mathbf{w}_2, \ldots, \mathbf{w}_r, \ldots, \mathbf{w}_m\}$ of $\mathbb{R}^n$. Then relative to these bases T is represented by

$$\left[\begin{array}{c|c} I_r & 0 \\ \hline 0 & 0 \end{array}\right],$$

so that for some non-singular matrices P, Q we have

$$Q^{-1}AP = \left[\begin{array}{c|c} I_r & 0 \\ \hline 0 & 0 \end{array}\right].$$

Evidently, if A is non-singular, then $r = n$.

3.6 Elementary matrices

A matrix obtained from the identity matrix by means of an elementary row or column operation is called an *elementary matrix*. There are thus three kinds of elementary matrix; typical examples are as follows. (Undisplayed entries are zero.)

$$\begin{bmatrix} 1 & 0 & 0 & 0 & 0 & 0 \\ & 0 & & & 1 & \\ & & 1 & & & \\ & & & 1 & & \\ & 1 & & & 0 & \\ & & & & & 1 \end{bmatrix} \begin{bmatrix} 1 & 0 & 0 & 0 & 0 & 0 \\ & 1 & & & & \\ & & 1 & & & \\ & & & 1 & & \\ & \lambda & & & 1 & \\ & & & & & 1 \end{bmatrix} \begin{bmatrix} 1 & 0 & 0 & 0 & 0 & 0 \\ \lambda & & & & & \\ & & 1 & & & \\ & & & 1 & & \\ & & & & 1 & \\ & & & & & 1 \end{bmatrix}$$

Recall that to perform an elementary row operation on a matrix A is the same as premultiplying A by the corresponding elementary matrix (one obtained from the identity by the same row operation).

It is easy to see that elementary matrices are non-singular, their inverses being elementary matrices of the same type.

Clearly column operations amount to basis changes; to see this note that a column operation on a matrix A is equivalent to post-multiplication, say to AE, but since E is non-singular E represents a basis change for which $P = E$. Similarly row operations amout to basis changes in the range space, since they take the form EA.

We know that any matrix A may be reduced by row and column operations to the echelon form

$$\left[\begin{array}{c|c} I_k & 0 \\ \hline 0 & 0 \end{array}\right].$$

We thus infer that for some non-singular matrices

$$Q^{-1}AP = \left[\begin{array}{c|c} I_k & 0 \\ \hline 0 & 0 \end{array}\right]$$

and in fact each of P and Q is a product of elementary matrices. Clearly, if A is non-singular then $k = n$ (otherwise $Q^{-1}APe_{k+1} = 0$, hence $A(Pe_{k+1}) = 0$, so $N(A) \neq \{0\}$). Thus for A non-singular we conclude $A = QP^{-1}$ and so A is itself a product of elementary matrices.

For a general matrix A we infer from the equation above that $k = \text{rank } A$, since that is the dimension of the range space of $Q^{-1}AP$. The latter matrix represents the same linear transformation as does A (though relative to different bases), so it must have identical rank.

We conclude our revision of rank by deducing that A and A^t have identical rank. First note that if P is non-singular then $(P^{-1})^t = (P^t)^{-1}$. Indeed, since $P \cdot P^{-1} = I$ we have, taking transposes, $(P^{-1})^t \cdot P^t = I$, and the result follows. But now

$$Q^{-1}AP = \left[\begin{array}{c|c} I_r & 0 \\ \hline 0 & 0 \end{array}\right],$$

where $r = \text{rank } A$. Taking transposes we obtain

$$P^t A^t (Q^t)^{-1} = \left[\begin{array}{c|c} I_r & 0 \\ \hline 0 & 0 \end{array}\right].$$

Now interpreting P^t and Q^t as the matrices corresponding to basis changes, we read off that $r = \text{rank } A^t$.

3.7 The orthogonal complement $S^\perp$

Recall that, by definition, a line segment ℓ is perpendicular to a plane P if it is perpendicular to *every* line in that plane (Figure 3.4). We generalise this as follows. If S is any set in $\mathbb{R}^n$ we say that $\mathbf{v}$ is orthogonal to S if for all $\mathbf{s}$ in S

$$\langle \mathbf{v}, \mathbf{s} \rangle = 0,$$

that is, $\mathbf{v}$ is perpendicular to every vector lying in S. For a given S the set of such vectors $\mathbf{v}$, is called the *orthogonal complement* of S and is denoted by $S^\perp$; that is,

$$S^\perp = \{\mathbf{v}: \langle \mathbf{v}, \mathbf{s} \rangle = 0 \text{ for all } \mathbf{s} \in S\}.$$

In Figure 3.5 we sketch three examples of $S^\perp$ when S is (a) a pair of vectors, (b) a line through $\mathbf{0}$ and (c) a plane through $\mathbf{0}$.
We note that $S^\perp$ is always a vector space, even if S is not a subspace. This is easily verified; if $\mathbf{u}$ and $\mathbf{v}$ are in $S^\perp$ then, for all $\mathbf{s}$ in S

$$\langle \alpha\mathbf{u} + \beta\mathbf{v}, \mathbf{s} \rangle = \alpha\langle \mathbf{u}, \mathbf{s} \rangle + \beta\langle \mathbf{v}, \mathbf{s} \rangle = 0.$$

Evidently $S \subseteq S^{\perp\perp}$, since if $\mathbf{s}$ is in S, then for every $\mathbf{v}$ in $S^\perp$ we have $\langle \mathbf{s}, \mathbf{v} \rangle = 0$. When S is a subspace (and in fact only then) the inclusion may be improved to an equality. The reason for this is the following. Suppose that $\{\mathbf{u}_1, \ldots, \mathbf{u}_k\}$ is an orthonormal basis for S and that $\{\mathbf{u}_{k+1}, \ldots, \mathbf{u}_{k+\ell}\}$ is an orthonormal basis for $S^\perp$. $\{\mathbf{u}_1, \ldots, \mathbf{u}_{k+\ell}\}$ is then an orthonormal basis for the whole space $\mathbb{R}^n$. Indeed, if this were not a basis we could extend the linearly independent set $\{\mathbf{u}_1, \ldots, \mathbf{u}_{k+\ell}\}$ to an orthonormal basis, say to $\{\mathbf{u}_1, \ldots, \mathbf{u}_{k+\ell}, \ldots, \mathbf{u}_n\}$. But, then $\mathbf{u}_n$ would be orthogonal to the vectors $\{\mathbf{u}_1, \ldots, \mathbf{u}_k\}$ and, since these vectors span S, $\mathbf{u}_n$ would also be orthogonal to the whole of S. Evidently, that puts $\mathbf{u}_n$ into $S^\perp$ and so makes it linearly dependent on $\{\mathbf{u}_{k+1}, \ldots, \mathbf{u}_{k+\ell}\}$, which is

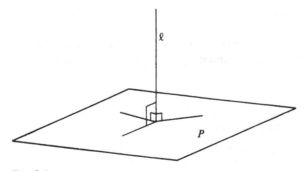

Fig. 3.4

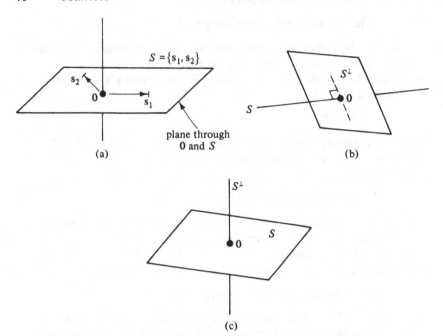

Fig. 3.5

absurd. Thus $\{\mathbf{u}_1, \ldots, \mathbf{u}_{k+\ell}\}$ is an orthonormal basis for $\mathbb{R}^n$ and we now read off the information that

$$\alpha_1 \mathbf{u}_1 + \cdots + \alpha_{k+\ell} \mathbf{u}_{k+\ell} \in S^{\perp\perp} \leftrightarrow \alpha_{k+1} = \alpha_{k+2} = \cdots = \alpha_{k+\ell} = 0.$$

3.8 The orthogonal complement of $N(A)$

If A is the 1×2 matrix (a_1, a_2) we see (Figure 3.6) that $N(A)$ is the line with equation

$$a_1 x_1 + a_2 x_2 = 0,$$

which has normal direction $(a_1, a_2)^t$. Thus $N(A)^{\perp}$ is the set of multiples of $(a_1, a_2)^t$; thus $N(A)^{\perp} = R(A^t)$. Similarly, if A is the 2×3 matrix

$$\begin{pmatrix} a_1 & a_2 & a_3 \\ b_1 & b_2 & b_3 \end{pmatrix}$$

we have $(x_1, x_2, x_3)^t \in N(A)$ if and only if

$$\left. \begin{aligned} a_1 x_1 + a_2 x_2 + a_3 x_3 &= 0 \\ b_1 x_1 + b_2 x_2 + b_3 x_3 &= 0 \end{aligned} \right\}.$$

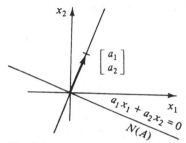

Fig. 3.6

Thus $(a_1, a_2, a_3)^t$ and $(b_1, b_2, b_3)^t$ are each orthogonal to every vector **x** in $N(A)$. But $R(A^t)$ here is the set of linear combinations of $(a_1, a_2, a_3)^t$ and $(b_1, b_2, b_3)^t$ and such vectors thus lie in $N(A)^\perp$. We now give a proof that this phenomenon occurs generally.

Theorem

For any real matrix A

$$R(A^t) = N(A)^\perp.$$

Proof. Since $N(A)^{\perp\perp} = N(A)$ it suffices to prove that $R(A^t)^\perp = N(A)$ or, equivalently, $R(A)^\perp = N(A^t)$.

Suppose then that $z \in R(A)^\perp$. Thus for every $y \in R(A)$, $\langle y, z \rangle = 0$. Hence for every vector **x**, since $A\mathbf{x} \in R(A)$, we have

$$0 = \mathbf{z}^t A \mathbf{x} = (A^t \mathbf{z})^t \mathbf{x} = \langle A^t \mathbf{z}, \mathbf{x} \rangle.$$

Thus $A^t \mathbf{z}$ is orthogonal to each and every vector **x**. Hence $A^t \mathbf{z} = 0$. Thus $\mathbf{z} \in N(A^t)$, i.e. $R(A)^\perp \subseteq N(A^t)$.

Now suppose $\mathbf{z} \in N(A^t)$. We show $\mathbf{z} \in R(A)^\perp$.

Consider any **y** in $R(A)$. Let $\mathbf{y} = A\mathbf{x}$. Then

$$A^t \mathbf{z} = 0 \Rightarrow 0 = \mathbf{x}^t A^t \mathbf{z} = (A\mathbf{x})^t \mathbf{z} = \mathbf{y}^t \mathbf{z} = \langle \mathbf{y}, \mathbf{z} \rangle,$$

so **z** is orthogonal to any **y** in $R(A)$. Thus $N(A^t) \subseteq R(A)^\perp$ and the theorem is proved.

Example

Prove that for a real $m \times n$ matrix A

$$\text{rank}\,(A^t A) = \text{rank}\,(A A^t) = \text{rank}\,(A).$$

Proof. In fact we show $R(AA^t) = R(A)$. This is equivalent to showing $N((AA^t)^t)^\perp = N(A^t)^\perp$ or $N(AA^t) = N(A^t)$.

If $z \in N(A^t)$ then $A^t z = 0$, so $AA^t z = 0$, giving $z \in N(AA^t)$. Thus $N(A^t) \subseteq N(AA^t)$.

If $z \in N(AA^t)$ then $AA^t z = 0$, so $z^t AA^t z = 0$ or $\| A z \|^2 = 0$. Hence $Az = 0$. Thus $N(AA^t) \subseteq N(A^t)$.

3.9   Rank of AB

Let A be $m \times k$ and B be $k \times n$. Now if $\mathbf{x} \in \mathbb{R}^n$ then $B\mathbf{x}$ is in $\mathbb{R}^k$ and $AB\mathbf{x}$ is in $\mathbb{R}^m$. We note that $R(AB)$ is the image under A of $R(B)$ whereas $R(A)$ is the image of *all* of $\mathbb{R}^k$ under A, see Figure 3.7.

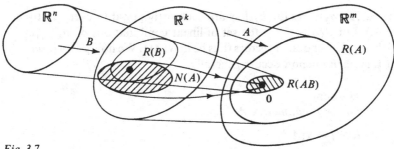

Fig. 3.7

We have

$$\text{rank}(AB) = \dim R(AB) \leqslant \dim R(A) = \text{rank}(A).$$

Also

$$\text{rank}(AB) = \text{rank}((AB)^t)$$
$$= \text{rank}(B^t A^t) \leqslant \text{rank}(B^t) = \text{rank}(B).$$

Thus

$$\text{rank}(AB) \leqslant \text{rank } A, \text{rank } B.$$

We can get a lower estimate on rank (AB) by considering the action of A on $R(B)$. Let T be the transformation from $R(B)$ to $\mathbb{R}^m$ where

$$T\mathbf{z} = A\mathbf{z} \quad \text{for } \mathbf{z} \in R(B).$$

Then

$$N(T) = N(A) \cap R(B).$$

Hence

$$\text{nullity}(T) \leqslant \text{nullity}(A) = k - \text{rank}(A).$$

Now

$$\text{rank}(T) = \text{rank}(AB),$$
$$\text{rank}(T) + \text{nullity}(T) = \dim R(B) = \text{rank}(B),$$
$$\text{rank}(B) - \text{rank}(AB) = \text{nullity}(T) \leqslant k - \text{rank}(A).$$

Thus

$$\text{rank}\,(A) + \text{rank}\,(B) - k \leqslant \text{rank}\,(AB).$$

Example

If $C = AB$ and both C and A have rank k, what is the rank of B?

Solution. By the first inequality, $k = \text{rank}\,C \leqslant \text{rank}\,B$. By the second inequality, $k + \text{rank}\,(B) - k \leqslant \text{rank}\,(C) = k$. Hence $\text{rank}\,(B) = k$.

3.10 Rank via subdeterminants

If A is any matrix, then a matrix obtained from A by the deletion of some of the rows and/or columns of A is called a submatrix of A.

Theorem

An $m \times n$ matrix A has rank r if and only if

(i)   *every square $(r + 1) \times (r + 1)$ submatrix is singular, and*

(ii) *at least **one** $r \times r$ submatrix is non-singular.*

Proof. Let A have rank r. Write $A = [\mathbf{a}_1, \ldots, \mathbf{a}_n]$ where $\mathbf{a}_1, \ldots, \mathbf{a}_n$ are the columns of A. We may suppose that $\mathbf{a}_1, \ldots, \mathbf{a}_r$ are the maximum number of linearly independent *rows* of $\{\mathbf{a}_1, \ldots, \mathbf{a}_n\}$. Choose r such rows to obtain an $r \times r$ non-singular submatrix.

We now drop the assumption that $\mathbf{a}_1, \ldots, \mathbf{a}_r$ are linearly independent, but continue to assume that A has rank r. To see that any $(r + 1) \times (r + 1)$ submatrix is singular, consider for example a submatrix B which is also a submatrix of $[\mathbf{a}_1, \ldots, \mathbf{a}_{r+1}]$. Now for some scalars $\alpha_1, \ldots, \alpha_{r+1}$ not all zero we have

$$\alpha_1 \mathbf{a}_1 + \cdots + \alpha_{r+1} \mathbf{a}_{r+1} = \mathbf{0}.$$

Striking out irrelevant rows in this relationship shows that B has rank at most r so is singular.

The theorem is now clear.

Example

The matrix A below has rank 2.

$$\begin{bmatrix} 1 & 1 & 2 \\ 2 & 1 & 1 \\ 5 & 1 & -2 \end{bmatrix}$$

Indeed $\det(A) = 0$, so all (i.e. all one of) 3×3 submatrices of A are singular, and

$$\begin{pmatrix} 1 & 1 \\ 2 & 1 \end{pmatrix}$$

is clearly non-singular (since $\det B = -1$).

3.11   Solubility of equations

We close this discussion of matrices with a trivial remark: the equation $Ax = b$ has a solution if and only if $b \in R(A)$. This is of course equivalent to demanding that $\text{rank}(A|b) = \text{rank}(A)$. We shall have more to say on this topic in the chapter on inverses (see Section 8.3).

3.12   Revision exercises

1. If

$$A = \begin{pmatrix} 1 & -2 \\ -2 & 3 \end{pmatrix}, \quad B = \begin{pmatrix} -2 & 1 \\ 1 & 1 \end{pmatrix}$$

form $(AB)^t$ and $B^t A^t$ and verify that these are equal. Use these examples show that each of the following assertions is *not* true in general
   (i) $AB = BA$,
   (ii) $(AB)^t = A^t B^t$,
   (iii) $A = A^t$ and $B = B^t$ implies $AB = (AB)^t$. (Note if $A = A^t$ we say that A is *symmetric*.)
  2. If A is symmetric show that $B^t AB$ is symmetric.
  3. If

$$B = \begin{pmatrix} 1 & 0 & -2 \\ -1 & 3 & 0 \end{pmatrix},$$

calculate $B^t B$ and BB^t and hence verify that these are both symmetric but *not* equal.

Revision of partitioned matrices: If

$$A = \begin{pmatrix} A_1 & A_2 \\ A_3 & A_4 \end{pmatrix}, \quad B = \begin{pmatrix} B_1 & B_2 \\ B_3 & B_4 \end{pmatrix}$$

where the A_i and B_i are submatrices, recall that

$$AB = \begin{pmatrix} A_1 B_1 + A_2 B_3 & A_1 B_2 + A_2 B_4 \\ A_3 B_1 + A_4 B_3 & A_3 B_2 + A_4 B_4 \end{pmatrix}.$$

4. If

$$A = (A_1 | A_2) = \begin{pmatrix} 4 & 3 & -2 & 1 & 4 \\ 2 & -5 & 6 & 3 & -1 \end{pmatrix}, \quad B = \begin{bmatrix} 0 & -1 & 3 \\ 2 & -1 & 6 \\ 5 & 2 & 1 \\ -3 & 4 & -1 \\ 2 & -1 & 2 \end{bmatrix}$$

calculate AB by straightforward multiplication and also by block manipulations (see the formula cited above).

5. If A_1, A_2, A_3 are non-singular matrices prove that

$$\begin{bmatrix} A_1 & 0 & 0 \\ 0 & A_2 & 0 \\ 0 & 0 & A_3 \end{bmatrix} = \begin{bmatrix} A_1^{-1} & 0 & 0 \\ 0 & A_2^{-1} & 0 \\ 0 & 0 & A_3^{-1} \end{bmatrix},$$

6. In $\mathbb{R}^3$, the basis is changed from $\{(1,0,0)^t, (0,1,0)^t, (0,0,1)^t\}$ to $\{(1,2,3)^t, (3,1,2)^t, (2,3,1)^t\}$. Express the new co-ordinates in terms of the old.

7. If new co-ordinates are given in terms of the old by the equation

$$\begin{bmatrix} X_1 \\ X_2 \\ X_3 \end{bmatrix} = \begin{bmatrix} 1 & 2 & 1 \\ 2 & 1 & 0 \\ 1 & 1 & 1 \end{bmatrix} \begin{bmatrix} x_1 \\ x_2 \\ x_3 \end{bmatrix}$$

and the old co-ordinates refer to the natural basis, what is the new basis?

8. In $\mathbb{R}^2$ new axes are obtained by rotating the old axes through $\pi/6$. Express the equation of the ellipse

$$x_1^2 + 2x_2^2 = 1$$

in terms of new co-ordinates X_1, X_2.

9. Explore the effect of multiplying the matrix

$$A = \begin{bmatrix} a & b & c \\ d & e & f \\ g & h & i \end{bmatrix}$$

on the left or right by the elementary matrices

$$\begin{bmatrix} 0 & 1 & 0 \\ 1 & 0 & 0 \\ 0 & 0 & 1 \end{bmatrix}, \quad \begin{bmatrix} 1 & 0 & 0 \\ 0 & 1 & 0 \\ \lambda & 0 & 1 \end{bmatrix}, \quad \begin{bmatrix} 1 & \lambda & 0 \\ 0 & 1 & 0 \\ 0 & 0 & 1 \end{bmatrix}, \quad \begin{bmatrix} 1 & 0 & 0 \\ 0 & \lambda & 0 \\ 0 & 0 & 1 \end{bmatrix}.$$

10. Let $(p_1, p_2, p_3)^t, (q_1, q_2, q_3)^t$ be non-zero vectors in $\mathbb{R}^3$. Explain why $(r_1, r_2, r_3)^t$ is orthogonal to both these vectors if and only if

$$\left. \begin{aligned} p_1 r_1 + p_2 r_2 + p_3 r_3 &= 0 \\ q_1 r_1 + q_2 r_2 + q_3 r_3 &= 0 \end{aligned} \right\}.$$

Solve these equations for $(r_1, r_2, r_3)'$ and deduce that the vector

$$\mathbf{r} = \begin{vmatrix} \mathbf{e}_1 & \mathbf{e}_2 & \mathbf{e}_3 \\ p_1 & p_2 & p_3 \\ q_1 & q_2 & q_3 \end{vmatrix}$$

is orthogonal to p, q. (Here $\mathbf{e}_1 = (1, 0, 0)'$, $\mathbf{e}_2 = (0, 1, 0)'$, $\mathbf{e}_3 = (0, 0, 1)'$.)

11. Using the last example find the orthogonal complement in $\mathbb{R}^3$ of the subspace S spanned by $(0, 0, -1)'$ and $(1, 2, 3)'$.

12. For each of the following matrices find $R(A)$, $N(A)$, $R(A')$, $N(A')$ and draw diagrams to illustrate these subspaces. In each case check that

(a) $R(A') = N(A)^{\perp}$ (b) $R(A)^{\perp} = N(A')$.

(i) $\begin{pmatrix} 1 & 1 \\ -1 & -1 \end{pmatrix}$ (ii) $\begin{pmatrix} 5 & 0 \\ 3 & 0 \end{pmatrix}$ (iii) $\begin{pmatrix} 1 & -2 \\ -3 & 6 \end{pmatrix}$ (iv) $\begin{pmatrix} 1 & 2 \\ 3 & 4 \end{pmatrix}$

13. Find the ranks of the following matrices. From each matrix pick out a set of linearly independent columns and a set of linearly independent rows which both contain the same number of vectors as the rank.

(i) $\begin{pmatrix} 2 & 4 & 6 \\ 1 & 2 & 3 \end{pmatrix}$ (ii) $\begin{bmatrix} 2 & 1 & 1 \\ 5 & 1 & -2 \\ 1 & 1 & 2 \end{bmatrix}$ (ii) $\begin{bmatrix} 2 \\ 3 \\ 1 \end{bmatrix}$

14. Show that the theorem of Section 3.8 remains true when A is a complex matrix provided that A' is replaced by A^*, the complex conjugate of the transpose. (All the theorems about rank are therefore also true for complex matrices if A^* replaces A'.)

15. If P is non-singular show that the two systems of linear equations

$$A\mathbf{x} = \mathbf{b}; \quad PA\mathbf{x} = P\mathbf{b}$$

have the same solution set.

In solving the equations $A\mathbf{x} = \mathbf{b}$ by successive elimination of the variables we are essentially reducing the augmented matrix $(A|\mathbf{b})$ to echelon form by a sequence of *row* operations. Explain this and what it has to do with the previous paragraph and the fact that

$$P(A|\mathbf{b}) = (PA|P\mathbf{b}).$$

16. Consider the n straight lines in $\mathbb{R}^2$ defined by

$$a_i x + b_i y = c_i \qquad\qquad (i = 1, 2, \ldots, n).$$

Explain why the condition that they should all pass through a common point is that the matrices

$(\mathbf{a}|\mathbf{b})$, $(\mathbf{a}|\mathbf{b}|\mathbf{c})$

should have the same rank (the $n \times 1$ column vectors $\mathbf{a}, \mathbf{b}$ and $\mathbf{c}$ are defined in the obvious way).

17. Prove that three points $(x_i, y_i)^t$ $(i = 1, 2, 3)$ are collinear (i.e. lie on the same line) if and only if the matrix

$$\begin{bmatrix} x_1 & y_1 & 1 \\ x_2 & y_2 & 1 \\ x_3 & y_3 & 1 \end{bmatrix}$$

has rank less than 3.

18. Show that AB is *singular* if A is $m \times n$ and B is $n \times m$ where $n < m$.

19. Show that AB has the same rank as A if B is non-singular.

20. Find a relationship between the dimensions of the solution subspaces of the two systems of linear equations

$$A\mathbf{x} = 0; \quad A^t\mathbf{y} = 0$$

and show that when A is a square matrix, they are the same.

21. Given a real $m \times n$ matrix A and an $n \times 1$ column vector $\mathbf{b}$, show that only *one* of the following systems can be consistent.

(i) $A\mathbf{x} = \mathbf{b}$ (ii) $A^t\mathbf{y} = 0$ and $\mathbf{y}^t\mathbf{b} \neq 0$.

(Hint: Use the theorem of Section 3.8.)

22. A subspace S of dimension $(n-2)$ in $\mathbb{R}^n$ is given as the solution set of the equations

$$\left. \begin{array}{l} \langle \mathbf{u}, \mathbf{x} \rangle = 0 \\ \langle \mathbf{v}, \mathbf{x} \rangle = 0 \end{array} \right\}.$$

Show that the general hyperplane containing S has for some λ the equation

$$\langle \mathbf{u} + \lambda\mathbf{v}, \mathbf{x} \rangle = 0.$$

How does this generalize to an *affine* subspace S of dimension $(n-2)$?

23. If $S \subseteq \mathbb{R}^n$ show that $\mathbf{x}$ is orthogonal to every element of S if and only if $\mathbf{x}$ is orthogonal to every element of Lin (S).

4

Projections

4.1 Direct sums

If X is an n dimensional vector space and $\{x_1, \ldots, x_n\}$ is a basis consider the subspaces

$$Y = \text{Lin}\{x_1, \ldots, x_k\},$$
$$Z = \text{Lin}\{x_{k+1}, \ldots, x_n\}.$$

If x is any element of X, then

$$x = \alpha_1 x_1 + \cdots + \alpha_k x_k + \alpha_{k+1} x_{k+1} + \cdots + \alpha_n x_n$$
$$= (\alpha_1 x_1 + \cdots + \alpha_k x_k) + (\alpha_{k+1} x_{k+1} + \cdots + \alpha_n x_n),$$

where $\alpha_1, \ldots, \alpha_n$ are unique. Thus x may be written uniquely as

$$x = y + z \quad \text{with} \quad y \in Y \quad \text{and} \quad z \in Z.$$

Generally, if Y and Z are subspaces of a vector space X such that any x in X may be written uniquely as

$$x = y + z \quad \text{with} \quad y \in Y \quad \text{and} \quad z \in Z,$$

then we say that X is a *direct sum* of Y and Z; symbolically this is rendered

$$X = Y \oplus Z.$$

Example
If S is a subset of $\mathbb{R}^n$ recall that $S^\perp$ is defined by

$$S^\perp = \{x : \langle x, s \rangle = 0 \quad \text{for all} \quad s \in S\}.$$

If S happens to be a subspace we have the result

$$\mathbb{R}^n = S \oplus S^\perp.$$

To see this suppose that $x = s + t = s' + t'$; then $0 = (s - s') + (t - t')$ and

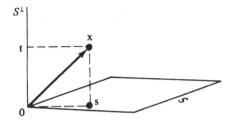

Fig. 4.1

since $(\mathbf{s} - \mathbf{s}') \in S$ and $(\mathbf{t} - \mathbf{t}') \in S^\perp$ we have by Pythagoras' theorem (see Figure 4.1) that

$$0 = \|\mathbf{s} - \mathbf{s}'\|^2 + \|\mathbf{t} - \mathbf{t}'\|^2,$$

so

$$0 = \mathbf{s} - \mathbf{s}' \quad \text{and} \quad 0 = \mathbf{t} - \mathbf{t}',$$

thus securing uniqueness. We have already shown that $S \cup S^\perp$ spans $\mathbb{R}^n$ (see Chapter 3, Section 3.7).

4.2 Projections

If $X = Y \oplus Z$ we define a mapping $P : X \to Y$ as follows. If $\mathbf{x}$ is in X then there is a unique $\mathbf{y}$ in Y and a unique $\mathbf{z}$ in Z such that $\mathbf{x} = \mathbf{y} + \mathbf{z}$. We let $P\mathbf{x}$ be $\mathbf{y}$. Thus

$$\mathbf{z} = (I - P)\mathbf{x}.$$

$P\mathbf{x}$ is called the *projection* of $\mathbf{x}$ *onto Y parallel* to Z. For a physical explanation of this terminology in three dimensions, consider an 'idealised' pin-hole projector which sends out a single beam of light (i.e. in the form of a straight line) and a two dimensional screen. The image (a point of light) on the screen obtained when the projector is placed at some point in space depends on both the position of the screen and the direction in which the projector emits the light beam. It projects from $\mathbf{x}$ onto the 'screen' Y parallel to a 'direction' Z. Now consider a one-dimensional screen (an infinitely thin line of screen material) and a projector which emits a sheet of light (two-dimensional emission). The sheet of light may be characterised as being parallel to a given two-dimensional space Z. We place the projector at a point $\mathbf{x}$ and the projected sheet lights up a single point on the screen (at the intersection of the light sheet and screen).

We claim that P is a linear transformation. This results from the uniqueness of the representation. Suppose $x = y + z$ and $x' = y' + z'$ with y and y' in Y and z and z' in Z. Then for any scalars α and β we have

$$\alpha x + \beta x' = \alpha(y + z) + \beta(y' + z')$$
$$= (\alpha y + \beta y') + (\alpha z + \beta z')$$

and so $P(\alpha x + \beta x')$ is the point $\alpha y + \beta y'$.

Note that for any z in Z, since $z = 0 + z$ we have $Pz = 0$; conversely, if $Px = 0$ then $x = (I - P)x$ and this is in Z by our earlier observation. Thus $N(P) = Z$.

Clearly the range of P is Y; now let y be in Y, then $y = y + 0$, so $Py = y$. But, for any x, Px is in Y, so $P(Px) = (Px)$. Thus

$$P^2 = P$$

and we refer to this relation by saying that P is *idempotent*.

4.3   Idempotents are projections

Let P be a linear mapping with $P^2 = P$; we are going to. show that P represents a projection. With this in mind, let $Y = $ range P and $Z = N(P)$. We claim that $X = Y \oplus Z$.

Given x consider $x - Px$. Clearly

$$P(x - Px) = Px - P^2x = Px - Px = 0.$$

Thus $z = x - Px \in Z$ and $y = Px \in Y$ and, of course, $x = y + z$. Next, we must check whether this representation for x is unique. Suppose therefore that $x = y' + z'$ with $y' \in Y$ and $z' \in Z$. Observe that

$$y - y' = z' - z$$

so

$$Py - Py' = Pz' - Pz = 0.$$

Put $y' = Px'$. Then

$$y = Px = P^2x = P(Px) = Py. = Py' = P(Px') = P^2x' = Px' = y'.$$

Hence also $z = z'$. Thus P projects onto Y parallel to Z.

4.4   Orthogonal projections

If S is a subspace of $\mathbb{R}^n$ the projection P onto S parallel to $S^\perp$ is referred to more simply as the *orthogonal projection onto S*.

Let us identify P with the matrix which represents it. Then $P^2 = P$; moreover, for any $\mathbf{x}$ and $\mathbf{y}$ in $\mathbb{R}^n$ we note that $P\mathbf{x}$ lies in S while $(I - P)\mathbf{y}$ lies in $S^\perp$ so the two vectors are orthogonal, i.e.

$$\langle P\mathbf{x}, (I - P)\mathbf{y} \rangle = 0 \quad \text{for any } \mathbf{x} \text{ and } \mathbf{y}.$$

Thus

$$\mathbf{x}^t P^t (I - P)\mathbf{y} = 0 \text{ for any } \mathbf{x} \text{ and } \mathbf{y}.$$

Taking for $\mathbf{x}$ the basis vector $\mathbf{e}_i$ (with 1 in position i and zeros elsewhere) and for $\mathbf{y}$ the vector $\mathbf{e}_j$ we obtain that the i,jth element of $P^t(I - P)$ satisfies

$$(P^t(I - P))_{ij} = 0.$$

Since this result is valid for all i and j we have

$$P^t(I - P) = 0$$

or $P^t = P^t P$. From this we deduce $P^{tt} = P^t P^{tt} = P^t$, so that P is symmetric. The argument above may be reversed (also we have $R(P)^\perp = N(P^t) = N(P)$). Thus orthogonal projections are precisely those which are represented by *symmetric idempotent* matrices.

4.5   Representation of an orthogonal projection

The easiest way to arrive at a formula for the orthogonal projection of a point onto a subspace S is to refer to an orthonormal basis for S. However, this is not necessarily the most convenient way to calculate the projection, as we shall see in the next section. If $\{\mathbf{e}_1, \mathbf{e}_2, \ldots, \mathbf{e}_k\}$ is an orthonormal basis for S the projection is

$$P\mathbf{x} = \langle \mathbf{x}, \mathbf{e}_1 \rangle \mathbf{e}_1 + \cdots + \langle \mathbf{x}, \mathbf{e}_k \rangle \mathbf{e}_k.$$

Indeed, if $\{\mathbf{e}_1, \ldots, \mathbf{e}_k, \mathbf{e}_{k+1}, \ldots, \mathbf{e}_n\}$ is an orthonormal basis for the whole of space then

$$x = \underbrace{(\langle x, \mathbf{e}_1 \rangle \mathbf{e}_1 + \cdots + \langle x, \mathbf{e}_k \rangle \mathbf{e}_k)}_{\text{in } S} + \underbrace{(\langle x, \mathbf{e}_{k+1} \rangle \mathbf{e}_{k+1} + \cdots + \langle x, \mathbf{e}_n \rangle \mathbf{e}_n)}_{\text{in } S^\perp}$$

and so the result follows.

4.6   Projection onto a column space

Let A be a matrix or size $m \times n$ and of rank n. We know that the rank of $A^t A$ is also n, and we note that $A^t A$ is square of size $n \times n$. It is natural to consider its inverse $(A'A)^{-1}$. The relation

$$(A'A)^{-1}A'A = I$$

is of interest, since it shows that the matrix $L = (A'A)^{-1}A'$ is a left inverse for A in the sense that $LA = I$. One is immediately tempted to study the matrix AL; now, clearly not much more can at first be said than that $ALAL = AL$ (playing on the fact that $LA = I$). Thus AL is an idempotent. A natural question is whether it is symmetric. Indeed it is. Thus $P = AL$ represents an orthogonal projection onto its own range. We determine that range. If $y = Px$ then $y = ALx$ so y is in $R(A)$. But if z is in $R(A)$ then $z = Ax'$, say, and $z = A(LA)x' = AL(Ax')$, so z is in the range of P.

We have thus arrived at a more useful formula for the determination of an orthogonal projection. For example if $\{x_1,\ldots,x_k\}$ is any basis for a subspace S of $\mathbb{R}^n$ form the matrix A whose columns are the vectors $\{x_1,\ldots,x_k\}$. Then A is $n \times k$ and of rank k. Now calculate P as above. This is less messy than the construction of an orthonormal basis for S out of $\{x_1,\ldots,x_k\}$.

Example

Let $S = \mathrm{Lin}(1, -1)'$. Find a matrix which represents orthogonal projection onto S.
Take

$$A = \begin{bmatrix} 1 \\ -1 \end{bmatrix}.$$

Then

$$A'A = (1, -1)\begin{bmatrix} 1 \\ -1 \end{bmatrix} = 2.$$

Thus

$$P = \begin{bmatrix} 1 \\ -1 \end{bmatrix}\frac{1}{2}(1, -1) = \frac{1}{2}\begin{bmatrix} 1 & -1 \\ -1 & 1 \end{bmatrix}.$$

Note that

$$P\begin{bmatrix} x \\ y \end{bmatrix} = \frac{1}{2}\begin{bmatrix} x - y \\ -(x - y) \end{bmatrix}.$$

We check this result (Figure 4.2) using the formula of Section 4.5.

$$Px = \left\langle x, \frac{1}{\sqrt{2}}(1, -1)' \right\rangle \frac{1}{\sqrt{2}}\begin{bmatrix} 1 \\ -1 \end{bmatrix} = \frac{1}{2}(x - y)\begin{bmatrix} 1 \\ -1 \end{bmatrix}.$$

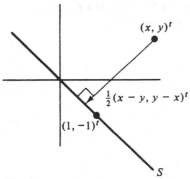

Fig. 4.2

4.7 Least squares analysis

We use projection matrices to find the best straight line through a given set of points which arise, say, as experimental data. For example, suppose that a car travels with constant velocity v starting at some point α. The distance y it has travelled by time t is measured for various values of t and tabulated thus:

t	0	3	5	8	10
y	2	5	6	9	11

The tabulated points are plotted in Figure 4.3. It soon transpires that the plotted points are not collinear, that is, if we seek values α and v so

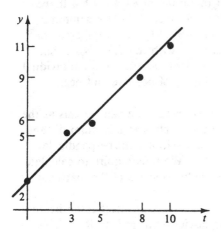

Fig. 4.3

that $y = \alpha + vt$ describes the car's motion we find that the equations to be satisfied by α and v shown below are inconsistent.

$$2 = \alpha \qquad\qquad (1)$$
$$5 = \alpha + 3v \qquad\qquad (2)$$
$$6 = \alpha + 5v \qquad\qquad (3)$$
$$9 = \alpha + 8v \qquad\qquad (4)$$
$$11 = \alpha + 10v \qquad\qquad (5)$$

or

$$\mathbf{b} = A\mathbf{x}, \quad \text{where} \quad \mathbf{b} = (2, 5, 6, 9, 11)^t$$

and $\mathbf{x} = (\alpha, v)^t$ while

$$A = \begin{bmatrix} 1 & 0 \\ 1 & 3 \\ 1 & 5 \\ 1 & 8 \\ 1 & 10 \end{bmatrix}.$$

For instance, the first two equations give $v = 1$ while the first and third equations give $v = 0.8$. Let us assume that experimental errors $r_1, \ldots, r_5$ have crept into the data. Then $\mathbf{b} - \mathbf{r} = A\mathbf{x}$ may give a consistent set of equations where $\mathbf{r} = (r_1, r_2, \ldots, r_5)^t$. However, we do not know $\mathbf{r}$, the vector of errors.

One technique for assessing α and v is to regard $\mathbf{r}$ as a variable and to seek values for $\mathbf{x}$ and for $\mathbf{r}$ satisfying the equation $A\mathbf{x} + \mathbf{r} = \mathbf{b}$ with the added requirement that $\mathbf{r}$ is selected in such a way that $\| \mathbf{r} \|$ is a minimum. In other words we are seeking α and v subject to the 'total' error Σr_i^2 being held down to a minimum consistent with the data. Note that our 'total' error is expressed as a convenient measure of the individual errors without permitting counterbalancing of positive and negative errors.

Since we are minimising $\| A\mathbf{x} - \mathbf{b} \|$, over all $\mathbf{x}$ we can see this as the problem of finding a point $A\mathbf{x}$ (in $R(A)$) which is at minimum distance to the point $\mathbf{b}$. But that evidently occurs at the foot of the perpendicular from $\mathbf{b}$ to the subspace $R(A)$ (Figure 4.4). We thus require to calculate the orthogonal projection P onto the column space of the matrix A precisely as in Section 4.6.

Let us just check the intuitively derived statement, viz. that the distance is minimised at the foot z of the perpendicular from $\mathbf{b}$ to $R(A)$. This follows from Pythagoras' theorem. Consider any other $\mathbf{y}$ in $R(A)$. Then,

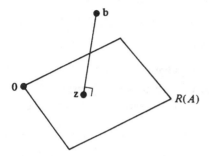

Fig. 4.4

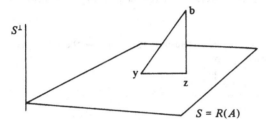

Fig. 4.5

since $\mathbf{z} - \mathbf{b}$ is orthogonal to $\mathbf{z} - \mathbf{y}$ we have (see Figure 4.5)

$$\|\mathbf{y} - \mathbf{b}\|^2 = \|\mathbf{y} - \mathbf{z}\|^2 + \|\mathbf{b} - \mathbf{z}\|^2$$
$$\geqslant \|\mathbf{b} - \mathbf{z}\|^2.$$

We return to the calculation. If $\mathbf{z}$ is at the foot of the perpendicular let $\mathbf{z} = A\mathbf{x}'$. Then $A\mathbf{x}' = P\mathbf{b} = AL\mathbf{b}$, where L is the left inverse of A as calculated in Section 4.6. But then $\mathbf{x}' = LA\mathbf{x}' = LAL\mathbf{b} = L\mathbf{b} = (A'A)^{-1}A'\mathbf{b}$.

In our example we have

$$\begin{bmatrix} 1 & 1 & 1 & 1 & 1 \\ 0 & 3 & 5 & 8 & 10 \end{bmatrix} \begin{bmatrix} 1 & 0 \\ 1 & 3 \\ 1 & 5 \\ 1 & 8 \\ 1 & 10 \end{bmatrix} = \begin{bmatrix} 5 & 26 \\ 26 & 198 \end{bmatrix}.$$

This has as inverse

$$\frac{1}{314} \begin{bmatrix} 198 & -26 \\ -26 & 5 \end{bmatrix}.$$

Also $A'\mathbf{b} = (33,227)'$. Thus

$$\mathbf{x}' = \frac{1}{314}(632, 277)' \approx (2.012, 0882)'$$

and so we have estimated (α, v) as $(2.012, 0.882)$.

In a later chapter we shall return to the subject of projections and their relation to various generalisations of the inverse of a matrix (such as the left inverse introduced above).

4.8   Exercises

1. Let S be the subspace of $\mathbb{R}^3$ spanned by $(1, 2, 3)$ and $(1, 1, -1)$. Find a matrix so that $P(x_1, x_2, x_3)'$ is the orthogonal projection of $(x_1, x_2, x_3)'$ onto S.

2. An input variable θ and a response variable y are related by a law of the form

$$y = a + b\cos^2\theta$$

with a, b being constants. Observation of y is subject to error; use the following data to estimate a and b by 'least squares'.

θ	0	$\pi/6$	$\pi/4$	$\pi/3$	$\pi/2$
y	4.1	3.4	2.7	2.1	1.6

3. Find the least squares fit for a curve of the form

$$y = \frac{m}{x} + c$$

given the following data points.

x	1	2	3	5
y	5	3	2	1

Why would it be wrong to suppose this was equivalent to the problem of fitting a curve of the form

$$z(=xy) = cx + m$$

through the data points (xy, x)?

4. The following table shows the amount of electricity, y, generated by a steam plant in the United Kingdom in the ten years, t, 1945–1954, and the amount x of coal used in the generation. Estimate a linear relationship between y, x and t by least squares.

t	1945	1946	1947	1948	1949	1950	1951	1952	1953	1954	
x	23	23	27	29	30	33	35	36	37	40	(10^6 tons)
y	37	41	43	47	49	55	60	62	65	72	(10^6)

5. Find the least squares solution to the system:

$$\left.\begin{array}{l} x_1 = a_1 \\ x_1 = a_2 \\ \vdots \\ x_1 = a_n \end{array}\right\}.$$

6. A variable y is linearly related to a variable x. k observations on y, all subject to error are available corresponding to each of the values $x_1, x_2, \ldots, x_n$. Denote by y_{ij} the jth observation corresponding to the value x_i. Show that the line fitted to such data by least squares is the same as that fitted to the data set $(x_i, \bar{y}_i)$ for $i = 1, 2, \ldots, n$ with

$$\bar{y}_i = \frac{1}{k} \sum_{j=1}^{k} y_{ij}.$$

7. Show that if $Ax = b$ is consistent then every solution of $A'Ax = A'b$ also solves the original system.

8. Suppose $S: V \to V$ is a linear transformation and X and Y are subspaces of V such that $S(X) \subseteq X$ and $S(Y) \subseteq Y$. Show that if

$$V = X \oplus Y$$

then

$$S(V) = S(X) \oplus S(Y).$$

9. Let X be a subspace of the space V. Let P denote the orthogonal projection onto X and let $Q = I - P$. Show that

$$\text{Lin}\{X \cup Y\} = \text{Lin}\{X \cup Q(Y)\}$$

for any subspace Y. Interpret this result on a diagram when X and Y are 1-dimensional subspaces of $\mathbb{R}^3$.

10. Let A and B be matrices of respective sizes $m \times k$ and $m \times j$. Let C be the partitioned matrix $(A|B)$. Denote by P_A the orthogonal projection onto $R(A)$ and let $Q_A = I - P_A$. Show that
   (i) $R(C)^\perp \subseteq R(A)^\perp$,
   (ii) $R(A) \subseteq Q_A(R(B))^\perp$,
   (iii) $R(C)^\perp \subseteq Q_A(R(B))^\perp$.
Interpret these results on a diagram where A and B are both 3×1 columns. [Hint: For (iii) use Question 9.]

11. With the same assumptions on A, B and C as in the previous question show that

$$P_C = P_A + P_{Q_A(B)}.$$

[Hint: Interpret this result geometrically and use the previous question.]

12. For any vector x in $\mathbb{R}^n$ show that $P = xx'$ is an orthogonal projection onto $\text{Lin}\{x\}$.

13. Verify that for $V = \mathbb{R}^n$ and L, M subspaces of V

$$V = L \oplus M \qquad\qquad (*)$$

holds if and only if

$$L \cap M = \{\mathbf{0}\} \quad \text{and} \quad \operatorname{Lin}(L \cup M) = V,$$

where $\operatorname{Lin}(S)$ is the smallest vector subspace containing S. Assuming (*) prove that

$$L^{\perp} \cap M^{\perp} = \{\mathbf{0}\}$$

and that

$$(\operatorname{Lin}(L^{\perp} \cup M^{\perp}))^{\perp} = \{\mathbf{0}\}.$$

Deduce that $L^{\perp} \oplus M^{\perp} = V$.

5

Spectral theory

5.1   Eigenvalues and eigenvectors – revision

We confine our attention for the moment to square matrices A. We
say that λ is an *eigenvalue* (characteristic root, latent root) of the
matrix A if and only if

$$A\mathbf{x} = \lambda\mathbf{x}$$

for some $\mathbf{x} \neq 0$. A vector $\mathbf{x} \neq 0$ which satisfies the equation is called
an *eigenvector* corresponding to the eigenvalue λ. If $\mathbf{x}$ is an
eigenvector then so is $\alpha\mathbf{x}$ for any $\alpha \neq 0$. A *normalised eigenvector* $\mathbf{x}$
satisfies $\| \mathbf{x} \| = 1$.

Let λ be an eigenvalue of the $n \times n$ matrix A, then

$$(A - \lambda I)\mathbf{x} = 0$$

has a non-zero solution so the matrix $A - \lambda I$ is singular and
therefore has a zero determinant, i.e.

$$|A - \lambda I| = 0.$$

We call $p(\lambda) = |A - \lambda I|$ the *characteristic polynomial* of A. It is a
polynomial of degree n if A is of size $n \times n$. Let us write

$$p(\lambda) = |A - \lambda I| = (-1)^n \lambda^n + p_{n-1}\lambda^{n-1} + \cdots + p_1\lambda + p_0.$$

Since this polynomial has n (possibly complex) roots $\lambda_1, \lambda_2, \ldots, \lambda_n$ we
have also the representation

$$p(\lambda) = (\lambda_1 - \lambda)(\lambda_2 - \lambda)\cdots(\lambda_n - \lambda).$$

Recall that some of the roots (or rather their values) may be
repeated in the listing $\lambda_1, \lambda_2, \ldots, \lambda_n$ and that the number of times a
root is repeated is called its *multiplicity*. Thus the square matrix A
has n eigenvalues, multiple roots being counted according to

multiplicity. It is sometimes useful to observe that

$$p_{n-1} = (-1)^{n-1}\{a_{11} + a_{22} + \cdots + a_{nn}\} = (-1)^{n-1}\operatorname{trace}(A),$$
$$p_0 = \det A,$$

so that

$$\operatorname{trace}(A) = \{\lambda_1 + \lambda_2 + \cdots + \lambda_n\}.$$

The collection of eigenvalues of A (also known as the *spectrum* of A) plays an important part in constructing a new basis relative to which the transformation A is represented by a simpler matrix. The hopeful simplification is of course a diagonal matrix with the eigenvalues as entries. The idea lying behind the simplification is that if eigenvectors are chosen as basis elements the transformation acts on each basis element by simply scaling it; the scaling factor here is of course the corresponding eigenvalue. More will be said on this point in the following section.

The choice of the term 'spectrum' is of interest. Recall that the passage of a ray of white light through a prism gives rise to a spectrum of rays of different colour. However, light of a single colour is transformed by the prism into the *same* colour, just as an eigenvector is transformed into itself, apart from a scaling factor. This is not just an analogy, the connection being in the way that the phenomenon of light is represented in physics.

5.2   Similarity

Recall that the matrix equation

$$\mathbf{y} = A\mathbf{x},$$

where A is $m \times n$ defines a linear transformation from $\mathbb{R}^n$ to $\mathbb{R}^m$, the natural bases being selected in these spaces.

By choosing new bases in $\mathbb{R}^n$ and $\mathbb{R}^m$ we can represent the *same* linear transformation by a matrix of the form

$$B = \left[\begin{array}{c|c} I_r & 0 \\ \hline 0 & 0 \end{array}\right].$$

If the new basis vectors in $\mathbb{R}^n$ are arranged as the columns of a matrix P and the new basis vectors in $\mathbb{R}^m$ are likewise arranged as the columns of Q, then as we have seen in the chapter on matrices the old and new co-ordinates are related by

$$\mathbf{x} = P\mathbf{X} \quad \text{and } \mathbf{y} = Q\mathbf{Y}.$$

Then $QY = APX$ and so the new representing matrix is $B = Q^{-1}AP$. We say that A and B are *equivalent* matrices.

If A is an $n \times n$ matrix, it defines a linear transformation of $\mathbb{R}^n$ into itself (with respect to the natural basis). In this case we are obviously interested in having both basis changes coincide, viz. $P = Q$. The question then arises: how simple can we make the matrix representation B of our linear transformation $y = Ax$ given that $P = Q$? If

$$B = P^{-1}AP,$$

we say that A and B are *similar*. What then is the simplest matrix B similar to A?

If A has n *linearly independent* eigenvectors the answer is easy. We simply take these eigenvectors to be our new basis (i.e. the columns $\mathbf{p}_1, \mathbf{p}_2, \ldots, \mathbf{p}_n$ of P). Write

$$\mathbf{y} = Y_1 \mathbf{p}_1 + Y_2 \mathbf{p}_2 + \cdots + Y_n \mathbf{p}_n,$$
$$\mathbf{x} = X_1 \mathbf{p}_1 + X_2 \mathbf{p}_2 + \cdots + X_n \mathbf{p}_n.$$

Then

$$
\begin{aligned}
Y_1 \mathbf{p}_1 + Y_2 \mathbf{p}_2 + \cdots + Y_n \mathbf{p}_n &= \mathbf{y} = A\mathbf{x} \\
&= X_1 A\mathbf{p}_1 + X_2 A\mathbf{p}_2 + \cdots + X_n A\mathbf{p}_n \\
&= X_1 \lambda_1 \mathbf{p}_1 + X_2 \lambda_2 \mathbf{p}_2 + \cdots + X_n \lambda_n \mathbf{p}_n,
\end{aligned}
$$

i.e.

$$
\begin{aligned}
Y_1 &= \lambda_1 X_1, \\
Y_2 &= \lambda_2 X_2, \\
&\vdots \\
Y_n &= \lambda_n X_n,
\end{aligned}
\quad \text{or} \quad
\begin{bmatrix} Y_1 \\ Y_2 \\ \vdots \\ Y_n \end{bmatrix}
=
\begin{bmatrix} \lambda_1 & & & \\ & \lambda_2 & & \\ & & \ddots & \\ & & & \lambda_n \end{bmatrix}
\begin{bmatrix} X_1 \\ X_2 \\ \vdots \\ X_n \end{bmatrix}.
$$

Thus with respect to the new basis $\mathbf{p}_1, \mathbf{p}_2, \ldots, \mathbf{p}_n$ the linear transformation $y = Ax$ is represented by a matrix which is *diagonal*, the diagonal entries being the eigenvalues of A. Equivalently:

$$
B = \begin{bmatrix} \lambda_1 & & & 0 \\ & \lambda_2 & & \\ & & \ddots & \\ 0 & & & \lambda_n \end{bmatrix} = P^{-1}AP,
$$

provided the columns of P are n linearly independent eigenvectors of A.

When does A have n linearly independent eigenvectors? This is easily shown to be the case when all the eigenvalues of A are *distinct*.

What happens if the eigenvalues are *not* distinct?

The results above remain valid provided that any eigenvalue of

multiplicity greater than 1, equal say to k, has k linearly independent
eigenvectors. Alas, this is *not* always the case. The dimension of the
eigenspace (the subspace spanned by the eigenvectors corresponding
to that eigenvalue) might well take any value between 1 and k.

A matrix which does *not* have n linearly independent eigenvectors
is called *defective*. It is *not* similar to a diagonal matrix. It may be
shown that such a matrix will be similar to a block diagonal matrix,
i.e. a partitioned matrix in the form (undisplayed entries are zero)

$$J = \begin{bmatrix} J_1 & & & \\ & J_2 & & \\ & & \cdots & \\ & & & J_r \end{bmatrix},$$

where each block is of either of the two Jordan types:

$$[\lambda], \quad \begin{bmatrix} \lambda & 1 & & & 0 \\ & \lambda & 1 & & \\ & & \ddots & & \\ & & & & 1 \\ & & & & \lambda \end{bmatrix}.$$

If a matrix is in block diagonal form with the blocks as above, we say
that the matrix is in *Jordan canonical form*.

5.3   Unitary and orthogonal transformations

In what follows we shall suppose that we are working in the space
$\mathbb{C}^n$. As usual if the space is $\mathbb{R}^n$, the assertions should be translated
according to Table 5.1.

Table 5.1

$\mathbb{R}^n$	$\mathbb{C}^n$
transpose A^t	conjugate transpose A^*
orthogonal matrix	unitary matrix
symmetric matrix	hermitian matrix

Consider again an $n \times n$ matrix A which defines the linear transformation

$$\mathbf{y} = A\mathbf{x}$$

from $\mathbb{C}^n$ to itself where it is understood that the natural basis is to be used.

We know that for a non-defective matrix A, we can choose a new basis $\mathbf{p}_1, \mathbf{p}_2, \ldots, \mathbf{p}_n$ (consisting of n linearly independent eigenvectors of A) with respect to which the linear transformation takes the form

$$\mathbf{Y} = B\mathbf{X},$$

where B is a diagonal matrix whose diagonal entries are the eigenvalues of the matrix A. Equivalently,

$$B = \begin{bmatrix} \lambda_1 & & & 0 \\ & \lambda_2 & & \\ & & \ddots & \\ 0 & & & \lambda_n \end{bmatrix} = P^{-1}AP$$

and $\mathbf{p}_1, \mathbf{p}_2, \ldots, \mathbf{p}_n$ are the columns of P.

The natural basis is orthonormal. Our natural greed for simple forms impels us to demand that our new basis also be orthonormal. That is to say we ask the question: for what matrices A can we find an *orthonormal* set of eigenvectors?

If $\mathbf{p}_1, \ldots, \mathbf{p}_n$ is an orthonormal set the matrix P will be called *unitary* (i.e. $P^* = P^{-1}$). In this case we have that

$$B = \begin{bmatrix} \lambda_1 & & & 0 \\ & \lambda_2 & & \\ & & \ddots & \\ 0 & & & \lambda_n \end{bmatrix} = P^*AP.$$

The answer is gratifyingly simple. A matrix has this property if and only if it is *normal*, i.e. $A^*A = AA^*$. (See Exercises 6.6.)

The most important class of normal matrices are the *hermitian matrices* which have the property that $A = A^*$. Other examples of normal matrices are unitary matrices ($PP^* = I = P^*P$) and skew-symmetric matrices ($A = -A^*$).

Hermitian matrices have the two further pleasant properties that all their eigenvalues are *real* and that eigenvectors corresponding to distinct eigenvalues are necessarily orthogonal. The proof of the first fact is easy. Suppose that

$$A\mathbf{x} = \lambda\mathbf{x},$$

then

$$\mathbf{x}^*A = \mathbf{x}^*A^* = (A\mathbf{x})^* = \bar{\lambda}\mathbf{x}^*.$$

Hence

$$\mathbf{x}^*(A\mathbf{x}) = \lambda\mathbf{x}^*\mathbf{x} = \lambda \|\mathbf{x}\|^2.$$

Also

$$\mathbf{x}^*(A\mathbf{x}) = (\mathbf{x}^*A)\mathbf{x} = \bar{\lambda}\mathbf{x}^*\mathbf{x} = \bar{\lambda}\|\mathbf{x}\|^2.$$

Therefore, since $\|\mathbf{x}\| \neq 0$,

$$\lambda = \bar{\lambda}$$

and so λ is *real*. In particular we draw attention to the result below:

The eigenvalues of a real symmetric matrix are all *real*.

As regards the second property, let $\lambda \neq \mu$ be distinct eigenvalues and let $\mathbf{p}$ and $\mathbf{q}$ be corresponding eigenvectors. We have

$$\lambda\mathbf{p}^*\mathbf{q} = (\lambda\mathbf{p})^*\mathbf{q} = (A\mathbf{p})^*\mathbf{q} = \mathbf{p}^*A\mathbf{q} = \mu\mathbf{p}^*\mathbf{q}.$$

Thus

$$(\lambda - \mu)\mathbf{p}^*\mathbf{q} = 0$$

and so $\mathbf{p}^*\mathbf{q} = 0$.

Example
Consider the real symmetric matrix

$$A = \begin{bmatrix} 1 & 3 & 0 \\ 3 & -2 & -1 \\ 0 & -1 & 1 \end{bmatrix}.$$

Thus

$$|A - \lambda I| = \begin{bmatrix} 1-\lambda & 3 & 0 \\ 3 & -2-\lambda & -1 \\ 0 & -1 & 1-\lambda \end{bmatrix} = -\lambda^3 + 13\lambda - 12.$$

One root of this is obviously $\lambda = 1$. Hence we have the factorisation for $p(\lambda)$ as $(1 - \lambda)(4 + \lambda)(\lambda - 3)$. Thus the eigenvalues are $1, -4, 3$ which are all *real*. They are distinct in this case so we expect the corresponding eigenvectors to be orthogonal. We have, solving $A\mathbf{x} = \lambda\mathbf{x}$ succesively for $\lambda = 1, -4, 3$:

$$(A - I)\mathbf{x} = \begin{bmatrix} 1-1 & 3 & 0 \\ 3 & -2-1 & -1 \\ 0 & -1 & 1-1 \end{bmatrix}\mathbf{x} = 0,$$

so $x_2 = 0$ and $x_3 = 3x_1$, i.e. $\mathbf{x} = (x_1, 0, 3x_1)^t = x_1(1, 0, 3)^t$. Thus the eigenspace for $\lambda = 1$ is spanned by $(1, 0, 3)^t$. Proceeding to $\lambda = -4$ we obtain

$$(A + 4I)\mathbf{x} = \begin{bmatrix} 1+4 & 3 & 0 \\ 3 & -2+4 & -1 \\ 0 & -1 & 1+4 \end{bmatrix} \mathbf{x} = 0,$$

so

$$x_2 = 5x_3 \quad \text{and} \quad -5x_1 = 3x_2 = 15x_3$$

i.e.

$$\mathbf{x} = (-3x_3, 5x_3, x_3)^t = x_3(-3, 5, 1)^t.$$

Thus the eigenspace for $\lambda = -4$ is spanned by $(-3, 5, 1)^t$. Finally for $\lambda = 3$ we have

$$(A - 3I)\mathbf{x} = \begin{bmatrix} 1-3 & 3 & 0 \\ 3 & -2-3 & -1 \\ 0 & -1 & 1-3 \end{bmatrix} \mathbf{x} = 0,$$

so that

$$x_2 = -2x_3 \quad \text{and} \quad 2x_1 = 3x_2 = -6x_3,$$

i.e.

$$\mathbf{x} = (-3x_3, -2x_3, x_3)^t = x_3(-3, -2, 1)^t.$$

The eigenspace to value 3 is spanned by $(3, 2, -1)^t$. Normalising the three vectors we have just calculated, gives the orthonormal set:

$$\mathbf{p}_1 = \frac{1}{\sqrt{10}} \begin{bmatrix} 1 \\ 0 \\ 3 \end{bmatrix}, \quad \mathbf{p}_2 = \frac{1}{\sqrt{35}} \begin{bmatrix} -3 \\ 5 \\ 1 \end{bmatrix}, \quad \mathbf{p}_3 = \frac{1}{\sqrt{14}} \begin{bmatrix} 3 \\ 2 \\ -1 \end{bmatrix}.$$

We may check orthogonality:

$$\langle \mathbf{p}_1, \mathbf{p}_2 \rangle = \frac{1}{\sqrt{350}} \{1 \cdot (-3) + 0 \cdot 5 + 3 \cdot 1\} = 0,$$

$$\langle \mathbf{p}_2, \mathbf{p}_3 \rangle = \frac{1}{7\sqrt{10}} \{(-3) \cdot 3 + 5 \cdot 2 + 1 \cdot (-1)\} = 0,$$

$$\langle \mathbf{p}_3, \mathbf{p}_1 \rangle = \frac{1}{\sqrt{140}} \{3 \cdot 1 + 2 \cdot 0 + (-1) \cdot 3\} = 0.$$

Hence

$$\begin{bmatrix} 1 & 0 & 0 \\ 0 & -4 & 0 \\ 0 & 0 & 3 \end{bmatrix} = P^* \begin{bmatrix} 1 & 3 & 0 \\ 3 & -2 & -1 \\ 0 & -1 & 1 \end{bmatrix} P,$$

where

$$P = \begin{bmatrix} \dfrac{1}{\sqrt{10}} & \dfrac{-3}{\sqrt{35}} & \dfrac{3}{\sqrt{14}} \\[2ex] 0 & \dfrac{5}{\sqrt{35}} & \dfrac{2}{\sqrt{14}} \\[2ex] \dfrac{3}{\sqrt{10}} & \dfrac{1}{\sqrt{35}} & \dfrac{-1}{\sqrt{14}} \end{bmatrix}.$$

Example

Consider the real symmetric matrix

$$A = \begin{bmatrix} 7 & -16 & -8 \\ -16 & 7 & 8 \\ -8 & 8 & -5 \end{bmatrix}.$$

We have

$$|A - \lambda I| = \begin{vmatrix} 7-\lambda & -16 & -8 \\ -16 & 7-\lambda & 8 \\ -8 & 8 & -5-\lambda \end{vmatrix} = \lambda^3 - 9\lambda^2 - 405\lambda - 2187.$$

Those interested may check that the roots are $\lambda_1 = 27$ and $\lambda_2 = \lambda_3 = -9$. The single root $\lambda_1 = 27$ yields without difficulty the eigenvector $\mathbf{q}_1 = (-2, 2, 1)^t$. The eigenvalue -9 has multiplicity 2. Since A is real and symmetric (and hence normal) the general theory assures us that:

(i)   The eigenvectors corresponding to -9 are all orthogonal to those corresponding to 27

(ii) We can find *two* orthonormal eigenvectors corresponding to -9 (since it has multiplicity *two*).

To find the eigenvectors corresponding to -9 we consider

$$(A + 9I)\mathbf{x} = \begin{bmatrix} 7+9 & -16 & -8 \\ -16 & 7+9 & 8 \\ -8 & 8 & -5+9 \end{bmatrix} \mathbf{x} = 0,$$

which system reduces to the one equation $2x_1 - 2x_2 - x_3 = 0$. The eigenvectors in this eigenspace are thus

$$(x_1, x_2, 2x_1 - 2x_2)^t = (x_1, 0, 2x_1)^t + (0, x_2, -2x_2)^t$$
$$= x_1(1, 0, 2)^t + x_2(0, 1, -2)^t$$

and therefore form together with the zero vector a two-dimensional space. The vectors $(1, 0, 2)^t$ and $(0, 1, -2)^t$ are a basis for this space.

We can find two orthonormal eigenvectors by using the Gram–Schmidt process. Take $\mathbf{q}_2 = (1, 0, 2)^t$ and $\mathbf{q}_3 = \beta(1, 0, 2)^t + \gamma(0, 1, -2)^t$, then

$$\langle \mathbf{q}_2, \mathbf{q}_3 \rangle = 5\beta - 4\gamma.$$

We want $\mathbf{q}_2$ and $\mathbf{q}_3$ to be orthogonal and so we choose $\beta = 4$ and $\gamma = 5$. Then $\mathbf{q}_3 = 4(1, 0, 2)^t + 5(0, 1, -2)^t = (4, 5, -2)^t$. An orthonormal set may now be obtained by normalising the chosen vectors. We have

$$\mathbf{p}_1 = \frac{1}{3} \begin{bmatrix} -2 \\ 2 \\ 1 \end{bmatrix}, \quad \mathbf{p}_2 = \frac{1}{\sqrt{5}} \begin{bmatrix} 1 \\ 0 \\ 2 \end{bmatrix}, \quad \mathbf{p}_3 = \frac{1}{\sqrt{45}} \begin{bmatrix} 4 \\ 5 \\ -2 \end{bmatrix}.$$

Thus

$$P^* \begin{bmatrix} 7 & -16 & -8 \\ -16 & 7 & 8 \\ -8 & 8 & -5 \end{bmatrix} P = \begin{bmatrix} 27 & 0 & 0 \\ 0 & -9 & 0 \\ 0 & 0 & -9 \end{bmatrix},$$

where

$$P = \begin{bmatrix} \dfrac{-2}{3} & \dfrac{1}{\sqrt{5}} & \dfrac{4}{\sqrt{45}} \\[2ex] \dfrac{2}{3} & 0 & \dfrac{5}{\sqrt{45}} \\[2ex] \dfrac{1}{3} & \dfrac{2}{\sqrt{5}} & \dfrac{-2}{\sqrt{45}} \end{bmatrix}.$$

5.4 Spectral decomposition

Let A be a normal matrix (usually hermitian, or real and symmetric). Then we know that

$$X^*AX = \Lambda = \begin{bmatrix} \lambda_1 & 0 & \cdots & 0 \\ 0 & \lambda_2 & & 0 \\ \vdots & & & \\ 0 & 0 & & \lambda_n \end{bmatrix},$$

where the columns of the unitary matrix X are an orthonormal set of eigenvectors $\mathbf{x}_1, \mathbf{x}_2, \ldots, \mathbf{x}_n$. To say that X is unitary means that

$$X^*X = XX^* = I.$$

Let us examine the implications of the equations $X^*X = I$ and $XX^* = I$ for the eigenvectors $x_1, x_2, \ldots, x_n$. We observe that the equation

$$I = X^*X = \begin{bmatrix} x_1^* \\ x_2^* \\ \vdots \\ x_n^* \end{bmatrix} [x_1, x_2 \cdots x_r] = \begin{bmatrix} x_1^*x_1 & \cdots & x_1^*x_n \\ & \vdots & \\ x_n^*x_1 & & x_n^*x_n \end{bmatrix}$$

simply confirms that $\{x_1, x_2, \ldots, x_n\}$ is orthonormal because $x_i^*x_j = \langle x_j, x_j \rangle$. But

$$I = XX^* = [x_1, x_2, \ldots, x_n] \begin{bmatrix} x_1^* \\ x_2^* \\ \vdots \\ x_n^* \end{bmatrix} = x_1x_1^* + \cdots + x_nx_n^*$$

is more interesting. Note that

$$E_i = x_i x_i^*$$

is an $n \times n$ matrix being an $(n \times 1) \times (1 \times n)$ matrix product.

The *spectral decomposition* of A is the formula

$$A = \lambda_1 x_1 x_1^* + \lambda_2 x_2 x_2^* + \cdots + \lambda_n x_n x_n^*.$$

The proof of this formula is trivial. We have that

$$\begin{aligned} \lambda_1 x_1 x_1^* + \lambda_2 x_2 x_2^* + \cdots + \lambda_n x_n x_n^* &= A x_1 x_1^* + A x_2 x_2^* + \cdots + A x_n x_n^* \\ &= A(x_1 x_1^* + x_2 x_2^* + \cdots + x_n x_n^*) \\ &= AI = A. \end{aligned}$$

The geometric interpretation of the matrices $E_i = x_i x_i^*$ is of interest (Figures 5.1 and 5.2). We have already shown the first of the following properties:

    (i) $E_1 + E_2 + \cdots + E_n = I$,

    (ii) $E_iE_j = 0 \ (i \neq j)$,

    (iii) $E_i^2 = E_i$.

To prove (ii) observe that $E_iE_j = (x_i x_i^*)(x_j x_j^*) = x_i(x_i^*x_j)x_j = x_i 0 x_j^* = 0$ provided that $i \neq j$. To prove (iii) we write

$$E_i^2 = (x_i x_i^*)(x_i x_i^*) = x_i(x_i^* x_i)x_i^* = x_i \cdot 1 \cdot x_i^* = x_i x_i^* = E_i,$$

Item (iii) says that E_i is an idempotent and hence it represents a projection. In fact E_i is the *orthogonal projection* of $\mathbb{C}^n$ onto the subspace spanned by x_i.

Since $\{x_1, x_2, \ldots, x_n\}$ is a basis for $\mathbb{C}^n$, any x in $\mathbb{C}^n$ may be expressed uniquely in the form

$$x = \alpha_1 x_1 + \alpha_2 x_2 + \cdots + \alpha_n x_n.$$

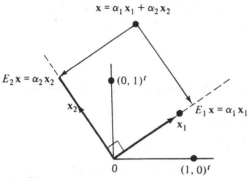

Fig. 5.1

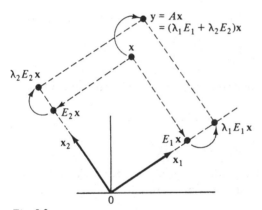

Fig. 5.2

Hence

$$E_i \mathbf{x} = \sum_j \alpha_j E_i \mathbf{x}_j = \sum_j \alpha_j \mathbf{x}_i (\mathbf{x}_i^* \mathbf{x}_j) = \alpha_i \mathbf{x}_i$$

5.5 Quadratic forms

A *quadratic form* is an expression

$$\mathbf{x}^* A \mathbf{x},$$

where A is a *hermitian matrix*.

Let the eigenvalues of A be $\lambda_1, \lambda_2, \ldots, \lambda_n$. These are all real. Let the corresponding orthonormal eigenvectors be $\mathbf{p}_1, \mathbf{p}_2, \ldots, \mathbf{p}_n$. If we change from the natural basis to the basis $\mathbf{p}_1, \mathbf{p}_2, \ldots, \mathbf{p}_n$, then the new co-ordinate vector $\mathbf{X}$ is related to the old co-ordinate vector $\mathbf{x}$ by the equation:

$$\mathbf{x} = P\mathbf{X},$$

where P is the matrix whose columns are $\mathbf{p}_1, \mathbf{p}_2, \ldots, \mathbf{p}_n$. The quadratic form then becomes

$$\mathbf{x}^* A \mathbf{x} = \mathbf{X}^* P^* A P \mathbf{X} = \mathbf{X}^* \Lambda \mathbf{X}$$
$$= \lambda_1 |X_1|^2 + \lambda_2 |X_2|^2 + \cdots + \lambda_n |X_n|^2,$$

i.e. the transformation $\mathbf{X} = P^* \mathbf{x}$ reduces the quadratic form to a 'sum of squares'.

Application

What sort of curve in $\mathbb{R}^2$ is $5x^2 + 4xy + 2y^2 = 1$? We observe that

$$5x^2 + 4xy + 2y^2 = (xy) \begin{bmatrix} 5 & 2 \\ 2 & 2 \end{bmatrix} \begin{bmatrix} x \\ y \end{bmatrix},$$

so that we are dealing with a quadratic form. We calculate the eigenvalues:

$$|A - \lambda I| = \begin{vmatrix} 5 - \lambda & 2 \\ 2 & 2 - \lambda \end{vmatrix} = (5 - \lambda)(2 - \lambda) - 4$$
$$= \lambda^2 - 7\lambda + 6 = (\lambda - 6)(\lambda - 1).$$

Hence is new co-ordinates the quadratic form reduces to $6X^2 + Y^2$, so the curve in question is an ellipse. For more information we need to calculate the eigenvectors

$$\lambda = 6 \qquad\qquad\qquad \lambda = 1$$

$$\begin{bmatrix} -1 & 2 \\ 2 & -4 \end{bmatrix} \begin{bmatrix} x \\ y \end{bmatrix} = \begin{bmatrix} 0 \\ 0 \end{bmatrix} \qquad \begin{bmatrix} 4 & 2 \\ 2 & 1 \end{bmatrix} \begin{bmatrix} x \\ y \end{bmatrix} = \begin{bmatrix} 0 \\ 0 \end{bmatrix}$$

or

$$x = 2y \qquad\qquad\qquad y = -2x.$$

Thus eigenvectors are of the form $\alpha(2, 1)^t$ and $\beta(1, -2)^t$. For an orthonormal set we take $\alpha = \beta = 1/\sqrt{5}$. Then

$$\begin{bmatrix} x \\ y \end{bmatrix} = \begin{bmatrix} \dfrac{2}{\sqrt{5}} & \dfrac{-1}{\sqrt{5}} \\ \dfrac{1}{\sqrt{5}} & \dfrac{2}{\sqrt{5}} \end{bmatrix} \begin{bmatrix} X \\ Y \end{bmatrix}.$$

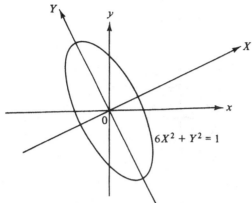

Fig. 5.3

Thus

$$\begin{bmatrix} X \\ Y \end{bmatrix} = \begin{bmatrix} \dfrac{2}{\sqrt{5}} & \dfrac{1}{\sqrt{5}} \\ \dfrac{-1}{\sqrt{5}} & \dfrac{2}{\sqrt{5}} \end{bmatrix} \begin{bmatrix} x \\ y \end{bmatrix}.$$

The *principal axes* of the ellipse are the X-axis given by $Y = 0$ (which in the old co-ordinates is given by $-x + 2y = 0$) and the Y-axis given by $X = 0$ (which in the old co-ordinates is given by $2x + y = 0$). See Figure 5.3.

5.6 Positive and negative definite forms

A hermitian matrix is *positive definite* if and only if all its eigenvalues are positive. If all the eigenvalues are non-negative, the matrix is *non-negative definite*. Similar definitions hold for negative definite and non-positive definite.

Suppose that A is positive definite then the eigenvalues satisfy $\lambda_1 > 0$, $\lambda_2 > 0, \ldots, \lambda_n > 0$ and so

$$\mathbf{x}^* A \mathbf{x} = \mathbf{X}^* \Lambda \mathbf{X} = \lambda_1 |X_1|^2 + \lambda_2 |X_2|^2 + \cdots + \lambda_n |X_n|^2 > 0$$

unless $X_1 = X_2 = \cdots = X_n = 0$ i.e. $\mathbf{X} = 0$. But then $\mathbf{x} = P\mathbf{X} = 0$. This result and similar ones are illustrated in Figure 5.4 and incorporated in Table 5.2.

Let A be positive definite. Then

$$|A| = \lambda_1 \lambda_2 \cdots \lambda_n > 0.$$

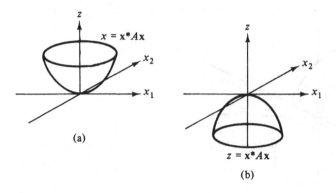

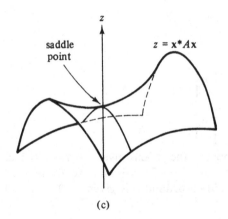

Fig. 5.4. (a) positive definite, (b) negative definite, (c) neither positive definite nor negative definite

Table 5.2

A positive definite	$\lambda_1 > 0, \lambda_2 > 0, \ldots, \lambda_n > 0$	$\mathbf{x}^* A \mathbf{x} > 0$ for $\mathbf{x} \neq \mathbf{0}$
A non-negative definite	$\lambda_1 \geqslant 0, \lambda_2 \geqslant 0, \ldots, \lambda_n \geqslant 0$	$\mathbf{x}^* A \mathbf{x} \geqslant 0$
A negative definite	$\lambda_1 < 0, \lambda_2 < 0, \ldots, \lambda_n < 0$	$\mathbf{x}^* A \mathbf{x} < 0$ for $\mathbf{x} \neq \mathbf{0}$
A non-positive definite	$\lambda_1 \leqslant 0, \lambda_2 \leqslant 0, \ldots, \lambda_n \leqslant 0$	$\mathbf{x}^* A \mathbf{x} \leqslant 0$

Further, if we partition A as shown,

$$A = \begin{bmatrix} & & & & \times \\ & A_{n-1} & & & \times \\ & & & & \times \\ & & & & \times \\ \hline \times & \times & \times & \times & \times \end{bmatrix},$$

then the matrix A_{n-1} is also positive definite because $\mathbf{x}^* A_{n-1}\mathbf{x} = \mathbf{X}^* A\mathbf{X} > 0$ ($\mathbf{x} \neq 0$) where $\mathbf{X}^* = (\mathbf{x}^*|0)$. It follows that $|A_{n-1}| > 0$.

The same argument shows that all the *principal minors* of a positive definite matrix are positive. A principal minor is, we recall, the determinant of a submatrix of A whose diagonal lies along the diagonal of A.

Less easy to prove, but much more useful is the converse result which is quoted below.

Theorem

Let A be a hermitian matrix:

$$A = \begin{bmatrix} a_{11} & a_{12} & \cdots & a_{1n} \\ a_{21} & a_{22} & & a_{2n} \\ \vdots & & & \\ a_{n1} & a_{n2} & & a_{nn} \end{bmatrix}.$$

Then:

(i) *A is positive definite if and only if*

$$a_{11} > 0, \quad \begin{vmatrix} a_{11} & a_{12} \\ a_{21} & a_{22} \end{vmatrix} > 0, \quad \begin{vmatrix} a_{11} & a_{12} & a_{13} \\ a_{21} & a_{22} & a_{23} \\ a_{31} & a_{32} & a_{33} \end{vmatrix} > 0, \ldots, |A| > 0$$

(ii) *A is negative definite if and only if*

$$a_{11} < 0, \quad \begin{vmatrix} a_{11} & a_{12} \\ a_{21} & a_{22} \end{vmatrix} > 0, \quad \begin{vmatrix} a_{11} & a_{12} & a_{13} \\ a_{21} & a_{22} & a_{23} \\ a_{31} & a_{32} & a_{33} \end{vmatrix} < 0, \ldots, (-1)^n|A| > 0$$

Note that (ii) follows from (i) because A is positive definite if and only if $-A$ is negative definite.

Example

Prove that the quadratic form below is positive definite.

$$(x\ y\ z)\begin{bmatrix} 1 & 1 & 0 \\ 1 & 4 & 2 \\ 0 & 2 & 3 \end{bmatrix}\begin{bmatrix} x \\ y \\ z \end{bmatrix}$$

We have that

$$1 > 0, \quad \begin{vmatrix} 1 & 1 \\ 1 & 4 \end{vmatrix} = 3 > 0, \quad \begin{vmatrix} 1 & 1 & 0 \\ 1 & 4 & 2 \\ 0 & 2 & 3 \end{vmatrix} = (12 + 0 + 0) - (4 + 3 + 0) = 5 > 0$$

5.7   Functions of matrices

Let A be a non-defective matrix. Then $A = P\Lambda P^{-1}$ where Λ is a diagonal matrix the entries of which are the eigenvalues of A and P is the matrix whose columns are the corresponding eigenvectors. We observe that

$$A^k = P\Lambda P^{-1}P\Lambda P^{-1}P\cdots\Lambda P^{-1}P\Lambda P^{-1}$$

$$= P\Lambda^k P^{-1} = P\begin{bmatrix} \lambda_1^k & & & \\ & \lambda_2^k & & \\ & & \ddots & \\ & & & \lambda_n^k \end{bmatrix}P^{-1}.$$

It follows that for any polynomial $f(x)$ we have

$$f(A) = P\begin{bmatrix} f(\lambda_1) & & & \\ & f(\lambda_2) & & \\ & & \ddots & \\ & & & f(\lambda_n) \end{bmatrix}P^{-1}.$$

If $f(x)$ is *not* a polynomial, we use the formula above to *define* $f(A)$. In particular, if A is non-negative definite, so that $\lambda_1 \geqslant 0, \lambda_2 \geqslant 0,\ldots,\lambda_n \geqslant 0$, then

$$A^{1/2} = P\begin{bmatrix} \lambda_1^{1/2} & & & \\ & \lambda_2^{1/2} & & \\ & & \ddots & \\ & & & \lambda_n^{1/2} \end{bmatrix}P^{-1}.$$

Observe that on this definition $A^{1/2}A^{1/2} = A$ and so our definition is consistent.

5.8 Spectral decomposition of a non-square matrix

Let A be a non-square matrix. Unless A is normal (i.e. $A^*A = AA^*$), the analysis of Section 5.4 does *not* apply. But all is not lost as we now explain.

We begin by considering the two square hermitian matrices A^*A and AA^*. Both these matrices are non-negative definite. The proof is simple; we have that

$$\mathbf{x}^*A^*A\mathbf{x} = (A\mathbf{x})^*A\mathbf{x} = \|A\mathbf{x}\|^2 \geqslant 0,$$
$$\mathbf{x}^*AA^*\mathbf{x} = (A^*\mathbf{x})^*A^*\mathbf{x} = \|A^*\mathbf{x}\|^2 \geqslant 0.$$

Hence all the eigenvalues of A^*A and AA^* are non-negative. We next show that A^*A and AA^* have precisely the *same* non-zero eigenvalues.

Theorem

Let A be any $m \times n$ matrix. Then the hermitian matrices A^*A and AA^* have the same *non-zero eigenvalues $\lambda_1, \lambda_2, \ldots, \lambda_k$.*

A corresponding orthonormal set $\{\mathbf{x}_1, \mathbf{x}_2, \ldots, \mathbf{x}_k\}$ of eigenvectors for A^*A and a corresponding orthonormal set $\{\mathbf{y}_1, \mathbf{y}_2, \ldots, \mathbf{y}_k\}$ of eigenvectors for AA^* may be chosen so that

$$\mathbf{x}_i = \frac{1}{\sqrt{\lambda_i}} A^*\mathbf{y}_i, \quad \mathbf{y}_j = \frac{1}{\sqrt{\lambda_j}} A\mathbf{x}_j.$$

Proof. Suppose that $A^*A\mathbf{x} = \lambda\mathbf{x}$ with $\mathbf{x} \neq 0$. Then $AA^*(A\mathbf{x}) = \lambda(A\mathbf{x})$. Hence each eigenvalue λ of A^*A is an eigenvalue of AA^* and $\mathbf{y} = A\mathbf{x}$ is a corresponding eigenvector, unless $A\mathbf{x} = 0$. But then $\lambda\mathbf{x} = A^*A\mathbf{x} = A^*0 = 0$ and so $\lambda = 0$.

This shows that each non-zero eigenvalue of A^*A is an eigenvalue of AA^*. The same argument proves the converse and so A^*A and AA^* have the same non-zero eigenvalues.

Let $\{\mathbf{x}_1, \mathbf{x}_2, \ldots, \mathbf{x}_k\}$ be an orthonormal set of eigenvectors corresponding to the non-zero eigenvalues $\lambda_1, \lambda_2, \ldots, \lambda_k$ of A^*A. Then we have seen that $\{A\mathbf{x}_1, A\mathbf{x}_2, \ldots, A\mathbf{x}_k\}$ are eigenvectors for the non-zero eigenvalues $\lambda_1, \lambda_2, \ldots, \lambda_k$ of AA^*.

Also,

$$(A\mathbf{x}_i)^*A\mathbf{x}_j = \mathbf{x}_i^*A^*A\mathbf{x}_j = \mathbf{x}_i^*\lambda_j\mathbf{x}_j = \begin{cases} \lambda_j & i = j, \\ 0, & i \neq j. \end{cases}$$

Thus the vectors $\{\mathbf{y}_1, \mathbf{y}_2, \ldots, \mathbf{y}_k\}$, where

$$\mathbf{y}_j = \frac{1}{\sqrt{\lambda_j}} A\mathbf{x}_j,$$

are an orthonormal set of eigenvectors corresponding to the non-zero eigenvalues $\lambda_1, \lambda_2, \ldots, \lambda_k$ of AA^*. Finally, observe that

$$A^* y_i = \frac{1}{\sqrt{\lambda_i}} A^* A x_i = \frac{\lambda_i}{\sqrt{\lambda_i}} x_i = \sqrt{\lambda_i} x_i.$$

The numbers $\sqrt{\lambda_1}, \sqrt{\lambda_2}, \ldots, \sqrt{\lambda_k}$ are called the *singular values* of the matrix A. Hence also the formula

$$A = \sum_{i=1}^{k} \sqrt{\lambda_i} y_i x_i^*$$

is called the *singular values decomposition* of the matrix A. The proof is as follows. Selecting $x_{k+1}, x_{k+2}, \ldots, x_n$ to be eigenvectors to the value zero in such a way that $\{x_1, x_2, \ldots, x_n\}$ is also an orthonormal set (cf. Section 5.4) we have

$$A = AI = A \sum_{i=1}^{n} x_i x_i^* = \sum_{i=1}^{n} (A x_i) x_i^* = \sum_{i=1}^{k} \sqrt{\lambda_i} y_i x_i^*.$$

(Observe that $A^* A x_i = 0$ implies $x_i^* A^* A x_i = 0$, i.e. $\| A x_i \|^2 = 0$ so $A x_i = 0$ and this explains the disappearance of terms beyond k in the summation.) The singular values decomposition may also be expressed in the alternative form

$$Y \begin{bmatrix} \lambda_1^{1/2} & & & \\ & \lambda_2^{1/2} & & \\ & & \ddots & \\ & & & \lambda_n^{1/2} \end{bmatrix} X^*$$

$$= \begin{bmatrix} | & | & \cdots & | \\ y_1 & y_2 & & y_k \\ | & | & & | \end{bmatrix} \begin{bmatrix} \lambda_1^{1/2} & & & \\ & \lambda_2^{1/2} & & \\ & & \ddots & \\ & & & \lambda_n^{1/2} \end{bmatrix} \begin{bmatrix} - & x_1^* & - \\ - & x_2^* & - \\ & \vdots & \\ - & x_k^* & - \end{bmatrix}.$$

The columns of Y and X are orthonormal and in particular, these matrices both have rank k.

Example
Consider

$$A = \begin{bmatrix} 0 & 1 \\ 1 & 0 \\ 1 & 1 \end{bmatrix}.$$

Then

$$A^*A = \begin{bmatrix} 0 & 1 & 1 \\ 1 & 0 & 1 \end{bmatrix}\begin{bmatrix} 0 & 1 \\ 1 & 0 \\ 1 & 1 \end{bmatrix}, \quad AA^* = \begin{bmatrix} 0 & 1 \\ 1 & 0 \\ 1 & 1 \end{bmatrix}\begin{bmatrix} 0 & 1 & 1 \\ 1 & 0 & 1 \end{bmatrix},$$

$$= \begin{bmatrix} 2 & 1 \\ 1 & 2 \end{bmatrix}. \qquad = \begin{bmatrix} 1 & 0 & 1 \\ 0 & 1 & 1 \\ 1 & 1 & 2 \end{bmatrix}.$$

Thus

$$\begin{vmatrix} 2-\lambda & 1 \\ 1 & 2-\lambda \end{vmatrix} \begin{aligned} &= (2-\lambda)^2 - 1 \\ &= \lambda^2 - 4\lambda + 3 \\ &= (\lambda - 3)(\lambda - 1). \end{aligned}$$

$$\begin{vmatrix} 1-\lambda & 0 & 1 \\ 0 & 1-\lambda & 1 \\ 1 & 1 & 2-\lambda \end{vmatrix} \begin{aligned} &= (1-\lambda)^2(2-\lambda) - 2(1-\lambda) \\ &= (1-\lambda)[(1-\lambda)(2-\lambda)-2] \\ &= (1-\lambda)\lambda(\lambda - 3). \end{aligned}$$

We now compute the eigenvectors corresponding to the eigenvalues $\lambda_1 = 1$ and $\lambda_2 = 3$ for the matrix A^*A and then use the theorem.

For $\lambda = 1$: $\begin{bmatrix} 2-1 & 1 \\ 1 & 2-1 \end{bmatrix}\begin{bmatrix} x \\ y \end{bmatrix} = 0,$

so $x + y = 0$ and the eigenvectors are $\alpha(1, -1)^t$.

For $\lambda = 3$: $\begin{bmatrix} 2-3 & 1 \\ 1 & 2-3 \end{bmatrix}\begin{bmatrix} x \\ y \end{bmatrix} = 0,$

so $x - y = 0$ and the eigenvectors are $\beta(1, 1)^t$.
Thus we may take

$$\mathbf{x}_1 = \frac{1}{\sqrt{2}}\begin{bmatrix} 1 \\ -1 \end{bmatrix} \quad \text{and} \quad \mathbf{x}_2 = \frac{1}{\sqrt{2}}\begin{bmatrix} 1 \\ 1 \end{bmatrix},$$

so that

$$\mathbf{y}_1 = \frac{1}{\sqrt{1}}\begin{bmatrix} 0 & 1 \\ 1 & 0 \\ 1 & 1 \end{bmatrix}\frac{1}{\sqrt{2}}\begin{bmatrix} 1 \\ -1 \end{bmatrix} = \frac{1}{\sqrt{2}}\begin{bmatrix} -1 \\ 1 \\ 0 \end{bmatrix}$$

and

$$\mathbf{y}_2 = \frac{1}{\sqrt{3}}\begin{bmatrix} 0 & 1 \\ 1 & 0 \\ 1 & 1 \end{bmatrix}\frac{1}{\sqrt{2}}\begin{bmatrix} 1 \\ 1 \end{bmatrix} = \frac{1}{\sqrt{6}}\begin{bmatrix} 1 \\ 1 \\ 2 \end{bmatrix}.$$

The singular values decomposition of A is now

$$A = \begin{bmatrix} 0 & 1 \\ 1 & 0 \\ 1 & 1 \end{bmatrix} = \sqrt{\lambda_1}\, y_1 x_1^* + \sqrt{\lambda_2}\, y_2 x_2^*$$

$$= \frac{\sqrt{1}}{2}\begin{bmatrix} -1 \\ 1 \\ 0 \end{bmatrix}(1,-1) + \frac{\sqrt{3}}{\sqrt{12}}\begin{bmatrix} 1 \\ 1 \\ 2 \end{bmatrix}(1,1)$$

$$= \frac{1}{2}\begin{bmatrix} -1 & 1 \\ 1 & -1 \\ 0 & 0 \end{bmatrix} + \frac{1}{2}\begin{bmatrix} 1 & 1 \\ 1 & 1 \\ 2 & 2 \end{bmatrix}.$$

Alternatively,

$$A = \begin{bmatrix} 0 & 1 \\ 1 & 0 \\ 1 & 1 \end{bmatrix} = \begin{bmatrix} \dfrac{-1}{\sqrt{2}} & \dfrac{1}{\sqrt{6}} \\ \dfrac{1}{\sqrt{2}} & \dfrac{1}{\sqrt{6}} \\ 0 & \dfrac{2}{\sqrt{6}} \end{bmatrix} \begin{bmatrix} 1 & 0 \\ 0 & \sqrt{3} \end{bmatrix} \begin{bmatrix} \dfrac{1}{\sqrt{2}} & -\dfrac{1}{\sqrt{2}} \\ \dfrac{1}{\sqrt{2}} & \dfrac{1}{\sqrt{2}} \end{bmatrix}.$$

5.9   Exercises

1. Find orthogonal matrices which reduce the following hermitian matrices to diagonal form.

$$\text{(i)}\ \begin{bmatrix} 0 & -1 \\ 1 & 0 \end{bmatrix} \quad \text{(ii)}\ \begin{bmatrix} 1 & 0 & 1 \\ 0 & 4 & 0 \\ 1 & 0 & 1 \end{bmatrix} \quad \text{(iii)}\ \begin{bmatrix} 53 & 4 & 1 \\ 4 & 38 & -4 \\ 1 & -4 & 53 \end{bmatrix}$$

$$[\text{Try } \lambda = 54]$$

2. Find the principle axes of the following quadratic forms
   (i) $10x^2 + 4xy + 7y^2 = 100$,
   (ii) $x^2 + 5xy - 11y^2 = 4$,
   (iii) $-x^2 - 2\sqrt{3}xz - 4yz + 3z^2 = 25$,
   (iv) $x^2 + 2y^2 + 6yz + 4z^2 = 1$.
Determine which are positive definite, etc.

3. Which of the following quadratic forms are positive definite, etc.
   (i) $x_1 x_2 + 2x_1 x_3 + 4x_1 x_4 + x_2 x_3 + x_4^2$,
   (ii) $x_1^2 + 2x_1 x_2 + x_2^2 + 2x_1 x_3 + 2x_2 x_3 + x_3^2$,
   (iii) $2x_1^2 + 8x_1 x_2 - 12x_1 x_3 + 7x_2^2 - 24x_2 x_3 + 15x_3^2$.

4. Write down explicitly the spectral decomposition for the matrix

$$\begin{bmatrix} 1 & 0 & 1 \\ 0 & 4 & 0 \\ 1 & 0 & 1 \end{bmatrix}.$$

5. Find a square matrix B with the property that

$$B^2 = \begin{bmatrix} 1 & 0 & 1 \\ 0 & 4 & 0 \\ 1 & 0 & 1 \end{bmatrix}.$$

6. Given the matrix

$$A = \begin{bmatrix} 2 & 0 & 1 \\ 0 & -2 & -1 \end{bmatrix}$$

find an orthonormal set of eigenvectors $\{x_1, x_2\}$ for the non-zero eigenvalues of A^*A and an orthonormal set of eigenvectors $\{y_1, y_2\}$ for the non-zero eigenvalues of AA^*. Write down the singular values decomposition for A. Describe the linear transformation represented by $y_1 x_1^*$ and $y_2 x_2^*$. Finally, express A in the form

$$A = Y \begin{bmatrix} \sqrt{\lambda_1} & 0 \\ 0 & \sqrt{\lambda_2} \end{bmatrix} X^*.$$

7. Prove that a square matrix A is singular if and only if it has a zero eigenvalue.

8. Explain why $\text{trace}(A) = \lambda_1 + \lambda_2 + \cdots + \lambda_n$ and $\det(A) = \lambda_1 \lambda_2 \cdots \lambda_n$, where $\lambda_1, \lambda_2, \ldots, \lambda_n$ are the eigenvalues of A.

9. Prove that if $\lambda_1, \lambda_2, \ldots, \lambda_k$ are distinct eigenvalues of a square matrix A then the corresponding eigenvectors $p_1, p_2, \ldots, p_k$ are linearly independent.

10. How are the eigenvalues of A^k related to those of A (where A is square)? If A is non-singular, how are the eigenvalues of A^{-1} related to those of A?

11. If A is normal, show that λ is an eigenvalue if and only if $\bar{\lambda}$ is an eigenvalue of A^*.

12. Prove that the eigenvalues of a unitary matrix have modulus 1.

13. If the eigenvalues of the normal matrix A are real, prove that A is hermitian.

14. Show that the eigenvalues of

$$\left[\begin{array}{c|c} A & C \\ \hline 0 & B \end{array} \right]$$

with A and B square are simply the eigenvalues of A and B.

15. Show that if A is positive definite hermitian and B is hermitian, then there exists a non-singular P such that

$$P^*AP = I$$
$$P^*BP = D$$

where D is diagonal. [Hint: consider $A^{-1/2}$.]

16. Show that a 2×2 matrix $A = (a_{ij})$ is non-negative definite if $\det(A) \geq 0$ and $a_{11} \geq 0$.

6

Reduction to upper triangular form

6.1 Upper triangular and block diagonal forms

In this section we explain how a square matrix A may be reduced by a similarity transformation to upper triangular form. It is of interest to note that using this process it is possible to make a further reduction (see for example D. Russell, *Mathematics of Finite Dimensional Control Systems*) by means of a similarity to the following block diagonal matrix

$$P^{-1}AP = \Lambda = \begin{bmatrix} \lambda_1 I_1 + N_1 & 0 & 0 \\ 0 & \lambda_2 I_2 + N_2 & \\ & & \ddots\ \lambda_k I_k + N_k \end{bmatrix},$$

where $\lambda_1, \lambda_2, \ldots, \lambda_k$ lists the distinct eigenvalues of A, I_1 is of size $m_1 \times m_1$ and m_1 is the multiplicity of λ_1, I_2 is of size $m_2 \times m_2$ and m_2 is the multiplicity of λ_2, etc. whereas $N_1, N_2, \ldots, N_k$ are matrices whose entries along the below the diagonal are zero. Thus Λ is upper-triangular. The Jordan canonical form is of course, a special case of the above form. The advantage of the present form is that a computational procedure is available, whereas for the Jordan form the calculations are very much more complicated.

6.2 Reduction to upper triangular form

For the time being let $\lambda_1, \lambda_2, \ldots, \lambda_n$ be the n eigenvalues of A listed in any order but with repetitions in cases of multiplicity. Let $\mathbf{v}_1$ be an eigenvector to value λ_1 and let us assume for the moment that $\mathbf{v}_1 = (v_{11}, v_{12}, \ldots)^t$ has $v_{11} \neq 0$. We will deal with the case $v_{11} = 0$ later.

Step 1 with $v_{11} \neq 0$

Let $P_1 = [\mathbf{v}_1, \mathbf{e}_2, \mathbf{e}_3, \ldots, \mathbf{e}_n]$, where as usual $\mathbf{e}_1, \mathbf{e}_2, \mathbf{e}_3, \ldots, \mathbf{e}_n$ denote the natural base vectors: $\mathbf{e}_1 = (1, 0, \ldots, 0)^t$ etc. in $\mathbb{R}^n$.

Since $v_{11} \neq 0$ the matrix P_1 has rank n and hence P_1^{-1} exists. (Observe that P_1 is in echelon form.) Now

$$AP_1 = A[\mathbf{v}_1, \mathbf{e}_2, \mathbf{e}_3, \ldots] = [\lambda_1 \mathbf{v}_1, \mathbf{a}_2, \mathbf{a}_3, \ldots]$$

$$= [\mathbf{v}_1, \mathbf{e}_2, \mathbf{e}_3, \ldots] \begin{bmatrix} \lambda_1 & c_{12} \cdots & c_{1n} \\ 0 & c_{22} & c_{2n} \\ \vdots & & \\ 0 & c_{n2} & c_{nn} \end{bmatrix},$$

where $\mathbf{a}_1, \mathbf{a}_2, \mathbf{a}_3, \ldots$ are the columns of the matrix A and the constants c_{ij} are so selected as to satisfy

$$\mathbf{a}_2 = c_{12}\mathbf{v}_1 + c_{22}\mathbf{e}_2 + \cdots + c_{n2}\mathbf{e}_n,$$
$$\mathbf{a}_3 = c_{13}\mathbf{v}_1 + c_{23}\mathbf{e}_2 + \cdots + c_{n3}\mathbf{e}_n,$$

etc.,

which is clearly possible since $\mathbf{v}_1, \mathbf{e}_2, \ldots, \mathbf{e}_n$ form a basis for $\mathbb{R}^n$. We thus have

$$P_1^{-1}AP_1 = \left[\begin{array}{c|c} \lambda_1 & \mathbf{d}_1 \\ \hline 0 & A_1 \end{array} \right] = \Lambda_1,$$

where A_1 is $(n-1) \times (n-1)$ and $\mathbf{d}_1$ is a row vector. Observe that

$$\det(\Lambda_1 - \lambda I) = (\lambda_1 - \lambda)\det(A_1 - \lambda I),$$

but

$$\det(P_1^{-1}AP_1 - \lambda I) = \det\{P_1^{-1}(A - \lambda I)P_1\}$$
$$= \det P_1^{-1} \det(A - \lambda I) \det P_1$$
$$= \det(A - \lambda I).$$

Hence the eigenvalues of A_1 are precisely $\lambda_2, \lambda_3, \ldots, \lambda_n$. This concludes the first step.

Step 1 with $v_{11} = 0$

In this case pick some other co-ordinate of $\mathbf{v}_1$, say v_{1j}, which is non-zero. This is clearly possible since $\mathbf{v}_1$ is itself non-zero (why?). We now take P_1 to be the same as before but with its jth column replaced by $\mathbf{e}_1$; thus

$$P_1 = [\mathbf{v}_1, \mathbf{e}_2, \mathbf{e}_3, \ldots, \mathbf{e}_{j-1}, \mathbf{e}_j, \mathbf{e}_{j+1}, \ldots, \mathbf{e}_n].$$

Just as before P_1 is of rank n (swopping the first and jth column will give an echelon form). The preceding argument is now once again applicable. (Check!)

Step 2

The next step is to apply the first step to the submatrix A_1 this time using the eigenvalue λ_2. Thus we obtain an $(n-1) \times (n-1)$ matrix say $\hat{P}_2$ such that

$$\hat{P}_2^{-1} A_1 \hat{P}_2 = \left[\begin{array}{c|c} \lambda_2 & \mathbf{d}_2 \\ \hline 0 & A_2 \end{array} \right] = \Lambda_2.$$

Now let P_2 be the bordered matrix

$$P_2 = \left[\begin{array}{c|c} 1 & 0 \\ \hline 0 & \hat{P}_2 \end{array} \right].$$

It may be checked that the inverse of this matrix is

$$P_2^{-1} = \left[\begin{array}{c|c} 1 & 0 \\ \hline 0 & \hat{P}_2^{-1} \end{array} \right]$$

and it is now easy to compute the result

$$P_2 P_1^{-1} A P_1 P_2 = \left[\begin{array}{c|c|c} \lambda_1 & \multicolumn{2}{c}{\mathbf{a}_1'} \\ \hline 0 & \lambda_2 & \mathbf{d}_2' \\ \hline 0 & 0 & A_3 \end{array} \right] = \Lambda_2.$$

At the next step we work on A_3 using λ_3 and first obtaining a matrix $\hat{P}_3$ of size $(n-2) \times (n-2)$; we then let P_3 be the bordered matrix

$$P_3 = \left[\begin{array}{c|c} I_2 & 0 \\ \hline 0 & \hat{P}_3 \end{array} \right].$$

where I_2 is a 2×2 identity matrix and 0 represents zero matrices of appropriate sizes. Again it may be checked that the inverse of this matrix is

$$P_3^{-1} = \left[\begin{array}{c|c} I_2 & 0 \\ \hline 0 & \hat{P}_3^{-1} \end{array} \right]$$

so the previous step may be mimicked.

Repeating this process $n-1$ times in all finally gives
$P_{n-1}^{-1} P_{n-2}^{-1} \cdots P_2^{-1} P_1^{-1} A P_1 P_2 \cdots P_n = \Lambda$, where Λ is upper triangular.

6.3   Example

$$A = \begin{bmatrix} 2 & 2 & -1 \\ -1 & -1 & 1 \\ -1 & -2 & 2 \end{bmatrix}.$$

Thus

$$\det(A - \lambda I) = \begin{vmatrix} 2-\lambda & 2 & -1 \\ -1 & -1-\lambda & 1 \\ -1 & -2 & 2-\lambda \end{vmatrix}$$

$$= -\lambda^3 + 3\lambda^2 - 3\lambda + 1 = -(\lambda - 1)^3.$$

Solving $(A - \lambda I) = 0$ for $\lambda = 1$ yields

$$\left. \begin{array}{l} 1x_1 + 2x_2 - x_3 = 0 \\ -x_1 - 2x_2 + x_3 = 0 \\ -x_1 - 2x_2 + x_3 = 0 \end{array} \right\}.$$

So the eigenspace consists of vectors of the form $(x_1, x_2, x_1 + 2x_2)^t$, i.e. is spanned by $(1, 0, 1)^t$ and $(0, 1, 2)^t$. Thus we may take

$$P_1 = \begin{bmatrix} 1 & 0 & 0 \\ 0 & 1 & 0 \\ 1 & 0 & 1 \end{bmatrix}, \quad \text{so that } P_1^{-1} = \begin{bmatrix} 1 & 0 & 0 \\ 0 & 1 & 0 \\ -1 & 0 & 1 \end{bmatrix}.$$

Now

$$P_1^{-1} A P_1 = \begin{bmatrix} 1 & 0 & 0 \\ 0 & 1 & 0 \\ -1 & 0 & 1 \end{bmatrix} \begin{bmatrix} 2 & 2 & -1 \\ -1 & -1 & 1 \\ -1 & -2 & 2 \end{bmatrix} \begin{bmatrix} 1 & 0 & 0 \\ 0 & 1 & 0 \\ 1 & 0 & 1 \end{bmatrix}$$

$$= \begin{bmatrix} 1 & 0 & 0 \\ 0 & 1 & 0 \\ -1 & 0 & 1 \end{bmatrix} \begin{bmatrix} 1 & 2 & -1 \\ 0 & -1 & 1 \\ 1 & -2 & 2 \end{bmatrix}$$

$$= \begin{bmatrix} 1 & 2 & -1 \\ 0 & -1 & 1 \\ 0 & -4 & 3 \end{bmatrix}.$$

Thus

$$A_1 = \begin{bmatrix} -1 & 1 \\ -4 & 3 \end{bmatrix}.$$

Solving $(A_1 - I)x = 0$ yields

$$\left. \begin{array}{l} -2x_1 + x_2 = 0 \\ -4x_1 + 2x_2 = 0 \end{array} \right\}.$$

The solution space is spanned by $(1, 2)^t$. We now take

$$\hat{P}_2 = \begin{bmatrix} 1 & 0 \\ 2 & 1 \end{bmatrix}, \quad \text{so that } \hat{P}_2^{-1} = \begin{bmatrix} 1 & 0 \\ -2 & 1 \end{bmatrix}.$$

We thus have (after bordering $\hat{P}_2$ to obtain P_2):

$$P_2^{-1}\Lambda_1 P_2 = \begin{bmatrix} 1 & 0 & 0 \\ 0 & 1 & 0 \\ 0 & -2 & 1 \end{bmatrix} \begin{bmatrix} 1 & 2 & -1 \\ 0 & -1 & 1 \\ 0 & -4 & 3 \end{bmatrix} \begin{bmatrix} 1 & 0 & 0 \\ 0 & 1 & 0 \\ 0 & 2 & 1 \end{bmatrix}$$

$$= \begin{bmatrix} 1 & 0 & 0 \\ 0 & 1 & 0 \\ 0 & -2 & 1 \end{bmatrix} \begin{bmatrix} 1 & 0 & -1 \\ 0 & 1 & 1 \\ 0 & 2 & 3 \end{bmatrix}$$

$$= \begin{bmatrix} 1 & 0 & -1 \\ 0 & 1 & 1 \\ 0 & 0 & 1 \end{bmatrix}.$$

Thus

$$P = P_1 P_2 = \begin{bmatrix} 1 & 0 & 0 \\ 0 & 1 & 0 \\ 1 & 0 & 1 \end{bmatrix} \begin{bmatrix} 1 & 0 & 0 \\ 0 & 1 & 0 \\ 0 & 2 & 1 \end{bmatrix} = \begin{bmatrix} 1 & 0 & 0 \\ 0 & 1 & 0 \\ 1 & 2 & 1 \end{bmatrix}.$$

6.4   Further examples of reduction to upper triangular form

1   An example with $v_{11} = 0$

We recall we have to pick $v_{1j} \neq 0$ (possible since $\mathbf{v}_1 \neq 0$) and then use

$$P_1 = (\mathbf{v}_1, \mathbf{e}_2, \dots, \mathbf{e}_{j-1}, \mathbf{e}_1, \mathbf{e}_{j+1}, \dots, \mathbf{e}_n).$$

For example, if

$$A = \begin{bmatrix} 2 & 2 & -1 \\ -1 & -1 & 1 \\ -1 & -2 & 2 \end{bmatrix},$$

then we have seen in Section 6.3 that $\lambda = 1$ is the only eigenvalue and that the eigenspace is generated by the vectors $(1, 0, 1)^t$ $(0, 1, 2)^t$. Suppose we take for $\mathbf{v}_1$ the vector

$$\begin{bmatrix} 0 \\ 1 \\ 2 \end{bmatrix}$$

which has $v_{11} = 0$. Let us follow through step 1 for this case.

Thus

$$P_1 = \begin{bmatrix} 0 & 1 & 0 \\ 1 & 0 & 0 \\ 2 & 0 & 1 \end{bmatrix},$$

hence

$$P_1^{-1} = -1 \begin{bmatrix} 0 & -1 & 0 \\ -1 & 0 & 0 \\ 0 & 2 & -1 \end{bmatrix}.$$

Thus

$$\Lambda_1 = \begin{bmatrix} 0 & 1 & 0 \\ 1 & 0 & 0 \\ 0 & -2 & 1 \end{bmatrix} \begin{bmatrix} 2 & 2 & -1 \\ -1 & -1 & 1 \\ -1 & -2 & 2 \end{bmatrix} \begin{bmatrix} 0 & 1 & 0 \\ 1 & 0 & 0 \\ 2 & 0 & 1 \end{bmatrix}$$

$$= \begin{bmatrix} 0 & 1 & 0 \\ 1 & 0 & 0 \\ 0 & -2 & 1 \end{bmatrix} \begin{bmatrix} 0 & 2 & -1 \\ 1 & -1 & 1 \\ 2 & -1 & 2 \end{bmatrix} = \begin{bmatrix} 1 & -1 & 1 \\ 0 & 2 & -1 \\ 0 & 1 & 0 \end{bmatrix}.$$

Now let

$$A_1 = \begin{pmatrix} 2 & -1 \\ 1 & 0 \end{pmatrix}.$$

The eigenvalues of A_1 are still $+1$, $+1$. Solving

$$(A_1 - I)\begin{pmatrix} a \\ b \end{pmatrix} = \mathbf{0}$$

we have

$$\left. \begin{array}{l} a - b = 0 \\ a - b = 0 \end{array} \right\}.$$

We take

$$\mathbf{v}_2 = \begin{pmatrix} 1 \\ 1 \end{pmatrix} \quad \text{so} \quad \hat{P}_2 = \begin{pmatrix} 1 & 0 \\ 1 & 1 \end{pmatrix}, \quad \hat{P}_2^{-1} = \begin{pmatrix} 1 & 0 \\ -1 & 1 \end{pmatrix}$$

$$\Lambda_2 = P_2^{-1}\Lambda_1 P_2 = \begin{bmatrix} 1 & 0 & 0 \\ 0 & 1 & 0 \\ 0 & -1 & 1 \end{bmatrix} \begin{bmatrix} 1 & -1 & 1 \\ 0 & 2 & -1 \\ 0 & 1 & 0 \end{bmatrix} \begin{bmatrix} 1 & 0 & 0 \\ 0 & 1 & 0 \\ 0 & 1 & 1 \end{bmatrix}$$

$$= \begin{bmatrix} 1 & 0 & 0 \\ 0 & 1 & 0 \\ 0 & -1 & 1 \end{bmatrix} \begin{bmatrix} 1 & 0 & 1 \\ 0 & 1 & -1 \\ 0 & 1 & 0 \end{bmatrix} = \begin{bmatrix} 1 & 0 & 1 \\ 0 & 1 & -1 \\ 0 & 0 & 1 \end{bmatrix}.$$

Here

$$P = P_1 P_2 = \begin{bmatrix} 0 & 1 & 0 \\ 1 & 0 & 0 \\ 2 & 0 & 1 \end{bmatrix} \begin{bmatrix} 1 & 0 & 0 \\ 0 & 1 & 0 \\ 0 & 1 & 1 \end{bmatrix} = \begin{bmatrix} 0 & 1 & 0 \\ 1 & 0 & 0 \\ 2 & 1 & 1 \end{bmatrix}.$$

2.   A 4 × 4 example

Using a similarity transformation reduce to upper triangular form

$$A = \begin{bmatrix} 4 & 1 & 0 & 1 \\ 2 & 4 & 0 & 0 \\ 0 & -1 & 4 & -1 \\ 0 & 0 & 2 & 4 \end{bmatrix}.$$

We have

$$p(\lambda) = \det(A - \lambda I)$$
$$= (4 - \lambda)^2 \{(4 - \lambda)^2 + 2\} - 1 \cdot 2 \cdot \{(4 - \lambda)^2 + 2\} - 1\{2(-2)\}$$
$$= (4 - \lambda)^4.$$

We find $\mathbf{v}_1$ with $(A - 4I)\mathbf{v}_1 = 0$. Say

$$\mathbf{v}_1 = \begin{bmatrix} x \\ y \\ z \\ t \end{bmatrix},$$

then

$$\left. \begin{array}{l} y \qquad\quad\ + t = 0 \\ 2x \qquad\qquad\quad = 0 \\ \quad -y \qquad\ -t = 0 \\ \qquad\quad 2z \qquad = 0 \end{array} \right\}$$

giving

$$\mathbf{v}_1 = \begin{bmatrix} 0 \\ 1 \\ 0 \\ -1 \end{bmatrix}$$

We take

$$P_1 = [\mathbf{v}_1 \ \mathbf{e}_1 \ \mathbf{e}_3 \ \mathbf{e}_4] = \begin{bmatrix} 0 & 1 & 0 & 0 \\ 1 & 0 & 0 & 0 \\ 0 & 0 & 1 & 0 \\ -1 & 0 & 0 & 1 \end{bmatrix},$$

$$P_1^{-1} = -1 \begin{bmatrix} 0 & -1 & 0 & 0 \\ -1 & 0 & 0 & 0 \\ 0 & 0 & -1 & 0 \\ 0 & -1 & 0 & -1 \end{bmatrix}.$$

Now

$$AP_1 = \begin{bmatrix} 0 & 4 & 0 & 1 \\ 4 & 2 & 0 & 0 \\ 0 & 0 & 4 & -1 \\ -4 & 0 & 2 & 4 \end{bmatrix} \text{ (interpreting } P_1 \text{ as column operations).}$$

Thus,

$$\Lambda_1 = P_1^{-1}AP_1 = \begin{bmatrix} 4 & 2 & 0 & 0 \\ 0 & 4 & 0 & 1 \\ 0 & 0 & 4 & -1 \\ 0 & 2 & 2 & 4 \end{bmatrix} \text{ (interpreting } P_1^{-1} \text{ as row operations).}$$

Thus

$$A_1 = \begin{bmatrix} 4 & 0 & 1 \\ 0 & 4 & -1 \\ 2 & 2 & 4 \end{bmatrix}$$

has eigenvalue 4, so we solve

$$(A_1 - 4I)\begin{bmatrix} a \\ b \\ c \end{bmatrix} = 0,$$

or

$$\left.\begin{array}{r} c = 0 \\ -c = 0 \\ 2a + 2b \quad\; = 0 \end{array}\right\}$$

giving

$$\mathbf{v}_2 = \begin{bmatrix} 1 \\ -1 \\ 0 \end{bmatrix}.$$

We now take

$$\hat{P}_2 = \begin{bmatrix} 1 & 0 & 0 \\ -1 & 1 & 0 \\ 0 & 0 & 1 \end{bmatrix} \quad \text{so that } \hat{P}_2^{-1} = \begin{bmatrix} 1 & 0 & 0 \\ 1 & 1 & 0 \\ 0 & 0 & 1 \end{bmatrix}.$$

Now

$$\Lambda_1 P_2 = \Lambda_1 \begin{bmatrix} 1 & 0 & 0 & 0 \\ 0 & 1 & 0 & 0 \\ 0 & -1 & 1 & 0 \\ 0 & 0 & 0 & 1 \end{bmatrix} = \begin{bmatrix} 4 & 2 & 0 & 0 \\ 0 & 4 & 0 & 1 \\ 0 & -4 & 4 & -1 \\ 0 & 0 & 2 & 4 \end{bmatrix} \quad \begin{array}{l}\text{(treating } P_2 \\ \text{as column} \\ \text{operations)}\end{array}$$

$$\Lambda_2 = P_2^{-1}\Lambda_1 P_2 = \begin{bmatrix} 1 & 0 & 0 & 0 \\ 0 & 1 & 0 & 0 \\ 0 & 1 & 1 & 0 \\ 0 & 0 & 0 & 1 \end{bmatrix} \Lambda_1 P_2 = \begin{bmatrix} 4 & 2 & 0 & 0 \\ 0 & 4 & 0 & 1 \\ 0 & 0 & 4 & 0 \\ 0 & 0 & 2 & 4 \end{bmatrix}$$

$$\text{(treating } P_2^{-1} \text{ as row operations)}$$

Now

$$A_2 = \begin{pmatrix} 4 & 0 \\ 2 & 4 \end{pmatrix}.$$

Solving

$$(A_2 - 4I)\begin{pmatrix} u \\ v \end{pmatrix} = 0$$

gives $u = 0$ so

$$\mathbf{v}_3 = \begin{pmatrix} 0 \\ 1 \end{pmatrix}.$$

Thus

$$P_3 = \begin{pmatrix} 0 & 1 \\ 1 & 0 \end{pmatrix} \quad \text{and} \quad P_3^{-1} = -1 \begin{pmatrix} 0 & -1 \\ -1 & 0 \end{pmatrix}$$

$$\Lambda_3 = P_3^{-1} \Lambda_2 \begin{bmatrix} 1 & 0 & 0 & 0 \\ 0 & 1 & 0 & 0 \\ 0 & 0 & 0 & 1 \\ 0 & 0 & 1 & 0 \end{bmatrix}$$

$$= \begin{bmatrix} 1 & 0 & 0 & 0 \\ 0 & 1 & 0 & 0 \\ 0 & 0 & 0 & 1 \\ 0 & 0 & 1 & 0 \end{bmatrix} \begin{bmatrix} 4 & 2 & 0 & 0 \\ 0 & 4 & 1 & 0 \\ 0 & 0 & 0 & 4 \\ 0 & 0 & 4 & 2 \end{bmatrix} = \begin{bmatrix} 4 & 2 & 0 & 0 \\ 0 & 4 & 1 & 0 \\ 0 & 0 & 4 & 2 \\ 0 & 0 & 0 & 4 \end{bmatrix}$$

$$P = P_1 P_2 P_3 = P_1 \begin{bmatrix} 1 & 0 & 0 & 0 \\ 0 & 1 & 0 & 0 \\ 0 & -1 & 1 & 0 \\ 0 & 0 & 0 & 1 \end{bmatrix} \begin{bmatrix} 1 & 0 & 0 & 0 \\ 0 & 1 & 0 & 0 \\ 0 & 0 & 0 & 1 \\ 0 & 0 & 1 & 0 \end{bmatrix}$$

$$= P_1 \begin{bmatrix} 1 & 0 & 0 & 0 \\ 0 & 1 & 0 & 0 \\ 0 & -1 & 0 & 1 \\ 0 & 0 & 1 & 0 \end{bmatrix} = \begin{bmatrix} 0 & 1 & 0 & 0 \\ 1 & 0 & 0 & 0 \\ 0 & -1 & 0 & 1 \\ -1 & 0 & 1 & 0 \end{bmatrix}$$

6.5   Application: simultaneous differential equations

We consider the following simultaneous system of differential equations

$$\frac{d}{dt} \begin{bmatrix} x_1 \\ x_2 \\ \vdots \\ x_n \end{bmatrix} = A \begin{bmatrix} x_1 \\ x_2 \\ \vdots \\ x_n \end{bmatrix} + \begin{bmatrix} g_1(t) \\ g_2(t) \\ \vdots \\ g_n(t) \end{bmatrix}$$

where $x_1, x_2, \ldots, x_n$ are functions of t and the matrix A consisting of constants is of size $n \times n$. We will write this in the form

$$\dot{\mathbf{x}} = A\mathbf{x} + \mathbf{g},$$

where $\dot{\mathbf{x}}$ denotes the column of derivatives of the components of $\mathbf{x}$. The system may easily be solved by first finding a similarity transformation reducing A to upper triangular form. Say

$$P^{-1}AP = B,$$

where B is in upper triangular form. Let us make the change of variable $\mathbf{x} = P\mathbf{z}$, i.e. $\mathbf{z} = P^{-1}\mathbf{x}$. Then since P is a constant matrix we have

$$P\dot{\mathbf{z}} = AP\mathbf{z} + \mathbf{g}.$$

or

$$\dot{\mathbf{z}} = P^{-1}AP\mathbf{z} + P^{-1}\mathbf{g}.$$

We have thus to solve the equation,

$$\dot{\mathbf{z}} = B\mathbf{z} + \mathbf{f},$$

where $\mathbf{f} = P^{-1}\mathbf{g}$ and B is upper triangular. This is particularly easy to solve when we remember that the last row of the equation involves only z_n (which can thus be solved immediately), the last but one only z_{n-1} and z_n (which by now is known), etc. Let us consider the matrix A of Section 6.3 and take $\mathbf{g} = \mathbf{0}$. Using the similarity computed there we have

$$\dot{\mathbf{z}} = \begin{bmatrix} 1 & 0 & -1 \\ 0 & 1 & 1 \\ 0 & 0 & 1 \end{bmatrix} \mathbf{z}$$

so $\dot{z}_3 = z_3$ hence $z_3 = K_1 e^t$ where K_1 is a constant. Thus

$$\dot{z}_2 = z_2 + z_3 = z_2 + K_1 e^t,$$

or

$$\dot{z}_2 - z_2 = K_1 e^t,$$

and using an integrating factor (see Part II, Chapter 20) we have

$$\frac{d}{dt}\{e^{-t}z_2\} = e^{-t}\dot{z}_2 - e^{-t}z_2 = K_1.$$

Thus on integrating

$$z_2 = (K_1 t + K_2)e^t,$$

where K_2 is a constant. Now

$$\dot{z}_1 = z_1 - z_3 = z_1 - K_1 e^t$$

and using an integrating factor again we have

$$\frac{d}{dt}\{e^{-t}z_1\} = e^{-t}\dot{z}_1 - e^{-t}z_1 = -K_1.$$

Thus on integrating we obtain

$$z_1 = -K_1 t e^t.$$

We now have the vector **z** and hence also $\mathbf{x} = P\mathbf{z}$.

6.6   Exercises

1. Find a similarity transformation for each of the following matrices which reduces it to upper triangular form:

$$\begin{bmatrix} 1 & -1 & 1 \\ 0 & 1 & 1 \\ 0 & -1 & 3 \end{bmatrix} \begin{bmatrix} 9 & 3 & -1 \\ 0 & 6 & 4 \\ 3 & 3 & 9 \end{bmatrix} \begin{bmatrix} 2 & 6 & 0 \\ -1 & -2 & -1 \\ 2 & 3 & 4 \end{bmatrix}$$

$$\begin{bmatrix} 3 & 1 & 2 \\ -1 & 1 & -1 \\ -1 & -1 & 0 \end{bmatrix} \begin{bmatrix} 2 & 0 & -4 \\ -1 & 2 & 4 \\ 5 & 2 & 8 \end{bmatrix} \begin{bmatrix} 3 & -3 & -3 \\ 2 & 8 & 2 \\ 1 & 1 & 7 \end{bmatrix}$$

2. Show that a diagonal matrix is normal.
3. Show that a unitary matrix Q exists so that Q^*AQ is upper triangular.
[Hint: In the standard procedure rescale $\mathbf{v}_1$ to unit length, apply Gram–Schmidt to $\mathbf{v}_1, \mathbf{e}_2, \ldots, \mathbf{e}_n$ obtaining an orthonormal system $\mathbf{v}_1, \mathbf{u}_2, \ldots, \mathbf{u}_n$ and use these to construct P_1.]
4. For any unitary matrix P show that if A is normal then P^*AP is normal. Show also that if P is a unitary matrix and P^*AP is normal then A is normal.
5. If B is upper triangular and normal show that B is a diagonal matrix.
[Hint. Consider the (i,i) entry of B^*B and of BB^*. Show successively, starting with the first row, that all rows of B are zero off the diagonal.]
6. Show that if A is normal then A may be reduced by a unitary matrix to a diagonal matrix. (Hint: first reduce A to upper triangular form.]

7

Reduction to tridiagonal form

We present a method due to Householder for reducing a *symmetric* matrix to tridiagonal form by means of an orthogonal transformation. The reduced matrix has zero entries in all positions except possibly those on: the main diagonal, the superdiagonal (i.e. the diagonal immediately above) and the subdiagonal (the diagonal below). This form has some important properties useful in the location of the eigenvalues of the original matrix. No such method seems available for non-symmetric (or non-hermitian) matrices.

7.1   Reflection matrices

Let $\mathbf{x}$ be any vector in $\mathbb{R}^n$ and let $\mathbf{e}_1 = (1,0,\ldots,0)^t$ be as usual, the first basis vector of the natural basis. Now let H be the hyperplane which internally bisects the angle between $\mathbf{x}$ and $\mathbf{e}_1$ and has a *unit* normal vector $\mathbf{v}$ co-planar with $\mathbf{x}$ and $\mathbf{e}_1$ (we show how to find $\mathbf{v}$ later in Section 7.4). Extend $\mathbf{e}_1$ by a factor of $\|\mathbf{x}\|$. Noting that $\mathbf{x} - \mathbf{e}_1\|\mathbf{x}\|$ is parallel to $\mathbf{v}$ (see the illustration in Figure 7.1),

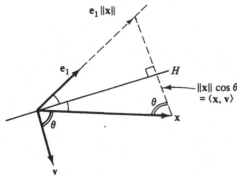

Fig. 7.1

we have

$$\mathbf{x} - \mathbf{e}_1 \| \mathbf{x} \| = (2 \| \mathbf{x} \| \cos \theta) \mathbf{v} = 2 \langle \mathbf{x}, \mathbf{v} \rangle \mathbf{v}.$$

We re-write the formula

$$\mathbf{e}_1 \| \mathbf{x} \| = \mathbf{x} - 2\mathbf{v} \langle \mathbf{v}, \mathbf{x} \rangle = \mathbf{x} - 2\mathbf{v}\mathbf{v}^t \mathbf{x}$$
$$= (I - 2\mathbf{v}\mathbf{v}^t)\mathbf{x}.$$

Thus $H = (I - 2\mathbf{v}\mathbf{v}^t)$ transforms $\mathbf{x}$ to $\mathbf{e}_1 \| \mathbf{x} \|$, i.e. H lines $\mathbf{x}$ up on the first basis element and so reflects across the bisector. H is called a *reflection matrix*.

We now show that H is both symmetric and orthogonal so that in fact it is self-inverse. Symmetry is clear from the formula. $H^2 = I$ may be verified as follows

$$H^t H = HH = (I - 2\mathbf{v}\mathbf{v}^t)(I - 2\mathbf{v}\mathbf{v}^t)$$
$$= I - 4\mathbf{v}\mathbf{v}^t + 4\mathbf{v}(\mathbf{v}^t\mathbf{v})\mathbf{v}^t$$
$$= I$$

(because $\mathbf{v}^t\mathbf{v} = 1$).

Thus $H^{-1} = H$ and so, assuming A is a symmetric matrix, HAH will also be a symmetric matrix similar to A. It will therefore have the same eigenvalues as A.

Although the matrix H was constructed with a view to reflecting the vector $\mathbf{x}$ onto $\mathbf{e}_1 \| \mathbf{x} \|$ we ought also to note how H transforms other vectors $\mathbf{z}$. Since we have

$$\mathbf{y} = H\mathbf{z} = (I - 2\mathbf{v}\mathbf{v}^t)\mathbf{z}$$
$$= \mathbf{z} - 2\langle \mathbf{v}, \mathbf{z} \rangle \mathbf{v}$$
$$= \mathbf{z} - (2 \| \mathbf{z} \| \cos \psi)\mathbf{v},$$

where ψ is the angle between $\mathbf{v}$ and $\mathbf{z}$ the argument developed for $\mathbf{x}$ now shows that $\mathbf{y}$ is the 'mirror image' of $\mathbf{z}$ in the hyperplane H with normal $\mathbf{v}$ (see Figure 7.2).

7.2 Householder's method

The reduction of a *symmetric* matrix A of size $n \times n$ proceeds by a sequence of steps. The first step is to define a vector $\mathbf{x}_1$ in $\mathbb{R}^{n-1}$ to be the first column of A with its first element deleted, thus

$$\mathbf{x}_1 = \begin{bmatrix} a_{21} \\ a_{31} \\ \vdots \\ a_{n1} \end{bmatrix}, \quad \text{where} \quad A = (a_{ij}).$$

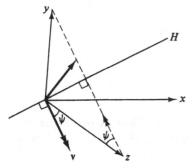

Fig. 7.2

Now let H_1 be a reflection matrix reflecting $\mathbf{x}_1$ across the angle bisector between $\mathbf{x}_1$ and the first basis element in $\mathbb{R}^{n-1}$. Observe that

$$B_1 = \left[\begin{array}{c|c} 1 & \mathbf{0} \\ \hline \mathbf{0} & H_1 \end{array}\right]\left[\begin{array}{c|c} a_{11} & \mathbf{x}_1^t \\ \hline \mathbf{x}_1 & A_{n-1} \end{array}\right]\left[\begin{array}{c|c} 1 & \mathbf{0} \\ \hline \mathbf{0} & H_1 \end{array}\right]$$

$$= \left[\begin{array}{c|c} a_{11} & \mathbf{x}_1^t H_1 \\ \hline H_1\mathbf{x}_1 & H_1 A_{n-1} H_1 \end{array}\right] = \left[\begin{array}{c|c} a_{11} & b_1 0 \cdots 0 \\ \hline \begin{array}{c} b_1 \\ 0 \\ \vdots \\ 0 \end{array} & H_1 A_{n-1} H_1 \end{array}\right].$$

We have thus gained a row of zeros from the third element onwards in the first row and similarly in the first column.

The next step is to perform the same operations on the smaller matrix $H_1 A_{n-1} H_1$ thus introducing still more zeros. More precisely, a reflection matrix H_2 of size $(n-2) \times (n-2)$ is constructed by reference to the vector $\mathbf{x}_2$ consisting of the first column of $H_1 A_{n-1} H_1$ with its first entry deleted. One then verifies that

$$B_2 = \left[\begin{array}{c|c} I_2 & 0 \\ \hline 0 & H_2 \end{array}\right] B_1 \left[\begin{array}{c|c} I_2 & 0 \\ \hline 0 & H_2 \end{array}\right]$$

(where I_2 is the 2×2 identity matrix) is similar to B_1 and has the same first row and first column as B_1. Thus the upper left-hand corner identity I_2 has the effect of keeping unaltered the arrangement obtained in the first step. Since $H_2 H_1 A_{n-1} H_1 H_2$ has a first column and first row containing zeros everywhere except the first two entries, B_2 has the form

$$\left[\begin{array}{cc|ccc}a_1 & b_1 & 0 & 0 & 0 & \cdots \\ b_1 & a_2 & b_2 & 0 & 0 \\ \hline 0 & b_2 \\ 0 & 0 & & A_{n-2} \\ & \cdots \end{array}\right],$$

where $a_1 = a_{11}$, etc.

The general step should now be clear: working with reflection matrices H_k on successively smaller submatrices A_{n-k} we set

$$B_k = \left[\begin{array}{c|c} I_k & 0 \\ \hline 0 & H_k \end{array}\right] B_{k-1} \left[\begin{array}{c|c} I_k & 0 \\ \hline 0 & H_k \end{array}\right].$$

After $n - 2$ steps we will have achieved the desired matrix:

$$\left[\begin{array}{ccccc} a_1 & b_1 & 0 & 0 & 0 & \cdots \\ b_1 & a_2 & b_2 & 0 & 0 \\ 0 & b_2 & a_3 & b_3 & 0 \\ \vdots & & & & \\ 0 & & & & a_n \end{array}\right].$$

7.3 Evaluation of $p(\lambda)$

The first advantage of the tridiagonal form is in the computation of the characteristic polynomial

$$p(\lambda) = \det\left[\begin{array}{ccccc} a_1 - \lambda & b_1 & 0 & 0 & 0 & \cdots \\ b_1 & a_2 - \lambda & b_2 & 0 & 0 \\ 0 & b_2 & a_3 - \lambda & b_3 & 0 \\ \vdots & & & & \\ 0 & & & & a_n - \lambda \end{array}\right].$$

Expanding successively the upper left-hand corner subdeterminants of size $1 \times 1, 2 \times 2, 3 \times 3$, etc., which we denote by $p_k(\lambda)$ (for $k = 1, 2, 3, \ldots$), we observe the following pattern:

$$p_0(\lambda) = 1 \quad \text{(included for later convenience)}$$
$$p_1(\lambda) = (a_1 - \lambda)$$
$$p_2(\lambda) = (a_2 - \lambda)(a_1 - \lambda) - b_1^2$$
$$= (a_2 - \lambda)p_1(\lambda) - b_1^2 p_0(\lambda)$$

$$p_3(\lambda) = (a_3 - \lambda)p_2(\lambda) - b_2^2 p_1(\lambda)$$
$$\vdots$$
$$p_k(\lambda) = (a_2 - \lambda)p_{k-1}(\lambda) - b_{k-1}^2 p_{k-2}(\lambda).$$

Of course $p_n(\lambda)$ is $p(\lambda)$.

Now a second and more important property of the tridiagonal form is the fact that the above sequence of polynomials is what is called a *Sturm sequence*, meaning that:

(a)   the zeros of all the polynomials are real;
(b)   the zeros of $p_k(\lambda)$ and $p_{k+1}(\lambda)$ interlace (see Figure 7.3);
(c)   the number of zeros of $p_n(\lambda)$ which are less than λ_0 is the same as the number of changes of sign in the sequence:

$$1, p_1(\lambda_0), p_2(\lambda_0), p_3(\lambda_0), \ldots, p_n(\lambda_0).$$

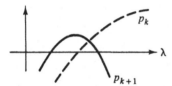

Fig. 7.3

Any zero values appearing in this list should be suppressed for the purposes of counting sign changes.

7.4   Finding a bisecting hyperplane and its normal

We begin by working in the two-dimensional subspace spanned by e_1 and x. Let u be the unit vector

$$u = \frac{x}{\|x\|}.$$

Observe that the vectors $e_1 \pm u$ are angle bisectors for the angle between e_1 and x, the choice of 'plus' giving the internal bisector and 'minus' the external bisector (cf. Figure 7.4).

Now normalise the vectors $w = e_1 \pm u$; we obtain

$$\|w\|^2 = \|(1 \pm u_1, \pm u_2, \ldots, \pm u_n)^t\|^2$$
$$= (1 \pm 2u_1 + u_1^2) + u_2^2 + \cdots + u_n^2$$
$$= 2 \pm 2u_1 \quad \text{(since } u \text{ is a unit vector)}.$$

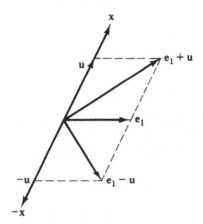

Fig. 7.4

If $\mathbf{v}$ is the unit vector $\mathbf{v} = \mathbf{w}/\|\mathbf{w}\|$, we have

$$v_1^2 = \frac{(1 \pm u_1)^2}{2(1 \pm u_1)} = \tfrac{1}{2}(1 \pm u_1)$$

and of course

$$v_j = \frac{\pm u_j}{\sqrt{\{2(1 \pm u_1)\}}} = \frac{\pm u_j}{2v_1}.$$

Suppose we take the *minus* sign, then $\mathbf{v}$ is the external bisector and defines the normal direction to a hyperplane bisecting the angle between $\mathbf{e}_1$ and $\mathbf{x}$. If, on the other hand, we take the *plus* sign, then $\mathbf{v}$ is the internal bisector and defines the normal direction to a hyperplane bisecting the angle between $\mathbf{e}_1$ and $-\mathbf{x}$. Either sign thus gives a reflection matrix. Clearly (why?) only the minus sign secures $H\mathbf{x} > 0$.

Example
Reduce to tridiagonal form the following matrix.

$$\begin{bmatrix} 3 & 1 & 0 & 0 & 0 \\ 1 & 3 & 0 & 0 & 0 \\ 0 & 0 & 3 & 1 & 1 \\ 0 & 0 & 1 & 3 & 1 \\ 0 & 0 & 1 & 1 & 3 \end{bmatrix}$$

We only have to work on

$$\left[\begin{array}{c|cc} 3 & 1 & 1 \\ \hline 1 & 3 & 1 \\ 1 & 1 & 3 \end{array}\right].$$

Call this matrix A_1. Our first step is to construct a reflection matrix for the vector $\mathbf{x} = (1, 1)^t$. Normalizing we have $\mathbf{u} = (1/\sqrt{2})(1, 1)^t$. Let us use the formulas developed above choosing the negative sign. Thus we have:

$$v_1^2 = \frac{1}{2}\left(1 - \frac{1}{\sqrt{2}}\right) = \alpha^2, \text{ say.}$$

Hence

$$\mathbf{v} = \left[\begin{array}{c} \alpha \\ \dfrac{-1}{2\sqrt{2\alpha}} \end{array}\right] \text{ and so } \mathbf{v}\mathbf{v}^t = \left[\begin{array}{cc} \alpha^2 & \dfrac{-1}{2\sqrt{2}} \\ \dfrac{-1}{2\sqrt{2}} & \dfrac{1}{8\alpha^2} \end{array}\right],$$

so

$$H_1 = I - 2\mathbf{v}\mathbf{v}^t = \left[\begin{array}{cc} 1 - 2\alpha^2 & \dfrac{1}{\sqrt{2}} \\ \dfrac{1}{\sqrt{2}} & 1 - \dfrac{1}{4\alpha^2} \end{array}\right].$$

Now we may calculate that

$$\frac{1}{\alpha^2} = \frac{2\sqrt{2}}{(\sqrt{2}-1)} \cdot \frac{\sqrt{2}+1}{\sqrt{2}+1} = \frac{4 + 2\sqrt{2}}{1}$$

$$H_1 = \left[\begin{array}{cc} \dfrac{1}{\sqrt{2}} & \dfrac{1}{\sqrt{2}} \\ \dfrac{1}{\sqrt{2}} & -\dfrac{1}{\sqrt{2}} \end{array}\right],$$

and

$$H_1 A_1 H_1 = \left[\begin{array}{cc} 4 & 0 \\ 0 & 2 \end{array}\right],$$

so that letting

$$H = \begin{bmatrix} 1 & 0 & \\ 0 & 1 & \\ \hline & & H_1 \end{bmatrix},$$

we have HAH equal to

$$\begin{bmatrix} 3 & 1 & 0 & 0 & 0 \\ 1 & 3 & 0 & 0 & 0 \\ 0 & 0 & 3 & \sqrt{2} & 0 \\ 0 & 0 & \sqrt{2} & 4 & 0 \\ 0 & 0 & 0 & 0 & 2 \end{bmatrix}.$$

7.5   Exercises

1. Find an orthogonal matrix for each of the following matrices which reduces it to tridiagonal form.

$$\begin{bmatrix} 3 & 0 & 1 \\ 0 & 5 & 2 \\ 1 & 2 & 4 \end{bmatrix} \quad \begin{bmatrix} 1 & 0 & 0 & 0 \\ 0 & -\sqrt{3} & \sqrt{3} & 1 \\ 0 & \sqrt{3} & \sqrt{3} & -2 \\ 0 & 1 & -2 & \sqrt{3} \end{bmatrix} \quad \begin{bmatrix} 1 & 0 & 0 & 0 \\ 0 & 2 & 3 & 4 \\ 0 & 3 & 1 & 0 \\ 0 & 4 & 0 & 1 \end{bmatrix}$$

$$\begin{bmatrix} 1 & 0 & 0 & 0 \\ 0 & 1 & \sqrt{2} & 2 \\ 0 & \sqrt{2} & 2 & \sqrt{2} \\ 0 & 2 & \sqrt{2} & 3 \end{bmatrix} \quad \begin{bmatrix} 1 & \sqrt{2} & 0 & 0 \\ \sqrt{2} & 2 & \sqrt{3} & 1 \\ 0 & \sqrt{3} & 2 & -\sqrt{3} \\ 0 & 1 & -\sqrt{3} & 4 \end{bmatrix} \quad \begin{bmatrix} 2 & 4 & & & \\ 4 & 2 & & & \\ & & 4 & 2 & 2 \\ & & 2 & 7 & -1 \\ & & 2 & -1 & -1 \end{bmatrix}$$

In each case determine how many positive eigenvalues the matrices have.

2. Why are the roots of the sequence $p_k(\lambda)$ all real (see Section 7.3)?

3. Examine why the roots of the Sturm sequence interlace.

4. Verify for the case of 3×3 matrices the assertion that the number of eigenvalues to the left of λ_0 equals the number of sign changes in the Sturm sequence.

5. Show that in the case of 2×2 matrices the reflection matrix H is equal to

$$\begin{bmatrix} u_1 & u_2 \\ u_2 & -u_1 \end{bmatrix}.$$

6. A tridiagonal matrix T has all its diagonal entries equal to a and its sub- and super-diagonal entries equal to b (with $b \neq 0$).

Show for some constant $A_\lambda \neq 0$ dependent on λ, that the Sturm sequence $p_0(\lambda), p_1(\lambda), \ldots, p_n(\lambda)$ of the matrix $T - \lambda I$ consisting of its principal subdeterminants in ascending order is given by

$$p_r(\lambda) = A_\lambda(t_1^{r+1} - t_2^{r+1}), \quad (r = 0, 1, \ldots, n)$$

where t_1 and t_2 are the roots of the quadratic

$$t^2 - (a - 1)t + b^2.$$

Hence show that the eigenvalues of T are of the form

$$\lambda = a + 2b \cos \frac{m\pi}{n+1} \quad (m = 1, \ldots, n).$$

(Hint: Make use of the $(n+1)^{st}$ roots of unity: $\exp\{i(2m\pi)/(n+1)\}$.)

7. Develop the argument of Section 7.2 for the general step. In particular check that bordering the matrix H_k as indicated in the step from B_{k-1} to B_k ensures that the work done at the previous step is not altered, viz. the leftmost $n \times (k-1)$ submatrix and the uppermost $(k-1) \times n$ submatrix of B_{k-1} are equal to the corresponding submatrices of B_k.

8. Show that if α is a root of the polynomial

$$p(x) \equiv x^n + a_{n-1}x^{n-1} + a_{n-2}x^{n-2} + \cdots + a_1 x + a_0$$

then $|\alpha| \leqslant 1 + A$ where $A = \max\{|a_i| : 0 \leqslant i < n\}$.

(Hint: $p(\alpha) = 0$ implies $|\alpha|^n \leqslant A\{|\alpha|^n - 1\}/\{|\alpha| - 1\}$.)

8

Inverses

In this chapter we shall be concerned with matrices that are not necessarily square. If A is of size $m \times n$ we shall be interested in generalizing the notion of an inverse. Recall that for the existence of an ordinary inverse it is necessary that A be square (i.e. $m = n$). An $n \times n$ matrix B is then an ordinary inverse for A if and only if $AB = BA = I$. We then write $B = A^{-1}$. An $n \times n$ matrix A has an ordinary inverse A^{-1} if and only if it is non-singular for which various criteria are given in Chapter 3. In particular, it is necessary and sufficient that its determinant be non-zero or that A have full rank n. If A is not square of if A is a singular matrix, it cannot have an ordinary inverse. Instead we must look for matrices which satisfy only some of the properties of an ordinary inverse. Two immediate candidates are: a *left inverse*, that is a matrix L of size $n \times m$ such that

$$LA = I_n,$$

where I_n is an $n \times n$ identity matrix, and a *right inverse*, that is a matrix R of size $n \times m$ satisfying

$$AR = I_m.$$

An even more general notion will be considered in Section 8.4, where we seek a matrix B of size $n \times m$ satisfying

$$ABA = A.$$

Ordinary inverses have an important role to play in the construction of left, right and generalised inverses.

8.1 Calculation of ordinary inverses I – standard techniques (revision)

Only a non-singular, square matrix can have an ordinary inverse. If A is non-singular it can be reduced by a sequence of elementary row operations to the identity matrix I. If the same sequence of row operations is carried out on the matrix I instead, the result will be the matrix A^{-1}. To see this let $E_1, E_2, E_3, \ldots, E_k$ be the elementary matrices corresponding to the elementary row operations (cf. Chapter 3 on matrices), then

$$E_k E_{k-1} \cdots E_1 A = I,$$

so

$$A = E_1^{-1} \cdots E_k^{-1}$$

and

$$A^{-1} = E_k \cdots E_1.$$

With an alternative technique in mind, we shall use the notation A_{ij} to denote the determinant of that submatrix of A which is obtained by deleting the ith row and the jth column of A. The *adjugate matrix* of A (we prefer this to the term 'adjoint' which is also used for something else) is the matrix B whose i, jth entry is $(-1)^{i+j} A_{ji}$, thus

$$B = \begin{bmatrix} +A_{11} & -A_{12} & +A_{13} & \cdots \\ -A_{21} & +A_{22} & -A_{23} & \cdots \\ +A_{31} & +A_{32} & +A_{33} & \cdots \\ & & \cdots & \end{bmatrix}^t = \begin{bmatrix} +A_{11} & -A_{21} & +A_{31} & \cdots \\ -A_{12} & +A_{22} & -A_{32} & \cdots \\ +A_{13} & -A_{23} & +A_{33} & \cdots \\ & & \cdots & \end{bmatrix}.$$

The numbers A_{ij} are called the *minors* of A and the numbers $(-1)^{i+j} A_{ij}$ are called the *cofactors*. We have

$$(AB)_{ij} = \sum_k a_{ik} b_{kj} = \sum_k (-1)^{k+j} a_{ik} A_{jk}.$$

Now observe that if $i = j$ this formula is the same as that obtained by expanding $\det A$ by its jth row. Thus $(AB)_{ii} = \det A$. On the other hand, if $i \neq j$, the formula is that obtained by replacing the ith row of A by its jth row and expanding the determinant of the resulting matrix by its ith row. Since a determinant with two equal rows is zero, it follows that $(AB)_{ij} = 0$ when $i \neq j$. We conclude that $AB = (\det A) I$. Hence A is non-singular if and only if $\det A$ is non-zero and

$$A^{-1} = \frac{1}{\det A} B.$$

This is the *Cramer rule*.

8.2   Calculation of ordinary inverses II – exotic techniques

(i)   We sketch a technique for inverting a matrix given in partitioned form. Suppose A and P are invertible matrices.

$$\text{If } A = \begin{bmatrix} P & Q \\ R & S \end{bmatrix}, \quad \text{then } A^{-1} = \begin{bmatrix} X & -P^{-1}QW \\ -WRP^{-1} & W \end{bmatrix},$$

where $W = (S - RP^{-1}Q)^{-1}$ and $X = P^{-1} + P^{-1}QWRP^{-1}$. This may readily be verified by multiplying the two matrices. The formula is useful for reducing the size of matrices requiring inversion. Note the special case when Q is a column:

$$\text{if } A = \left[\begin{array}{c|c} P & \mathbf{u} \\ \hline \mathbf{v}^t & \alpha \end{array} \right], \quad \text{then } A^{-1} = \left[\begin{array}{c|c} H & \mathbf{p} \\ \hline \mathbf{q}^t & \alpha \end{array} \right].$$

Here $\mathbf{u}$ and $\mathbf{v}$ are $n \times 1$ column vectors, α is a scalar and H, $\mathbf{p}$ and $\mathbf{q}$ are given by the formulas

$$H = A^{-1} + \alpha A^{-1}\mathbf{u}\mathbf{v}^t A^{-1},$$
$$\alpha = (\alpha - \mathbf{v}^t A^{-1}\mathbf{u})^{-1},$$
$$\mathbf{p} = -\alpha A^{-1}\mathbf{u},$$
$$\mathbf{q}^t = -\alpha \mathbf{v}^t A^{-1}.$$

(ii) Suppose that H, A and D are non-singular matrices and that

$$H = A + BDC.$$

Then

$$H^{-1} = A^{-1} - A^{-1}B(D^{-1} + CA^{-1}B)^{-1}CA^{-1}.$$

Again there is an important special case to note. If $H = A + \mathbf{b}\mathbf{b}^t$ then

$$H^{-1} = A^{-1} - \frac{A^{-1}\mathbf{b}\mathbf{c}^t A^{-1}}{1 + \mathbf{c}^t A^{-1}\mathbf{b}},$$

where $\mathbf{b}$ and $\mathbf{c}$ are column vectors. This idea is known as *tearing*.

8.3   Left and right inverses

Let A be an $m \times n$ matrix with real entries. We shall relate the existence of left and right inverses to a consideration of the system of equations

$$A\mathbf{x} = \mathbf{b}.$$

Theorem
The following are equivalent:
(i)   *A has a right inverse, e.g. $A^t(AA^t)^{-1}$;*
(ii) *$A\mathbf{x} = \mathbf{b}$ has a solution for every $\mathbf{b}$;*
(iii) *A has rank m (i.e. its rank is the left side of $m \times n$).*

Proof. Suppose that A has a right inverse. Then for *any* $\mathbf{b}$ in $\mathbb{R}^m$ we have

$$\mathbf{b} = I\mathbf{b} = (AR)\mathbf{b} = A(R\mathbf{b}),$$

so that $\mathbf{x} = R\mathbf{b}$ is a solution to the equation $A\mathbf{x} = \mathbf{b}$.

Clearly (ii) implies (iii). (See Chapter 1, Section 1.5.) Now if A has rank m then also the $m \times m$ square matrix AA^t has rank m (see Chapter 3, Section 3.8). Hence the matrix $A^t(AA^t)^{-1}$ is well-defined and is a right inverse for A because $A\{A^t(AA^t)^{-1}\} = (AA^t)(AA^t)^{-1} = I$.

Theorem
The following are equivalent:
  (i) A *has a left inverse, e.g.* $(A^tA)^{-1}A^t$;
  (ii) $A\mathbf{x} = \mathbf{b}$ *has a **unique** solution whenever the system* $A\mathbf{x} = \mathbf{b}$ *is consistent*;
  (iii) A *has rank* n *(i.e. its rank is the right side of* $m \times n$).

Proof. Suppose A has a left inverse L and that the equation $A\mathbf{x} = \mathbf{b}$ has a solution. Then $L\mathbf{b} = LA\mathbf{x} = I\mathbf{x} = \mathbf{x}$. Thus the solution is necessarily $L\mathbf{b}$.

To see that (ii) implies (iii) note that the equation $A\mathbf{x} = 0$ is consistent since it has a trivial solution. Thus the unique solution is $\mathbf{x} = 0$. But this is to say that $N(A) = \{0\}$, i.e. that the nullity of A is zero. (See Chapter 3, Section 2.) But then the rank must be n.

If the rank of A is n then the rank of the $n \times n$ square matrix A^tA is also n. Hence the matrix $(A^tA)^{-1}A^t$ is well-defined and is a left inverse for A because $\{(A^tA)^{-1}A^t\}A = (A^tA)^{-1}(A^tA) = I$.

8.4   Generalised inverses

Let A be an arbitrary $m \times n$ matrix. Any $n \times m$ matrix B which satisfies

$$ABA = A$$

is called a *weak generalised inverse* of A. We shall use the notation A^g to denote any weak generalised inverse of A. In general a given matrix A will have *many* weak generalised inverses. For instance if C is a matrix such that $AC = 0$ and if B is a weak generalised inverse of A, then so also is $B + C$ provided C is $n \times m$.

Example
For any matrix A, a left inverse L or a right inverse R (when one or other exists) is also a weak generalised inverse. We have

$$ALA = A(LA) = AI = A,$$
$$ARA = (AR)A = IA = A.$$

Theorem

Suppose that $ABA = A$. Then:

(i) *AB projects $\mathbb{R}^m$ onto the column space $R(A)$ of A,*

(ii) *BA projects $\mathbb{R}^n$ parallel to the null space $N(A)$.*

Proof. We have the following equations:

$$(AB)^2 = (AB)(AB) = (ABA)B = AB,$$
$$(BA)^2 = (BA)(BA) = B(ABA) = BA.$$

So the two matrices AB and BA are idempotents and hence represent projections. (See Chapter 4.) To show that the range of AB is the same as that of A, observe that any vector $A\mathbf{x}$ may be rewritten as $(ABA)\mathbf{x} = AB(A\mathbf{x})$ and so is in the range of AB; evidently any vector $AB\mathbf{z}$ is in the form $A(B\mathbf{z})$ and so is in the range of A. Now for (ii) we verify that the null spaces of BA and of A are identical:

$$A\mathbf{x} = 0 \Rightarrow BA\mathbf{x} = 0$$
$$BA\mathbf{x} = 0 \Rightarrow ABA\mathbf{x} = 0 \Rightarrow A\mathbf{x} = 0.$$

This completes the proof.

Like left and right inverses, a generalised inverse A^g is intimately connected with the solubility of the equation $A\mathbf{x} = \mathbf{b}$. Observe that, if the latter system has a solution at all, $\mathbf{b}$ is in the range A. Hence since AB projects onto $R(A)$ the projection must leave $\mathbf{b}$ unchanged, i.e. $AB\mathbf{b} = \mathbf{b}$. Conversely, if this equation holds then $\mathbf{x} = B\mathbf{b}$ is a solution of $A\mathbf{x} = \mathbf{b}$. Thus a generalised inverse will give a solution whenever a solution exists. But the general solution of $A\mathbf{x} = \mathbf{b}$ takes the form $\mathbf{x}_0 + \mathbf{w}$ where $\mathbf{w}$ is any solution of the equation $A\mathbf{x} = 0$. Thus $\mathbf{w}$ lies in the null space of A. But $I - BA$ is a projection onto the null space of BA (by part (ii) of the theorem above). Thus $\mathbf{w}$ takes the form $(I - BA)\mathbf{z}$ for some $\mathbf{z}$. In conclusion we have:

The system $A\mathbf{x} = \mathbf{b}$ is consistent if and only if $\mathbf{b} = AA^g\mathbf{b}$. When consistent, its solutions are given by the formula:

$$\mathbf{x} = A^g\mathbf{b} + (I - A^g A)\mathbf{z}$$

where $\mathbf{z}$ is arbitrary.

The existence of these formulae does not imply that the use of generalised inverses is the optimal way for solving systems of linear equations. Usually this will not be the case.

As we have already noted the weak generalised inverse is not unique. Among the many generalised inverses there is one that deserves special attention. It is the inverse that makes AB and BA *orthogonal* projections onto their respective ranges and furthermore and satisfies $BAB = B$. This is called the *strong* generalised inverse. Some properties of this inverse are discussed in Questions 10, 11 and 12 of Exercises 8.6. Here we stop only to prove its uniqueness.

Suppose that both G and H satisfy the definition of a strong generalised inverse of A. Then we have working on the first two terms of GAG that

$$
\begin{aligned}
G &= (GA)G = A^t G^t G &&(\text{since } GA = (GA)^t)\\
&= (AHA)^t G^t G = (HA)^t A^t G^t G\\
&= HAA^t G^t G &&(\text{since } HA = (HA)^t)\\
&= HAGAG &&(\text{since } A^t G^t = (GA)^t = GA)\\
&= HAG.
\end{aligned}
$$

But the same argument can be followed using H for G and working on the last two terms of HAH. Thus

$$
\begin{aligned}
H &= H(AH) = HH^t A^t\\
&= HH^t(AGA)^t = HH^t A^t (AG)^t\\
&= HH^t A^t AG\\
&= H(AH)AG &&(\text{since } AH = (AH)^t = H^t A^t)\\
&= HAG.
\end{aligned}
$$

Hence $H = G$.

8.5   Computation of generalised inverses

We suggest three methods, the usefulness of each of which depends on circumstance. The first method is rather special.

Method I

Suppose that A is of rank k and can be partitioned as below:

$$
A = \left[\begin{array}{c|c} B & C \\ \hline D & E \end{array}\right],
$$

where B is square of size $k \times k$ and is *also* of rank k. Then we claim that the matrix

$$
A^g = \left[\begin{array}{c|c} B^{-1} & 0 \\ \hline 0 & 0 \end{array}\right]
$$

is a weak generalised inverse. To see this note first that

$$\left[\begin{array}{c|c} B & C \\ \hline D & E \end{array}\right] \left[\begin{array}{c|c} B^{-1} & 0 \\ \hline 0 & 0 \end{array}\right] \left[\begin{array}{c|c} B & C \\ \hline D & E \end{array}\right] = \left[\begin{array}{c|c} I & 0 \\ \hline DB^{-1} & 0 \end{array}\right] \left[\begin{array}{c|c} B & C \\ \hline D & E \end{array}\right]$$

$$= \left[\begin{array}{c|c} B & C \\ \hline D & DB^{-1}C \end{array}\right].$$

We need to see that $E = DB^{-1}C$. Let us write

$$\left[\begin{array}{c|c} P_1 & P_2 \\ \hline P_3 & P_4 \end{array}\right] \left[\begin{array}{c|c} I_k & 0 \\ \hline 0 & 0 \end{array}\right] \left[\begin{array}{c|c} Q_1 & Q_2 \\ \hline Q_3 & Q_4 \end{array}\right] = \left[\begin{array}{c|c} B & C \\ \hline D & E \end{array}\right],$$

where the first and third matrices in the line above are non-singular. This is possible (Chapter 3, Section 3.5) since the matrix A has rank k and the two matrices just mentioned represent inverses of the row and column operations required to bring A to reduced echelon form. We may now compute:

$$\left[\begin{array}{c|c} P_1 & 0 \\ \hline P_3 & 0 \end{array}\right] \left[\begin{array}{c|c} Q_1 & Q_2 \\ \hline Q_3 & Q_4 \end{array}\right] = \left[\begin{array}{c|c} P_1 Q_1 & P_1 Q_2 \\ \hline P_3 Q_1 & P_3 Q_2 \end{array}\right] = \left[\begin{array}{c|c} B & C \\ \hline D & E \end{array}\right].$$

Hence

$$DB^{-1}C = P_3 Q_1 (Q_1^{-1} P_1^{-1}) P_1 Q_2 = P_3 Q_2 = E.$$

We need to justify the fact that both P_1 and Q_1 are invertible. Now $k = \text{rank } B = \text{rank } P_1 Q_1 \leqslant \text{rank } P_1$. But P_1 is square of size $k \times k$, so we conclude its rank is k. Similarly for Q_1.

We remark that if the top left corner B of A is not of rank k then appropriate modifications can be carried out by column and row operations to make the new upper left corner have rank k (how and why?).

Method II

This is a method which in fact yields the (unique) strong generalised inverse. For the moment we take for granted that, if the $m \times n$ matrix A is of rank k, then we may write

$$A = BC,$$

where the $m \times k$ matrix B and the $k \times n$ matrix C are both of rank k. Assuming this has been done, we can compute a left inverse for B, say L, and a right inverse for C, say R. Now note that

$$A^g = RL$$

is a weak generalised inverse for A; indeed,

$$AA^g A = A(RL)A = BC(RL)BC = B(CR)(LB)C = BIIC = BC = A.$$

The cases $L = (B^t B)^{-1} B^t$ and $R = C^t(CC^t)^{-1}$ are particularly interesting. Then

$$AA^g = BCC^t(CC^t)^{-1}(B^t B)^{-1} B^t = B(B^t B)^{-1} B^t$$
$$A^g A = C^t(CC^t)^{-1}(B^t B)^{-1} B^t BC = C^t(CC^t)^{-1}C$$

are symmetric and it may also be checked that $A^g AA^g = A^g$ hence the strong generalised inverse is

$$A^G = C^t(CC^t)^{-1}(B^t B)^{-1} B^t.$$

We return to the question of writing A in the form BC. The matrix B is constructed so as to consist of precisely k linearly independent columns of A, where k is the rank of A. Thus B is of size $m \times k$ and of rank k. With this choice for B the equation $A = BC$ expresses the fact that the columns $a_1, a_2, \ldots, a_n$ of A are combinations of the columns $b_1, b_2, \ldots, b_k$ of B. To see this write

$$[a_1, a_2, \ldots, a_n] = [b_1, b_2, \ldots, b_k] \begin{bmatrix} c_{11} \cdots c_{1n} \\ \vdots \\ c_{k1} \quad c_{kn} \end{bmatrix}$$

and note that this matrix equation reduces to a system of equations with the cs as unknowns; for example we have

$$a_1 = c_{11} b_1 + c_{21} b_2 + \cdots + c_{k1} b_k.$$

Finally, notice that by the Example of Chapter 3, Section 3.9 the matrix C has rank k.

Example

$$A = \begin{bmatrix} 1 & 1 & 2 \\ 1 & 2 & 3 \\ 2 & 1 & 3 \end{bmatrix} = \begin{bmatrix} 1 & 1 \\ 1 & 2 \\ 2 & 1 \end{bmatrix} \begin{bmatrix} 1 & 0 & 1 \\ 0 & 1 & 1 \end{bmatrix} = BC.$$

It is obvious that the first two columns of A are linearly independent and that the third is their sum. Thus

$$\begin{bmatrix} 1 \\ 1 \\ 2 \end{bmatrix} = 1 \cdot \begin{bmatrix} 1 \\ 1 \\ 2 \end{bmatrix} + 0 \cdot \begin{bmatrix} 1 \\ 2 \\ 1 \end{bmatrix} \qquad \begin{bmatrix} 1 \\ 2 \\ 1 \end{bmatrix} = 0 \cdot \begin{bmatrix} 1 \\ 1 \\ 2 \end{bmatrix} + 1 \cdot \begin{bmatrix} 1 \\ 2 \\ 1 \end{bmatrix}$$

$$\begin{bmatrix} 2 \\ 3 \\ 2 \end{bmatrix} = 1 \cdot \begin{bmatrix} 1 \\ 1 \\ 2 \end{bmatrix} + 1 \cdot \begin{bmatrix} 1 \\ 2 \\ 1 \end{bmatrix}.$$

We may now calculate the strong generalised inverse of A. We have:

$$CC^t = \begin{bmatrix} 1 & 0 & 1 \\ 0 & 1 & 1 \end{bmatrix} \begin{bmatrix} 1 & 0 \\ 0 & 1 \\ 1 & 1 \end{bmatrix} = \begin{bmatrix} 2 & 1 \\ 1 & 2 \end{bmatrix}, \quad (CC^t)^{-1} = \frac{1}{3} \begin{bmatrix} 2 & -1 \\ -1 & 2 \end{bmatrix},$$

$$B^t B = \begin{bmatrix} 1 & 1 & 2 \\ 1 & 2 & 1 \end{bmatrix} \begin{bmatrix} 1 & 1 \\ 1 & 2 \\ 2 & 1 \end{bmatrix} = \begin{bmatrix} 6 & 5 \\ 5 & 6 \end{bmatrix}, \quad (B^t B)^{-1} = \frac{1}{11} \begin{bmatrix} 6 & -5 \\ -5 & 6 \end{bmatrix}.$$

Hence,

$$A^G = \begin{bmatrix} 1 & 0 \\ 0 & 1 \\ 1 & 1 \end{bmatrix} \frac{1}{3} \begin{bmatrix} 2 & -1 \\ -1 & 2 \end{bmatrix} \frac{1}{11} \begin{bmatrix} 6 & -5 \\ -5 & 6 \end{bmatrix} \begin{bmatrix} 1 & 1 & 2 \\ 1 & 2 & 1 \end{bmatrix}$$

$$= \frac{1}{33} \begin{bmatrix} 2 & -1 \\ -1 & 2 \\ 1 & 1 \end{bmatrix} \begin{bmatrix} 1 & -4 & 7 \\ 1 & 7 & -4 \end{bmatrix} = \frac{1}{33} \begin{bmatrix} 1 & -15 & 18 \\ 1 & 18 & -15 \\ 2 & 3 & 3 \end{bmatrix}.$$

Method III

Given the singular values decomposition of an $m \times n$ matrix A, it is easy to write down the value of A^G. If

$$A = \sum_1^k \sqrt{\lambda_i} \, \mathbf{y}_i \mathbf{x}_i^*$$

is the singular values decomposition then

$$B = A^G = \sum_1^k \frac{1}{\sqrt{\lambda_i}} \mathbf{x}_i \mathbf{y}_i^*.$$

We have that

$$ABA = \sum_1^k \sqrt{\lambda_r} \mathbf{y}_r \mathbf{x}_r^* \sum_1^k \frac{1}{\sqrt{\lambda_s}} \mathbf{x}_s \mathbf{y}_s^* \sum_1^k \sqrt{\lambda_t} \mathbf{y}_t \mathbf{x}_t^*$$

$$= \sum_r \sum_s \sum_t \frac{\sqrt{\lambda_r}\sqrt{\lambda_t}}{\sqrt{\lambda_s}} \mathbf{y}_r (\mathbf{x}_r^* \mathbf{x}_s)(\mathbf{y}_s^* \mathbf{y}_t) \mathbf{x}_t^*$$

$$= \sum_1^k \sqrt{\lambda_r} \mathbf{y}_r \mathbf{x}_r^* = A.$$

The orthonormality of the sets $\{x_1, \ldots, x_k\}$ and $\{y_1, \ldots, y_k\}$ has been used very strongly here. The matrix

$$AB = \sum_1^k \sqrt{\lambda_r} y_r x_r^* \sum_1^k \frac{1}{\sqrt{\lambda_s}} x_s y_s^* = \sum_1^k y_r y_r^*$$

and similarly the matrix

$$BA = \sum_1^k \frac{1}{\sqrt{\lambda_r}} x_r y_r^* \sum_1^k \sqrt{\lambda_s} y_s x_s^* = \sum_1^k x_r x_r^*$$

is symmetric and hence both represent orthogonal projections.

We note for reference that the orthogonal projection of $\mathbb{R}^m$ onto the range $R(A)$ of A is

$$\sum_1^k y_i y_i^*.$$

The orthogonal projection of $\mathbb{R}^n$ parallel to $N(A)$ is

$$\sum_1^k x_i x_i^*.$$

8.6   Exercises

1. Find the equations which u, v, w, x, y and z must satisfy in order that

$$\begin{bmatrix} 1 & -1 & 1 \\ 1 & 1 & 2 \end{bmatrix} \begin{bmatrix} u & v \\ w & x \\ y & z \end{bmatrix} = \begin{bmatrix} 1 & 0 \\ 0 & 1 \end{bmatrix},$$

hence find all the right inverses of the matrix

$$A = \begin{bmatrix} 1 & -1 & 1 \\ 1 & 1 & 2 \end{bmatrix}.$$

Show that this matrix has no left inverse.

2. Show that the matrix

$$A = \begin{bmatrix} 1 & -1 & 1 \\ 1 & 1 & 1 \end{bmatrix}$$

has no left inverse and no right inverse.

3. If A, B and $A + B$ are non-singular matrices prove that

$$(A^{-1} + B^{-1})^{-1} = A(A + B)^{-1} B = B(A + B)^{-1} A.$$

4. Calculate the inverse of the non-singular matrix

$$\begin{bmatrix} 0 & 1 & 2 \\ 2 & 0 & 1 \\ 1 & 2 & 0 \end{bmatrix}$$

by the following methods: (i) Cramer's rule, (ii) elementary operations.

5. Explain Cramer's rule for the solution of systems of linear equations on the basis of Section 8.1.

6. Explain why a weak generalised inverse of the matrix

$$A = \begin{bmatrix} 1 & 2 & 3 & 0 & 6 \\ 3 & 1 & 2 & 6 & 0 \\ 2 & 3 & 1 & 6 & 6 \\ 1 & 1 & 1 & 2 & 2 \\ 1 & 2 & 0 & 4 & 4 \\ 4 & 3 & 2 & 10 & 4 \end{bmatrix} \quad \text{is} \quad \frac{1}{18} \begin{bmatrix} -5 & 7 & 1 & 0 & 0 & 0 \\ 1 & -5 & 7 & 0 & 0 & 0 \\ 7 & 1 & -5 & 0 & 0 & 0 \\ 0 & 0 & 0 & 0 & 0 & 0 \\ 0 & 0 & 0 & 0 & 0 & 0 \end{bmatrix}.$$

What is the weak generalised inverse of the following matrix?

$$B = \begin{bmatrix} 6 & 1 & 2 & 0 & 3 \\ 0 & 3 & 1 & 6 & 2 \\ 6 & 2 & 3 & 6 & 1 \\ 2 & 1 & 1 & 2 & 1 \\ 4 & 1 & 2 & 4 & 0 \\ 4 & 4 & 3 & 10 & 2 \end{bmatrix}$$

7. Use the generalised inverse given in the last question to write down a formula for the solutions of the system

$$\begin{bmatrix} 1 & 2 & 3 & 0 & 6 \\ 3 & 1 & 2 & 6 & 0 \\ 2 & 3 & 1 & 6 & 6 \\ 1 & 1 & 1 & 2 & 2 \\ 1 & 2 & 0 & 4 & 4 \\ 4 & 3 & 2 & 10 & 4 \end{bmatrix} \begin{bmatrix} u \\ v \\ w \\ x \\ y \end{bmatrix} = \begin{bmatrix} 3 \\ 4 \\ 5 \\ 2 \\ 3 \\ 7 \end{bmatrix}.$$

8. Express the matrix

$$A = \begin{bmatrix} 1 & 1 & 0 & -1 \\ 0 & 1 & 1 & 1 \\ 1 & 1 & 0 & -1 \end{bmatrix}$$

in the form $A = BC$ where B and C are both of rank 2 and of sizes 3×2 and 2×4 respectively.

Hence calculate A^G and find the matrix which represents the orthogonal projection of $\mathbb{R}^3$ onto the column space of A.

9. If

$$A = \left[\begin{array}{c|c} B & C \\ \hline D & E \end{array}\right]$$

where A is of rank k and B is a non-singular $k \times k$ matrix prove that

$$A = \left[\frac{B}{D}\right][I|B^{-1}C].$$

Hence express A^G in terms of the matrices A, B, C and D.

10. Let G be the (strong) generalised inverse of Section 8.5, Method II. Show that $GAG = G$. Prove that:
   (i) AG projects $\mathbb{R}^n$ onto $R(A)$ parallel to $N(G)$,
   (ii) GA projects $\mathbb{R}^n$ onto $R(G)$ parallel to $N(A)$.

11. Suppose that $Ax = b$ is consistent. Explain why $x = A^G b$ is the solution nearest to the origin (i.e. the solution which minimises $\|x\|$).

12. Let A be an $m \times n$ matrix. Show that the general 'least squares solution' of the system $Ax = b$ is given by

$$x = A^G b + (I - A^G A)z.$$

13. Use the singular values decomposition of the matrix

$$A = \begin{bmatrix} 0 & 1 \\ 1 & 0 \\ 1 & 1 \end{bmatrix}$$

to calculate the value of A^G. Write down orthogonal projections of $\mathbb{R}^m$ onto $R(A)$ and of $\mathbb{R}^n$ parallel to $N(A)$.

14. Suppose that the real matrices A and B satisfy $AB^t B = 0$. Prove that $AB^t = 0$. [Hint: Consider the equation $C^t C = 0$.]

15. Let $(A^t A)^g$ be any weak generalised inverse of $A^t A$. Prove that $A(A^t A)^g A^t$ is then the orthogonal projection of $\mathbb{R}^m$ onto $R(A)$. [Hint: Use the last question.]

16. Suppose that A has a right inverse B and that B satisfies

$$R(B) = R(A^t). \tag{+}$$

Show that $B = A^t C$ for some $(m \times m)$ matrix C. Deduce that C is non-singular and conclude that there is exactly one right inverse for A satisfying $(+)$.

17. Show that if A has a right inverse then $R = A^T(AA^T)^{-1}$ is the strong generalized inverse of A and that $X = Rb$ is the solution of $Ax = b$ nearest the origin. [Hint: Use $\mathcal{N}(A) = \mathcal{R}(A^t)^\perp$, consider $u = Rb - z$ where z is any other solution and apply Pythagoras' theorem.]

18. Deduce from the last question that if A is any matrix then of all the least squares solutions of the possibly inconsistent system $Ax = b$ the one nearest the origin is $x = A^G b$. [Hint: If $A = BC$, replace $BCx = b$ by $Bz = b$ and $Cx = z$ where B has a left inverse and C has a right inverse.]

9

Convexity

9.1   Lines and line segments

The *line* which passes through two vectors **a** and **b** is the set of all **x** such that

$$\mathbf{x} = \alpha\mathbf{a} + \beta\mathbf{b}$$

where $\alpha + \beta = 1$ (Figure 9.1). The *line segment* (or chord) joining **a** and **b** is the set of all **x** which satisfy

$$\mathbf{x} = \alpha\mathbf{a} + \beta\mathbf{b}$$

where $\alpha + \beta = 1$, $\alpha \geqslant 0$ and $\beta \geqslant 0$ (Figure 9.2).

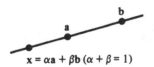

$\mathbf{x} = \alpha\mathbf{a} + \beta\mathbf{b}\ (\alpha + \beta = 1)$

Fig. 9.1

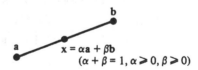

$\mathbf{x} = \alpha\mathbf{a} + \beta\mathbf{b}$
$(\alpha + \beta = 1, \alpha \geqslant 0, \beta \geqslant 0)$

Fig. 9.2

9.2   Convex sets

A set is *convex* if and only if whenever the set contains **a** and **b** it then also contains the line segment which joins them (Figure 9.3).

Example
Prove that the n-dimensional ball

$$B = \{\mathbf{x} : \|\mathbf{x} - \mathbf{a}\| < r\}$$

is convex in $\mathbb{R}^n$.

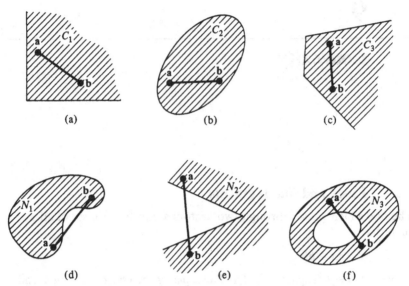

Fig. 9.3. (a)–(c) convex sets, (d)–(f) non-convex sets

Suppose that **x** and **y** lie in *B*. We have to show that $\alpha\mathbf{x} + \beta\mathbf{y}$ lies in *B* provided that $\alpha + \beta = 1$, $\alpha \geqslant 0$ and $\beta \geqslant 0$. But

$$
\begin{aligned}
\| \alpha\mathbf{x} + \beta\mathbf{y} - \mathbf{a} \| &= \| \alpha(\mathbf{x} - \mathbf{a}) + \beta(\mathbf{y} - \mathbf{a}) \| \\
&\leqslant |\alpha| \, \|(\mathbf{x} - \mathbf{a})\| + |\beta| \, \|(\mathbf{y} - \mathbf{a})\| \\
&= \alpha \, \|(\mathbf{x} - \mathbf{a})\| + \beta \, \|(\mathbf{y} - \mathbf{a})\| \\
&< \alpha r + \beta r = r.
\end{aligned}
$$

so $\alpha\mathbf{x} + \beta\mathbf{y}$ lies in *B* (Figure 9.4).

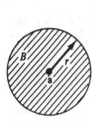

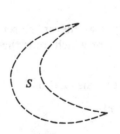

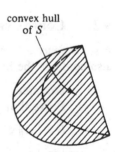

Fig. 9.4   Fig. 9.5

9.3   Convex hull

Given a set S its *convex hull* is the smallest convex set which contains S (Figure 9.5). We will see below that such a smallest convex set exists and that it is possible to characterise the points which lie in it. For this we need the following notion.

9.4   Convex combinations

A *convex combination* of a set of vectors $\{x_1, x_2, \ldots, x_k\}$ is an expression of the form

$$y = \alpha_1 x_1 + \alpha_2 x_2 + \cdots + \alpha_k x_k$$

in which $\alpha_1 + \alpha_2 + \cdots + \alpha_k = 1$, $\alpha_1 \geqslant 0$, $\alpha_2 \geqslant 0, \ldots, \alpha_k \geqslant 0$.

A convex combination of the set $\{x_1, x_2, \ldots, x_k\}$ can be identified geometrically with a point in the convex hull of $\{x_1, x_2, \ldots, x_k\}$. It may also be interpreted physically: suppose weights $\alpha_1, \ldots, \alpha_k$ are distributed at positions $x_1, \ldots, x_k$ in a plane made of infinitely thin carboard (so that the cardboard is itself weightless), then the centre of gravity of this plane, that is the position at which it would balance on a needle point, will be at $\alpha_1 x_1 + \cdots + \alpha_k x_k$ (Figure 9.6).

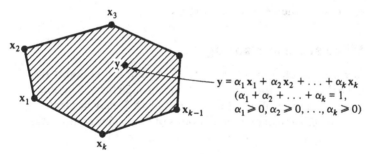

Fig. 9.6

The convex hull of a set S may be characterized as the set of all convex combinations of points taken from that set. In particular a convex set contains all convex combinations of its points. We sketch a proof. It is an easy exercise to check that the set of convex combinations of members of S is a convex set. We need to show also that if a convex set C contains S then it also contains all convex combinations of members of S. If the combination is of *two* members of S it will be in C by the definition of convexity. If the combination is of *three* members of S we show how to reduce the problem of membership of C back to combinations of two points of S. Say the combination is

$$\mathbf{y} = \alpha_1 \mathbf{s}_1 + \alpha_2 \mathbf{s}_2 + \alpha_3 \mathbf{s}_3$$

where $\alpha_1 + \alpha_2 + \alpha_3 = 1, \alpha_1 \geqslant 0, \alpha_2 \geqslant 0$, and $\alpha_3 \geqslant 0$ and $\mathbf{s}_1, \mathbf{s}_2, \mathbf{s}_3$ are points of S. Since not all the coefficients are zero we may assume without loss of generality that $\alpha_1 + \alpha_2 \neq 0$. Put

$$\mathbf{y}_1 = \frac{\alpha_1}{\alpha_1 + \alpha_2} \mathbf{s}_1 + \frac{\alpha_2}{\alpha_1 + \alpha_2} \mathbf{s}_2.$$

We have arranged for the coefficients here to sum to unity, so $\mathbf{y}_1$ is a member of C by the convexity of C. But likewise we see that because $\mathbf{y} = (\alpha_1 + \alpha_2)\mathbf{y}_1 + \alpha \mathbf{s}_3$ the latter point also belongs to C. The reduction argument generalises. For instance, if we consider a combination of *four* points, say

$$\mathbf{y} = \alpha_1 \mathbf{s}_1 + \alpha_2 \mathbf{s}_2 + \alpha_3 \mathbf{s}_3 + \alpha_4 \mathbf{s}_4,$$

with the usual assumptions, then we put

$$\mathbf{y}_1 = \frac{\alpha_1}{\alpha_1 + \alpha_2 + \alpha_3} \mathbf{s}_1 + \frac{\alpha_2}{\alpha_1 + \alpha_2 + \alpha_3} \mathbf{s}_2 + \frac{\alpha_3}{\alpha_1 + \alpha_2 + \alpha_3} \mathbf{s}_3$$

which is a combination of three members of S so is in C by our previous argument. We leave the details to the reader.

9.5   Affine and linear analogues

An *affine set* may be characterised by the fact that, if it contains the points $\mathbf{a}$ and $\mathbf{b}$ then it contains the straight line through them. The *affine hull* of a set is the smallest affine set which contains S. It may be identified with the set of all *affine combinations* of points of S, an affine combination of the vectors $\{\mathbf{x}_1, \mathbf{x}_2, \ldots, \mathbf{x}_k\}$ being an expression of the form

$$\mathbf{y} = \alpha_1 \mathbf{x}_1 + \alpha_2 \mathbf{x}_2 + \cdots + \alpha_k \mathbf{x}_k$$

in which $\alpha_1 + \alpha_2 + \cdots + \alpha_k = 1$.

We already know what a linear combination is. Instead of a 'linear set' we usually speak of a 'subspace'. Instead of a 'linear hull of S' we usually speak of the 'subspace spanned by S', or the 'linear span of S'.

9.6   Dimension of convex sets

Given a convex set S we define its dimension to be that of its affine hull A. Thus a triangular lamina in $\mathbb{R}^3$ has dimension 2 since its affine hull is a plane (Figure 9.7).

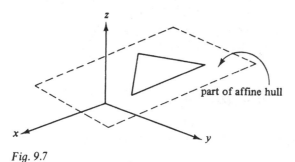

Fig. 9.7

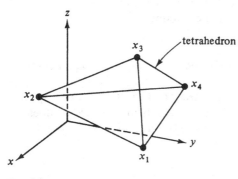

Fig. 9.8

A *simplex* in $\mathbb{R}^n$ is the convex hull of any $n + 1$ points which do not lie in a hyperplane (i.e. an affine set of dimension $n - 1$). Thus a simplex in $\mathbb{R}^n$ has dimension n. In $\mathbb{R}^3$ a simplex is a tetrahedron (Figure 9.8).

The convex hull of an arbitrary finite collection of points is called a *convex polytope*. This notion generalises convex polygons and polyhedra.

9.7 Some properties of convex sets

Bodily movement of a convex set around in space will obviously not affect its convexity. This fact is contained in the following more general theorem.

Theorem
If S and T are convex, then for scalars λ and μ so is the set $\lambda S + \mu T$ which consists of all vectors of the form $\lambda \mathbf{s} + \mu \mathbf{t}$ where $\mathbf{s}$ is in S and $\mathbf{t}$ is in T.

Proof. Let **a** and **b** be points in $\lambda S + \mu T$ and let $\alpha + \beta = 1$ with $\alpha \geqslant 0$ and $\beta \geqslant 0$. We have that for some s_1, s_2 in S and t_1, t_2 in T

$$\mathbf{a} = \lambda \mathbf{s}_1 + \mu \mathbf{t}_1, \quad \mathbf{b} = \lambda \mathbf{s}_2 + \mu \mathbf{t}_2.$$

Hence·

$$\alpha \mathbf{a} + \beta \mathbf{b} = \alpha(\lambda \mathbf{s}_1 + \mu \mathbf{t}_1) + \beta(\lambda \mathbf{s}_2 + \mu \mathbf{t}_2)$$
$$= \lambda(\alpha \mathbf{s}_1 + \beta \mathbf{s}_2) + \mu(\alpha \mathbf{t}_1 + \beta \mathbf{t}_2)$$
$$= \lambda \mathbf{s} + \mu \mathbf{t}.$$

Thus $\alpha \mathbf{s} + \beta \mathbf{t}$ lies in the set $\lambda S + \mu T$ (the point $\mathbf{s} = \alpha \mathbf{s}_1 + \beta \mathbf{s}_2$ lies in S because S is convex; similarly for $\mathbf{t} = \alpha \mathbf{t}_1 + \beta \mathbf{t}_2$).

The *intersection* of a collection of sets is the set of points which belong to every one of the sets in the collection.

Theorem
The intersection of any collection of convex sets is again convex.

Proof. Let **a** and **b** belong to the intersection. Then **a** and **b** belong to each of the sets in the collection. Since these sets are convex, so does $\alpha \mathbf{a} + \beta \mathbf{b}$ provided $\alpha + \beta = 1$, $\alpha \geqslant 0$ and $\beta \geqslant 0$. Thus $\alpha \mathbf{a} + \beta \mathbf{b}$ belongs to the intersection. (See Figure 9.9.)

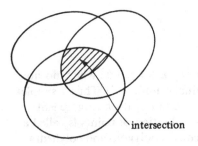

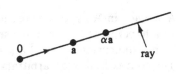

Fig. 9.9   *Fig. 9.10*

9.8   Cones

A *ray* (with vertex **0**) in the direction of **a** is the set of all **x** of the form

$$\mathbf{x} = \alpha \mathbf{a}$$

where $\alpha \geqslant 0$ (Figure 9.10).

A *cone* is a set with the property that if it contains **a** then it

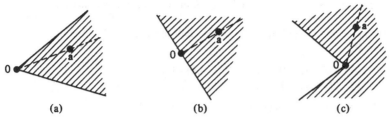

Fig. 9.11. (a), (b) convex cones, (c) non-convex cone

contains the ray through **a**. A *convex cone* is a cone which is a convex set (Figure 9.11).

A set S is a convex cone if and only if whenever **a** and **b** are in S, then so is

$$\alpha\mathbf{a} + \beta\mathbf{b}$$

provided $\alpha \geqslant 0$ and $\beta \geqslant 0$.

The smallest convex cone containing a given set S is the set of all expressions of the form

$$\mathbf{y} = \alpha_1\mathbf{x}_1 + \alpha_2\mathbf{x}_2 + \cdots + \alpha_k\mathbf{x}_k$$

where $\alpha_1 \geqslant 0$, $\alpha_2 \geqslant 0, \ldots, \alpha_k \geqslant 0$ and $\{\mathbf{x}_1, \mathbf{x}_2, \ldots, \mathbf{x}_k\}$ is any finite set of points of S.

9.9 Exercises

1. Which of the following sets in $\mathbb{R}^2$ are convex and which are not?

$$\{(x, y)^t : x > 0\} \qquad \{(x, y)^t : x = y\}$$
$$\{(x, y)^t : x^2 + 2y^2 < 1\} \quad \{(x, y)^t : x^2 + 2y^2 \leqslant 1\}$$
$$\{(x, y)^t : x^2 + 2y^2 > 1\} \quad \{(x, y)^t : x^2 + 2y^2 \geqslant 1\}$$
$$\{(x, y)^t : y \geqslant x^2\} \qquad \{(x, y)^t : y > \sin x\}$$

2. The same as question 1 for the following sets.

$$\{(x, y, z)^t : x > 0, y > 0\} \quad \{(x, y, z)^t : x^2 + 2y^2 + 3z^2 < 1\}$$
$$\{(x, y, z)^t : z > x^2 + 2y^2\} \quad \{(x, y, z)^t : z \leqslant x^2 + 2y^2\}$$

3. The non-negative orthant Θ of $\mathbb{R}^n$ is the set

$$\Theta = \{(x_1, x_2, \ldots, x_n)^t : x_1 \geqslant 0, x_2 \geqslant 0, \ldots, x_n \geqslant 0\}.$$

Prove that Θ is a convex cone.
4. Prove that any affine set is convex.
5. If S is a convex set in $\mathbb{R}^n$ and T is the set of all **y** of the form $\mathbf{y} = A\mathbf{x}$

where A is an $m \times n$ matrix and $\mathbf{x}$ lies in S, prove that T is a convex set in $\mathbb{R}^m$.

6. Determine the dimension of the convex hull of the following sets of points in $\mathbb{R}^3$.

(i) $(0,0,0)^t, (0,0,1)^t, (0,1,0)^t, (1,0,0)^t, (1,1,1)^t$
(ii) $(0,0,1)^t, (0,1,0)^t, (0,1,1)^t, (0,2,2)^t$.

7. Prove that the set of all $(x_1, x_2, \ldots, x_n)^t$ which satisfy

$$\left.\begin{array}{l} a_{11}x_1 + a_{12}x_2 + \cdots + a_{1n}x_n \leqslant b_1 \\ a_{21}x_1 + a_{22}x_2 + \cdots + a_{2n}x_n \leqslant b_2 \\ \vdots \\ a_{m1}x_1 + a_{m2}x_2 + \cdots + a_{mn}x_n \leqslant b_m \end{array}\right\}$$

is convex.

8. Sketch the following sets in $\mathbb{R}^2$:

(i) $(3,3)^t + A$ (ii) $A + B$ (iii) $A - B$

where

$$A = \{(x,y)^t : x^2 + y^2 < 1\} \quad \text{and} \quad B = \{(x,y)^t : 0 < x < 1, 0 < y < 1\}.$$

9.10   Further exercises

1. Show that

$$\{\mathbf{a}\} + \text{conv}(B) \subseteq \text{conv}(\{\mathbf{a}\} + B)$$

and that

$$A + \text{conv}(B) \subseteq \text{conv}(A + B).$$

Infer that

$$\text{conv}(A) + \text{conv}(B) \subseteq \text{conv}(A + B)$$

and deduce that equality obtains.

2. Sketch the convex hulls of the following sets.

(a) $\{(2,2)^t\} \cup \{(x,y)^t : x = y^2 - 1\}$,
(b) $\{(2,2)^t\} \cup \{(x,y)^t : x^2 = y^2 + 1 \ \& \ x \geqslant 0\}$
(c) $\{(0,0)^t\} \cup \{(x, x^2+1)^t : -\infty < x < \infty\}$
(d) $\left\{\left(x, \dfrac{1}{x^2+1}\right)^t : -\infty < x < \infty\right\}$
(e) $\{(x,y)^t : 0 \leqslant x \leqslant 2, y = 4x^2 - x^4\}$
(f) $\{(0,0)^t\} \cup \{(x,y)^t : (x-3)^2 + y^2 = 1\}$
(g) $\{(\theta^2 \cos\theta, \theta \sin\theta)^t : 0 \leqslant \theta \leqslant 5\pi/4\}$

(h) $\{(0, y, 0)^t : -\infty < y < \infty\} \cup \{(1, 0, 1)^t, (0, 0, 1)^t\}$

(i) $\{(0, y, 0)^t : -\infty < y < \infty\} \cup \{(x, 0, 1)^t : -\infty < x < \infty\}$

3. If $\mathbf{a}, \mathbf{b}, \mathbf{c}$ are three vectors in $\mathbb{R}^n$ and

$$\mathbf{x} = \alpha \mathbf{a} + \beta \mathbf{b} + \gamma \mathbf{c}$$

is a combination of $\mathbf{a}, \mathbf{b}, \mathbf{c}$ find a formula in terms of $\mathbf{a}, \mathbf{b}, \alpha, \beta$ for a point $\mathbf{z}$ in conv $\{\mathbf{a}, \mathbf{b}\}$ so that $\mathbf{x}$ is a convex combination of $\mathbf{z}$ and $\mathbf{c}$.
(Hint: Draw a triangle with vertices at $\mathbf{a}, \mathbf{b}, \mathbf{c}$ and consider the line through $\mathbf{c}$ and $\mathbf{x}$.)
4. If $\mathbf{x}_0, \mathbf{x}_1, \ldots, \mathbf{x}_{n+1}$ are vectors in $\mathbb{R}^n$ show by considering the $n+1$ vectors

$$\mathbf{x}_1 - \mathbf{x}_0, \mathbf{x}_2 - \mathbf{x}_0, \ldots, \mathbf{x}_{n+1} - \mathbf{x}_0$$

that there are constants $\alpha_0, \alpha_1, \ldots, \alpha_n$ not all zero satisfying

$$\alpha_0 \mathbf{x}_0 + \alpha_1 \mathbf{x}_1 + \cdots + \alpha_{n+1} \mathbf{x}_{n+1} = 0,$$

and

$$\alpha_0 + \alpha_1 + \cdots + \alpha_{n+1} = 0.$$

Letting $I = \{i : \alpha_i > 0\}$ show that

$$\text{conv} \{\mathbf{x}_i : i \in I\} \cap \text{conv} \{\mathbf{x}_j : j \notin I\}.$$

(Hint: Consider the previous question.)
5. Prove *Radon's* Theorem: If $X = \{\mathbf{x}_1, \ldots, \mathbf{x}_m\}$ is a finite set of vectors in $\mathbb{R}^n$ such that $m \geq n + 2$ then X can be partitioned into two sets X_1 and X_2 such that $\text{conv}(X_1) \cap \text{conv}(X_2) \neq \emptyset$. (Hint: use the method of the last question.)
6. Prove *Helly's* Theorem: If $A_1, \ldots, A_m$ are convex subsets of $\mathbb{R}^n$ with $m \geq n + 1$ such that every intersection of $n + 1$ of these sets is non-empty, then the intersection $\bigcap_{i=1}^m A_i$ is non-empty. (Hint: Apply induction. Suppose true for a given $m \geq n + 1$ and consider $A_1, \ldots, A_{m+1}$. Pick x_i in $A_1 \cap \cdots \cap A_{i-1} \cap A_{i+1} \cap \cdots \cap A_{m+1}$ and, using the last question, show that $x \in \text{conv}(X_1) \cap \text{conv}(X_2)$ lies in the intersection by assuming that $X_1 = \{x_1, \ldots, x_j\}$.)

10

The separating hyperplane theorem

10.1 Hyperplane and halfspaces

Recall that an affine set of dimension $n-1$ in $\mathbb{R}^n$ is called a *hyperplane*. A hyperplane is therefore defined by one linear equation

$$v_1 x_1 + v_2 x_2 + \cdots + v_n x_n = p,$$

provided not all the coefficients $v_1, v_2, \ldots, v_n$ are zero. Alternatively, we can write this in the form

$$\langle \mathbf{x}, \mathbf{v} \rangle = p,$$

in which case the vector $\mathbf{v} = (v_1, v_2, \ldots, v_n)^t$ may be regarded as the *normal* to the hyperplane.

Any hyperplane H defines two *half-spaces*. One of these is defined by the inequality

$$\langle \mathbf{x}, \mathbf{v} \rangle \geqslant p.$$

This is the half-space towards which the normal $\mathbf{v}$ points. The other half-space is defined by the inequality

$$\langle \mathbf{x}, \mathbf{v} \rangle \leqslant p.$$

To pass from the interior of one half-space to the interior of the other it is necessary to cross the hyperplane $\langle \mathbf{x}, \mathbf{v} \rangle = p$.

10.2 Separation

A hyperplane H *separates* two sets A and B if they lie in different half-spaces determined by H (Figure 10.1).

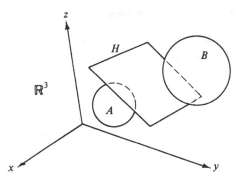

Fig. 10.1

Examples

(i) The hyperplane $3x + 4y = 2$ in $\mathbb{R}^2$ separates the points $(0, 0)^t$ and $(1, 1)^t$ (Figure 10.2). Observe that

$$3{\cdot}0 + 4{\cdot}0 \leqslant 2,$$

but

$$3{\cdot}1 + 4{\cdot}1 \geqslant 2.$$

(ii) The hyperplane $3x + 4y = 2$ in $\mathbb{R}^2$ separates the sets A and B defined respectively by the inequalities

$$(x - 3)^2 + (y - 9/2)^2 \leqslant 25,$$
$$(x + 7/3)^2 + (y + 4)^2 \leqslant 25,$$

(the sets A and B are both discs which touch the line $3x + 4y = 2$). See Figure 10.3.

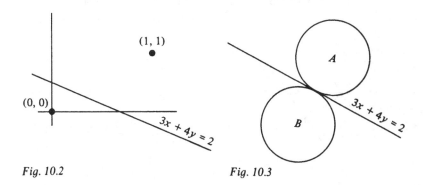

Fig. 10.2   Fig. 10.3

10.3   Separating hyperplanes and convex sets

This section depends on certain results from analysis, but we shall not make a fuss about these. We give a geometrical argument which will at least make for plausibility.

Theorem
*Let B be a non-empty convex set in $\mathbb{R}^n$ and let **a** be any point of $\mathbb{R}^n$ which does not lie in the interior of B. Then there is a hyperplane H which separates **a** and B.*

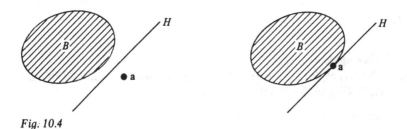

Fig. 10.4

Proof. See Figure 10.4 for an illustration of the theorem. We suppose first that B is *compact*, i.e. *closed* and bounded (closed in that B contains all of its boundary). We also suppose to begin with that **a** is *not* a boundary point of B. Then by our assumptions on B there is a point **b** in B which is *closest* to **a**, i.e.

$$\min_{y \in B} \| \mathbf{a} - \mathbf{y} \| = \| \mathbf{a} - \mathbf{b} \|$$

and let d be this minimum distance (cf. Figure 10.5). Now let H be the hyperplane through **a** which is normal to the vector $\mathbf{v} = \mathbf{b} - \mathbf{a}$. Then H

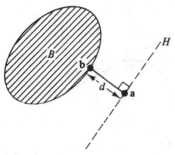

Fig. 10.5

is determined by the equation

$$\langle \mathbf{v}, \mathbf{x} - \mathbf{a} \rangle = 0.$$

If it were true that B contained a point $\mathbf{b}_1$, on the opposite side of H from $\mathbf{b}$, then it would also be true that there was a point $\mathbf{b}_2$ on the line segment joining $\mathbf{b}$ to $\mathbf{b}_1$ *closer* to $\mathbf{a}$ than $\mathbf{b}$. But, since B is convex, $\mathbf{b}_2$ would belong to B and thus $\mathbf{b}$ could not be the closest point of B to $\mathbf{a}$ (cf. Figure 10.6).

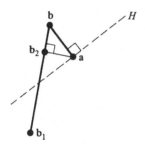

Fig. 10.6

Now suppose that $\mathbf{a}$ is a boundary point of B; then it must be the limit of a sequence of points $\{\mathbf{a}_n\}$ outside B which are *not* boundary points of B. Associated with each point $\mathbf{a}_n$ is a separating hyperplane H_n given by an equation

$$\langle \mathbf{v}_n, \mathbf{x} - \mathbf{a}_n \rangle = 0.$$

If the normals $\mathbf{v}_n$ are chosen to be of *unit* length then the sequence $\{\mathbf{v}_n\}$ is bounded (lies on the unit sphere, in fact) and hence has a convergent subsequence say with limit $\mathbf{v}$, as in Figure 10.7. (This is by the Bolzano–Weierstrass theorem of analysis; see for example K.G. Binmore,

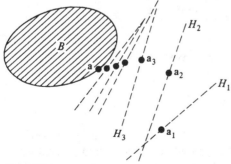

Fig. 10.7

Mathematical Analysis.) The hyperplane defined by

$$\langle \mathbf{v}, \mathbf{x} - \mathbf{a} \rangle = 0$$

then separates **a** and *B*. We are now in a position to prove:

Theorem of the separating hyperplane

*Let A and B be two non-empty **convex** sets with no points in common. Then there exists a hyperplane H which separates A and B.*

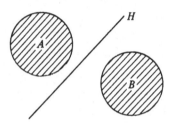

Fig. 10.8

Proof. The theorem is illustrated in Figure 10.8. Consider the set $A - B$ consisting of all vectors of the form $\mathbf{a} - \mathbf{b}$ where **a** lies in *A* and **b** lies in *B*. This set is convex and the point **0** does *not* belong to the set, for if it did there would be point $\mathbf{a}_0$ in *A* and $\mathbf{b}_0$ in *B* such that

$$\mathbf{0} = \mathbf{a}_0 - \mathbf{b}_0,$$

i.e.

$$\mathbf{a}_0 = \mathbf{b}_0,$$

so *A* and *B* would have a common point, a contradiction.

Now by the last theorem we can find a hyperplane *H* through **0** which has the whole of $A - B$ on one side. Suppose that *H* is given by

$$\langle \mathbf{v}, \mathbf{x} \rangle = 0$$

and $A - B$ lies in the half-space determined by

$$\langle \mathbf{v}, \mathbf{x} \rangle \geqslant 0.$$

Then, for each **a** in *A* and **b** in *B*, we have

$$\langle \mathbf{v}, \mathbf{a} - \mathbf{b} \rangle \geqslant 0,$$

i.e.

$$\langle \mathbf{v}, \mathbf{a} \rangle \geqslant \langle \mathbf{v}, \mathbf{b} \rangle.$$

The last inequality says that for any **b** in *B* the number $\langle \mathbf{v}, \mathbf{b} \rangle$ is a lower bound for the set of numbers $T = \{\langle \mathbf{v}, \mathbf{a} \rangle : \mathbf{a} \in A\}$. If *T* had a minimum it would then be true that $\min T \geqslant \langle \mathbf{v}, \mathbf{b} \rangle$. Since **b** was arbitrary we could then conclude that min T is an upper bound for the set of numbers $S = \{\langle \mathbf{v}, \mathbf{b} \rangle : \mathbf{b} \in B\}$. If the set S had a maximum we would then be able to say that

$$\max S \leqslant \min T.$$

We would then infer that any number p between these two is simultaneously greater than or equal to each member of S and smaller than or equal to each member of T. This, as we show below, would provide us with a separating hyperplane. See Figure 10.9.

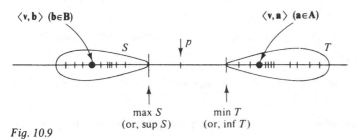

Fig. 10.9

Unfortunately not every set which is bounded below has a minimum, for example $\{(1/n) : n = 1, 2, 3, \dots\}$ does not. However, every set Z which is bounded from below does have a greatest lower bound. For example 0 is the greatest lower bound of the set $\{(1/n) : n = 1, 2, 3, \dots\}$. This greatest lower bound of Z is called its *infimum* and is denoted by inf Z. Evidently we have for every z in Z that

$$\inf Z \leqslant z.$$

The inf is often a good substitute when a minimum is not available. Of course if the minimum exists it is equal to the infimum. Similarly, a set Z which is bounded from above has a least upper bound known as its *supremum*, denoted sup Z. Again we have for any z in Z that

$$z \leqslant \sup Z.$$

When a set has a maximum the supremum will be equal to it. The supremum is likewise a good substitute when the maximum does not exist. In this connection see Part II Section 1.3.

Returning to our main argument, even if T does not have a minimum the set of lower bounds of T has a greatest member, namely inf T. Since for any $\mathbf{b}$ in B, $\langle \mathbf{v}, \mathbf{b} \rangle$ is among the lower bounds of T, it must be the case that $\langle \mathbf{v}, \mathbf{b} \rangle$ is less than or equal to the greatest of all the lower bounds. Thus

$$\langle \mathbf{v}, \mathbf{b} \rangle \leqslant \inf T.$$

Thus inf T is an upper bound for the set S and hence is greater than or equal to the smallest of all the upper bounds, i.e.

$$\sup S \leqslant \inf T.$$

Now, if p satisfies

$$\sup S \leqslant p \leqslant \inf T,$$

we conclude that it is simultaneously greater than or equal to each member of S and smaller than or equal to each member of T. In other words:

$$\langle \mathbf{v}, \mathbf{a} \rangle \geqslant p \geqslant \langle \mathbf{v}, \mathbf{b} \rangle$$

for any $\mathbf{a}$ in A and any $\mathbf{b}$ in B. Hence the hyperplane determined by

$$\langle \mathbf{x}, \mathbf{v} \rangle = p$$

separates A and B. The set A lies in the half-space $\langle \mathbf{v}, \mathbf{x} \rangle \geqslant p$ and B in the half-space $\langle \mathbf{v}, \mathbf{x} \rangle \leqslant p$.

Note. The conditions of the separating hyperplane theorem may be relaxed, and this is sometimes important.

The theorem remains true if we only assume that A and B are non-empty convex sets whose *relative interiors* have no points in common. To find the relative interior of a convex set, first find its affine hull. The points of the relative interior are then those points which are not boundary points *relative to the affine hull*. See Figure 10.10.

Some such condition on the sets A and B is clearly necessary. For example, if A and B are two non-parallel hyperplanes then their interiors (relative to the whole space $\mathbb{R}^n$) have no points in common but they cannot be separated. See Figure 10.11.

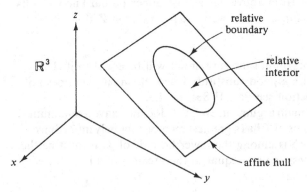

Fig. 10.10

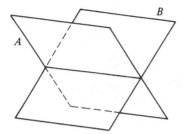

Fig. 10.11

A stronger but less complicated condition under which the theorem is still true is the following. The theorem of the separating hyperplane holds if A is a non-empty convex set while B is a convex set having a non-empty interior which does not have points in common with A.

10.4 Supporting hyperplanes

Let S be a non-empty convex set in $\mathbb{R}^n$. Let $\mathbf{a}$ be a boundary point of S. If H is a hyperplane through $\mathbf{a}$ then it determines two half-spaces. If S lies entirely inside one of these half-spaces, then H is called a *supporting hyperplane* at $\mathbf{a}$. (The half-space in which S lies is called a *supporting half-space*.) See Figure 10.12.

The theorem of the separating hyperplane assures the existence of at least one supporting hyperplane at each boundary point $\mathbf{a}$. If there is only one supporting hyperplane at $\mathbf{a}$, it is sensible to call it the *tangential hyperplane* at $\mathbf{a}$.

Given a supporting hyperplane

$$\langle \mathbf{v}, \mathbf{x} - \mathbf{a} \rangle = 0$$

to the non-empty convex set S at the boundary point $\mathbf{a}$, let us choose

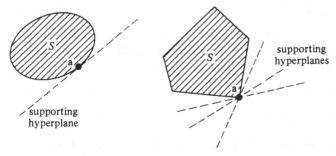

Fig. 10.12

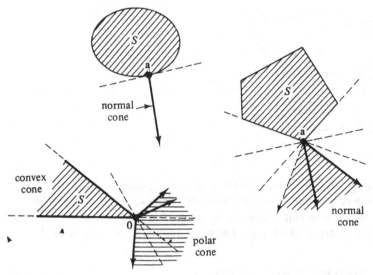

Fig. 10.13

the normal **v** to point away from S (so that S lies in the half-space $\langle \mathbf{v}, \mathbf{x} - \mathbf{a} \rangle \leqslant 0$).

The convex cone generated by such normals to the supporting hyperplanes at **a** is called the *normal cone* at **a**. (We usually draw this normal cone with its vertex transferred to the point **a**.)

If S is a convex cone to start with, the normal cone at its vertex **0** is called the *polar cone* to S. See Figure 10.13.

10.5   Extreme points

Recall that a set S is convex if and only if for any **a** and **b** in the set S the point

$$\mathbf{x} = \alpha \mathbf{a} + \beta \mathbf{b}$$

is also in the set S provided that $\alpha \geqslant 0$, $\beta \geqslant 0$ and $\alpha + \beta = 1$.

A point **x** in a convex set S is an *extreme point* of S if **x** *cannot* be expressed in the form

$$\mathbf{x} = \alpha \mathbf{a} + \beta \mathbf{b}$$

where **a** and **b** are in S $\alpha, \beta \geqslant 0$ and $\alpha + \beta = 1$ unless $\alpha = 0$ or $\beta = 0$. See Figure 10.14.

Geometrically this just means that **x** cannot be in the middle of any line segment which joins two points of S.

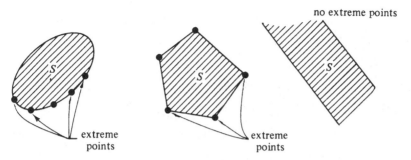

Fig. 10.14

It is obvious that every extreme point must be a boundary point. Open sets contain none of their boundary points and hence cannot have any extreme points at all. When discussing extreme points it is therefore sensible to confine attention to *closed, convex sets*. A closed set is one which contains *all* its boundary points.

Observe, however, that not every boundary point of a closed, convex set is an extreme point of the set. See Fig. 10.15.

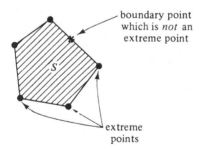

Fig. 10.15

A closed convex set which contains the whole of at least one line is called a *closed convex cylinder*. Closed convex cylinders may also be characterised as those closed convex sets which have no extreme points at all. See Figure 10.16.

Note that a half-space is an example of a closed, convex cylinder and hence has no extreme points.

We conclude this section with two results which are geometrically 'obvious' (Their proofs are not quite so obvious.)

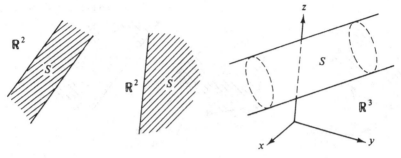

Fig. 10.16

Theorem

   Let S be a closed convex set which is **not** a cylinder. Then every supporting hyperplane to S contains an extreme point of S.

(The idea of the proof is that, if H did not contain an extreme point, it would have to contain a whole line of points of S, thus S would be a cylinder. Try to prove this in $\mathbb{R}$, then in $\mathbb{R}^2$ and then in $\mathbb{R}^3$ to see the shape of the proof. Compare Figure 10.17.)

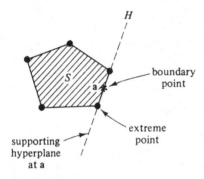

Fig. 10.17

Theorem

   A closed and bounded convex set in $\mathbb{R}^n$ is the convex hull of its extreme points.

(Note that this result is false for unbounded closed convex sets. For example, a closed convex cylinder has *no* extreme points. The theorem is illustrated in Figure 10.18.)

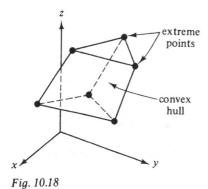

Fig. 10.18

10.6   Extremal rays

A closed convex cone *C* has only one extreme point (unless it is a cylinder, in which case it has no extreme points at all). The extreme point is its vertex at the origin. Little profit can therefore be expected from the study of the extreme points of a convex cone. We consider instead *extremal rays*.

If a closed convex cone is sliced by an appropriate hyperplane, then the rays through the extreme points of the convex set formed by the intersection of the cone and the hyperplane will be the extremal rays. (Cf. Figure 10.19.)

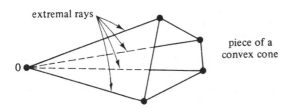

Fig. 10.19

10.7   Exercises

1. Sketch the following subsets of $\mathbb{R}^2$ and decide whether or not they are convex.

$$\left.\begin{array}{r} x + 2y \leqslant 1 \\ y - 2x \geqslant -2 \\ x + 5 \geqslant 0 \end{array}\right\} \qquad \left.\begin{array}{r} x + y \geqslant 0 \\ x - y \leqslant 0 \\ x \geqslant 0 \end{array}\right\}$$

$$\left.\begin{array}{c} xy \leqslant 1 \\ x \geqslant 0 \end{array}\right\} \qquad \left.\begin{array}{c} y \geqslant e^x \end{array}\right.$$

$$\left.\begin{array}{c} x + 3y \geqslant 1 \\ x \leqslant 1 - 3y \end{array}\right\} \qquad \left.\begin{array}{c} y \geqslant x^2 \\ y \leqslant x \\ x \geqslant 0 \end{array}\right\}$$

For each convex set, identify its extreme points (where they exist) and, in each case, find all supporting hyperplanes at an extreme point of your choice. Also find the normal cone at this point.

2. Find a separating hyperplane in $\mathbb{R}^2$ for the convex sets A and B determined by

$$\left.\begin{array}{c} xy \geqslant 1 \\ x \geqslant 0 \end{array}\right\} \qquad \left.\begin{array}{c} xy \leqslant -1 \\ x \geqslant 0 \end{array}\right\}$$

3. Let S be the convex hull of the points $(1, 2)^t$, $(2, 1)^t$, $(2, 2)^t$ and $(3, 2)^t$ in $\mathbb{R}^2$. Let C be the convex cone generated by these points. Determine the extreme points of S and the extremal rays of C.

4. Explain why the supporting hyperplane to a convex cone must contain the vertex $\mathbf{0}$.

5. Justify the assertion that a closed convex set is the intersection of its supporting half-spaces.

6. Given that the set of linear inequalities

$$\left.\begin{array}{c} a_{11}x_1 + a_{12}x_2 + \cdots + a_{1n}x_n \leqslant b_1 \\ a_{21}x_1 + a_{22}x_2 + \cdots + a_{2n}x_n \leqslant b_2 \\ \vdots \\ a_{m1}x_1 + a_{m2}x_2 + \cdots + a_{mn}x_n \leqslant b_m \\ x_1 \geqslant 0; \quad x_2 \geqslant 0; \ldots; x_n \geqslant 0 \end{array}\right\}$$

has at least one solution, show that the set of all solutions is convex and has at least one extreme point.

7. With reference to Question 2 of Section 9.10, find the extreme points of the convex hulls of the sets (a)–(i).

11

Linear inequalities

11.1 Notation

In this chapter we use extensively some notation for a partial ordering of column vectors. We list for reference the following terminology for an $n \times 1$ column vector $\mathbf{x} = (x_1, x_2, \ldots, x_n)^t$.

$\mathbf{x} > 0$ $\mathbf{x}$ positive $x_1 > 0, x_2 > 0, \ldots, x_n > 0$

$\mathbf{x} \geqslant 0$ $\mathbf{x}$ non-negative $x_1 \geqslant 0, x_2 \geqslant 0, \ldots, x_n \geqslant 0$

$\mathbf{x} \gneqq 0$ $\mathbf{x}$ semi-positive $\mathbf{x} \geqslant 0$ and $\mathbf{x} \neq 0$

The notation in this area is not yet standardised. For example, some authors use the notation $\mathbf{x} \gg 0$ for $\mathbf{x} > 0, \mathbf{x} \geqslant 0$ for $\mathbf{x} \geqslant 0$ and $\mathbf{x} > 0$ for $\mathbf{x} \gneqq 0$.

Entirely non-standard is the following notation which we find convenient:

$$\mathbf{x} \gneqq_j 0 \Leftrightarrow x_1 \geqslant 0, x_2 \geqslant 0, \ldots, x_{j-1} \geqslant 0, x_j > 0, x_{j+1} \geqslant 0, \ldots, x_n \geqslant 0.$$

For example

$$(0, 1, 2)^t \gneqq_2 (0, 0, 0)^t.$$

Example
Prove that
(1) $\langle \mathbf{u}, \mathbf{v} \rangle = \mathbf{u}^t \mathbf{v} > 0$ for every $\mathbf{u} \gneqq_j 0$
    if and only if
(2) $\mathbf{v} \gneqq_j 0$.

Proof. First suppose (2) holds. So let $\mathbf{u}$ and $\mathbf{v}$ be such that

$$\mathbf{u} \gneqq_j 0 \quad \text{and} \quad \mathbf{v} \gneqq_j 0$$

then $\langle \mathbf{u}, \mathbf{v} \rangle = \mathbf{u}^t \mathbf{v} = u_1 v_1 + u_2 v_2 + \cdots + u_n v_n \geqslant u_j v_j > 0$, so that (1) holds.

Now suppose that (1) holds. By taking $\mathbf{u} = \mathbf{e}_i + \delta \mathbf{e}_j = (0, 0, \ldots, 1, 0, \ldots, \delta, 0, \ldots, 0)'$, where δ is a small positive number, we obtain a vector which is strictly positive at its jth co-ordinate as required in (1). So, $\mathbf{u}'\mathbf{v} = v_i + \delta v_j > 0$. Now take limits as δ tends to zero to obtain $v_i \geqslant 0$. Also taking $\mathbf{u} = \mathbf{e}_j$ we obtain $\mathbf{u}'\mathbf{v} = v_j > 0$. Thus $\mathbf{v}$ is semi-positive as required in (2).

11.2 Orthants

The set $\{\mathbf{x} : \mathbf{x} \geqslant \mathbf{0}\}$ is called the *non-negative orthant*. The set

$$\{\mathbf{x} : \mathbf{x} \gneqq \mathbf{0}\}$$

is the non-negative orthant with the point $\mathbf{0}$ removed. The set

$$\{\mathbf{x} : \mathbf{x} \gneqq_j \mathbf{0}\}$$

is the non-negative orthant with one of its faces removed. The *positive orthant* is the non-negative orthant with all its faces removed, i.e. $\{\mathbf{x} : \mathbf{x} > \mathbf{0}\}$.

The non-negative orthant in $\mathbb{R}^n$ is an example of a closed, convex cone (with vertex $\mathbf{0}$). The other orthants mentioned are all convex sets.

11.3 Solution sets of linear inequalities

Let A be an $m \times n$ matrix. Recall that the solution set of $A\mathbf{x} = \mathbf{b}$ is an affine set (possibly empty). The solution set of $A\mathbf{x} = \mathbf{0}$ is a subspace of $\mathbb{R}^n$ (and always contains $\mathbf{x} = \mathbf{0}$). The solution set of the system of inequalities

$$A\mathbf{x} \leqslant \mathbf{b}$$

is a *closed convex* set with a finite number of extreme points. To see this, let the rows of A be $\mathbf{a}_1', \mathbf{a}_2', \ldots, \mathbf{a}_m'$. Then the system $A\mathbf{x} \leqslant \mathbf{b}$ becomes

$$\left. \begin{array}{l} \langle \mathbf{a}_1, \mathbf{x} \rangle = \mathbf{a}_1' \mathbf{x} \leqslant b_1 \\ \langle \mathbf{a}_2, \mathbf{x} \rangle = \mathbf{a}_2' \mathbf{x} \leqslant b_2 \\ \cdots\cdots\cdots\cdots\cdots \\ \langle \mathbf{a}_m, \mathbf{x} \rangle = \mathbf{a}_m' \mathbf{x} \leqslant b_m \end{array} \right\}.$$

The solution set is therefore the intersection of a finite number of closed half-spaces.

Observe that the solution set may turn out to be empty. Evidently the solution set of the system $A\mathbf{x} \leqslant \mathbf{0}$ always contains $\mathbf{x} = \mathbf{0}$. It is a closed convex cone with a finite number of extremal rays.

More complicated systems can be reduced to those above. For example, the system

$$\left.\begin{array}{r}Ax \leqslant b \\ x \leqslant 0\end{array}\right\}$$

may be written in the form

$$\left(\frac{-A}{I}\right) x \geqslant \left(\frac{-b}{0}\right).$$

Similarly the system

$$\left.\begin{array}{r}Ax = b \\ x \geqslant 0\end{array}\right\}$$

may be re-written in the form

$$\begin{bmatrix} A \\ -A \\ I \end{bmatrix} x \geqslant \begin{bmatrix} b \\ -b \\ 0 \end{bmatrix}.$$

11.4 Duality

The basic duality result that we shall prove is the following. One and only one of the two systems of inequalities

$$\left.\begin{array}{r}Ax = 0 \\ x \geqslant_j 0\end{array}\right\} \qquad\qquad A^t y \geqslant_j 0$$

has a solution. This is illustrated in Figure 11.1. The first system has no

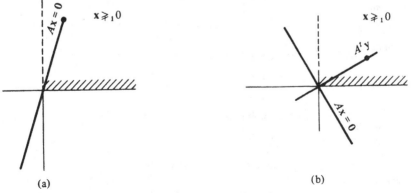

(a)   (b)

Fig. 11.1. Either $Ax = 0$ meets the quadrant – (a) – or else $R(A^t)$ does – (b)

solution if and only if the two non-empty convex sets

$$\mathscr{A} = \{\mathbf{x}:A\mathbf{x} = \mathbf{0}\} = N(A)$$
$$\mathscr{B} = \{\mathbf{x}:\mathbf{x} \geqslant_j \mathbf{0}\}$$

have no point in common. But this happens if and only if there is a separating hyperplane. Such a hyperplane must contain the subspace $\mathscr{A}$. Thus, for some vector $\mathbf{v}$ (normal to the hyperplane) we have

$$\langle \mathbf{v}, \mathbf{a} \rangle = 0 \quad \text{if} \quad A\mathbf{a} = \mathbf{0},$$
$$\langle \mathbf{v}, \mathbf{b} \rangle > 0 \quad \text{if} \quad \mathbf{b} \geqslant_j \mathbf{0}.$$

But $\langle \mathbf{v}, \mathbf{a} \rangle = 0$ for each $\mathbf{a}$ in $\mathscr{A}$ if and only if $\mathbf{v}$ lies in $\mathscr{A}^\perp$. But $\mathscr{A}^\perp = N(A)^\perp = R(A')$. Thus $\mathbf{v} = A'\mathbf{y}$ for some $\mathbf{y}$ in $\mathbb{R}^m$.

Also $\langle \mathbf{v}, \mathbf{b} \rangle > 0$ for each $\mathbf{b} \geqslant 0$ such that $b_j > 0$. So by the earlier result in Section 11.1 $\mathbf{v} \geqslant 0$ and $v_j > 0$ as required.

We conclude that the system

$$\left. \begin{array}{r} A\mathbf{x} = \mathbf{0} \\ \mathbf{x} \geqslant_j \mathbf{0} \end{array} \right\}$$

has no solution if and only if the system

$$A'\mathbf{y} \geqslant_j \mathbf{0}$$

does have a solution.

From this preliminary work we can deduce the following useful duality results (and many others besides – see the exercises). Of each of the following pairs of systems in the table below, one and only one has a solution.

References	Primal system	Dual system
Lemma 1	$\left. \begin{array}{r} A\mathbf{x} = \mathbf{0} \\ \mathbf{x} \geqslant \mathbf{0} \end{array} \right\}$	$A'\mathbf{y} > \mathbf{0}$
Lemma 2	$\left. \begin{array}{r} A\mathbf{x} = \mathbf{0} \\ \mathbf{x} > \mathbf{0} \end{array} \right\}$	$A'\mathbf{y} \geqslant \mathbf{0}$
Lemma 3	$\left. \begin{array}{r} A\mathbf{x} \leqslant \mathbf{0} \\ \mathbf{x} \geqslant \mathbf{0} \end{array} \right\}$	$\left. \begin{array}{r} A'\mathbf{y} > \mathbf{0} \\ \mathbf{y} > \mathbf{0} \end{array} \right\}$
Farkas' lemma	$\left. \begin{array}{r} A\mathbf{x} = \mathbf{b} \\ \mathbf{x} \geqslant \mathbf{0} \end{array} \right\}$	$\left. \begin{array}{r} A'\mathbf{y} \geqslant \mathbf{0} \\ \mathbf{b}'\mathbf{y} < \mathbf{0} \end{array} \right\}$
Lemma 4	$\left. \begin{array}{r} A\mathbf{x} \leqslant \mathbf{b} \\ \\ \mathbf{x} \geqslant \mathbf{0} \end{array} \right\}$	$\left. \begin{array}{r} A'\mathbf{y} \geqslant \mathbf{0} \\ \mathbf{y} \geqslant \mathbf{0} \\ \mathbf{b}'\mathbf{y} > \mathbf{0} \end{array} \right\}$

Proofs

Lemmas 1 and 2. Both lemmas require very similar ideas so we prove only Lemma 1. Suppose the primal system has no solution, then for each j the system

$$\left. \begin{array}{l} A\mathbf{x} = \mathbf{0} \\ \mathbf{x} \geqslant_j \mathbf{0} \end{array} \right\}$$

has no solution. Thus for each j there is a solution, call it $\mathbf{Y}(j)$ for the system

$$A^t \mathbf{y} \geqslant_j \mathbf{0}. \tag{*}$$

Consider the vector $\mathbf{z} = \mathbf{Y}(1) + \mathbf{Y}(2) + \cdots + \mathbf{Y}(n)$. Then, for any j, the jth co-ordinate of $A^t \mathbf{z}$ is $A^t \mathbf{Y}(1)_j + A^t \mathbf{Y}(2)_j + \cdots + A^t \mathbf{Y}(j)_j + \cdots + A^t \mathbf{Y}(n)_j$ which is greater than or equal to $A^t \mathbf{Y}(j)_j$ (why?) and this is positive since $\mathbf{Y}(j)$ solves (*). Thus $A^t \mathbf{z} > 0$. Thus $\mathbf{z}$ solves the dual system. Conversely, suppose $\mathbf{z}$ solves the dual system then, for every j, $\mathbf{z}$ solves the system (*) and hence for each j the system

$$\left. \begin{array}{l} A\mathbf{x} = \mathbf{0} \\ \mathbf{x} \geqslant_j \mathbf{0} \end{array} \right\}$$

has *no* solution. But, this says that the primal system has no solution.

Lemma 3. We start with the dual system:

$$\left. \begin{array}{l} A^t \mathbf{y} > \mathbf{0} \\ \mathbf{y} > \mathbf{0} \end{array} \right\} \quad viz. \quad \begin{pmatrix} A^t \\ I \end{pmatrix} \mathbf{y} > \mathbf{0}.$$

By Lemma 1 this has no solution if and only if the following system has no solutions:

$$\left. \begin{array}{l} (A \mid I)\mathbf{X} = \mathbf{0} \\ \mathbf{X} \geqslant \mathbf{0} \end{array} \right\}.$$

Write $\mathbf{X}^t = (\mathbf{x}^t \mid \mathbf{z}^t)$ where $\mathbf{x}$ and $\mathbf{z}$ and $n \times 1$ column vectors. Then the system reduces to

$$\left. \begin{array}{l} A\mathbf{x} = -\mathbf{z} \\ \mathbf{x} \geqslant \mathbf{0} \\ \mathbf{z} \geqslant \mathbf{0} \\ \text{and } (\mathbf{x} \mid \mathbf{z}) \neq \mathbf{0} \end{array} \right\} \qquad \text{i.e.} \quad \left. \begin{array}{l} A\mathbf{x} \leqslant \mathbf{0} \\ \mathbf{x} \geqslant \mathbf{0} \end{array} \right\}.$$

The last step requires explanation. The system on the left cannot have a solution with $\mathbf{x} = \mathbf{0}$; for if $\mathbf{x} = \mathbf{0}$ then $\mathbf{z} = -A\mathbf{x} = A \cdot \mathbf{0} = \mathbf{0}$ and hence

$\mathbf{X} = \mathbf{0}$, a contradiction. Thus any solution of the left system will certainly solve the system on the right. Conversely, any solution of the system on the right generates a solution of the system on the left. It suffices to set $\mathbf{z}$ at the value of $-A\mathbf{x}$ and then since $A\mathbf{x} \leqslant 0$ we have $\mathbf{z} \geqslant 0$.

Farkas' lemma. Start with the dual system:

$$\left.\begin{array}{l} A'\mathbf{y} \geqslant \mathbf{0} \\ \mathbf{b'y} < \mathbf{0} \end{array}\right\} \qquad\qquad \begin{pmatrix} A' \\ -\mathbf{b'} \end{pmatrix}\mathbf{y} \not\geqslant_{n+1} \mathbf{0}.$$

This has *no* solution if and only if

$$\left.\begin{array}{r} (A|-\mathbf{b})\mathbf{X} = \mathbf{0} \\ \mathbf{X} \not\geqslant_{n+1} \mathbf{0} \end{array}\right\}$$

has a solution. Write $\mathbf{X}' = (\mathbf{x}'|z)$ where this time $\mathbf{x}$ is an $n \times 1$ column vector and z is a *scalar*. The system now becomes:

$$\left.\begin{array}{r} A\mathbf{x} = \mathbf{b}z \\ \mathbf{x} \geqslant \mathbf{0} \\ z > 0 \end{array}\right\}.$$

If this has a solution then so does the system

$$\left.\begin{array}{r} A\mathbf{x} = \mathbf{b} \\ \mathbf{x} \geqslant \mathbf{0} \end{array}\right\}$$

(simply divide the former system's $\mathbf{x}$-solution by $z \neq 0$). Conversely, if this last system has a solution then so does the preceding one (take $z = 1$).

Lemma 4. Again we start with the dual system:

$$\left.\begin{array}{l} A'\mathbf{y} \geqslant \mathbf{0} \\ \mathbf{y} \geqslant \mathbf{0} \\ -\mathbf{b'y} > \mathbf{0} \end{array}\right\} \quad \text{i.e.} \quad \begin{bmatrix} A' \\ I \\ -\mathbf{b'} \end{bmatrix}\mathbf{y} \not\geqslant_{m+n+1} \mathbf{0}.$$

This has no solution if and only if

$$\left.\begin{array}{r} (A|I|-\mathbf{b})\mathbf{X} = \mathbf{0} \\ \mathbf{X} \not\geqslant_{m+n+1} \mathbf{0} \end{array}\right\}.$$

has a solution. Write $\mathbf{X}' = (\mathbf{x}'|\boldsymbol{\xi}'|z)$ where $\mathbf{x}$ and $\boldsymbol{\xi}$ are $n \times 1$ and $m \times 1$ column vectors respectively and z is a *scalar*. The system now becomes

$$\left.\begin{array}{r} A\mathbf{x} + \boldsymbol{\xi} = \mathbf{b}z \\ \mathbf{x} \geqslant \mathbf{0} \\ \boldsymbol{\xi} \geqslant \mathbf{0} \\ z > 0 \end{array}\right\} \quad \text{i.e.} \quad \left.\begin{array}{r} A\mathbf{x} \leqslant \mathbf{b}z \\ \mathbf{x} \geqslant \mathbf{0} \\ z > 0 \end{array}\right\}.$$

The system on the right has a solution if and only if

$$\left. \begin{array}{r} A\mathbf{x} \leqslant \mathbf{b} \\ \mathbf{x} \geqslant \mathbf{0} \end{array} \right\}$$

has a solution (as in the previous lemma: division by z in one direction and put $z = 1$ for the converse).

11.5   Exercises

1. Determine which of the following assertions are true and which are false.

$$\begin{bmatrix} 1 \\ 2 \\ 4 \end{bmatrix} > \begin{bmatrix} 1 \\ 2 \\ 3 \end{bmatrix} \qquad\qquad \begin{bmatrix} 1 \\ 2 \\ 4 \end{bmatrix} \not\geqslant \begin{bmatrix} 1 \\ 2 \\ 3 \end{bmatrix}$$

$$\begin{bmatrix} 1 \\ 2 \\ 3 \end{bmatrix} \geqslant \begin{bmatrix} 1 \\ 2 \\ 3 \end{bmatrix} \qquad\qquad \begin{bmatrix} 1 \\ 2 \\ 4 \end{bmatrix} \not\geqslant_2 \begin{bmatrix} 1 \\ 2 \\ 3 \end{bmatrix}$$

2. Prove the following:
    (i) $\mathbf{u}^t\mathbf{v} > 0$ for each $\mathbf{u} \geqslant \mathbf{0}$ with $\mathbf{u} \neq \mathbf{0}$ if and only if $\mathbf{v} > \mathbf{0}$;
    (ii) $\mathbf{u}^t\mathbf{v} > 0$ for each $\mathbf{u} > \mathbf{0}$ if and only if $\mathbf{v} \geqslant \mathbf{0}$ and $\mathbf{v} \neq \mathbf{0}$.
3. What is the obvious definition of the *column cone* of a matrix A? If $\mathbf{b} \neq \mathbf{0}$, explain why the system

$$\left. \begin{array}{r} A\mathbf{x} = \mathbf{b} \\ \mathbf{x} \geqslant \mathbf{0} \end{array} \right\}$$

has a solution if and only if $\mathbf{b}$ lies in the column cone of A.

4. Sketch and describe the solution sets of the following systems.

(i)
$$\left. \begin{array}{r} 2x + 3y \leqslant 1 \\ x \geqslant 0 \\ y \geqslant 0 \end{array} \right\}$$

(ii)
$$\left. \begin{array}{r} 2x - 3y \geqslant 0 \\ x \geqslant 0 \\ y \geqslant 0 \end{array} \right\}$$

(iii)
$$\left. \begin{array}{r} 2x + 3y \geqslant 1 \\ 2x - 3y \geqslant 4 \\ x \geqslant 10 \\ y \geqslant -5 \end{array} \right\}$$

(iv)
$$\left. \begin{array}{r} x + y \geqslant 0 \\ 3x + 3y \leqslant 0 \end{array} \right\}$$

5. One of the solution sets in Question 4 is a bounded convex set. Find its extreme points and hence write down a formula for the solutions.

6. One of the solution sets in Question 4 is an unbounded convex set, but not a cone. Find the convex hull of its extreme points.

7. One of the solution sets in Question 4 is a cylinder. Which?

8. The second of the solution sets in Question 4 is a convex cone. Find points on its extremal rays and hence write down a formula for the solutions.

9. Use the Farkas lemma to show that one and only one of the systems:

$$Ax \leqslant b \qquad \left.\begin{array}{l} A'y = 0 \\ y \geqslant 0 \\ b'y < 0 \end{array}\right\}$$

has a solution. [Hint: $A^y y = 0 \Leftrightarrow A'y \geqslant 0$ and $-A'y \geqslant 0$.]

10. Use Question 9 to show that of the two systems:

$$Ax = b \qquad \left.\begin{array}{l} A'y = 0 \\ b'y \neq 0 \end{array}\right\}$$

one and only one has a solution. Give a simpler proof based directly on the fact that $R(A) = N(A')^\perp$.

11. Use Farkas' lemma to find the dual to:

$$\left.\begin{array}{l} Px = x \\ x \geqslant 0 \\ e'x = 1 \end{array}\right\}$$

where $P = [p_{ij}]$ is a square matrix of size $n \times n$ and $e = (1,1,\ldots,1)'$. If for each j

$$\sum_{i=1}^{n} p_{ij} = 1$$

and all the entries p_{ij} are non-negative show that the dual cannot be solved.

[Hint: For any z

$$\sum_{i=1}^{n} p_{ij} \leqslant z_m,$$

where $z_m = \max\{z_1,\ldots,z_n\}$.]

12

Linear programming and game theory

12.1 Introduction

We discuss only some of the theoretical aspects of linear programming. One practical technique is outlined in the next chapter but short of that we refer the reader to courses on operational research.

The *primal linear programming* problem is to find

$$\max \mathbf{c}^t \mathbf{x}$$

subject to the constraints

$$A\mathbf{x} \leqslant \mathbf{b},$$
$$\mathbf{x} \geqslant 0.$$

The associated *dual linear programming* problem is to find

$$\min \mathbf{b}^t \mathbf{y}$$

subject to the constraints

$$A^t \mathbf{y} \geqslant \mathbf{c},$$
$$\mathbf{y} \geqslant \mathbf{0}.$$

(Here A is an $m \times n$ matrix, $\mathbf{b}$ an $m \times 1$ column vector and $\mathbf{c}$ an $n \times 1$ column vector.)

The connection between these problems will be discussed later on. For the moment we consider the terminology and elementary theory of the primal problem.

An $n \times 1$ column vector which satisfies the constraints

$$A\mathbf{x} \leqslant \mathbf{b},$$
$$\mathbf{x} \geqslant \mathbf{0},$$

is called a *feasible solution*. We shall use F to denote the set of all feasible solutions. The set F may be empty (i.e. there may be *no* solutions, the inequalities $A\mathbf{x} \geqslant \mathbf{b}, \mathbf{x} \geqslant \mathbf{0}$ having no common solution). In this case the problem is of little interest.

A feasible solution which maximises the *objective function* $\mathbf{c}'\mathbf{x}$ is called an *optimal solution*. It may be that $\mathbf{c}'\mathbf{x}$ is unbounded on the set F. In this case there will be *no* optimal solutions.

We know that F is a closed convex set with a finite number of extreme points. Also F is not a cylinder, being a subset of the orthant $\{\mathbf{x}:\mathbf{x} \geqslant \mathbf{0}\}$. Hence it contains at least one extreme point (see Figure 12.1).

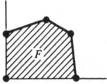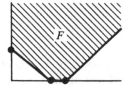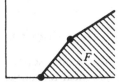

Fig. 12.1

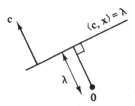

Fig. 12.2. Provided $\|\mathbf{c}\| = 1$

Given λ, the equation

$$\mathbf{c}'\mathbf{x} = \langle \mathbf{c}, \mathbf{x} \rangle = \lambda$$

defines a hyperplane. The geometric interpretation of the vector $\mathbf{c}$ and the scalar λ are indicated in Figure 12.2 (see also Chapter 2, Section 5).

It is important to recall that, if the hyperplane is moved parallel to itself in the direction of the vector $\mathbf{c}$, then the value of λ *increases*.

We are seeking the maximum value of λ such that the hyperplane $\langle \mathbf{c}, \mathbf{x} \rangle = \lambda$ cuts the set F.

Figure 12.3(a) indicates a situation in which *no* such maximum exists (the objective function $\mathbf{c}'\mathbf{x} = \langle \mathbf{c}, \mathbf{x} \rangle$ being unbounded above).

Given that a maximum value λ^* of λ *does* exist, then $\langle \mathbf{c}, \mathbf{x} \rangle = \lambda^*$ cannot pass through an interior point of F. (If it did, then we could find a value of λ a little larger than λ^* and $\langle \mathbf{c}, \mathbf{x} \rangle = \lambda$ would still cut F.)

It follows that $\langle \mathbf{c}, \mathbf{x} \rangle = \lambda^*$ is a *supporting hyperplane* to F. Hence it

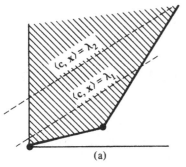 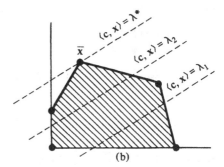

Fig. 12.3. (a) $\lambda_1 < \lambda_2$, (b) $\lambda_1 < \lambda_2 < \lambda^*$

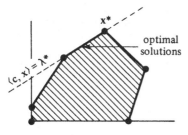 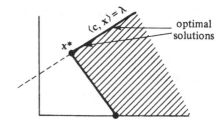

Fig. 12.4

passes through an extreme point of F. Such an extreme point $\mathbf{x}^*$ is therefore an optimal solution (cf. Figure 12.4).

From the above discussion, it is clear that we need only examine the extreme points of F in seeking an optimal solution. In particular, given that optimal solutions exist, we can find an optimal solution by evaluating the objective function at each extreme point to see at which it is the largest. This idea is the foundation for various numerical techniques for solving linear programming problems.

12.2 The duality theorem

Recall the primal and dual linear programming problems

Maximise $\mathbf{c}^t\mathbf{x}$	*Minimise* $\mathbf{b}^t\mathbf{y}$
subject to	*subject to*
$A\mathbf{x} \leqslant \mathbf{b}$,	$A^t\mathbf{y} \geqslant \mathbf{c}$,
$\mathbf{x} \geqslant \mathbf{0}$.	$\mathbf{y} \geqslant \mathbf{0}$.

The dual assumes the same form as the primal problem if we consider the problem of maximising $(-\mathbf{b})'\mathbf{y}$ subject to

$$(-A)'\mathbf{y} \leqslant -\mathbf{c},$$
$$\mathbf{y} \geqslant \mathbf{0}.$$

The reason for considering the two problems together is the following theorem.

Duality theorem

Suppose that both the primal and dual problems have feasible solutions. Then both have optimal solutions and

$$\max \mathbf{c}'\mathbf{x} = \min \mathbf{b}'\mathbf{y}.$$

Proof. Since the primal problem is feasible

$$A\mathbf{x} \leqslant \mathbf{b},$$
$$\mathbf{x} \geqslant \mathbf{0},$$

has a solution.

By Lemma 4 of Chapter 11, it follows that

$$\left.\begin{array}{r} A'\mathbf{y} \geqslant \mathbf{0} \\ \mathbf{y} \geqslant \mathbf{0} \\ \mathbf{b}'\mathbf{y} < 0 \end{array}\right\} \tag{1}$$

has no solution.

Next consider the system

$$\left.\begin{array}{r} A\mathbf{x} \leqslant \mathbf{b} \\ \mathbf{x} \geqslant \mathbf{0} \\ \mathbf{c}'\mathbf{x} \geqslant \lambda \end{array}\right\} \quad \text{i.e.} \quad \left(\begin{array}{c} A \\ -\mathbf{c}' \end{array}\right)\mathbf{x} \leqslant \left(\begin{array}{c} \mathbf{b} \\ -\lambda \end{array}\right) \\ \mathbf{x} \geqslant \mathbf{0} \right\}. \tag{2}$$

By Lemma 4 again, this system has *no* solution if and only if the system

$$\left.\begin{array}{r} (A'|\mathbf{c})\mathbf{Y} \geqslant \mathbf{0} \\ \mathbf{Y} \geqslant \mathbf{0} \\ (\mathbf{b}'|-\lambda)\mathbf{Y} < 0 \end{array}\right\} \quad \text{i.e.} \quad \left.\begin{array}{r} A'\mathbf{y} \geqslant \mathbf{c}z \\ \mathbf{y} \geqslant \mathbf{0} \\ z \geqslant 0 \\ \mathbf{b}'\mathbf{y} < \lambda z \end{array}\right\} \tag{3}$$

has a solution. Here

$$\mathbf{Y} = \begin{bmatrix} \mathbf{y} \\ \overline{z} \end{bmatrix}$$

where $\mathbf{y}$ is $m \times 1$ and z is a scalar. Since the system (1) is insoluble, there cannot be a solution to (3) with $z = 0$. Hence we may suppose that $z > 0$ in (3). Thus (3) is equivalent to

$$\left. \begin{aligned} A^t\mathbf{y} &\geqslant \mathbf{c} \\ \mathbf{y} &\geqslant \mathbf{0} \\ \mathbf{b}^t\mathbf{y} &< \lambda \end{aligned} \right\}. \tag{4}$$

To summarise: system (2) is insoluble if and only if system (4) is soluble.

We have yet to use the fact that the dual problem has a feasible solution, i.e. that

$$A^t\mathbf{y} \geqslant \mathbf{c},$$
$$\mathbf{y} \geqslant \mathbf{0},$$

has a solution. This implies that system (4) has a solution provided that λ is sufficiently large. Hence (2) has *no* solution if λ is sufficiently large. But this simplies that $\mathbf{c}^t\mathbf{x}$ is bounded above for $\mathbf{x}$ satisfying $A\mathbf{x} \leqslant \mathbf{b}$ and $\mathbf{x} \geqslant \mathbf{0}$. Hence

$$M = \max \mathbf{c}^t\mathbf{x}$$

exists. A similar argument establishes the existence of

$$m = \min \mathbf{b}^t\mathbf{y}.$$

Now observe that

$$\lambda > M \Leftrightarrow (2)\text{ insoluble} \Leftrightarrow (3)\text{ soluble} \Leftrightarrow \lambda > m.$$

It follows that $M = m$ and the theorem is proved.

12.3 Economic interpretation

Suppose that the co-ordinates of the $n \times 1$ *production vector* $\mathbf{x}$ represent quantites of processed goods. The co-ordinates of the $m \times 1$ *resource vector* z represent quantities of raw materials. To produce $\mathbf{x}$ one consumes z where

$$z = A\mathbf{x}.$$

An entrepreneur can sell his products at fixed prices represented by the co-ordinates of the price vector $\mathbf{c}$. He seeks to *maximise* his *revenue*

$$\mathbf{c}^t\mathbf{x}$$

subject to the constraints

$$A\mathbf{x} \leqslant \mathbf{b},$$
$$\mathbf{x} \geqslant \mathbf{0},$$

where **b** represents his *stock* of raw materials. (The production process is assumed to involve negligible cost.)

Now consider the problem of an auditor who has to assign a value to the stock **b**. He wishes to value the stock as low as is consistent with the fact that it can be converted into processed goods and sold at price **c**. Thus the auditor seeks to *minimise*

$$\mathbf{b}^t\mathbf{y}$$

subject to the constraints

$$A^t\mathbf{y} \geqslant \mathbf{c},$$
$$\mathbf{y} \geqslant 0.$$

The co-ordinates of the $m \times 1$ column vector **y** are the 'prices' to be assigned to the various raw materials stocked. Since these prices are for book-keeping purposes, we call **y** the *shadow price vector*.

The constraint $A^t\mathbf{y} \geqslant \mathbf{c}$ should be explained. We cannot value the resource vector $\mathbf{z} = A\mathbf{x}$ at less than the amount for which we can sell the corresponding production **x**. Hence

$$\mathbf{y}^t\mathbf{z} \geqslant \mathbf{c}^t\mathbf{x},$$

i.e.

$$\mathbf{y}^t A\mathbf{x} \geqslant \mathbf{c}^t\mathbf{x}.$$

But this inequality must hold for each **x**. We can therefore apply it with $\mathbf{x} = (\delta, 0, 0, \ldots, 0)^t, \mathbf{x} = (0, \delta, 0, \ldots, 0)^t$ etc., where δ is a small enough number, and hence obtain that $\mathbf{y}^t A \geqslant \mathbf{c}^t$, i.e.

$$A^t\mathbf{y} \geqslant \mathbf{c}.$$

A co-ordinate of $A^t\mathbf{y}$ may be interpreted as the cost of producing a unit of the corresponding processed good, given that raw materials are to be valued at shadow prices **y**.

If the optimal shadow price vector $\mathbf{y}^*$ is unique, then a small change $\delta\mathbf{b}$ in the stock vector **b** will leave $\mathbf{y}^*$ unchanged. (See Figure 12.5, where it is clear that for small enough $\delta\mathbf{b}$ the hyperplane with normal $\mathbf{b} + \delta\mathbf{b}$ continues to support the feasible set of the dual problem at $\mathbf{y}^*$.) But the change in **b** will imply a change $\delta\mathbf{x}^*$ in the optimal production $\mathbf{x}^*$. By the duality theorem

$$\mathbf{c}^t(\mathbf{x}^* + \delta\mathbf{x}^*) = (\mathbf{b} + \delta\mathbf{b})^t\mathbf{y}^*,$$
$$\mathbf{c}^t\mathbf{x}^* = \mathbf{b}^t\mathbf{y}^*.$$

Hence

$$\mathbf{c}^t\delta\mathbf{x}^* = \delta\mathbf{b}^t\mathbf{y}^*.$$

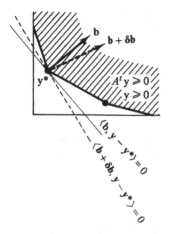

Fig. 12.5

It follows that if extra stock is bought at the shadow price levels $\mathbf{y}^*$ then the cost will be exactly balanced by the revenue on the extra production. Thus $\mathbf{y}^*$ represents the price levels below which it becomes worth purchasing extra stock.

We end this section with some remarks about the nature of the optimal solutions $\mathbf{x}^*$ and $\mathbf{y}^*$.

The expression

$$\mathbf{y}^t(\mathbf{b} - A\mathbf{x}) = \langle \mathbf{y}, \mathbf{b} - A\mathbf{x} \rangle$$

may be interpreted as the value (at price $\mathbf{y}$) of the stock which is left unused in the production of $\mathbf{x}$.

Equally, we may consider

$$\mathbf{x}^t(A^t\mathbf{y} - \mathbf{c}) = \langle \mathbf{x}, A^t\mathbf{y} - \mathbf{c} \rangle$$

which is the loss involved in producing $\mathbf{x}$ given that the stock is valued at price $\mathbf{y}$.

Since $A\mathbf{x} \leqslant \mathbf{b}$ we have that $\mathbf{y}^t A\mathbf{x} \leqslant \mathbf{y}^t\mathbf{b}$. Also, $A^t\mathbf{y} \geqslant \mathbf{c}$, so $\mathbf{x}^t A^t\mathbf{y} \geqslant \mathbf{x}^t\mathbf{c}$. By the duality theorem $\mathbf{y}^{*t}\mathbf{b} = \mathbf{x}^{*t}\mathbf{c}$. It follows that

$$\mathbf{y}^{*t}\mathbf{b} = \mathbf{x}^{*t}\mathbf{c} \leqslant \mathbf{x}^{*t}A^t\mathbf{y}^* = \mathbf{y}^{*t}A\mathbf{x}^* = \mathbf{y}^{*t}\mathbf{b}.$$

We have shown that $\mathbf{y}^{*t}\mathbf{b} = \mathbf{y}^{*t}A\mathbf{x}^*$ i.e.

$$\langle \mathbf{y}^*, \mathbf{b} - A\mathbf{x}^* \rangle = 0.$$

Since $\mathbf{y}^* \geqslant 0$ and $\mathbf{b} - A\mathbf{x}^* \geqslant 0$, it follows that

$$(\mathbf{b} - A\mathbf{x}^*)_i > 0 \Leftrightarrow y_i^* = 0$$

– i.e the shadow price of a raw material which is not entirely used up in producing the optimal production vector $\mathbf{x}^*$ is zero.

Using the same argument as above, it can also be shown that

$$\langle \mathbf{x}^*, A'\mathbf{y}^* - \mathbf{c} \rangle = 0.$$

Since $\mathbf{x}^* \geqslant 0$ and $A'\mathbf{y}^* \geqslant \mathbf{c}$, it follows that

$$(A'\mathbf{y}^* - \mathbf{c})_i > 0 \Leftrightarrow x_i^* = 0$$

– i.e. if the price of the raw material required to produce one unit of a given processed good exceeds the revenue in selling it, then none of that good will be produced at the optimum.

12.4   The Lagrangian

The *Lagrangian* $L(\mathbf{x}, \mathbf{y})$ for the primal linear programming problem:

Maximise $\mathbf{c}'\mathbf{x}$

subject to the constraints

$$\left.\begin{array}{l} A\mathbf{x} \leqslant \mathbf{b} \\ \mathbf{x} \geqslant 0 \end{array}\right\}$$

is defined by

$$L(\mathbf{x}, \mathbf{y}) = \langle \mathbf{c}, \mathbf{x} \rangle + \langle \mathbf{y}, \mathbf{b} - A\mathbf{x} \rangle \quad (\mathbf{x} \geqslant 0, \mathbf{y} \geqslant 0).$$

If we write the dual linear programming problem in the form:

Maximise $-\mathbf{b}'\mathbf{y}$

subject to the constraints

$$-A'\mathbf{y} \leqslant -\mathbf{c}$$
$$\mathbf{y} \geqslant 0$$

we obtain the Lagrangian $L'(\mathbf{x}, \mathbf{y})$

$$L'(\mathbf{x}, \mathbf{y}) = \langle -\mathbf{b}, \mathbf{y} \rangle + \langle \mathbf{x}, -\mathbf{c} + A'\mathbf{y} \rangle$$
$$= -\{\langle \mathbf{c}, \mathbf{x} \rangle + \langle \mathbf{y}, \mathbf{b} - A\mathbf{x} \rangle\} = -L(\mathbf{x}, \mathbf{y}) \quad (\mathbf{x} \geqslant 0, \mathbf{y} \geqslant 0).$$

We are interested in the *saddle points* of Lagrangians. The Lagrangian just defined being a very simple function (linear in both arguments) we can give a simplified definition of a saddle point. We say that $(\mathbf{x}^*, \mathbf{y}^*)$ is a saddle point for the Lagrangian $L(\mathbf{x}, \mathbf{y})$ provided that $\mathbf{x}^* \geqslant 0$ and $\mathbf{y}^* \geqslant 0$ and

$$L(\mathbf{x}, \mathbf{y}^*) \leqslant L(\mathbf{x}^*, \mathbf{y}^*) \leqslant L(\mathbf{x}^*, \mathbf{y})$$

for each $\mathbf{x} \geqslant 0$ and $\mathbf{y} \geqslant 0$ (see Figure 12.6).

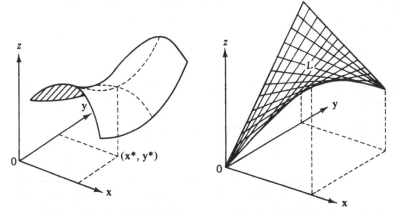

Fig. 12.6. *Left: a saddle point on a general surface; right: a ruled surface with a saddle point*

Using the results of the last section one can show the following theorem.

Theorem

The vectors x^* and y^* are optimal for the primal and dual problems if and only if (x^*, y^*) is a saddle point for $L(x, y)$.

12.5 Game theory

We confine our attention to two-person, zero-sum games. For such a game one requires two *strategy sets* X and Y and a *payoff function $K(x, y)$*.

Player I chooses a strategy x from the set X and Player II a strategy y from the set Y. Player I then receives a reward of $K(x, y)$ and Player II a reward of $-K(x, y)$.

If Player I *knows* that Player II will select strategy y^*, then Player I will choose his strategy x^* so that

$$K(x^*, y^*) = \max_x K(x, y^*)$$

in order to maximise his gain.

Suppose Player II *knows* that his strategy choice will become known to Player I, he therefore has little option but to choose his strategy y^* (if he can) so that

$$K(x^*, y^*) = \max_x K(x, y^*) = \min_y \max_x K(x, y)$$

in order to minimise his loss. Similar considerations with Players I and II reversed would lead to the conclusion that

$$K(x^*, y^*) = \min_y K(x, y^*) = \max_x \min_y K(x, y).$$

We are faced with the question: when does there exist a pair (x^*, y^*) for which

$$K(x^*, y^*) = \max_x \min_y K(x, y) = \min_y \max_x K(x, y).$$

If such a pair (x^*, y^*) exists, then it is an *equilibrium point* for the game in the sense that if the game is played over and over again, it will profit neither player to deviate from the equilibrium strategy, unless the other does so as well. The number $K(x^*, y^*)$ is the *value* of the game.

If we restrict our attention to the case when the strategy sets X and Y may be regarded as closed convex sets of $\mathbb{R}^n$ and $\mathbb{R}^m$ respectively $K(\mathbf{x}, \mathbf{y})$ is convex in $\mathbf{x}$ and concave in $\mathbf{y}$ (see Section 15.1), then (x^*, y^*) is an equilibrium point if and only if it is a *saddle-point* for $K(\mathbf{x}, \mathbf{y})$ – i.e.

$$K(\mathbf{x}, \mathbf{y}^*) \leqslant K(\mathbf{x}^*, \mathbf{y}^*) \leqslant K(\mathbf{x}^*, \mathbf{y}).$$

Any linear programming situation therefore provides an example of a game. The two players are the entrepreneur and the auditor and the payoff function is the Lagrangian

$$L(\mathbf{x}, \mathbf{y}) = \langle \mathbf{c}, \mathbf{x} \rangle + \langle \mathbf{y}, \mathbf{b} - A\mathbf{x} \rangle.$$

This can be regarded as the 'value' of the company. The entrepreneur seeks to make it large by his choice of $\mathbf{x}$. The auditor seeks to make it small by his choice of $\mathbf{y}$. As we know an equilibrium $(\mathbf{x}^*, \mathbf{y}^*)$ exists provided that both the primal and the dual problems have feasible solutions.

12.6   Finite matrix games

The most important result in game theory is von Neumann's celebrated Minimax theorem which asserts that every finite matrix game has an equilibrium point $(\mathbf{x}^*, \mathbf{y}^*)$ (provided 'mixed strategies' are allowed).

For a finite matrix game we require an $m \times n$ matrix A. Player I has first of all m *pure strategies* from which he may choose. Similarly Player II has n pure strategies.

The pay-off to Player I if he chooses his jth pure strategy, while Player II chooses his kth pure strategy is a_{jk}.

A *strategy* (sometimes called a 'mixed strategy') for Player I is a vector

$\mathbf{p} = (p_1, p_2, \ldots, p_m)^t$ for which $p_1 \geqslant 0, p_2 \geqslant 0, \ldots, p_m \geqslant 0$ and

$$p_1 + p_2 + \cdots + p_m = 1.$$

This choice of strategy is interpreted to mean that the jth stategy is chosen with *probability* p_j. The pure strategies are therefore identified with the vectors $(1, 0, \ldots, 0)^t, (0, 1, \ldots, 0)^t$, etc.

Similarly a strategy for Player II is a vector $\mathbf{q} = (q_1, q_2, \ldots, q_n)^t$ for which $q_1 \geqslant 0, q_2 \geqslant 0, \ldots, q_n \geqslant 0$ and

$$q_1 + q_2 + \cdots + q_n = 1.$$

The *expected reward* for Player I if the strategies chosen are $\mathbf{p}$ and $\mathbf{q}$ is equal to

$$\mathbf{p}^t A \mathbf{q} = \sum_{j=1}^{m} \sum_{k=1}^{n} a_{jk} p_j q_k$$

and we therefore define the payoff function $K(\mathbf{p}, \mathbf{q})$ by

$$K(\mathbf{p}, \mathbf{q}) = \mathbf{p}^t A \mathbf{q}.$$

Von Neumann's theorem is then the assertion that $K(\mathbf{p}, \mathbf{q})$ has a saddle point. This can be proved directly from the theorem of the Separating Hyperplane. It is perhaps more instructive to explain first how each player is faced with a linear programming problem and then to deduce von Neumann's theorem from the duality theorem.

Consider the finite matrix game with payoff matrix

$$A = \begin{bmatrix} a_{11} & a_{12} & \cdots & a_{1n} \\ a_{21} & a_{22} & & a_{2n} \\ \vdots & & & \\ a_{m1} & a_{m2} & & a_{mn} \end{bmatrix}.$$

We shall suppose that each of the entries of A is non-negative. (This can be ensured by adding one and the same constant, say c, to each entry of A, an activity which will not affect the equilibrium strategies $\mathbf{x}^*$ and $\mathbf{y}^*$ of the two players as it is equivalent to demanding an admission fee of c from Player II before the game is played.)

Player I can ensure that his expected payoff is at least as large as λ provided that he can find $(p_1, p_2, \ldots, p_m)^t$ such that

$$a_{11} p_1 + a_{21} p_2 + \cdots + a_{m1} p_m \geqslant \lambda$$

(Expected gain if II chooses first pure strategy)

$$a_{12} p_1 + a_{22} p_2 + \cdots + a_{m2} p_m \geqslant \lambda$$

$$\cdots \cdots \cdots \cdots \cdots \cdots \cdots$$

$$a_{1n} p_1 + a_{2n} p_2 + \cdots + a_{mn} p_m \geqslant \lambda$$

(1a)

$$a_{1n}p_1 + a_{2n}p_2 + \cdots + a_{mn}p_m \geqslant \lambda$$
$$p_1 + \quad p_2 + \cdots + \quad p_m = 1$$
$$p_1 \geqslant 0$$
$$\qquad p_2 \geqslant 0$$
$$\cdots$$
$$\qquad\qquad p_m \geqslant 0.$$

$$(1b)$$

If $\mathbf{e} = (1, 1, 1, \ldots, 1)^t$ we can write this system in the form

$$A^t\mathbf{p} \geqslant \lambda\mathbf{e}$$
$$\mathbf{e}^t\mathbf{p} = 1 \quad\Big\}.$$
$$\mathbf{p} \geqslant 0$$

$$(2)$$

Observe that whatever Player II's choice of strategy $\mathbf{q}$ (subject to $\mathbf{e}^t\mathbf{q} = 1$ and $\mathbf{q} \geqslant 0$) we have that the expected payoff for Player I is

$$\mathbf{p}^t A\mathbf{q} = \mathbf{q}^t A^t\mathbf{p} \geqslant \lambda\mathbf{q}^t\mathbf{e} = \lambda.$$

Observe also that the set of all vectors of the form $A^t\mathbf{p}$ where $\mathbf{e}^t\mathbf{p} = 1$ and $\mathbf{p} \geqslant 0$ is the set of all convex combinations of the columns of A^t – i.e. it is the convex hull of the rows of A. Since the entries of A are non-negative, this set lies in the non-negative orthant and so a $\lambda > 0$ can be found to satisfy (2) for a suitable choice of $\mathbf{p}$. Indeed if $\mathbf{p}$ is any fixed vector of the required form we may inspect the left-hand sides of (1a) and denoting by $\lambda(\mathbf{p})$ the least value occurring, we see that (1) is satisfied when $\lambda = \lambda(\mathbf{p})$. Player I thus seeks to maximize $\lambda(\mathbf{p})$ over all permissible vectors $\mathbf{p}$. See Figure 12.7.

Alternatively we may regard Player I as seeking to maximise λ subject to the constraints (2). We put $\mathbf{y} = \mathbf{p}/\lambda$. Then Player I seeks to minimise

$$\frac{1}{\lambda} = \frac{1}{\lambda}\mathbf{e}^t\mathbf{p} = \mathbf{e}^t\mathbf{y}$$

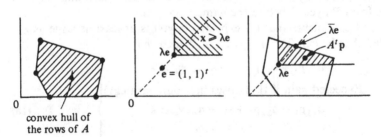

convex hull of
the rows of A

Fig. 12.7

subject to the constraints

$$A'\mathbf{y} \geqslant \mathbf{e},$$
$$\mathbf{y} \geqslant 0.$$

Similar arguments lead to Player II seeking to minimise μ subject to the constraints

$$\left.\begin{array}{l} A\mathbf{q} \leqslant \mu\mathbf{e} \\ \mathbf{e}'\mathbf{q} = 1 \\ \mathbf{q} \geqslant 0 \end{array}\right\}. \tag{3}$$

We put $\mathbf{x} = \mathbf{q}/\mu$. Then Player II seeks to maximise

$$\frac{1}{\mu} = \frac{1}{\mu}\mathbf{e}'\mathbf{q} = \mathbf{e}'\mathbf{x}$$

subject to the constraints

$$A\mathbf{x} \leqslant \mathbf{e}$$
$$\mathbf{x} \geqslant 0.$$

Thus Player II seeks to solve a feasible primal problem and Player I a feasible dual. By the duality theorem both have optimal solutions $\mathbf{x}^*$ and $\mathbf{y}^*$ and, moreover,

$$\frac{1}{\lambda^*} = \mathbf{e}'\mathbf{x}^* = \mathbf{e}'\mathbf{y}^* = \frac{1}{\mu^*}$$

and thus both players can solve their problems simultaneously.

[Note that we have used $\mathbf{e}$ to denote ambiguously both the $m \times 1$ vector associated with Player I and the $n \times 1$ vector associated with Player II consisting entirely of ones.]

We now briefly develop the parallel approach using the separating hyperplane theorem.

We begin by interpreting (2) geometrically.

Let

$$\mathfrak{A} = \{A'\mathbf{p} : \mathbf{p} \geqslant 0 \text{ and } \mathbf{e}'\mathbf{p} = 1\}$$

which we have noted is the convex hull of the rows of A. For any λ let

$$K(\lambda) = \{\mathbf{x} : \mathbf{x} \geqslant \lambda\mathbf{e}\} = \lambda\mathbf{e} + \{\mathbf{z} : \mathbf{z} \geqslant 0\}.$$

Thus $K(\lambda)$ is the translate by $\lambda\mathbf{e}$ of the non-negative orthant. It is thus a convex cone (see Chapter 10). Now Player I seeks the largest value of λ such that

$$K(\lambda) \cap \mathfrak{A} \neq 0. \tag{4}$$

Such a value evidently exists (why?). If λ^* is this largest value let $\mathbf{p}^*$ with $\mathbf{e}^t\mathbf{p}^* = 1$ satisfy

$$A^t\mathbf{p}^* \in K(\lambda^*). \tag{5}$$

Thus

$$A^t\mathbf{p}^* \geqslant \lambda^*\mathbf{e}$$

and so for any mixed strategy $\mathbf{q}$

$$\mathbf{p}^{*t}A\mathbf{q} \geqslant \lambda^*\mathbf{e}^t\mathbf{q} = \lambda^*, \tag{6}$$

so Player I has an expected reward of at least λ^* no matter what strategy Player II employs.

Now $\mathfrak{A}$ is disjoint from the interior of $K(\lambda^*)$ for otherwise we could find $\lambda > \lambda^*$ so that (4) holds. Hence there is a separating hyperplane between $\mathfrak{A}$ and $K(\lambda^*)$. Evidently $A^t\mathbf{p}^*$ is by (5) on the hyperplane which thus supports $K(\lambda^*)$. In this case the hyperplane must also pass through the apex of the cone $K(\lambda^*)$ (see Question 11 of Exercise 12.8) and so its equation is

$$\langle \mathbf{v}, \mathbf{x} - \lambda^*\mathbf{e} \rangle = 0$$

for some non-zero vector $\mathbf{v}$. We assume $\mathbf{v}$ is so chosen that $K(\lambda^*)$ is in the half-space

$$\langle \mathbf{v}, \mathbf{x} \rangle \geqslant \lambda^*\mathbf{v}^t\mathbf{e}.$$

It now follows that for any $\mathbf{z} \geqslant 0$, since $\lambda^*\mathbf{e} + \mathbf{z}$ is in $K(\lambda^*)$ that

$$\langle \mathbf{v}, \lambda^*\mathbf{e} + \mathbf{z} \rangle \geqslant \lambda^*\mathbf{v}^t\mathbf{e}$$

or

$$\lambda^*\mathbf{v}^t\mathbf{e} + \mathbf{v}^t\mathbf{z} \geqslant \lambda^*\mathbf{v}^t\mathbf{e}.$$

Thus for any $\mathbf{z} \geqslant 0$

$$\mathbf{v}^t\mathbf{z} \geqslant 0,$$

from which it follows that $\mathbf{v} \geqslant 0$ and so $\mathbf{v}^t\mathbf{e} > 0$. Put

$$\mathbf{q}^* = \frac{\mathbf{v}}{\mathbf{v}^t\mathbf{e}} \geqslant 0.$$

Then $\mathbf{e}^t\mathbf{q}^* = 1$ so $\mathbf{q}^*$ is a mixed strategy available to Player II. But for any $\mathbf{p}$, since $A^t\mathbf{p}$ is in $\mathfrak{A}$ we have

$$\langle \mathbf{v}, A^t\mathbf{p} \rangle \leqslant \lambda^*\mathbf{v}^t\mathbf{e}$$

hence

$$\mathbf{v}^t\mathbf{e}\mathbf{q}^*A^t\mathbf{p} \leqslant \lambda^*\mathbf{v}^t\mathbf{e}$$

or

$$\mathbf{p}^t A \mathbf{q}^* \leqslant \lambda^*. \tag{7}$$

Thus by employing $\mathbf{q}^*$ Player II can ensure that his loss is not greater than λ^*. Together (6) and (7) prove that $(\mathbf{p}^*, \mathbf{q}^*)$ is an equilibrium.

12.7   Example

Consider the matrix game with payoff

$$\text{I} \begin{cases} 1 \\ 2 \end{cases} \begin{array}{|c|c|} \hline 0 & 1 \\ \hline 3 & -1 \\ \hline \end{array}$$

$$\underbrace{ 1 2 }$$

$$\text{II}$$

(Thus if I chooses pure strategy 2 and II chooses pure strategy 1 the payoff to I is 3 and to II is -3.)

From the analysis above we know that at the equilibrium point Player I has maximised λ and Player II has minimised μ subject to

$$\left.\begin{array}{l} A^t\mathbf{p} \geqslant \lambda\mathbf{e} \\ \mathbf{e}^t\mathbf{p} = 1 \\ \mathbf{p} \geqslant 0 \end{array}\right\} \quad (1) \qquad \left.\begin{array}{l} A\mathbf{q} \leqslant \mu\mathbf{e} \\ \mathbf{e}^t\mathbf{q} = 1 \\ \mathbf{q} \geqslant 0 \end{array}\right\} \quad (2)$$

For system (1) we are interested in the convex hull of the rows of A; for the system (2) in the convex hull of the columns of A. We refer now to Figure 12.8.

The convex hull of the rows of A comprises the line segment L joining the points $(0, 1)^t$ and $(3, -1)^t$. The line l through the two points has equation

$$\frac{x_2 - 1}{-1 - 1} = \frac{x_1 - 0}{3 - 0}$$

i.e.

$$2x_1 + 3x_2 = 3.$$

To find λ^* in this case we first solve simultaneously the last equation with $x_1 = x_2$. Then

$$5x_1 = 3.$$

Since this gives a point of the line *segment* L (rather than just of l) we have for some $\mathbf{p}^*$ that

$$\tfrac{3}{5}(1, 1)^t = A^t\mathbf{p}^*.$$

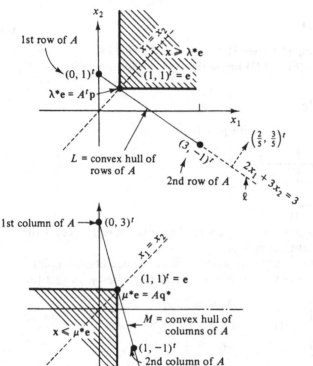

Fig. 12.8

We note that for any $\mathbf{q}$

$$\mathbf{p}^{*t}A\mathbf{q} = \mathbf{q}^t A^t \mathbf{p}^* = \tfrac{3}{5}\mathbf{q}^t(1, 1)^t$$
$$= \tfrac{3}{5},$$

hence Player I can secure an expected reward of $\tfrac{3}{5}$.

Before finding $\mathbf{p}^*$ we repeat the calculation above for the system (2).

The convex hull of the columns of A comprises the line segment M joining the points $(1, -1)^t$ and $(0, 3)^t$. The line through these two points thus has equation

$$\frac{x_2 - 3}{-1 - 3} = \frac{x_1 - 0}{1 - 0}$$

i.e.

$$4x_1 + x_2 = 3.$$

To find μ^* in this case we first solve simultaneously the last equation with $x_1 = x_2$. Then

$$5x_1 = 3.$$

Again this gives a point of the line *segment M* itself and so for some $\mathbf{q}^*$ it is true that

$$\tfrac{3}{5}(1, 1)^t = A\mathbf{q}^*.$$

Again too we note that for any $\mathbf{p}$

$$\mathbf{p}^t A\mathbf{q}^* = \tfrac{3}{5}\mathbf{p}^t(1, 1)^t = \tfrac{3}{5}.$$

Hence Player II can ensure that his expected loss no matter what Player I does is *not* greater than $\tfrac{3}{5}$. We have thus discovered that $\lambda^* = \tfrac{3}{5} = \mu^*$ and the *value* of the game is $\tfrac{3}{5}$ – i.e. the expected payoff for Player I at the equilibrium is $\tfrac{3}{5}$. (Von Neumann's theorem assures us that $\lambda^* = \mu^*$, and we observe that this is borne out by our calculations.)

Finally we find $\mathbf{p}^*$ and $\mathbf{q}^*$. One way to do this is to refer to A^{-1} *assuming this exists*. In our case, since the rows are linearly independent, we have an inverse and

$$A = \begin{bmatrix} 0 & 1 \\ 3 & -1 \end{bmatrix} \quad A^{-1} = \frac{1}{3}\begin{bmatrix} 1 & 1 \\ 3 & 0 \end{bmatrix} \quad (A^t)^{-1} = \frac{1}{3}\begin{bmatrix} 1 & 3 \\ 1 & 0 \end{bmatrix}.$$

We obtain that

$$\mathbf{p}^* = \frac{3}{5}\frac{1}{3}\begin{bmatrix} 1 & 3 \\ 1 & 0 \end{bmatrix}\begin{bmatrix} 1 \\ 1 \end{bmatrix}, \qquad \mathbf{q}^* = \frac{3}{5}\frac{1}{3}\begin{bmatrix} 1 & 1 \\ 3 & 0 \end{bmatrix}\begin{bmatrix} 1 \\ 1 \end{bmatrix}$$

$$= \begin{bmatrix} \tfrac{4}{5} \\ \tfrac{1}{5} \end{bmatrix} \qquad\qquad = \begin{bmatrix} \tfrac{2}{5} \\ \tfrac{3}{5} \end{bmatrix}$$

i.e. Player I choses pure strategy 1 with probability $\tfrac{4}{5}$ and pure strategy 2 with probability $\tfrac{1}{5}$. Player II on the other hand choses his pure strategy 1 with probability $\tfrac{2}{5}$ and pure strategy 2 with probability $\tfrac{3}{5}$.

12.8 Exercises

1. Solve the following linear programming problem. Maximise $z = x + y$ subject to the constraints

$$\left.\begin{array}{l} x + 2y \leqslant 1 \\ y - 2x \geqslant -2 \\ 2x - 1 \leqslant 0 \end{array}\right\} \quad \left.\begin{array}{l} x \geqslant 0 \\ y \geqslant 0 \end{array}\right\}.$$

2. Write down the dual of the linear programming problem. Minimise $u = x + 2y - z$ subject to the constraints

$$\left.\begin{array}{r} x + 2y - 2z \geqslant 1 \\ 2x - y \qquad \geqslant 1 \end{array}\right\} \quad \left.\begin{array}{l} x \geqslant 0 \\ y \geqslant 0 \\ z \geqslant 0 \end{array}\right\}$$

and hence compute the minimum. Find also the *maximum* of u subject to the constraints (if such exists).

3. Two products x and y pass through machine operations (I, II, III and IV) in being manufactured. The machine times in hours per unit are

	I	II	III	IV
x	2	4	3	1
y	$\frac{1}{4}$	2	2	45

The total time available on machines I, II, III and IV respectively is 45, 100, 300 and 50 hours. Product x sells for £6 and product y for £4 a unit.

What combination of x and y should be produced to maximise revenue?

4. An entrepreneur can produce manufactured goods x and y which each sell at a price of £1 per unit. To manufacture x and y respectively he requires quantities u and v of raw materials, where

$$\left.\begin{array}{l} u = x + 2y \\ v = 2x + y \end{array}\right\}.$$

If he begins with a stock of 3 units of raw material u and 4 units of raw material v, at what prices for the raw materials will it pay him to buy a little more?

5. Consider the problem: Maximise $\mathbf{c}'\mathbf{x}$ subject to the constraints

$$\left.\begin{array}{l} A\mathbf{x} = \mathbf{b} \\ \mathbf{x} \geqslant \mathbf{0} \end{array}\right\}.$$

Show that the dual problem is to minimise $\mathbf{b}'\mathbf{z}$ subject to the simple constraints

$$A'\mathbf{z} \geqslant \mathbf{c}.$$

6. Suppose that $\mathbf{x}^* > \mathbf{0}$ is optimal for the standard primal problem. If the rows of A are $\mathbf{a}_1', \mathbf{a}_2', \ldots, \mathbf{a}_m'$ explain why

$$\mathbf{c} = \sum_{y_i^* > 0} y_i^* \mathbf{a}_i.$$

This equation implies that $\mathbf{c}$ lies in the normal cone at $\mathbf{x}^*$ to the feasible set of the primal problem. Explain this.

7. Find the equilibrium strategies and value of the finite matrix game with

$$A = \begin{bmatrix} -1 & 0 \\ 2 & 1 \end{bmatrix}.$$

8. Find the equilibrium strategy for Player II and the value of the finite matrix game with

$$A = \begin{bmatrix} 1 & 0 & 4 \\ 0 & -1 & -5 \end{bmatrix}.$$

9. Explain why chess may be regarded as a two-person, zero-sum game.

10. Write down the Lagrangian for the primal linear programming problem introduced in Section 12.4. This has a saddle point at $(\mathbf{x}^*, \mathbf{y}^*)$. To what does this assertion reduce when $\mathbf{x} = \mathbf{q}/\mu, \mathbf{y} = \mathbf{p}/\lambda, \mathbf{x}^* = \mathbf{q}^*/\mu, \mathbf{y}^* = \mathbf{p}^*/\lambda$? What is the significance (if any) of your answer?

11. The hyperplane $\mathbf{v}^t\mathbf{x} = c$ is known to support the non-negative orthant $\{\mathbf{x} : \mathbf{x} \geqslant 0\}$. Assume that $\mathbf{v}^t\mathbf{x} \leqslant c$ for all $\mathbf{x} \geqslant 0$. Assume further that for some fixed $\mathbf{z} \geqslant 0, \mathbf{v}^t\mathbf{z} = c$. Show by considering the sequence of points $\mathbf{z}, 2\mathbf{z}, 3\mathbf{z}, \ldots, n\mathbf{z}, \ldots$ which are all in the orthant that $c \leqslant 0$. Deduce that the hyperplane passes through the origin, i.e. that $c = 0$.

12. Find a vector of the form $(\lambda, \lambda, \lambda)^t$ belonging to the convex hull C of the three points

$$(4, 1, 2)^t, \quad (1, 3, 1)^t, \quad (2, 2, 4)^t$$

and prove that for this λ one of the players in the zero-sum game with payoff matrix

$$\begin{bmatrix} 4 & 1 & 2 \\ 1 & 3 & 2 \\ 2 & 1 & 4 \end{bmatrix}$$

can ensure that the expected loss is λ. Find a normal vector to the plane in $\mathbb{R}^3$ separating C from the translate by $(\lambda, \lambda, \lambda)^t$ of the positive quadrant. How can this normal vector be used to show that λ is the most favourable possible expected loss for the aforesaid player?

13

The simplex method

We describe an algorithm due to van Dantzig for solving the linear programme

$$\max \mathbf{c}^t \mathbf{x}$$

subject to

$$A\mathbf{x} \leqslant 0, \qquad\qquad\qquad\qquad (*)$$

and

$$\mathbf{x} \geqslant 0. \qquad\qquad\qquad\qquad (**)$$

We already know that if the feasible set F is bounded, the optimal $\mathbf{x}$ is an extreme point of F. Since the extreme points of F form a finite set (this is geometrically obvious – see Section 11.3 for an indication of a proof), it would in principle suffice to check the value of $\mathbf{c}^t \mathbf{x}$ at each extreme point. The algorithm not only locates the extreme points for us, but presents them in such an order that the value of the *objective function* $\mathbf{c}^t \mathbf{x}$ increases. In the example illustrated in Figure 13.1 we have:

$$A = \begin{bmatrix} 1 & 2 \\ 1 & 1 \\ 4 & 2 \end{bmatrix} \quad \text{and} \quad \mathbf{b} = \begin{bmatrix} 10 \\ 6 \\ 19 \end{bmatrix}.$$

Observe that the lines $\mathbf{c}^t \mathbf{x} = 2x_1 + 3x_2 = \lambda$ for λ increasing pass through the sequence of points $(0,0)$, $(0,5)$ and $(2,4)$ which are all connected successively by edges on the boundary of the feasible set. The algorithm when started at $(0,0)$ will present $(0,5)$ for evaluation, then $(2,4)$ and stop there since that is where the maximum occurs.

13.1 The key idea

There is a very easy way to locate algorithmically an extreme point other than the origin. Search along one of the co-ordinate axes. The advantage

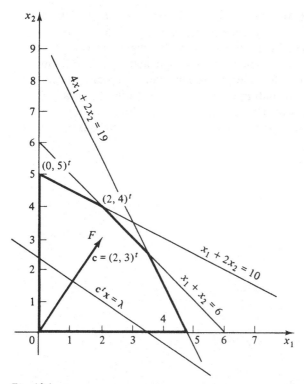

Fig. 13.1

of this idea is that all but one variable is zero. Since the objective function $2x_1 + 3x_2$ increases fastest with x_2 (per unit increase) we search the x_2-axis (where $x_1 = 0$). To find the extreme point consider the constraining inequalities. With $x_1 = 0$ they take a very simple form. We have from (*) that:

$$x_1 + 2x_2 \leqslant 10 \quad \text{becomes } 2x_2 \leqslant 10 \text{ or } x_2 \leqslant 5, \tag{1}$$

$$x_1 + x_2 \leqslant 6 \quad \text{becomes } x_2 \leqslant 6, \tag{2}$$

$$4x_1 + 2x_2 \leqslant 19 \quad \text{becomes } 2x_2 \leqslant 19 \text{ or } x_2 \leqslant 9.5 \tag{3}$$

The extreme point occurs at the largest value of x_2 which satisfies all the constraints. That, clearly, occurs at the most restrictive constraint $x_2 \leqslant 5$. We have thus reached the extreme point $(0, 5)$. The procedure is very simple indeed, and evidently works when we are confronted with an *n* variable problem since on an axis all variables except one are zero.

Now either the optimum is $(0, 5)$ or else we need to proceed to a further extreme point. We thus have to ask ourselves whether there is an equally easy way to find another extreme point at which the objective function is greater than $2·0 + 3·5 = 15$. Van Dantzig's idea was to construct a new co-ordinate system with its origin at $(0, 5)$ and with axes along the boundary of the feasible set, the aim being to repeat the same process as above for calculating the next extreme point. Naturally the co-ordinate system has to have oblique axes, but then our earlier calculation nowhere used the orthogonality of the natural axes.

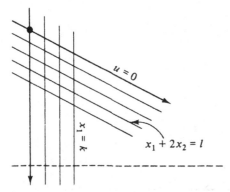

Fig. 13.2. New co-ordinate through (0,5)

We set about determining the form of the co-ordinate system with axes as illustrated in Figure 13.2. The parallels in this system are given by

$$x_1 = k$$

and by

$$x_1 + 2x_2 = l.$$

for constants k and l.

Observe that we know the equation of the next edge as the one which corresponds to the most restrictive constraint identified at the last step viz. (1).

The new co-ordinate axis is obtained when $l = 10$. Introduce a new variable u. We would like this new axis to have equation $u = 0$. We can arrange for that to happen with either of two choices:

$$u = 10 - (x_1 + 2x_2) \quad \text{or} \quad u = x_1 + 2x_2 - 10.$$

Which should it be?

Since we would want points in the feasible set to satisfy $u \geqslant 0$ (because the co-ordinate variables in the original statement of the problem were also non-negative) we are led to select

$$u = 10 - (x_1 + 2x_2).$$

Having determined that the new co-ordinate system has axes $u = 0$ and $x_1 = 0$ we need to reformulate the original problem using the variables x_1 and u instead of x_1 and x_2. This we can do by employing the equation which defines u to eliminate all reference to x_2 in all the other constraints.

Performing this elimination first on (*) (viz. $A\mathbf{x} \leqslant \mathbf{b}$) we have

$$-u \leqslant 0 \qquad \text{or} \quad u \geqslant 0, \tag{1'}$$

$$x_1 + \tfrac{1}{2}\{10 - u - x_1\} \leqslant 6 \qquad \text{or} \quad \tfrac{1}{2}x_1 - \tfrac{1}{2}u \leqslant 1, \tag{2'}$$

$$4x_1 + \{10 - u - x_1\} \leqslant 10 \qquad \text{or} \quad 3x_1 - u \leqslant 9, \tag{3'}$$

(and we notice that $u \geqslant 0$ as pre-arranged). Next we transform the vector inequality (**) viz. $\mathbf{x} \geqslant 0$ which now reads:

$$x_1 \geqslant 0,$$
$$x_2 = \tfrac{1}{2}\{10 - u - x_1\} \geqslant 0,$$
$$\text{i.e. } x_1 + u \leqslant 5. \tag{1''}$$

(Notice that the latter is really a restatement of (1) without using x_2.)

Next we observe that the objective function itself is transformed as follows

$$2x_1 + 3x_2 = 2x_1 + \tfrac{3}{2}\{10 - u - x_1\}$$
$$= 15 + \tfrac{1}{2}x_1 - \tfrac{3}{2}u.$$

The constant occurring in the last line has to equal the value of the objective function at the extreme point (0, 5). To see this note that $x_1 = 0$ and $u = 0$ defines that extreme point, so on substituting these values in the objective function a value of 15 must be returned. Thus *the constant term records the objective value at the current extreme point.* Note also that 15 is not the maximum we are seeking since the objective function may be increased by setting $u = 0$ and then allowing x_1 to increase (subject of course to the constraints).

We thus know that we must look for a further extreme point and have ended up with the problem:
maximise

$$15 + \tfrac{1}{2}x_1 - \tfrac{3}{2}u$$

subject to

$$\tfrac{1}{2}x_1 + \tfrac{1}{2}u \leqslant 5, \tag{4}$$

$$\tfrac{1}{2}x_1 - \tfrac{1}{2}u \leqslant 1, \tag{5}$$

$$3x_1 - u \leqslant 9, \tag{6}$$

with $x_1 \geqslant 0$ and $u \geqslant 0$.

(Notice that (5) and (6) are (2') and (3') while (4) arises from (1").)

This is once again in the form of the original problem and we may take $u = 0$ and seek the most restrictive constraint. The constraints this time are as follows.

(4) becomes $\frac{1}{2}x_1 \leqslant 5$,

(5) becomes $\frac{1}{2}x_1 \leqslant 1$,

(6) becomes $3x_1 \leqslant 9$,

and the most restrictive as far as maximising x_1 is concerned is (5). Thus at the next extreme point we have $u = 0$ and $x_1 = 2$. The objective function value here is $15 + \frac{1}{2} \cdot 2 = 16$, certainly an improvement on the last value returned.

The next step is obvious. We attempt a further increase in the objective function by the same trick. We use the second constraint (5) to introduce a new co-ordinate variable, call it v, and we have:

$$v = 1 - \tfrac{1}{2}x_1 + \tfrac{1}{2}u$$

(so that the extreme point is now represented by $u = 0$ and $v = 0$). We ought now to rewrite the constraints eliminating x_1 in favour of v. Before doing so, let us first re-write the objective function in terms of our new variables. We have

$$15 + \tfrac{1}{2}x_1 - \tfrac{3}{2}u = 15 - \tfrac{3}{2}u + \{1 - v + \tfrac{1}{2}u\}$$
$$= 16 - u - v.$$

This last expression has the significant property that *all variables have negative coefficients*. Since the variables are all non-negative ($u \geqslant 0, v \geqslant 0$) the objective function cannot take a value greater than 16 in the feasible set. We have thus solved the linear programme.

13.2   Systematic approach

Now that the pattern of activity has emerged we set about performing the calculation in a systematic way. This is done by means of an array of numbers comprising the coefficients and the constant terms which we have been considering above. The array is known as a *tableau*. Before we can introduce it we explain the philosophy of the notation. Instead of introducing the letters u, v in an *ad hoc* way, we note that their role is to convert inequalities into equations and that they in effect take up the 'slack' in the inequalities. A systematic way to do this is to *extend* the original list of variables in the problem by introducing one *slack variable* into each inequality at the outset, thus:

$$x_1 + 2x_2 + x_3 \qquad = 10,$$
$$x_1 + x_2 + x_4 = 6,$$
$$4x_1 + 2x_2 + x_5 = 19.$$

The feasible set is now given in terms of the extended list of variables by:

$$x_1, x_2, x_3, x_4, x_5 \geqslant 0,$$

subject to the equations above.

Note that $u = x_3$ and $v = x_4$. Note also that the edges of the feasible set F (Figure 13.3) are respectively given by

$$x_1 = 0, \quad x_2 = 0, \quad x_3 = 0, \quad x_4 = 0, \quad x_5 = 0.$$

Extreme points in a *two* variable problem are characterised by the fact that *precisely two* of the extended list of variables are zero. In official parlance the variables that are non-zero at an extreme point are called the *basic variables* of the extreme point and the others are called *non-basic*. Thus the non-basic variables are those that refer to the 'new co-ordinate frame' which was constructed through the extreme point as origin. On the other hand the basic variables are those that evaluate slack. In other words they are the 'generalised slack variables'.

Secondly we give the objective function a name, say M. Thus:

$$M = 2x_1 + 3x_2$$
$$= 15 + \tfrac{1}{2}x_1 - \tfrac{3}{2}u$$
$$= 16 - u - v.$$

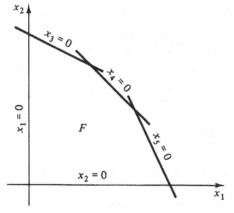

Fig. 13.3

13.3   The general setting

In general we can expect an objective function of n variables $x_1, x_2, \ldots, x_n$, thus:

$$M = c^t x,$$

and m linear inequalities

$$Ax \leqslant b,$$

where for the time being we shall suppose that $b \geqslant 0$. The latter may be converted to equations by the introduction of one slack variable per inequality: $x_{n+1}, x_{n+2}, \ldots, x_{n+m}$. We then select one variable which is to be increased as in the original example. For the sake of argument let this be x_1. (In the example it was x_2.) Setting $x_2 = 0$, $x_3 = 0, \ldots, x_n = 0$ we have that as x_1 increases the point $(x_1, 0, \ldots, 0, 0)$ moves along a straight line segment or edge of the feasible set until one of the constraining inequalities becomes *active* (viz. the inequality is actually an equation). Since to begin with all the slack variables are non-negative (usually in fact they are positive), the situation just envisaged brings one of the slack variables to *zero*. (We hope that only one slack variable becomes zero). Say this is x_{n+1}. We change to an oblique co-ordinate system in which the reference variables (the non-basic variables) are $x_2, \ldots, x_n, x_{n+1}$ and the basic variables which measure slack relative to the new variables are $x_1, x_{n+2}, \ldots, x_{n+m}$. The process is then repeated by selecting another non-basic variable for increase.

13.4   Simplex tableaux

Our original argument may be re-written in the tabular form below

M	x_1	x_2	x_3	x_4	x_5	b	Ratios
0	1	②	1	0	0	10	$10/2 = \underline{5}$
0	1	1	0	1	0	6	6/1
0	4	2	0	0	1	19	$19/2 = 9.5$
1	-2	$\underline{-3}$	0	0	0	0	

where the last line records the equation $M - 2x_1 - 3x_2 = 0$. Our technique is now to

— select the column corresponding to the most negative entry on the bottom line
— calculate the ratios of the entries in the b column to the corresponding coefficient in the selected column

—circle that coefficient in the selected column which corresponds to the least ratio. The circled entry is known as the *pivot*.

This done we use the pivotal row (the row in which the pivot lies) to eliminate by *row operations* all other entries in the selected column. (This is known as Gauss–Jordan elimination.) The result is another tableau:

M	x_1	x_2	x_3	x_4	x_5	b	Ratios
0	1/2	1	1/2	0	0	5	$5/(1/2) = 10$
0	①/②	0	$-1/2$	1	0	1	$1/(1/2) = \underline{2}$
0	3	0	-1	0	1	9	$9/3 = 3$
1	$-1/2$	0	3/2	0	0	15	

Repeating this process yields

M	x_1	x_2	x_3	x_4	x_5	b
0	0	1	$+1$	-1	0	4
0	1	0	-1	$+2$	0	2
0	0	0	2	-6	1	3
1	0	0	1	1	0	16

Since there are no negative entries in the bottom row we have completed our task.

The ratio calculations are of course related to our earlier determination of the most restrictive constraint. The most restrictive corresponds to the least ratio. To see this refer to the first tableau and note that if x_1 is set equal to zero the positivity of the slack variables implies the same inequalities as in Section 13.1.

We have for $x_1 = 0$ that:

$$2x_2 + x_3 = 10, \quad \text{so since } x_3 \geqslant 0 \quad \text{hence } 2x_2 \leqslant 10 \quad \text{or} \quad x_2 \leqslant 5,$$
$$x_2 + x_4 = 6, \quad \text{so since } x_4 \geqslant 0 \quad \text{hence } x_2 \leqslant 6,$$
$$2x_2 + x_5 = 19, \quad \text{so since } x_5 \geqslant 0 \quad \text{hence } 2x_2 \leqslant 19 \quad \text{or} \quad x_2 \leqslant 9.5,$$

hence the constraints on x_2 are obtained by taking the ratio of the constant appearing on the right to the coefficient of x_2. But that is what we calculated at the side of the tableau.

Remarks

1. Observe that if a constant appearing in the b column is *positive* and the coefficient in the pivotal column is negative then the ratio should be ignored. To see why observe that a constraint

like

$$-5x_2 \leqslant 10$$

is automatically satisfied when $x_2 \geqslant 0$. Likewise if the constant is *zero* and the coefficient in the pivotal column is *negative* the ratio is ignored. However a constraint like

$$5x_2 \leqslant 0$$

can only be satisfied for $x_2 \geqslant 0$ if $x_2 = 0$. The ratio here is *not* ignored.

Evidently, if there are no ratios to consider (as a result of the above rule concerning negative ratios) then the maximum of $c'x$ is $+\infty$.

2. We have assumed throughout that $x = 0$ satisfies $Ax \leqslant b$ since $b \geqslant 0$. We will consider in the next section how to solve the linear programme when the origin is not in the feasible set.

3. It is usual to omit the first column (under M) from the tableaux since it is never altered. Similarly it is unnecessary to write down the variable names in the tableau. Further simplifications are possible, for instance: omitting the identity submatrix in the tableau (the coefficients at the basic variables create an identity array) at the expense of slight notational changes. We will not pursue this abbreviation since it requires care in remembering which are the basic variables. In any event involved calculations are best consigned to the computer.

4. Degeneracy problems occur if some of the constraints happen to be redundant (in the presence of other constraints). Let us follow through what happens to our initial example when we use

$$A = \begin{bmatrix} 1 & 2 \\ 1 & 1 \\ 4 & 2 \end{bmatrix} \quad \text{and} \quad b = \begin{bmatrix} 10 \\ 6 \\ 10 \end{bmatrix}$$

Here the constraint $x_1 + 2x_2 \leqslant 10$ is redundant. See Figure 13.4. The effect on the tableau calculation is as follows. In the first tableau we have two choices for a pivot.

						Ratios
1	②	1	0	0	10	$10/2 = \underline{5}$
1	1	0	1	0	6	$6/1$
4	②	0	0	1	10	$10/2 = \underline{5}$
-2	$\underline{-3}$	0	0	0	0	

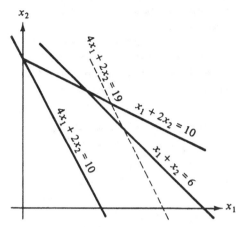

Fig. 13.4

Let us be obdurate and take the pivot in the first row (despite the fact that our picture tells us to ignore the first constraint altogether). Then we have

						Ratios
1/2	1	1/2	0	0	5	$5/(1/2) = 10$
1/2	0	−1/2	1	0	1	$1/(1/2) = 2$
③	0	−1	0	1	0	$0/3 = \underline{0}$
−1/2	0	3/2	0	0	15	

Note that the second tableau tells us to follow the edge $x_5 = 0$ in search of the next extreme point. (We had of course just been searching for an extreme point along $x_3 = 0$ when we opted for the pivot in the first row). Now note the objective value in the third tableau:

0	1	4/6	0	−1/6	5
0	0	−1/3	1	−1/6	1
1	0	−1/3	0	1/3	0
0	0	4/3	0	1/6	15

It is the same value as before. The reason is clear: we have changed edges without changing extreme point. But since we have arrived at the maximum the calculation has not been upset.

It is possible (in more general cases) for the calculation to cycle round a number of edges without moving on to the optimum extreme point. However, practitioners point out that cycling is rarely encountered in the real world, so we prefer not to take this question any further.

13.5   Calculation of a basic solution for *b* non-positive

Consider the modified problem illustrated in Figure 13.5:

maximise

$$2x_1 + 3x_2$$

subject to

$$x_1 + 2x_2 \leqslant 10,$$
$$x_1 + x_2 \geqslant 6, \quad (\text{i.e.} \ -x_1 - x_2 \leqslant -6),$$
$$4x_1 + 2x_2 \leqslant 19.$$

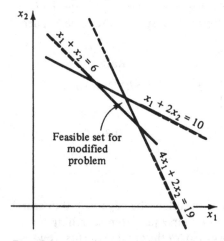

Fig. 13.5

As a first step we rewrite the inequalities as equations with slack variables just as before. Note that the middle equation has been written so as to keep x_4 non-negative just like the other slack variables.

$$\left.\begin{array}{rl} x_1 + 2x_2 + x_3 & = 10 \\ x_1 + x_2 \quad - x_4 & = 6 \\ 4x_1 + x_2 \quad + x_5 & = 19 \\ x_1, x_2, x_3, x_4, x_5 & \geqslant 0 \end{array}\right\}. \tag{1}$$

This approach has a serious obstacle in that $x_1 = x_2 = 0$ does not satisfy (1) since x_4 would need to be -6 which is not non-negative. So we create a further variable known as an *artificial variable* and consider the system:

$$
\left.
\begin{aligned}
x_1 + 2x_2 + x_3 \qquad\qquad\quad &= 10 \\
x_1 + \ x_2 + \quad -x_4 \ \ +x_6 &= 6 \\
4x_1 + 2x_2 + \qquad\quad x_5 \qquad &= 19 \\
x_1, x_2, x_3, x_4, x_5, x_6 \geqslant 0 &
\end{aligned}
\right\}
\qquad (2)
$$

This has the non-negative solution $x_1 = x_2 = x_4 = 0$, $x_3 = 10$, $x_6 = 6$, $x_5 = 19$. Our aim now is to find a solution of the system (2) with $x_6 = 0$. We therefore attack the problem of finding a solution of (2) with x_6 at a minimum. This leads us to solve the auxiliary problem:
maximise

$$-x_6$$

subject to

$$
\begin{aligned}
x_1 + 2x_2 + x_3 \qquad\qquad\quad &= 10, \\
x_1 + \ x_2 + \quad -x_4 \ \ +x_6 &= 6, \\
4x_1 + 2x_2 + \qquad\quad x_5 \qquad &= 19, \\
x_1, \ldots, x_6 \geqslant 0. &
\end{aligned}
$$

The linear programme above will have as its solution max $= 0$ if and only if the original problem has a feasible solution. One might at first be tempted to believe that the tableau for the auxiliary problem should be as below.

1	2	1	0	0	0		10
1	1	0	-1	0	1		6
4	2	0	0	1	0		19
0	0	0	0	0	1		0

However, the slack variables in this problem happen to be x_3, x_5, x_6 but the bottom line does not conform to the requirement that the objective function (of the auxiliary problem) must be expressed in the non-basic variables of the current extreme point viz. x_1, x_2, x_4. To counter this we quickly eliminate x_6 from the bottom line using the second row (since that row defines x_6 in terms of the non-basic variables x_1, x_2, x_4). This we now do. The tableau below also shows the pivot and additionally carries the objective function of the original problem (in brackets). Reasons for this will soon become apparent.

1	②1	0	0	0	10	10/2 = 5
1	1	0	−1	0	1	6	6/1
4	2	0	0	1	0	19	19/2 = 9.5

$$
\begin{array}{cccccc|c}
1 & ② & 1 & 0 & 0 & 0 & 10 \\
1 & 1 & 0 & -1 & 0 & 1 & 6 \\
4 & 2 & 0 & 0 & 1 & 0 & 19 \\
\hline
-1 & \underline{-1} & 0 & 1 & 0 & 0 & -6 \\
(-2 & -3 & 0 & 0 & 0 & 0 & 0)
\end{array}
\qquad
\begin{array}{l}
10/2 = 5 \\
6/1 \\
19/2 = 9.5
\end{array}
$$

We now have (applying Gauss–Jordan elimination to the bracketted row too):

$$
\begin{array}{cccccc|c}
1/2 & 1 & 1/2 & 0 & 0 & 0 & 5 \\
① & 0 & -1/2 & -1 & 0 & 1 & 1 \\
3 & 0 & -1 & 0 & +1 & 0 & 9 \\
\hline
-1/2 & 0 & 1/2 & 1 & 0 & 0 & -1 \\
(-1/2 & 0 & 3/2 & 0 & 0 & 0 & 15)
\end{array}
\qquad
\begin{array}{l}
5/(1/2) = 10 \\
1/(1/2) = \underline{2} \\
9/3 = 3
\end{array}
$$

and finally

$$
\begin{array}{cccccc|c}
0 & 1 & 1 & 1 & 0 & -1 & 4 \\
1 & 0 & -1 & -2 & 0 & 2 & 2 \\
0 & 0 & 2 & 6 & 1 & -6 & 3 \\
\hline
0 & 0 & 0 & 0 & 0 & 1 & 0 \\
(0 & 0 & 1 & -1 & 0 & 1 & 16)
\end{array}
$$

This gives $x_1 = 2, x_2 = 4, x_5 = 3, x_3 = x_4 = x_6 = 0$. Since $x_6 = 0$ we consider what happens if we suppress the x_6 column in our last tableau and use the bracketted row in the place of the current bottom row. The answer is clear. Since setting $x_6 = 0$ in (2) gives back the system (1), we obtain a tableau for the original problem corresponding to the extreme point $x_1 = 2, x_2 = 4, x_5 = 3, x_3 = x_4 = 0$. Thus the tableau is written in a form where x_3 and x_4 are the non-basic variables.

$$
\begin{array}{ccccc|c}
0 & 1 & 1 & 1 & 0 & 4 \\
1 & 0 & -1 & -2 & 0 & 2 \\
0 & 0 & 2 & 6 & 1 & 3 \\
\hline
0 & 0 & 1 & -1 & 0 & 16
\end{array}
$$

We leave it to the reader to complete the calculation from here.

13.6 Exercises

1. A plant makes two products A, B each needing to be processed by three machines. Time spent in process on each machine and hours available are itemised in the table

	Product A	B	Hrs available per week
Machine 1.	2	1	70
2.	1	1	40
3.	1	3	90

The profit on product A is \$2, on product B, it is \$3 per item.

(a) If profit is to be maximised, what amounts of each product should be made?

(b) The option has arisen of hiring an extra person to run the machines. If he/she can operate the machines for up to $10, 10, 20$ hrs per week (respectively) up to what wage is it profitable to employ him/her?

(c) If a person is employed what is in fact the most economical way to distribute his 40 hour week between the three machines? (You will probably have to do this one from first principles.)

2. Maximise $3x + 2y + z$ subject to

$$6x + 3y + 2z \leqslant 10$$
$$2x + y + 2z \leqslant 6$$
$$6x - 6y + z \leqslant 6$$
$$x, y, z \geqslant 0.$$

[Remember to disregard 'negative ratios'.]

3. Maximise $x + y + z$ subject to

$$3x + 3y + 4z \leqslant 12$$
$$4x + 6y + 3z \leqslant 12$$
$$3x - 6y + z \leqslant 3$$
$$x, y, z \geqslant 0.$$

When selecting the pivot note that for a positive c the relation $-cx_i \leqslant 0$ is always satisfied in the feasible set, so disregard the ratio $0/-c$; on the other hand $cx_i \leqslant 0$ is satisfied in the feasible set only for $x_i = 0$ so the ratio $0/+c$ is *not* to be disregarded.

13.7 Further exercises

1. (From Vajda.) At a signal-controlled intersection a two-way road meets three one-way roads as illustrated in Figure 13.6. Permitted movements a, b, c, d, e can

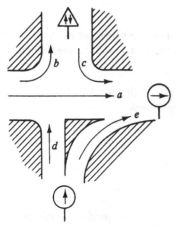

Fig. 13.6

be paired together by appropriate green signals being phased in the traffic lights. The average flow is tabulated below and possible phasing is (a, b), (d, e), (a, c), (b, c), (b, e) (check this!). The green time has to be allocated in such a way that the smallest number of vehicles allowed through on green via a, b, c, d, or e is maximised.

What porportion of green time should be given to the various phasings?

a	b	c	d	e
7	3	15	12	18

2. (Multiple solutions.) In the final tableau, if a *zero* appears in a column corresponding to a non-basic variable then we may obtain yet another tableau using that column to locate a further pivot. Evidently, the Gauss–Jordan elimination leaves the value of the objective function unchanged. Why does this correspond to multiple solutions?

3. Examine the relationship of the tableaux for a given problem (regarded as primal) to its dual.

4. (Tie-breaking.) If two 'pivotal ratios' (i.e. ratios examined to determine the pivot) are equal, then a zero will appear in the last column of the next tableau. What is the geometric significance of this and why will this mean that the tableau next after this may well have the same value for the objective value? Use geometric reasoning to suggest a tie-breaking rule between competing ratios. (Instead of a zero variable, use a very small value and examine ratios of pivotal column elements against corresponding elements from each of the other columns.)

5. (Checking rule.) Against each row i of your tableau place the coefficient α_i of the basic variable which it had in the original expression for the objective function. Do likewise for each column j corresponding to the non-basic variables. If d_j is

the entry in column j in the bottom row then

$$\sum \alpha_i c_i - \beta_j = - d_j$$

where c_is are indicated in the tableau below:

		β_j		
α_1	...	c_1	...	
		$\vdots$		
α_m	...	c_m	...	
		d_j		

14

Partial derivatives (revision)

14.1   Tangent plane to a surface

The tangent plane to the surface $z = f(x, y)$ at the point (X, Y, Z) where $Z = f(X, Y)$ takes the form

$$z - Z = l(x - X) + m(y - Y).$$

To obtain the numbers l and m we use calculus as follows. The plane $y = Y$ determines a section of the surface as illustrated in Figure 14.1. We work in that plane. The section of surface is, of course, a curve whose equation is $z = f(x, Y)$. This is a function of the one variable x. At the point $x = X$ the function has gradient

$$\lim_{h \to 0} \frac{f(X + h, Y) - f(X, Y)}{h}$$

and this is of course the partial derivative

$$\frac{\partial f}{\partial x}(X, Y).$$

The tangent plane's trace on $y = Y$ is a tangent line to the curve $z = f(x, Y)$; its equation is, evidently, $z - Z = l(x - X)$, i.e. its gradient is l and that must equal the derivative obtained above. Repeating this analysis in the plane $x = X$, we arrive at the conclusion that the tangent plane has equation

$$z - Z = \frac{\partial f}{\partial x} \cdot (x - X) + \frac{\partial f}{\partial y} \cdot (y - Y).$$

14.2   The directional derivative

The function $\phi(t) = f(X + tu_1, Y + tu_2)$ arises when we take a vertical section of the surface $z = f(x, y)$ by a plane through $(X, Y, 0)$ parallel to

178

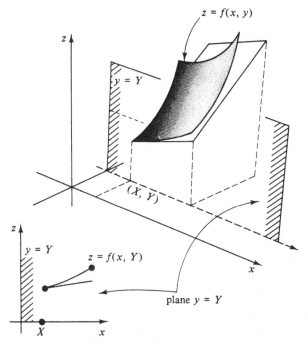

Fig. 14.1

the vector $\mathbf{u} = (u_1, u_2)^t$, as illustrated in Figure 14.2. The curve thus traced has slope $\phi'(0)$ at the point (X, Y, Z). The value of this slope can be obtained by using the chain rule. We have

$$\frac{d\phi}{dt} = \frac{\partial f}{\partial x}\frac{dx}{dt} + \frac{\partial f}{\partial y}\frac{dy}{dt}$$

$$= \frac{\partial f}{\partial x}u_1 + \frac{\partial f}{\partial y}u_2 = \nabla f \cdot \mathbf{u},$$

where ∇f is the row matrix

$$\left(\frac{\partial f}{\partial x}, \frac{\partial f}{\partial y}\right).$$

The slope obtained above is known as the *directional derivative of* $f(x, y)$ in direction $\mathbf{u}$.

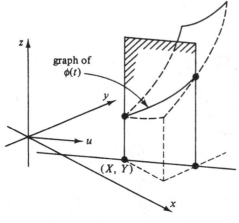

Fig. 14.2

14.3   The derivative *Df*

We claim that ∇f may properly be regarded as 'the derivative of f'. To say such a thing requires us to review the meaning attached to the usual derivative

$$\frac{dz}{dx}$$

of a function $z = \phi(x)$ of one variable. $\phi'(X)$ is a number that gives the best coefficient (or, multiplier) m for which the expression

$$z = Z + m(x - X)$$

is a good approximation to the function $z = \phi(x)$ near $x = X$ (Figure 14.3).

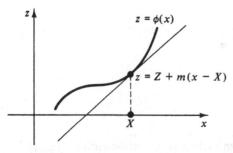

Fig. 14.3

It is 'best' in the sense that the error is of order $(x - X)^2$ (compare Taylor's theorem). Now note that if $x = X + h$ and $y = Y + k$

$$f(X + h, Y + k) - f(X, Y) = f(X + h, Y + k) - f(X, Y + k)$$
$$+ f(X, Y + k) - f(X, Y)$$

$$\approx \frac{\partial f}{\partial x} h + \frac{\partial f}{\partial y} k = \nabla f \cdot \begin{bmatrix} h \\ k \end{bmatrix},$$

so that ∇f is the 'best' matrix multiplier M in the linear approximation

$$z = Z + M \begin{bmatrix} x - X \\ y - Y \end{bmatrix}.$$

So it is sensible to think of ∇f as 'Df'.

Now consider the vector-valued function

$$\mathbf{z} = (z_1, z_2)^t = (f_1(x, y), \quad f_2(x, y))^t = \mathbf{f}(x, y).$$

Repeating the earlier analysis for each co-ordinate variable z_1 and z_2 gives

$$\mathbf{f}(X + h, Y + k) - \mathbf{f}(X, Y) \approx \left(\frac{\partial f_1}{\partial x} h + \frac{\partial f_1}{\partial y} k, \quad \frac{\partial f_2}{\partial x} h + \frac{\partial f_2}{\partial y} k \right)^t.$$

This last expression may be re-written as

$$\begin{bmatrix} \dfrac{\partial f_1}{\partial x} & \dfrac{\partial f_1}{\partial y} \\[2mm] \dfrac{\partial f_2}{\partial x} & \dfrac{\partial f_2}{\partial y} \end{bmatrix} \begin{bmatrix} h \\ k \end{bmatrix}.$$

So it is natural to *define* the derivative Df in this case to be the matrix

$$\begin{bmatrix} \dfrac{\partial f_1}{\partial x} & \dfrac{\partial f_1}{\partial y} \\[2mm] \dfrac{\partial f_2}{\partial x} & \dfrac{\partial f_2}{\partial y} \end{bmatrix}$$

since it is the best matrix multiplier M in the approximation

$$\mathbf{f}(X + h, Y + k) - \mathbf{f}(X, Y) \approx M \begin{bmatrix} h \\ k \end{bmatrix}.$$

Consider what meaning should be attached to the term $D^2 f$ when, as initially, we are concerned with the real-valued function $z = f(x, y)$. We evidently have

$$Df = \left(\frac{\partial f}{\partial x}, \frac{\partial f}{\partial y} \right).$$

So treating the first and second co-ordinates as though they were the functions f_1 and f_2 we obtain that $D^2f = D(Df)$ is:

$$\begin{bmatrix} \dfrac{\partial^2 f}{\partial x^2} & \dfrac{\partial^2 f}{\partial y\,\partial x} \\[2ex] \dfrac{\partial^2 f}{\partial x\,\partial y} & \dfrac{\partial^2 f}{\partial y^2} \end{bmatrix}.$$

One could proceed to obtain $D^3f,\ldots$ but we would have to agree on some convention regarding its representation, since matrices would not suffice. One spectacular idea would be to introduce 'solid' matrices (Figure 14.4). Less spectacular, and easier to handle, would be to re-write

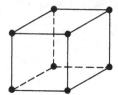

Fig. 14.4

D^2f as

$$\left(\frac{\partial^2 f}{\partial x^2}, \frac{\partial^2 f}{\partial y\,\partial y}, \frac{\partial^2 f}{\partial x\,\partial y}, \frac{\partial^2 f}{\partial^2 y} \right)$$

and to continue from there. Fortunately, we have no need for $D^3f,\ldots$ in this course (or for tensors, which would be the appropriate mathematical tool).

Example
Compute the directional derivative of the directional derivative of $f(x, y)$ when both are in the same direction **u**.

Solution. We have

$$D(Df \cdot \mathbf{u}) \cdot \mathbf{u} = \frac{\partial}{\partial x}(Df \cdot \mathbf{u})u_1 + \frac{\partial}{\partial y}(Df \cdot \mathbf{u})u_2$$

$$= \frac{\partial}{\partial x}\left\{\frac{\partial f}{\partial x}u_1 + \frac{\partial f}{\partial y}u_2\right\}u_1 + \frac{\partial}{\partial y}\left\{\frac{\partial f}{\partial x}u_1 + \frac{\partial f}{\partial y}u_2\right\}u_2$$

$$= \frac{\partial^2 f}{\partial x^2}u_1^2 + 2\frac{\partial^2 f}{\partial x \partial y}u_1 u_2 + \frac{\partial^2 f}{\partial^2 y}u_2^2,$$

$$= (u_1, u_2)D^2 f\begin{bmatrix} u_1 \\ u_2 \end{bmatrix}.$$

This result will be useful to us when we study convex functions.

14.4 Exercises

1. A function $f(x_1, \ldots, x_n)$ is said to be homogeneous of degree r if

$$f(kx_1, \ldots, kx_n) = k^r f(x_1, \ldots, x_n).$$

Show that in this case the partial derivatives $\partial f/\partial x_i$ are homogeneous of degree $r - 1$.

2. Show that $f(x, y)$ is homogeneous of degree r if and only if it is of the form $x^r F(y/x)$ for some function $F(t)$.

3. Derive Euler's Equation for a differentiable homogeneous function of degree r, $f(x, y)$, viz.

$$rf(x, y) = x\frac{\partial f}{\partial x} + y\frac{\partial f}{\partial y}.$$

Conclude that if r is an integer and f is r-fold differentiable, then

$$r! f(x, y) = \sum_{t=0}^{r} \binom{r}{t} x^{r-t} y^t \frac{\partial^r f}{\partial x^{r-t} \partial y^t}.$$

How do both equations generalise to n variables?
[Remark. If a function f satisfies Euler's Equation then it is necessarily homogeneous of degree r. See Exercises 20.4, question 7(a).]

4. Suppose the function $f(x, y, u, v)$ is homogeneous in x, y of degree r and is homogeneous in u, v of degree s i.e.

$$f(kx, ky, u, v) = k^r f(x, y, u, v),$$
$$f(x, y, ku, kv) = k^s f(x, y, u, v).$$

Show that

$$(r - s)f = \left(u\frac{\partial}{\partial u} + v\frac{\partial}{\partial v}\right)\left(x\frac{\partial f}{\partial u} + y\frac{\partial f}{\partial v}\right)$$

$$- \left(x\frac{\partial}{\partial u} + y\frac{\partial}{\partial v}\right)\left(u\frac{\partial f}{\partial x} + v\frac{\partial f}{\partial y}\right).$$

15

Convex functions

15.1 Convex and concave functions

Let X be a convex subset of $\mathbb{R}^n$ and let $y = f(\mathbf{x})$ be a real-valued function defined on X. We say that f is *convex* on X if the *epigraph*, i.e. the set of those points $(\mathbf{x}, y)$ in $\mathbb{R}^{n+1}$ with

$$y \geqslant f(\mathbf{x})$$

for $\mathbf{x}$ in the set X, is convex. The function $y = f(\mathbf{x})$ is *concave* if $y = -f(\mathbf{x})$ is convex. See Figure 15.1.

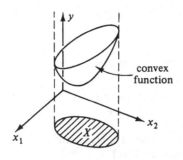
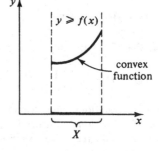
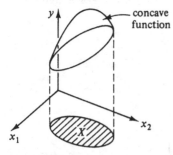
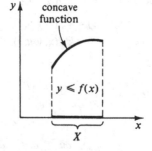

Fig. 15.1

184

It is obvious that a function $y = f(\mathbf{x})$ is convex if and only if all *chords* drawn through two points of the graph lie above the graph. Similarly, chords to a concave function lie below the graph. See Figure 15.2.

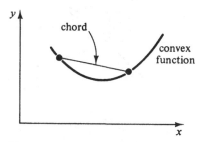

chord

convex function

Fig. 15.2

This geometric idea can be put into analytic language and yields the useful fact that a function f is convex on the set X if and only if

$$f(\alpha\mathbf{x}_1 + \beta\mathbf{x}_2) \leqslant \alpha f(\mathbf{x}_1) + \beta f(\mathbf{x}_2)$$

for each $\alpha \geqslant 0$ and each $\beta \geqslant 0$ satisfying $\alpha + \beta = 1$ and for each $\mathbf{x}_1$ and $\mathbf{x}_2$ in the set X. See Figure 15.3.

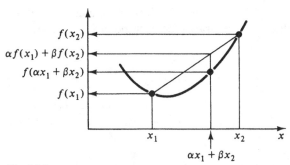

$f(x_2)$

$\alpha f(x_1) + \beta f(x_2)$

$f(\alpha x_1 + \beta x_2)$

$f(x_1)$

x_1 x_2 x

$\alpha x_1 + \beta x_2$

Fig. 15.3

For concave functions the inequality is replaced by

$$f(\alpha\mathbf{x}_1 + \beta\mathbf{x}_2) \geqslant \alpha f(\mathbf{x}_1) + \beta f(\mathbf{x}_2).$$

15.2 Examples

The functions (i) $y = x^2$ (ii) $y = e^x$ and (iii) $y = 3x_1^2 + 4x_2^2$ are all convex. The functions (iv) $y = -(x-1)^2$ and (v) $y = \log x \, (x > 0)$ are *concave*. See Figure 15.4.

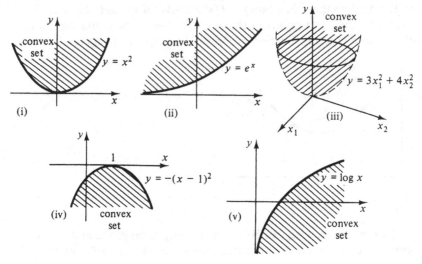

Fig. 15.4

15.3   Differentiable convex functions

If the function $y = f(x)$ is differentiable at each point of a convex set X, then a tangent hyperplane can be drawn for each point ξ in the set X (cf. Figure 15.5).

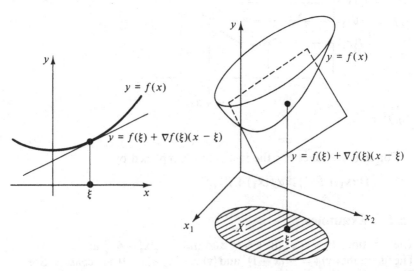

Fig. 15.5

For the function to be *convex* it is clear that each such tangent hyperplane must be a *supporting hyperplane* to the set of points above the graph. The condition

$$f(\mathbf{x}) - f(\xi) \geqslant \nabla f(\xi)(\mathbf{x} - \xi)$$

for each $\mathbf{x}$ and ξ in the set X is therefore necessary and sufficient that f be convex on X.

Now suppose that $y = f(\mathbf{x})$ is twice differentiable on X. Let $\mathbf{a}$ and $\mathbf{b}$ lie in X and let $\mathbf{u}$ be a unit vector pointing from $\mathbf{a}$ to $\mathbf{b}$. Write $\mathbf{b} = \mathbf{a} + t\mathbf{u}$. From the condition above we have

$$\nabla f(\mathbf{a})\mathbf{u} \leqslant \frac{f(\mathbf{b}) - f(\mathbf{a})}{t} \leqslant \nabla f(\mathbf{b})\mathbf{u}$$

since

$$f(\mathbf{b}) - f(\mathbf{a}) \geqslant \nabla f(\mathbf{a}) \cdot \mathbf{u}t$$

and

$$f(\mathbf{a}) - f(\mathbf{b}) \geqslant - \nabla f(\mathbf{b}) \cdot \mathbf{u}t.$$

See Figure 15.6.

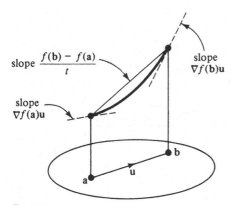

slope $\dfrac{f(\mathbf{b}) - f(\mathbf{a})}{t}$

slope $\nabla f(\mathbf{b})\mathbf{u}$

slope $\nabla f(\mathbf{a})\mathbf{u}$

Fig. 15.6

Thus $\nabla f(\mathbf{x})\mathbf{u} = Df(\mathbf{x})\mathbf{u}$ *increases* as $\mathbf{x}$ moves in the direction of $\mathbf{u}$. Its directional derivative in this direction must therefore be non-negative, i.e.

$$D\{Df(\mathbf{x})\mathbf{u}\}\mathbf{u} = \mathbf{u}^t D^2 f(\mathbf{x})\mathbf{u} \geqslant 0.$$

This is true for *all* unit vectors $\mathbf{u}$. Hence f is convex on X if and only if the matrix

$$D^2 f(\mathbf{x})$$

is *non-negative definite* for each $\mathbf{x}$ in X.

15.4   Examples

(i) Consider the function

$$y = \exp(18x^3),$$

we have

$$D\{\exp(18x^3)\} = 18 \cdot 3 \cdot x^2 \exp(18x^3)$$
$$D^2\{\exp(18x^3)\} = 18 \cdot 6 \cdot x \exp(18x^3) + 18 \cdot 3 \cdot x^2 18 \cdot 3x^2 \exp(18x^3)$$
$$= 18 \cdot 3 \exp(18x^3)(2x + 18 \cdot 3x^4).$$

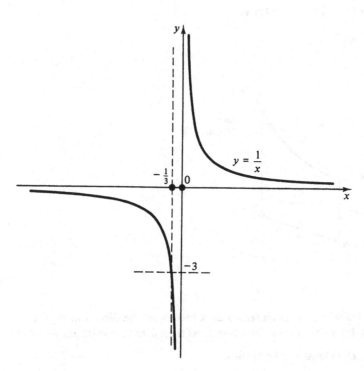

Fig. 15.7

The condition for convexity is

$$2x + 18 \cdot 3x^4 \geqslant 0$$

$$\frac{1}{x^3} + 27 \geqslant 0$$

$$\frac{1}{x} \geqslant -3.$$

Outside the region $-1/3 < x < 0$ the function is therefore convex (cf. Figure 15.7). Inside this region $D^2\{\exp(18x^3)\} \leqslant 0$ and so the function is concave. See Figure 15.8.

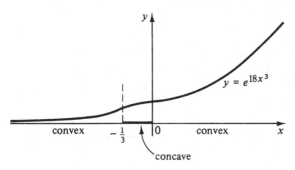

Fig. 15.8

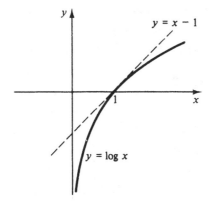

Fig. 15.9

(ii)   Consider the function $y = \log x$ $(x > 0)$.

$$D\{\log x\} = 1/x$$

$$D^2\{\log x\} = -\frac{1}{x^2} \qquad (x > 0)$$

Thus $y = \log x$ is concave for $x > 0$. The tangent line when $x = 1$ is given by $y = (x - 1)$. Note that

$$\log x \leqslant x - 1$$

and thus $y = (x - 1)$ is a supporting hyperplane to the set of those (x, y) satisfying $y \leqslant \log x$ and $x > 0$. See Figure 15.9.

(iii)   Consider the function $y = f(x_1, x_2) = x_1^3 + x_1 x_2 + x_2^2$. We have

$$\frac{\partial f}{\partial x_1} = 3x_1^2 + x_2 \qquad \frac{\partial f}{\partial x_2} = x_1 + 2x_2$$

$$\begin{bmatrix} \dfrac{\partial^2 f}{\partial x_1^2} & \dfrac{\partial^2 f}{\partial x_1 \partial x_2} \\[2mm] \dfrac{\partial^2 f}{\partial x_1 \partial x_2} & \dfrac{\partial^2 f}{\partial x_2^2} \end{bmatrix} = \begin{bmatrix} 6x_1 & 1 \\ 1 & 2 \end{bmatrix}$$

and the determinant of this matrix is $12x_1 - 1$. For $D^2 f$ to be non-negative definite or non-positive definite we need (cf. Section 5.6) that $12x_1 - 1 \geqslant 0$, i.e.

$$x_1 \geqslant \tfrac{1}{12}.$$

Consequently $6x_1 > 0$ and hence the function is convex in this region, but it is neither convex nor concave elsewhere.

(iv)   For what values of p and q is the function

$$y = \frac{1}{x_1^p} + \frac{1}{x_2^q}$$

(a) convex, (b) concave in the region $x_1 > 0$, $x_2 > 0$?
We have

$$\frac{\partial f}{\partial x_1} = -px_1^{-p-1}, \quad \frac{\partial f}{\partial x_2} = -x_2^{-q-1}$$

$$\begin{bmatrix} \dfrac{\partial^2 f}{\partial x_1^2} & \dfrac{\partial^2 f}{\partial x_1 \partial x_2} \\[2mm] \dfrac{\partial^2 f}{\partial x_1 \partial x_2} & \dfrac{\partial^2 f}{\partial x_2^2} \end{bmatrix} = \begin{bmatrix} p(p+1)x_1^{-p-2} & 0 \\ 0 & q(q+1)x_2^{-q-2} \end{bmatrix}$$

For the function to be convex *or* concave we need

$$\det D^2 f = pq(p + 1)(q + 1)x_1^{-p-2}x_2^{-q-2} \geq 0,$$

i.e.

$$pq(p + 1)(q + 1) \geq 0.$$

Given that this holds the function is convex if $p(p + 1) \geq 0$. The condition for convexity is therefore

$$\left.\begin{array}{l} p(p + 1) \geq 0 \\ \\ q(q + 1) \geq 0 \end{array}\right\} \quad \text{i.e.} \quad \begin{array}{c} (p \geq 0 \quad \text{or} \quad p \leq -1) \\ \text{and} \\ (q \geq 0 \quad \text{or} \quad q \leq -1). \end{array}$$

The condition for concavity is that $pq(p + 1)(q + 1) \geq 0$ and $p(p + 1) \leq 0$ i.e.

$$\left.\begin{array}{l} p(p + 1) \leq 0 \\ \\ q(q + 1) \leq 0 \end{array}\right\} \quad \text{i.e.} \quad \begin{array}{c} (-1 \leq p \leq 0) \\ \text{and} \\ (-1 \leq q \leq 0). \end{array}$$

15.5 Exercises

1. Explain why a function cannot be convex on the set X unless X is a convex set.
2. Where are the following functions (a) convex, (b) concave?

    (i) $y = e^{\sqrt{x}}$ $(x \geq 0)$ (ii) $y = \log(1 + x^2)$ (x real)

    (iii) $\dfrac{1}{1 + x^2}$ (x real)   (iv) $y = x^3$ (x real)

    (v) $f(x, y, z) = x^2 + 2y^2 + 3z^2$ $((x, y, z) \text{ in } \mathbb{R}^3)$

    (vi) $f(x, y, z) = -x^2 - 2y^2 - 3z^2 + 7x + 4y$ $((x, y, z) \text{ in } \mathbb{R}^3)$

    (vii) $f(x, y) = 3x^2 + 2xy + 4y^2 + x + 3y + 5$ $((x, y) \text{ in } \mathbb{R}^2)$

    (viii) $f(x, y) = \log(xy)$ $(x > 0 \quad \text{and} \quad y > 0)$

3. Let

$$f(x, y) = \exp\{-x^2 - y^2\}.$$

Calculate $D^2 f$ and find the region X in which $\det D^2 f \geq 0$. Also calculate f_{xx} and f_{yy} and find the regions in which these are non-negative. Prove that f is concave on X.

    Find the equation of the supporting hyperplane at the point

$$(\tfrac{1}{3}, \tfrac{1}{2}, \exp(-13/36))$$

to the convex set defined by

$$z \leq \exp\{-x^2 - y^2\}$$

where (x, y) lies in X.
4. An affine function is given by $f(\mathbf{x}) = A\mathbf{x} + \mathbf{b}$. Prove that it is both convex and concave.

5. Under what conditions is the real-valued function defined on $\mathbb{R}^n$ by the formula

$$f(\mathbf{x}) = \mathbf{x}^t A \mathbf{x}$$

(where A is symmetric) (a) convex, (b) concave?

6. A function $y = f(\mathbf{x})$ is differentiable and concave on $\mathbb{R}^n$. If

$$\nabla f(\boldsymbol{\xi}) = 0$$

explain why $f(\mathbf{x}) \leqslant f(\boldsymbol{\xi})$ for all $\mathbf{x}$ in $\mathbb{R}^n$.

7. Let $y = f(\mathbf{x})$ be convex on a subset X of $\mathbb{R}^n$. Justify *Jensen's Inequality* which asserts that

$$f(\alpha_1 \mathbf{x}_1 + \alpha_2 \mathbf{x}_2 + \cdots + \alpha_n \mathbf{x}_n) \leqslant \alpha_1 f(\mathbf{x}_1) + \alpha_2 f(\mathbf{x}) + \cdots + \alpha_n f(\mathbf{x}_n), \text{ provided}$$
$\alpha_1 \geqslant 0, \alpha_2 \geqslant 0, \ldots, \alpha_n \geqslant 0, \alpha_1 + \alpha_2 + \cdots + \alpha_n + 1$ and $x_1, x_2, \ldots, x_n$ all lie in X.
[Hint: See Section 9.4]

15.6   Further exercises

1. Show that if $g(\mathbf{x})$ and $h(\mathbf{x})$ are concave functions defined on a convex set $C \subseteq \mathbb{R}^n$ then the function

$$m(\mathbf{x}) = \min \{g(\mathbf{x}), h(\mathbf{x})\}$$

is concave. Is $M(\mathbf{x})$ defined by

$$M(\mathbf{x}) = \max \{g(\mathbf{x}), h(\mathbf{x})\}$$

concave?

2. Show that $1/t$ is convex for $t > 0$ and deduce that if $f(\mathbf{x})$ is a concave function defined on $\mathbb{R}$ with $f(\mathbf{x}) > 0$ for all $\mathbf{x}$ then $1/f(\mathbf{x})$ is convex.

3. Show that if $f(\mathbf{x})$ is a convex function defined on a convex set C then for $0 < p < q < r$

$$\frac{f(\mathbf{x} + q\mathbf{u}) - f(\mathbf{x} + p\mathbf{u})}{q - p} \geqslant \frac{f(\mathbf{x} + p\mathbf{u}) - f(\mathbf{x})}{p}$$

and

$$\frac{f(\mathbf{x} + r\mathbf{u}) - f(\mathbf{x} + q\mathbf{u})}{r - q} \geqslant \frac{f(\mathbf{x} + p\mathbf{u}) - f(\mathbf{x})}{p}.$$

4. If C is a closed convex set in $\mathbb{R}^3$ show that the set

$$D = \{(x, y)^t : (x, y, z)^t \in C \text{ for some } z\}$$

is convex and that the function g defined by
$$g(x, y) = \min \{z : (x, y, z)^t \in C\}$$

for $(x, y)^t \in D$ is a convex function.

5. If $f(\mathbf{x})$ and $g(\mathbf{x})$ are convex functions defined on C show that $\alpha f(\mathbf{x}) + \beta g(\mathbf{x})$ is convex for $\alpha, \beta \geqslant 0$.

6. If $f(\mathbf{x})$ is a convex function on C and $\phi : \mathbb{R} \to \mathbb{R}$ is increasing and convex show that $\phi(f(\mathbf{x}))$ is convex.

7. Let C be a closed, bounded, non-empty convex set; show that the

support function

$$f(\mathbf{x}) = \max\{\mathbf{v}^t\mathbf{x} : \mathbf{v} \in C\}$$

is convex. Give a geometric interpretation of this function.

8. The Minkowski functional of a closed, bounded, non-empty convex set is defined by

$$p(\mathbf{x}) = \{r \geqslant 0 : \mathbf{x} \in rC\}.$$

Show that the function is convex. Compute the value of p when C is the elliptical disc

$$\frac{x^2}{a^2} + \frac{y^2}{b^2} \leqslant 1.$$

9. If C is a closed, non-empty, convex set its distance function is defined by

$$d(\mathbf{x}) = \min\{\|\mathbf{x} - \mathbf{c}\| : \mathbf{c} \in C\}.$$

Show that d is convex. (Note its occurrence in the proof of the separating hyperplane theorem.)

10. Use the convexity of the function e^t and Jensen's inequality to prove that, for $x_1, \ldots, x_n \geqslant 0$ and $\alpha_1, \alpha_2, \ldots, \alpha_n \geqslant 0$ with $\alpha_1 + \alpha_2 + \cdots + \alpha_n = 1$ it is the case that

$$x_1^{\alpha_1} x_2^{\alpha_2} \cdots x_n^{\alpha_n} \leqslant \alpha_1 x_1 + \cdots + \alpha_n x_n.$$

11. Using the result of the last question show that for $p > 1$

$$\sum_1^n |x_i y_i| \leqslant \left[\sum_1^n |x_i|^p\right]^{1/p} \cdot \left[\sum_1^n |y_i|^q\right]^{1/q},$$

where

$$\frac{1}{p} + \frac{1}{q} = 1.$$

(Hint: Replace x_i by x_i/S where S denotes the sum of the pth powers.) This result is known as Hölder's inequality.

12. Deduce from Hölder's inequality the following inequality due to Minkowski

$$\left[\sum_1^n (|x_i| + |y_i|)^p\right]^{1/p} \leqslant \left[\sum_1^n |x_i|^p\right]^{1/p} + \left[\sum_1^n |y_i|^p\right]^{1/p}.$$

What does this have to do with the triangle inequality?

13. A real-valued function $f(\mathbf{x})$ defined on a convex subset C of $\mathbb{R}^n$ is called *quasiconvex* if for any $\mathbf{u}$ and $\mathbf{v}$ in C we have

$$f(\alpha\mathbf{u} + (1 - \alpha)\mathbf{v}) \leqslant \max\{f(\mathbf{u}), f(\mathbf{v})\} \tag{+}$$

whenever $0 \leqslant \alpha \leqslant 1$. Show that if f is convex then it is also quasiconvex. Show that f is quasiconvex if and only if each of the level sets

$$\{\mathbf{x} : f(\mathbf{x}) \leqslant \lambda\}$$

is convex. Show that an increasing function is quasiconvex and give an example of a quasiconvex function which is not convex.

14. State and prove a generalization of the Jensen inequality appropriate to quasiconvex functions. If $C = \text{conv} \{x_1, \ldots, x_n\}$ and f is a quasiconvex function defined on C show that f has a maximum over C at one of the points $x_1, \ldots, x_n$. Deduce that the maximum occurs at an extreme point of C.

15. If a real-valued function f defined on a convex set satisfies $(+)$ with strict inequality it is said to be strongly quasiconvex. Show that if a strongly quasiconvex function f defined on a convex set C has a local minimum at a point of C then it has a global minimum at the point.

Remark

A pseudoconvex function is a real-valued function f defined on a convex set $C \subseteq \mathbb{R}^n$ satisfying $f(x) \geqslant f(\xi)$ whenever $\nabla f(\xi) \cdot (x - \xi) \geqslant 0$. Thus every differentiable convex function is pseudoconvex. Every pseudoconvex function is in fact quasiconvex.

16. If f and g are convex functions defined on $\mathbb{R}^n$ let $f * g$ denote the function defined by

$$f * g(x) = \inf \{ f(u) + g(v) : u + v = x \}.$$

Show that $f * g$ is convex.

17. If $g : \mathbb{R}^n \to \mathbb{R}$ is a convex function show that

$$h(x, y) = g(x) + g(y)$$

is a convex function. Show also that a solution of the problem

$$\min h(x, y)$$

subject to

$$x + y = 2a$$

is provided by

$$x = y = a.$$

18. Show that in the special case of a function of two variables, $f(x, y)$ is convex if and only if

(i) $\partial^2 f / \partial x^2 \geqslant 0$, and
(ii) $\det (D^2 f) \geqslant 0$.

Deduce that the following two functions are concave

(a) $f(x, y) = x^\alpha y^\beta$, where $0 \leqslant \alpha, \beta, 0 \leqslant \alpha + \beta \leqslant 1$.
(b) $f(x, y) = (\alpha x^\gamma + \beta y^\gamma)^{1/\gamma}$, where $0 \leqslant \alpha, \beta, \alpha + \beta = 1$ and $\gamma \leqslant 1$.

16

Non-linear programming

16.1 Geometric considerations

Consider the problem of evaluating

$$M = f(\mathbf{x}) = \max_{\mathbf{x} \in S} f(\mathbf{x})$$

where S is a closed and bounded set in $\mathbb{R}^n$ and f is differentiable.

If the maximum is attained at an interior point as indicated in Figure 16.1 there is not much more to say about the optimum

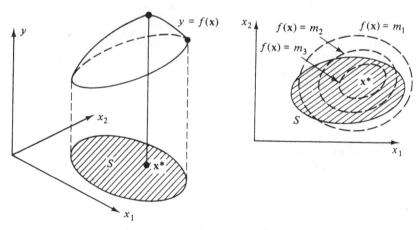

Fig. 16.1. $m_1 < m_2 < m_3 < M$

point $\mathbf{x}^*$ than that

$$\nabla f(\mathbf{x}^*) = 0.$$

What can be said if the optimum point $\mathbf{x}^*$ lies on the boundary of the set S? Consider the illustrations in Figure 16.2 and 16.3.

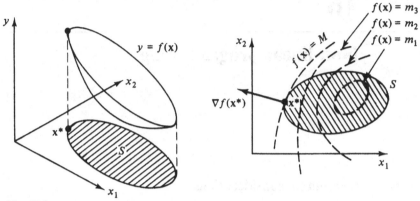

Fig. 16.2

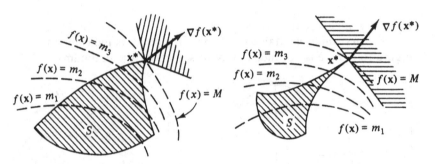

Fig. 16.3

The important point to note is that, in each case, the vector
$\nabla f(\mathbf{x}^*)$ points into the *normal cone* of the set S at the point $\mathbf{x}^*$.

16.2   Constraints

In practice, the set S is often specified by a number of
constraints – i.e. S is the set of all $\mathbf{x}$ for which the constraints

$$\left.\begin{array}{l} g_1(\mathbf{x}) \geqslant 0 \\ g_2(\mathbf{x}) \geqslant 0 \\ \vdots \\ g_k(\mathbf{x}) \geqslant 0 \end{array}\right\}$$

are all satisfied.

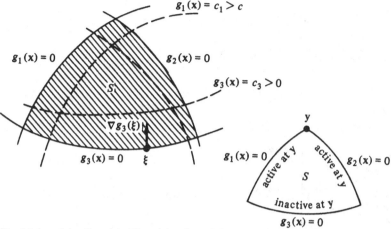

Fig. 16.4. $g_1(y) = 0$, $g_2(y) = 0$, $g_3(y) > 0$

A constraint g_i is said to be *active* at a point y in S (Figure 16.4) if and only if

$$g_i(\mathbf{y}) = 0.$$

Let y be a point in the set S at which the active constraints are $g_1, g_2, \ldots, g_i$. Consider a path which leaves y and enters the set S. Let u be the unit vector which is tangent to the path at y. As x leaves y in the direction u, the functions $g_1(\mathbf{x}), \ldots, g_i(\mathbf{x})$ cannot decrease. Hence we must have that

$$\left.\begin{aligned} \nabla g_1(\mathbf{y})\mathbf{u} &\geq 0 \\ \nabla g_2(\mathbf{y})\mathbf{u} &\geq 0 \\ &\vdots \\ \nabla g_j(\mathbf{y})\mathbf{u} &\geq 0 \end{aligned}\right\} \tag{1}$$

(since these are the directional derivatives in the direction of u).

If it is true that conditions (1) imply that the unit vector u is tangent to a path which enters S, then we say that the *constraint qualification* is satisfied (cf. Figure 16.5).

The constraint qualification only fails to hold in exceptional circumstances like that indicated in Figure 16.6. In particular, the constraint qualification is always satisfied if the gradient vectors

$$\nabla g_1(\mathbf{y}), \nabla g_2(\mathbf{y}), \ldots, \nabla g_k(\mathbf{y})$$

are *linearly independent*.

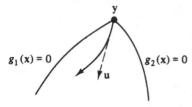

Fig. 16.5

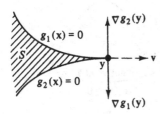

Fig. 16.6. $\nabla g_1(y)v = 0$, $\nabla g_2(y)v = 0$. *Observe that* **v** *does not point into S*

16.3 Kuhn–Tucker theorem

The general non-linear programming problem is to find $\mathbf{x}^*$ such that

$$f(\mathbf{x}^*) = M = \max f(\mathbf{x}) \tag{2}$$

where $\mathbf{x}$ is subject to the constraints

$$\left.\begin{array}{l} g_1(\mathbf{x}) \geqslant 0 \\ g_2(\mathbf{x}) \geqslant 0 \\ \vdots \\ g_k(\mathbf{x}) \geqslant 0 \end{array}\right\}. \tag{3}$$

An $\mathbf{x}$ satisfying conditions (3) is said to be *feasible*. A feasible $\mathbf{x}^*$ satisfying condition (2) is said to be *optimal* (cf. Chapter 12).

We shall always suppose that our functions are all differentiable and that the constraint qualification holds.

Kuhn–Tucker theorem

Suppose that $\mathbf{x}^*$ *is optimal for the general non-linear programming problem above. Then the following conditions hold:*

(i) $\mathbf{x}^*$ *is feasible,*

(ii) *there exist* $\lambda_1 \geqslant 0, \lambda_2 \geqslant 0, \ldots, \lambda_k \geqslant 0$ *such that*

$$\lambda_1 g_1(\mathbf{x}^*) = \lambda_2 g_2(\mathbf{x}^*) = \cdots = \lambda_k g_k(\mathbf{x}^*) = 0,$$

(iii) $\nabla f(\mathbf{x}^*) + \lambda_1 \nabla g_1(\mathbf{x}^*) + \cdots + \lambda_k \nabla g_k(\mathbf{x}^*) = 0.$

The conditions above are known as the *Kuhn–Tucker conditions*. Of these (i) is trivial and (ii) simply says that $\lambda_i = 0$ unless the constraint g_i is active at $\mathbf{x}^*$. Given this information (iii) may then be interpreted as the assertion that $\nabla f(\mathbf{x}^*)$ points into the normal cone at $\mathbf{x}^*$ (cf. Figure 16.7).

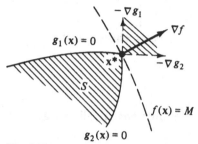

Fig. 16.7

This fact evidently depends on the directions of the various gradients and so ultimately on the form of (2) and (3). In this respect the mnemonic "Maximize your objective and think positive" may be helpful.

Proof of the Kuhn–Tucker theorem. Let $\mathbf{x}^*$ be optimal and let the active constraints at $\mathbf{x}^*$ be $g_1, g_2, \ldots, g_j$. Suppose that

$$\left.\begin{array}{l} \nabla g_1(\mathbf{x}^*)\mathbf{u} \geqslant 0 \\ \nabla g_2(\mathbf{x}^*)\mathbf{u} \geqslant 0 \\ \vdots \\ \nabla g_j(\mathbf{x}^*)\mathbf{u} \geqslant 0 \end{array}\right\} \quad \text{i.e.} \quad \begin{bmatrix} \text{---} & \nabla g_1 & \text{---} \\ \text{---} & \nabla g_2 & \text{---} \\ & \vdots & \\ \text{---} & \nabla g_j & \text{---} \end{bmatrix}\mathbf{u} \geqslant 0 \cdot \tag{4}$$

Assuming the constraint qualification, the vector $\mathbf{u}$ is tangent to a path which leaves $\mathbf{x}^*$ and enters the set S. Since $f(\mathbf{x})$ is largest when $\mathbf{x} = \mathbf{x}^*$, it follows that

$$\nabla f(\mathbf{x}^*)\mathbf{u} \leqslant 0,$$

for all $\mathbf{u}$ satisfying (4) – i.e. for no $\mathbf{u}$ satisfying (4) is it true that

$$-\nabla f(\mathbf{x}^*)\mathbf{u} < 0. \tag{5}$$

We now appeal to the Farkas lemma with $A = [\nabla g_1^t, \ldots, \nabla g_j^t]$ and $\mathbf{b}^t = -\nabla f(\mathbf{x}^*)$. This asserts that one and only one of the systems

$$\left.\begin{array}{r} A\boldsymbol{\lambda} = \mathbf{b} \\ \boldsymbol{\lambda} \geqslant \mathbf{0} \end{array}\right\}, \qquad\qquad \left.\begin{array}{r} A'\mathbf{u} \geqslant \mathbf{0} \\ \mathbf{b}'\mathbf{u} < \mathbf{0} \end{array}\right\}, \tag{6}$$

has a solution. But we have seen that (4) and (5) are contradictory, i.e. the second system of (6) is insoluble. We deduce from the first the existence of $\lambda_1 \geqslant 0, \lambda_2 \geqslant 0, \ldots, \lambda_j \geqslant 0$ such that

$$
\begin{bmatrix} \vert & \vert & & \vert \\ \nabla g_1 & \nabla g_2 & \cdots & \nabla g_j \\ \vert & \vert & & \vert \end{bmatrix} \begin{bmatrix} \lambda_1 \\ \lambda_2 \\ \lambda_j \end{bmatrix} = - \begin{bmatrix} \vert \\ \nabla f \\ \vert \end{bmatrix}. \tag{7}
$$

This completes the proof ($\lambda_{j+1}, \lambda_{j+2}, \ldots, \lambda_k$ being chosen so that $\lambda_{j+1} = \lambda_{j+2} = \cdots = \lambda_k = 0$).

16.4   Concave programming

A *concave programme* is one in which the functions are all concave.

Theorem

*For a **concave** programme, the Kuhn–Tucker conditions imply that* $\mathbf{x}^*$ *is optimal.*

Proof. Since the functions $g_1, g_2, \ldots, g_k$ are concave, the set S of feasible $\mathbf{x}$ is convex (why?). Given any $\mathbf{x}$ in S, the line segment joining $\mathbf{x}$ and $\mathbf{x}^*$

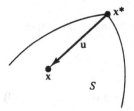

Fig. 16.8

therefore lies in the set S (cf. Figure 16.8). Put $\mathbf{u} = \mathbf{x} - \mathbf{x}^*$. Then

$$
\left. \begin{aligned} \nabla g_1(\mathbf{x}^*)\mathbf{u} &\geqslant 0 \\ \nabla g_2(\mathbf{x}^*)\mathbf{u} &\geqslant 0 \\ &\vdots \\ \nabla g_j(\mathbf{x}^*)\mathbf{u} &\geqslant 0 \end{aligned} \right\} \quad \text{i.e.} \quad \begin{bmatrix} \underline{\quad} & \nabla g_1 & \underline{\quad} \\ \underline{\quad} & \nabla g_2 & \underline{\quad} \\ & \vdots & \\ \underline{\quad} & \nabla g_j & \underline{\quad} \end{bmatrix} \mathbf{u} \geqslant 0
$$

where $g_1, g_2, \ldots, g_j$ are the active constraints at $\mathbf{x}^*$, i.e. (4) above holds. The Kuhn–Tucker conditions (7) (by way of the Farkas lemma (6)) therefore imply that (4) and (5) are contradictory. Hence

$$
\nabla f(\mathbf{x}^*)\mathbf{u} \leqslant 0.
$$

But f is *concave*, consequently

$$f(\mathbf{x}) - f(\mathbf{x}^*) \leqslant \nabla f(\mathbf{x}^*)\mathbf{u} = \nabla f(\mathbf{x}^*)(\mathbf{x} - \mathbf{x}^*) \leqslant 0.$$

Thus $f(\mathbf{x}) \leqslant f(\mathbf{x}^*)$ for each feasible $\mathbf{x}$.

16.5   Notes on the Kuhn–Tucker conditions

(i) Just as with linear programming, a duality theory can be developed for the general non-linear problem. (In the case of concave programmes the analogy is very close.) In particular the numbers $\lambda_1, \lambda_2, \ldots, \lambda_k$ can be interpreted as *shadow prices* once the problem has been suitably formulated.

(ii) The 'Lagrange multipliers theorem' asserts that if $\mathbf{x}^*$ is optimal for the programme:

$$\text{maximise } f(\mathbf{x})$$

subject to the constraints

$$\left.\begin{aligned} g_1(\mathbf{x}) &= 0 \\ g_2(\mathbf{x}) &= 0 \\ &\vdots \\ g_k(\mathbf{x}) &= 0 \end{aligned}\right\},$$

then there exist real numbers $\lambda_1, \lambda_2, \ldots, \lambda_k$ such that

$$\nabla f(\mathbf{x}^*) + \lambda_1 \nabla g_1(\mathbf{x}^*) + \cdots + \lambda_k \nabla g_k(\mathbf{x}^*) = 0. \tag{8}$$

Formal manipulation of the Kuhn–Tucker conditions for the given system of equation constraints rewritten as a system of inequalities, viz. $g_1(\mathbf{x}) \geqslant 0$, $-g_1(\mathbf{x}) \geqslant 0, \ldots$, etc., gives the equation (8). (Why?) Unfortunately these rewritten constraints do not satisfy the constraint qualification (Section 16.2). Nonetheless by analogy the numbers in the Kuhn–Tucker conditions are also called *Lagrange multipliers*.

(iii) Note that we did not need to assume the constraint qualification in proving our theorem on concave programming.

(iv) In applications the constraints often take the form

$$\left.\begin{aligned} g_1(\mathbf{x}) &\geqslant 0 \\ g_2(\mathbf{x}) &\geqslant 0 \\ &\vdots \\ g_m(\mathbf{x}) &\geqslant 0 \end{aligned}\right\} \text{ and } \begin{aligned} x_1 &\geqslant 0 \\ x_2 &\geqslant 0 \\ &\vdots \\ x_n &\geqslant 0 \end{aligned}$$

If $\mathbf{x}^*$ is optimal for maximising $f(\mathbf{x})$ subject to these constraints, then there exist $\lambda_1 \geqslant 0, \lambda_2 \geqslant 0, \ldots, \lambda_m \geqslant 0$ and $\mu_1 \geqslant 0, \mu_2 \geqslant 0, \ldots, \mu_n \geqslant 0$ such that

$$\lambda_1 g_1(\mathbf{x}^*) = \cdots = \lambda_m g_m(\mathbf{x}^*) = \mu_1 x_1^* = \mu_2 x_2^* = \cdots = \mu_n x_n^* = \mathbf{0}$$

and

$$\nabla f(\mathbf{x}^*) + \lambda_1 \nabla g_1(\mathbf{x}^*) + \cdots + \lambda_m \nabla g_m(\mathbf{x}^*) + (\mu_1, \mu_2, \ldots, \mu_n) = \mathbf{0}.$$

Note that $\nabla x_1 = (1, 0, \ldots, 0)$. Hence if $\mathbf{x}^*$ is optimal, there exist $\lambda_1 \geqslant 0, \lambda_2 \geqslant 0, \ldots, \lambda_m \geqslant 0$ such that

$$\lambda_1 g_1(\mathbf{x}^*) = \cdots = \lambda_m g_m(\mathbf{x}^*) = 0$$

and

$$\nabla f(\mathbf{x}^*) + \lambda_1 \nabla g_1(\mathbf{x}^*) + \cdots + \lambda_m \nabla g_m(\mathbf{x}^*) \leqslant 0.$$

16.6   A worked example

Find the minimum of

$$z = x^2 + y^2 - 4x$$

subject to

$$x^2 + 4y^2 \leqslant 1,$$
$$x + 2y \geqslant 1.$$

From Figure 16.9 it is clear that the answer is -3 which is attained at

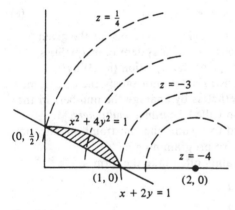

Fig. 16.9

the point $(1, 0)$. We shall, however, arrange the problem as a concave programme and use the Kuhn–Tucker conditions. Consider then:

maximise

$$\{4x - x^2 - y^2\}$$

subject to

$$1 - x^2 - 4y^2 \geqslant 0,$$
$$x + 2y - 1 \geqslant 0.$$

Then (x, y) is optimal, provided that it is feasible and $\lambda \geqslant 0$ and $\mu \geqslant 0$ exist for which

$$\lambda(1 - x^2 - 4y^2) = \mu(x + 2y - 1) = 0 \tag{1}$$

and

$$(4 - 2x, -2y) + \lambda(-2x, -8y) + \mu(1, 2) = 0. \tag{2}$$

Equation (2) yields that

$$\left. \begin{array}{l} 4 - 2x = 2\lambda x - \mu \\ -2y = 8\lambda y - 2\mu \end{array} \right. \text{ i.e. } \left. \begin{array}{l} 4 - 2x = 2\lambda x - \mu \\ -y = 4\lambda y - \mu \end{array} \right\}. \tag{3}$$

Suppose that $\lambda = 0$. From equation (3) we obtain that $y = \mu = -4 + 2x$. But this is inconsistent with the feasibility of (x, y). Hence $\lambda \neq 0$.

Suppose that $\mu = 0$. From equation (3), we have

$$4 - 2x = 2\lambda x$$
$$-y = 4\lambda y \quad \text{i.e.} \quad y = 0 \quad \text{or} \quad \lambda = -\tfrac{1}{4}.$$

If the latter $4 - 2x = 2(-\tfrac{1}{4})x$, and $x = \tfrac{8}{3}$. But $x = \tfrac{8}{3}$ is inconsistent with the feasibility of (x, y) whereas $y = 0$ is consistent only if $(x, y) = (1, 0)$. Thus, either $(x, y) = (1, 0)$ or else $\lambda \neq 0$ and $\mu \neq 0$. But then from equation (1)

$$1 - x^2 - 4y^2 = 0$$
$$x + 2y - 1 = 0$$

and hence $(x, y) = (1, 0)$ or $(x, y) = (0, \tfrac{1}{2})$.

At $(x, y) = (1, 0)$ equations (3) become

$$\left. \begin{array}{l} 2 = 2\lambda - \mu \\ 0 = -\mu \end{array} \right\} \text{ i.e. } \left. \begin{array}{l} \lambda = 1 \geqslant 0 \\ \mu = 0 \geqslant 0 \end{array} \right\}$$

and so $(1, 0)$ is the *optimal point*.

At $(x, y) = (0, \tfrac{1}{2})$, the equations (3) become

$$\left. \begin{array}{l} 4 = -\mu \\ -\tfrac{1}{2} = 2\lambda - \mu \end{array} \right\} \text{ i.e. } \left. \begin{array}{l} \mu = -4 < 0 \\ \lambda = -\tfrac{9}{4} < 0 \end{array} \right\}$$

and the Kuhn–Tucker conditions are not met.

16.7   Exercises

1. Sketch the region in $\mathbb{R}^2$ determined by the constraints

$$\left.\begin{array}{r} 1-x-y\geqslant0 \\ 5-x^2-y^2\geqslant0 \\ x\geqslant0 \end{array}\right\}.$$

Use the Kuhn–Tucker conditions to obtain the condition satisfied by ∇g at the point $\xi=(2,-1)^t$ if the function g has a maximum at ξ subject to the given constraints. Explain the geometric significance of your condition and sketch ∇g in your diagram.

2. Discuss the problem of evaluating

$$\min\{x^{-p}+y^{-q}\}$$

subject to the constraints

$$\left.\begin{array}{r} ax+by\leqslant c \\ 0\leqslant x\leqslant A \\ 0\leqslant y\leqslant B \end{array}\right\}.$$

3. Write down the Kuhn–Tucker conditions for the problem of evaluating

$$\max\{\log(xy)\}$$

subject to the constraints

$$\left.\begin{array}{r} x^2+2y^2\leqslant4 \\ x^2+2y^2\leqslant6x-5 \\ y\leqslant\tfrac{1}{2}x \end{array}\right\}.$$

Hence solve the problem.

4. What is the region determined by the concave constraints

$$\left.\begin{array}{r} 1-x^2-(y-1)^2\geqslant0 \\ y\leqslant0 \end{array}\right\}$$

Comment on this region and the 'constraint qualification'.

16.8   Further exercises

1. Find the maximum of the function

$$f(x,y,z)=x^2+2y^2+3z^2$$

subject to

$$\begin{array}{r} x^2+y^2+z^2=1, \\ x+y+z=0. \end{array}$$

Is your optimum still optimal when the equations are replaced by the symbol $\leqslant$?

2. Repeat question 1 with
$$f = x^2 + y^2 + z^2$$
and with the constraints
$$x^2 + y^2 + 4z^2 = 1,$$
$$x + 3y + 2z = 0.$$

3. Repeat question 1 with
$$f = xyz$$
and the constraints
$$x + y + z = 5,$$
$$xy + yz + zx = 8,$$
but adding $x, y, z \geqslant 0$ for the problem with inequalities.

4. Minimise $x^2 + y^2 + z^2$ subject to
$$xyz = c,$$
$$y = kx,$$
$$x, y, z > 0,$$
$(k \neq 0 \neq c)$.

5. Maximise xy^2z^2 subject to
$$x^3 + y^2 + z = 1,$$
$$x, y, z > 0.$$

6. Minimise $x + y$ subject to
$$\left[\frac{a}{x}\right]^2 + \left[\frac{b}{y}\right]^2 = 1.$$

What does this problem say about the possibility of carrying a ladder round a corner of two corridors, with widths a and b?

7. Write down the Kuhn–Tucker conditions for the problem of maximising
$$x^p y \quad (p > 0)$$
subject to
$$\left.\begin{array}{r} y \leqslant 2x \\ xy \leqslant 2 \\ x \leqslant 2y \\ x, y \geqslant 0 \end{array}\right\}.$$

Deduce from these conditions a condition on p which ensures that the optimal point $(\bar{x}, \bar{y})$ is (i) $(1, 2)$, (ii) $(2, 1)$.

8. Determine the part of the boundary on which the problem

$$\text{Max } x^p y^q$$

subject to

$$x^2 + (y + 1)^2 \leqslant 5$$
$$(x + 2)^2 + y^2 \leqslant 10$$
$$x, y \leqslant 0$$

with $p > 0$, $q > 0$ has a solution. How does the answer depend on the ratio p/q?

9. Use the convexity of

$$x^k + y^k + x^k \quad (k \geqslant 1)$$

and the Kuhn–Tucker theorem to prove that for $k \geqslant 1$

$$\frac{1}{3}(x^k + y^k + z^k) \geqslant \left(\frac{x + y + z}{3}\right)^k.$$

What does Jensen's inequality have to say about this result?

10. Compute the Kuhn–Tucker conditions when the objective function to be maximized is the quadratic

$$\mathbf{c}^t \mathbf{x} + \tfrac{1}{2} \mathbf{x}^t B \mathbf{x}$$

with B a symmetric matrix and the constraints are

$$A \mathbf{x} \leqslant \mathbf{b}.$$

Solve the problem when $c = (-1, 3)^T$, $b = (7, 8)^T$ and

$$A = \begin{bmatrix} 1 & 4 \\ 3 & 2 \end{bmatrix}, \quad B = \begin{bmatrix} -1 & 1 \\ 1 & -1 \end{bmatrix}.$$

11. (a) Find the minimum and maximum of the function

$$x^4 + y^4 + z^4$$

subject to the constraints

$$\left.\begin{array}{c} x^2 + y^2 + z^2 = 1 \\ x + y + z = 1 \end{array}\right\}.$$

(b) Check whether the maximum obtained in part (a) also solves the programme

maximize

$$x^4 + y^4 + z^4$$

subject to

$$\left.\begin{array}{c} x^2 + y^2 + z^2 \leqslant 1 \\ x + y + z \leqslant 1 \end{array}\right\}.$$

II ADVANCED CALCULUS

17

The integration process

17.1 Integration as area

In elementary treatments it is customary to explain the symbol

$$\int f(x)\,dx$$

as an anti-derivative, i.e. denoting any function $F(x)$ such that

$$\frac{d}{dx}F(x) = f(x).$$

For example, we might write $\int x\,dx = (x^2/2)$, since $(d/dx)(x^2/2) = x$. The notation is ambiguous because, of course, $(d/dx)(x^2/2 + c) = x$, where c is any constant. Indeed, quite generally, if $F_1(x)$ and $F_2(x)$ are two functions satisfying for some $f(x)$

$$\frac{d}{dx}F_1(x) = f(x) = \frac{d}{dx}F_2(x),$$

then

$$\frac{d}{dx}(F_1(x) - F_2(x)) = 0.$$

But it is known that if a function $G(x)$ has derivative identically zero, then $G(x)$ is a constant. For this reason, if we are to have a clear conscience, we should write

$$\int f(x)\,dx = F(x) + c.$$

But suppose we are to make sense of the symbol

$$\int \exp(-x^2)\,dx.$$

Now although $\exp(-x^2)$ is an elementary function, in that it is made up from the basic functions log, exp, polynomials, there is no elementary function whose derivative is $\exp(-x^2)$. In this case we are stuck: no 'formula' for $F(x)$ can be found. To get unstuck we have to appeal to the geometric interpretation of an integral as an area under a curve.

Let us briefly recall this. Let $f(x)$ be a real-valued function. In general, the domain of definition of $f(x)$ need not be the whole of $\mathbb{R}$ so we shall assume that $f(x)$ is defined only on some open interval $a < x < b$.

Pick c in this interval and, for x satisfying:

$$\left. \begin{array}{l} a < x < b \\ c < x \end{array} \right\},$$

let

$A(x) = $ area under the curve between the c and x abscissae.

The usual argument now runs as follows. For small enough h

$$A(x + h) - A(x) \approx f(x) \cdot h$$

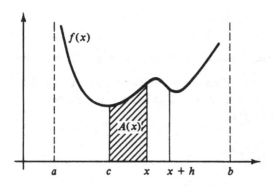

Fig. 17.1

(see Figure 17.1) so that

$$\lim_{h \to 0} \frac{A(x + h) - A(x)}{h} = f(x).$$

We shall be able to justify this formula more precisely at the end of Section 17.5. For the moment we have the plausible results

$$\frac{d}{dx} A(x) = f(x)$$

and

$$A(x) = \int_c^x f(t)\,dt.$$

So the answer to our initial question is to recognise that we should create a new function defined by means of area; we may then denote it by

$$\int_0^x \exp(-t^2)\,dt.$$

This function has been tabulated just like logarithms. Actually, a multiple of the function above is more useful in applications, viz.

$$\frac{2}{\sqrt{\pi}} \int_0^x \exp(-t^2)\,dt$$

and this has been christened the 'error function', denoted erf(x). We shall now turn our attention to the calculation of area, something that our definition above takes for granted.

17.2   A calculation

Find

$$I = \int_0^1 \exp(-t^2)\,dt.$$

The practical problem is to approximate to the answer and to compute, for a start, the area of the shaded region in Figure 17.2. That area is, of

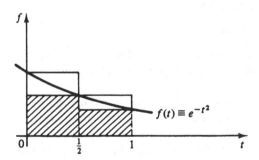

Fig. 17.2

course, an under-estimate. We obtain

$$I_1^- = \tfrac{1}{2} f(\tfrac{1}{2}) + \tfrac{1}{2} f(1).$$

But an approximation is no use, unless we know the extent to which we are in error. So we shall next over-estimate the area by using the larger unshaded rectangles. We now obtain

$$I_1^+ = \tfrac{1}{2}f(\tfrac{1}{2}) + \tfrac{1}{2}f(0).$$

The numbers work out to be $f(0) = 1$, $f(1) = 0.367\,879\,4$, $f(1/2) = 0.778\,800\,7$, so that

$$I_1^- = 0.573\,340\,1 < I < I_2^+ = 0.889\,400\,3.$$

Not a very gratifying result, considering that the two estimates do not agree even in the first place of decimal. We can, of course, use a larger number of narrower rectangles (as in Figure 17.3). To cut a long story

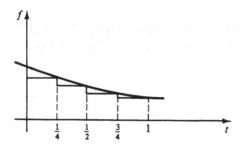

Fig. 17.3

short, it is not until we use a base width of 2^{-5} that we get even the first place of decimal agreeing in both the under-estimate and the over-estimate. The numbers are at that stage (order 5):

$$I_5^- = 0.736\,887\,4 < I < I_5^+ = 0.756\,641\,2.$$

In Sections 17.7 and 17.8 we shall see that, rather than increase the number of rectangles, it would be wiser to opt for an alternative approximation process.

The above calculation begs a very important question. How do we know that the process of under-estimating and over-estimating area will 'eventually' yield 'the right answer'; that is, an answer up to a desired degree of accuracy when enough subdivisions of $[0, 1]$ are made. In the example to hand it would not be very difficult to prove that the under-estimates and over-estimates will agree in as many places of decimal as we wish provided we use narrow enough bases for our rectangles. We need, however, to guarantee this more generally and this is the specific aim of the next two sections.

17.3 The Riemann definition

We start off by giving a formal definition of the definite integral

$$\int_a^b f(x)\,dx,$$

when the function is non-negative (so that its graph stays above the x-axis). We do not stop to consider functions of variable sign; they are dealt with in the obvious way by summing the positive/negative contributions arising from the subintervals where the integrand is positive/negative. The definition builds on the procedure of the last section.

By a *partition* of the interval $[a,b]$ we shall mean a finite set

$$P = \{x_0, x_1, x_2, \ldots, x_n\}$$

where $a = x_0 < x_1 < \cdots < x_n = b$. With its help we can obtain the lower estimate by reference to rectangles below the curve (Figure 17.4). Proceed

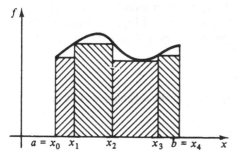

Fig. 17.4

as follows. In each of the subintervals $[x_0 x_1]$, $[x_1, x_2]$, $[x_2, x_3], \ldots,$ $[x_{n-1}, x_n]$ we calculate the minimum value of the function. More precisely: in each subinterval we calculate the maximum of the lower bounds on the function in that subinterval. The reason is that we have to cater for the possibility that a minimum is not attained by the function. See Exercise 17.20 question 6. We use the symbol inf to denote this greatest lower bound; see also Section 10.3. For $1 \leqslant r \leqslant n$, let

$$m_r = \inf\{f(x) : x_{r-1} \leqslant x \leqslant x_r\},$$

then let

$$L(P) = \sum_{r=1}^{n} m_r(x_r - x_{r-1}).$$

It now seems plausible to define the area under the curve as the smallest number greater than all the lower estimates $L(P)$ as P ranges over all partitions P. The symbol sup is used to denote the smallest among these upper bounds (known as the *supremum*). Here again it might well be that there is no 'attained' largest lower estimate for the area. We thus have

$$\int_a^b f(x)dx = \sup_P L(P).$$

Two questions immediately arise: is the number defined here finite, and, would the same number be obtained if we looked at over-estimates? Certainly, if f is bounded, say by M, we have

$$L(P) = \sum_{r=1}^{n} m_r(x_r - x_{r-1}) \leqslant \sum_{r=1}^{n} M(x_r - x_{r-1})$$

$$= M \sum_{r=1}^{n} (x_r - x_{r-1}) = M(b - a),$$

which settles the first question. The upper estimates (over-estimates) would be defined by refering in each subinterval $[x_{r-1}, x_r]$ to the least of the numbers bounding f from above in the subinterval $[x_{r-1}, x_r]$. Thus, if

$$M_r = \sup\{f(x) : x_{r-1} \leqslant x \leqslant x_r\}$$

(where we have again used the sup), the upper estimate would be

$$U(P) = \sum_{r=1}^{n} M_r(x_r - x_{r-1}).$$

The largest number which is below all the over-estimates is denoted by

$$\inf_P U(P).$$

This is also a candidate for the area under the curve. To show that the two are equal, amounts to the same question that we asked at the end of Section 17.2. The answer is 'yes' provided the integrand $f(x)$ is continuous. The next section studies this notion. See also Appendix A.

17.4   Continuity

Consider carefully the behaviour at and close to $x = 0$ of the functions graphed in Figure 17.5.

In (a) and (b) the value of $f(x)$ either side of 0 is very nearly $f(0)$ provided x is close enough to 0. Another way to say this is that the limiting value of $f(x)$ as x approaches 0 from either side is $f(0)$. If this is

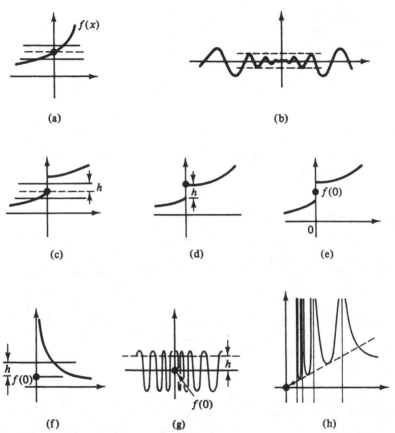

Fig. 17.5. (a) Near x = 0 the graph is almost straight. (b) The oscillations decrease either side of x = 0. (c) A jump discontinuity: values to the right at least h away from f(0). (d) This time the discontinuity is at left. (e) Two-sided jump. (f) Infinite jump. (g) There are x·s arbitrarily close to 0 where f(x) is h away from f(0). (h) Infinitely many infinite jumps.

so, we say that f is continuous at 0.

In all the other cases illustrated we can pick arbitrarily small intervals round 0 in which the function value $f(x)$ is at some x at least a predetermined value h away from $f(0)$. (Compare the illustrations.) The function is discontinuous and is said to have an oscillation of at least h at 0.

The precise definition of 'continuity at x_0' which follows simply says: no matter what hurdle we set the function to jump, it will be unable to jump as high as we ask anywhere in some particular vicinity of x_0.

Definition
    Let $f :(a, b) \to \mathbb{R}$ be a function. Let $a < x_0 < b$. We say that f is
continuous *at x_0 if for every $\varepsilon > 0$ there is a $\delta > 0$ such that for any x satisfying*

$$a < x < b \quad \& \quad x_0 - \delta < x < x_0 + \delta$$

we have

$$|f(x) - f(x_0)| \leqslant \varepsilon.$$

We further say that f is **continuous** *if it is continuous at every point in its domain of definition.*

17.5   What the definition means

Let us explain the definition above carefully and in detail, because the ideas involved here are common to a number of later definitions. The order of events in the definition is crucial. First of all, a positive number is assumed given. This is ε. Its purpose is to give meaning to the words '$f(x)$ is close to $f(x_0)$'. The rest of the definition then says that we are able to *control* the closeness of $f(x)$ to $f(x_0)$ by constraining the position of x through requiring it to lie at most δ either side of x_0.

To illustrate this point concretely imagine the problem of accurately computing π^2. (Here the function we are concerned with is $f(x) = x^2$ and x_0 will be π.) Our problem is that we do not actually *know* the value of π, but we can compute it to any desired degree of accuracy. So, in order to compute π^2 we need to use instead of x_0 an appropriate approximation x. How close must x be to x_0? Evidently, we *first* need to know what is the target accuracy in the answer for $f(x_0)$. If we required 20-figure accuracy we would want our computed answer $f(x)$ to be within at most 10^{-21} away from the actual answer. Writing $\varepsilon = 10^{-21}$ we require to use an x which will guarantee that

$$|f(x) - f(x_0)| \leqslant \varepsilon.$$

The *second* item of information which we require, and this can only be known once ε has been given, is 'how close to x_0 must x be'. The answer to that question must take the form 'to within so and so' and this has to be assessed somehow (see below). Say we assess it to be δ. When we have found this δ we can then assert that:

> Given the target accuracy ε for approximating $f(x_0)$ by $f(x)$ we can find δ so that when x lies to within δ of x_0, the target accuracy is achieved.

Another way of visualising our discussion is this. We are given an idealised calculator which computes x^2 but: (i) it will print the answer only to a finite number of decimal positions as output on a tape and we are allowed to tell the machine how many decimal positions we require; (ii) we can read in x as an input only to a finite number of decimal positions though we are allowed to select to how many. Our problem is to have π^2 accurately. The solution takes us through the following steps.

(1) Specify the target output accuracy (ε); this is equivalent to specifying the number of decimal positions on the output side.

(2) Discover whether it is possible that the actual input π may be replaced by an approximation (truncation of its decimal expansion) without necessarily incurring loss of accuracy at the output side.

(3) Hence compute the input accuracy (δ) required of the approximation to π to guarantee the targetted output accuracy.

The formal definition above for the continuity of f at x_0 says precisely the same thing, only symbolically. Indeed the fact that we can find δ i.e. the knowledge that we can obtain π^2 to within ε, upon using any approximation to π (which itself is to within δ of π) amounts to saying that the function $f(x) = x^2$ is continuous at π. (It is of course also continuous at all other locations for x_0, not just for $x_0 = \pi$.) And that is of course why continuity is so important.

We will now show that such a δ may be found (see Figure 17.6).

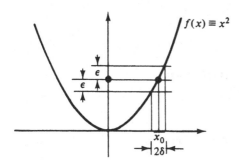

Fig. 17.6

Observe that

$$(x^2 - x_0^2) = (x - x_0)(x + x_0).$$

Suppose that $x_0 > 1$. To begin with we constrain x so that $x_0 - 1 < x < x_0 + 1$ then

$$x + x_0 \leqslant 2x_0 + 1$$

and
$$|(x^2 - x_0^2)| \leqslant |(x - x_0)| \, |(x + x_0)|$$
$$\leqslant |(x - x_0)|(2x_0 + 1).$$

This will be less than ε provided we ensure both that $\delta \cdot (2x_0 + 1) < \varepsilon$ and that $\delta < 1$. (Why?) We leave it to the reader to complete the argument for other values of x_0.

Illustrative use of definitions: anti-differentiation

Armed with the notion of continuity and the Riemann definition of integration we are in a position to give a proper proof (promised in Section 17.1) that differentiation is the opposite of integration. Let $f(t)$ be a continuous function defined on $[a, b]$. Put for x in $[a, b]$

$$F(x) = \int_a^x f(t) dt.$$

Let ξ be in (a, b) and suppose ε is given. By continuity at ξ there is a $\delta > 0$ such that

$$|f(x) - f(\xi)| \leqslant \varepsilon$$

for any x with $|x - \xi| < \delta$. Let $0 < |h| < \delta/2$. First suppose $h > 0$. Now for x in $[\xi, \xi + h]$ we have that

$$f(\xi) - \varepsilon \leqslant f(x) \leqslant f(\xi) + \varepsilon,$$

so, integrating the inequality from ξ to $\xi + h$

$$(f(\xi) - \varepsilon)h \leqslant \int_\xi^{\xi+h} f(t) dt \leqslant (f(\xi) + \varepsilon)h.$$

Thus since $h > 0$

$$f(\xi) - \varepsilon \leqslant \frac{F(\xi + h) - F(\xi)}{h} \leqslant f(\xi) + \varepsilon,$$

or

$$-\varepsilon \leqslant \frac{F(\xi + h) - F(\xi)}{h} - f(\xi) \leqslant + \varepsilon.$$

A similar inequality obtains for $0 > h > -\delta/2$. This implies that the limiting value of

$$\frac{F(\xi + h) - F(\xi)}{h}$$

is $f(\xi)$.

17.6 Which functions are continuous

The practical answer is twofold.

First of all we note that the polynomials, e^x and $\log x$ (the latter for $0 < x < \infty$) are continuous. Furthermore, if $f(x)$ and $g(x)$ are continuous in (a, b) then so also are

$$f(x) + g(x), \quad cf(x), \quad f(x)g(x).$$

So also is $f(x)/g(x)$ provided that in (a, b) $g(x)$ is never zero. But note that $1/x$ is discontinuous in any interval containing 0. Moreover if the values of the continuous function $f(x)$ all lie in the range of the continuous function $h(x)$ then the composite function $h(f(x))$ is also continuous. In other words if we combine continuous functions we obtain continuous functions except when we attempt to divide by zero.

Secondly, if in doubt about continuity plot the graph of the function and inspect whether there are any discontinuities.

17.7 Numerical integration

We saw in Section 17.2 that the calculation of area by upper and lower estimates in terms of rectangles could be a slow process. In this section we present two easy techniques which will often lead to more accurate results with less calculating effort.

(i) Trapezoidal rule

Instead of drawing rectangles construct trapezoids using the chords joining sucessive points on the the curve, as illustrated in Figure 17.7.

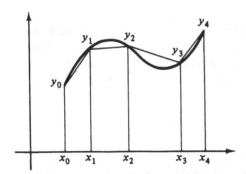

Fig. 17.7

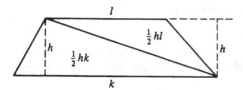

Fig. 17.8

Recall that the area of the trapezoid shown in Figure 17.8 is

$$\tfrac{1}{2}h \cdot (l + k).$$

We thus obtain the estimate

$$I = \sum_{r=0}^{n} \tfrac{1}{2}h(y_r + y_{r-1}),$$

where we have assumed a *spacing* of h throughout the partition. Note that

$$I = \tfrac{1}{2}h \cdot \{(y_0 + y_1) + (y_1 + y_2) + (y_2 + y_3) + \cdots (y_{n-1} + y_n)\},$$

so:

$$I = \tfrac{1}{2}h \cdot \left\{ (y_0 + y_n) + 2 \cdot \sum_{r=1}^{n-1} y_r \right\}.$$

Evidently, an approximation is not much use unless we can guarantee its order of accuracy. Fortunately it may be shown that the error committed *in each of the intervals* $[x_{r-1}, x_r]$ amounts to

$$-\frac{h^3}{12} f''(\xi_r),$$

where ξ_r is some point in the interval $[x_{r-1}, x_r]$ and it is assumed that the second order derivative $f''(x)$ is continuous. The significance of this formula in estimating the accuracy of a computation using the rule above will become apparent in the example of the next section. We now introduce a second method.

(ii)   Simpson's rule

In the trapezoidal rule we were in effect replacing arcs of the curve $f(x)$ by chords, that is by linear approximations. We can go one better by using a second order (parabolic) approximation. For this we must assume that the partition has equal spacing of width h and that n is even. We can then arrange the subintervals in pairs which are contiguous thus:

$$[x_0, x_1], [x_1, x_2]; \; [x_2, x_3], [x_3, x_4]; \ldots; [x_{n-2}, x_{n-1}], [x_{n-1}, x_n].$$

For given r (odd), let us temporarily take axes through the point $(x_r, 0)$ (cf. Figure 17.9). Writing x, y for our *new* variables the equation of the

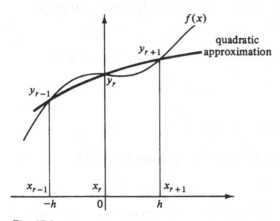

Fig. 17.9

quadratic curve approximating to the curve of f in the short range $[-h, h]$ will be, say,

$$y = Ax^2 + Bx + C.$$

Then we have

$$I_r = \int_{-h}^{h} y \, dx = \left[\frac{Ax^3}{3} + \frac{Bx^2}{2} + Cx \right]_{-h}^{h}$$

$$= \frac{2Ah^3}{3} + 2Ch.$$

But observe that, since the curve is assumed to pass through the points $(-h, y_{r-1}), (0, y_r), (h, y_{r-1})$, we have on substituting:

$$y_r = C,$$
$$y_{r+1} = Ah^2 + Bh + y_r,$$
$$y_{r-1} = Ah^2 - Bh + y_r,$$

from which we may deduce the values of A and B. Substituting in the calculation for I, above we obtain

$$I_r = \frac{h}{3}\{y_{r+1} + y_{r-1} - 2y_r + 6y_r\}$$

$$= \frac{h}{3}\{y_{r-1} + 4y_r + y_{r+1}\}.$$

Now summing over consecutive pairs of intervals of the partition we have the following estimate:

$$I = \frac{h}{3}\{(y_0 + 4y_1 + y_2) + (y_2 + 4y_3 + y_4) + (y_4 + 4y_5 + y_6) + \cdots\}$$

or

$$I = \frac{h}{3}\{(y_0 + y_n) + 4(y_1 + y_3 + y_5 + \cdots + y_{n-1}) + 2(y_2 + y_4 + \cdots + y_{n-2})\}.$$

This is known as Simpson's rule. As before we need to know the accuracy of this estimate. It may be proved that the error committed in the interval $[x_{r-1}, x_{r+1}]$ is

$$-\frac{h^5}{90} f^{(4)}(\bar{\xi}_r)$$

where $\bar{\xi}_r$ is some point in the interval $[x_{r-1}, x_{r+1}]$ and it is assumed that the fourth order derivative $f^{(4)}(x)$ is continuous. The proof of this formula is much trickier than that in the trapezoidal rule. For a discussion of this and of higher order approximations (the Newton–Coates quadrature formulas) see, for example, Hildebrand's *Introduction to Numerical Analysis*. We pass now to an illustrative example.

17.8   Example

Calculate

$$\int_0^1 \exp(-x^2)dx.$$

We begin by calculating the derivatives.

$$f(x) = \exp(-x^2) \quad f'(x) = -2x\exp(-x^2)$$

$$f''(x) = (4x^2 - 2)\exp(-x^2)$$
$$f^{(3)}(x) = (-8x^3 + 12x)\exp(-x^2)$$
$$f(x)^{(4)} = (16x^4 - 48x^2 + 12)\exp(-x^2)$$

We may examine the behaviour of the quartic factor in $f^{(4)}(x)$ by considering the graph of $y = 16z^2 - 48z + 12$ sketched in Figure 17.10. From it we see that $f^{(4)}$ is decreasing in $[0, 1]$. The largest that

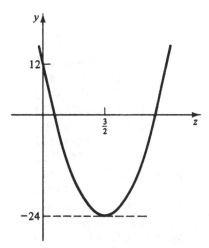

Fig. 17.10. Graph of $y = 16z^2 - 48z + 12$

$f^{(4)}(x)$ can be in absolute value is therefore either $|f^{(4)}(0)|$ or $|f^{(4)}(1)|$. A rough calculation shows the maximum absolute value to be at 0. So a bound on the error per pair of intervals in the Simpson rule will be

$$\frac{h^5}{90} \cdot f^{(4)}(0) = \frac{2}{15} h^5.$$

As regards the trapezoidal rule we note that, since $f^{(3)}(x)$ is positive in $[0, 1]$, f'' is increasing and so in absolute value is greatest either at 0 or at 1. Here a quick calculation reveals that the maximum of $|f''|$ occurs at 0. We may conclude that the error in any one interval of the trapezoidal calculation is in absolute value no greater than

$$\frac{h^3}{12} \cdot f''(0) = \tfrac{1}{6}h^3.$$

Let us take $h = 2^{-5}$ so that there are 32 subintervals or 16 pairs of contiguous intervals. Then we have:

$$\text{accuracy in Simpson formula} \leqslant 16 \cdot 2^{-25} \cdot \left(\tfrac{2}{15}\right) \leqslant 6 \cdot 36 \times 10^{-8}$$

$$\text{accuracy in trapezoidal rule} \leqslant 32 \cdot 5^{-15} \cdot \left(\tfrac{1}{6}\right) \leqslant 1 \cdot 63 \times 10^{-4}$$

We now calculate the estimate for the integral. Since $h = 2^{-5}$ we are assuming $n = 32$.

$$y_0 = f(0) = 1$$
$$y_{32} = f(1) = 0.367\,879\,4$$

$$\overline{y_0 + y_{32} = 1.367\,879\,4}$$

	$y_1 = f(\ 1/32) = 0.990\,239$
$y_2 = f(\ 2/32) = 0.996\,101\,3$	$y_3 = f(\ 3/32) = 0.991\,249$
$y_4 = f(\ 4/32) = 0.984\,496\,4$	$y_5 = f(\ 5/32) = 0.975\,881\,5$
$y_5 = f(\ 6/32) = 0.965\,454\,5$	$y_7 = f(\ 7/32) = 0.953\,275\,3$
$y_8 = f(\ 8/32) = 0.939\,413$	$y_9 = f(\ 9/32) = 0.923\,946\,1$
$y_{10} = f(10/32) = 0.906\,960\,6$	$y_{11} = f(11/32) = 0.888\,550\,2$
$y_{12} = f(12/32) = 0.868\,815$	$y_{13} = f(13/32) = 0.847\,860\,5$
$y_{14} = f(14/32) = 0.825\,797$	$y_{15} = f(15/32) = 0.802\,738\,2$
$y_{16} = f(16/32) = 0.778\,800\,7$	$y_{17} = f(17/32) = 0.754\,102\,8$
$y_{18} = f(18/32) = 0.728\,763\,3$	$y_{10} = f(19/32) = 0.702\,901\,1$
$y_{20} = f(20/32) = 0.676\,633\,8$	$y_{21} = f(21/32) = 0.650\,077\,2$
$y_{22} = f(22/32) = 0.623\,344\,3$	$y_{23} = f(23/32) = 0.596\,544\,4$
$y_{24} = f(24/32) = 0.569\,782\,8$	$y_{25} = f(25/32) = 0.543\,159\,8$
$y_{26} = f(26/32) = 0.516\,770\,5$	$y_{27} = f(27/32) = 0.490\,704$
$y_{28} = f(28/32) = 0.465\,043\,1$	$y_{29} = f(29/32) = 0.439\,864\,2$
$y_{30} = f(30/32) = 0.415\,236\,8$	$y_{31} = f(31/32) = 0.391\,223\,3$

Sum even $\quad= 11.261\,413$ Sum odd $\quad= 11.951\,102$

The trapezoidal estimate is thus

$$\tfrac{1}{2} \cdot \tfrac{1}{32} \cdot \{1.367\,879\,4 + 2 \cdot (11.261\,413 + 11.951\,101)\}$$
$$= 0.746\,764\,1 \pm 0.000\,163.$$

On the other hand the Simpson rule gives

$$\tfrac{1}{3} \cdot \tfrac{1}{32} \cdot \{1.367\,879\,4 + 4 \cdot (11.951\,101) + 2 \cdot (11.261\,413)\}$$
$$= 0.746\,824 \pm 0.000\,000\,063\,6.$$

The latter result is thus exact to six figures (within the accuracy guaranteed by the hand calculator used in this calculation).

17.9 The Riemann–Stieltjes integral

For reasons that will soon become apparent it will be useful to introduce a generalization of the definition presented in Section 17.3. The generalisation gives meaning to the symbol

$$\int_a^b f(x)\,d\alpha(x)$$

where $\alpha(x)$ is also a function which is defined on $[a, b]$. Before giving the definition we shall motivate it in two ways.

(i) Corrugated area

Somewhere in Wonderland is a cardboard mill which presses mashed raw material pulp into sheets. However, neither Rabbit nor his friends and relations know how to lubricate the rollers.
(Acknowledgements to Winnie the Pooh.) The cardboard therefore comes out in corrugated form (see Figure 17.11). It is possible to record how

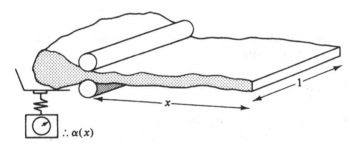

Fig. 17.11

much mass has been turned into a length x of the sheet (the width fortunately remains constant). Call this $\alpha(x)$. Now Alice, for reasons best known to herself, has cut a piece of cardboard so that one edge resembles the graph of the function $f(x)$ as in Figure 17.12. Had the cardboard been of uniform thickness and unit linear density, the card's mass would have been given by the integral

$$\int_a^b f(x)\,dx.$$

But the mass of sheet between length x and length $x + h$ is evidently

$$\alpha(x + h) - \alpha(x) \equiv \delta\alpha(x).$$

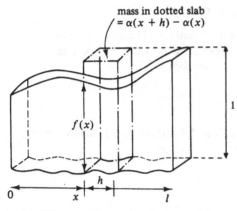

mass in dotted slab
$= \alpha(x + h) - \alpha(x)$

$f(x)$

1

h

0   x l

Fig. 17.12

So for h small the corresponding part of the cut-out has mass

$$f(x)\{\alpha(x + h) - \alpha(x)\}.$$

We are thus concerned with the limiting value of the sum $\Sigma f(x)\delta\alpha(x)$ which might naturally be written

$$\int_0^l f(x)d\alpha(x).$$

We may therefore think of this as 'corrugated area'.

Trite as this story is, it needs to be taken a little further. So far we are tacitly modelling a situation where $\alpha(x)$ is an increasing function of x. We should also make room for functions $\alpha(x)$ which might not be continuous. We therefore need to refer to Mr Tomkins in Wonderland (this time acknowledgements to Professor George Gamow) who knows both about matter and anti-matter (negative mass). He can also tell us that matter ($=$ energy) is available only in discrete lumps. So, for instance the jump function illustrated in Figure 17.13 corresponds to the inclusion of a particular quantum of mass in the cardboard sheet not earlier than at length l_0. We close this discussion by noting that if the function $\alpha(x)$ were differentiable then instead of writing

$$f(x)\{\alpha(x + h) - \alpha(x)\}$$

for the mass of a section, we could write

$$f(x)\alpha'(x)h.$$

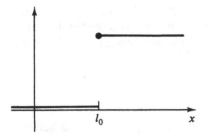

Fig. 17.13. Quantum jump!

This is because

$$\frac{\alpha(x + h) - \alpha(x)}{h} \approx \alpha'(x).$$

Thus the required mass of cardboard in this instance would be the limiting sum

$$\Sigma f(x)\alpha'(x)\delta x,$$

where we have written δx for the increment h. Our more general integral here reduces to

$$\int_0^l f(x)\alpha'(x)dx.$$

In a later section we shall consider what happens if α is not differentiable. In the meantime we can regard $\alpha(x)$ as being a 'weighting function', since it corresponds to uneven mass/weight distribution.

(ii)   Motivation from probability theory

Consider a large population of individuals each of whom posseses a certain amount of income. Choose units appropriately so that the total income is 1. Let

$$\alpha(x) = \text{proportion of people possessing an amount} \leq x \text{ of income.}$$

We can use α to estimate the income of a randomly selected individual as follows. Divide the whole income range $[0, 1]$ into income brackets:

x_0	x_1	$\cdots$	x_{r-1}	x_r	$\cdots$	x_n
0						1

Our randomly chosen individual falls in the income bracket $x_{r-1} < x \leqslant x_r$, with probability $\alpha(x_r) - \alpha(x_{r-1})$ and thus has, with this probability, at least x_{r-1}. Now we sum over all the income brackets and we obtain an 'expected' income of at least

$$\sum_{r=1}^{n} x_{r-1} \{\alpha(x_r) - \alpha(x_{r-1})\}.$$

More generally, if we are interested not in the expected income but in the expected savings of a randomly selected individual, we might first obtain the function $f(x)$ which describes the amount saved given an income level x and then 'argue' (this is a bit fudged) that the expected savings come to 'about':

$$\sum_{r=1}^{n} f(x_{r-1}) \{\alpha(x_r) - \alpha(x_{r-1})\}.$$

This expression looks very much like the Riemann sums of Section 17.3 except that

(i) m_r or M_r is replaced by an intermediate value $f(x_{r-1})$;
(ii)   the incremental term $(x_r - x_{r-1})$ has been replaced by-$(\alpha(x_r) - \alpha(x_{r-1}))$.

Our intention is to take some sort of a limit over all sums by passing to *fine enough* partitions and calling the limit

$$\int_0^1 f(x) d\alpha(x).$$

In the example above note that because the population is finite, the precise expected income or savings will be arrived at with a fine enough partition.

What is fudged in the formula for the savings is whether we should work with $f(x_{r-1})$ or $f(x_r)$ or even some intermediate value $f(t_r)$ with $x_{r-1} \leqslant t_r \leqslant x_r$. For a continuous function f these three possibilities will be approximately equal (for a fine enough spacing). The matter is far from clear when f is discontinuous and will be carefully examined in Section 17.16.

Finally, observe how $\alpha(x)$ provides a 'weighting' of possible saving levels $f(x)$ (according to their probability). We can thus think of

$$\int_0^1 f(x) d\alpha(x)$$

as a 'weighted' integration process.

17.10 Limits over partitions

The Riemann–Stieltjes integral of f with respect to the weighting function α over $[a, b]$ (or, *integrator*) is defined by reference to the approximating sums

$$S(P, \mathbf{t}) = \sum_{r=1}^{n} f(t_r)\{\alpha(x_r) - \alpha(x_{r-1})\},$$

where P is the partition $\{x_0, x_1, \ldots, x_n\}$, and $\mathbf{t}$ is a vector $(t_1, t_2, \ldots, t_n)$ which satisfies, for each r, $x_{r-1} \leqslant t_r \leqslant x_r$. Our aim is to take limits as P gets progressively finer. This presupposes that we are able to attach significance to the three terms: finer, progressively, limits. The last term concerns accuracy as in Section 17.6. The first term is easy and we attend to it first.

Definition

We say that the partition $Q = \{y_0, y_1, \ldots, y_m\}$ is finer than the partition $P = \{x_0, x_1, \ldots, x_n\}$, if Q contains all the points of P and, possibly, some others. Thus, in set notation $P \subseteq Q$.

The term requiring most attention is 'progressively'. Progress subsumes direction. It will be helpful to consider the role of direction in two natural cases where limit processes occur.

Case (i) 'Limit down a sequence'

One possible way to compute $\sqrt{2}$ (from first principles!) is to use the iteration formula:

$$x_{n+1} = 1 + \frac{1}{1 + x_n},$$

starting with $x_1 = 1$. We obtain successively $x_2 = 1.5$ $x_3 = 1.4$ $x_4 = 1.416\,66\cdots x_5 = 1.413\,44\cdots x_6 = 1.4142$. Clearly, if the limit is x (assuming the limit exists), we have

$$x = 1 + \frac{1}{1 + x}$$

and hence

$$(x-1)(x+1) = 1 \quad \text{i.e.} \quad x^2 = 2.$$

Here the sequence of approximations will be 'progressively' more accurate in the sense that for a given target degree of accuracy (e.g. 4 figure accuracy) we can find a position in the sequence starting at which all further approximations have the required accuracy. In symbols: there is a (starting) subscript n_0 such that for all $n \geqslant n_0$ the terms x_n have the target accuracy. We see from the above calculations that for 4 figure accuracy the starting position is $n = 6$.

The notion of direction in the case discussed is derived from the natural order relation $\leqslant$. We note that the term 'progressively' makes use of a 'starting position' appropriate to a given target accuracy.

Case (ii) 'Limit along a curve'

This time we consider the graph of a function f such as is given in Figure 17.14. We talk about the limiting value of f as t 'approaches' 0

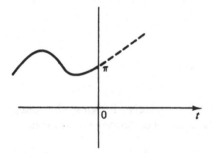

Fig. 17.14

from the left. In the illustration $f(t)$ approaches π as t approaches 0 from the left. What we mean is that $f(t)$ is a good approximation to π when t is close enough to 0. More precisely, if we wish to achieve a given target accuracy in approximating π by $f(t)$ we need to specify a starting position t_0 and require that for all $t \geqslant t_0$ the value $f(t)$ is to within the required accuracy of π.

Again we were guided by an ordering relation of the t-axis and by a 'starting position'.

Conclusion. We may regard the calculation of $S(P)$ (we omit reference to t for simplicity) as though it were indexed by P in analogy to case (i).

The notion of direction will be provided by the relation 'is finer than' (by analogy to 'is greater than'). This is graphed in the picture below.

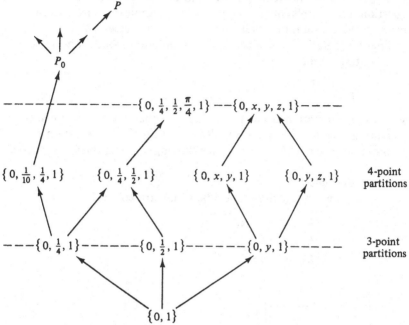

Partitions of [0,1]. Arrow indicates 'refined to' and provides a notion of direction for the limit process

To say that $S(P)$ is progressively closer to some limiting value we will need to say that for a given target accuracy (measured say by ε) there is a starting position P_0 such that, for all P finer than P_0, $S(P)$ is to within ε of I. Now let us make this official.

Definition
We say that the integral of f with respect to α over $[a,b]$ *exists if: there is a number I such that to **any** choice of* $\varepsilon > 0$ *there corresponds a partition* P_0 *with the property that all partitions P finer than* P_0 *satisfy*

$$|S(P,t) - \mathrm{I}| < \varepsilon$$

and this happens no matter how we select t. (Of course t is subject to $x_{r-1} \leqslant t_r \leqslant x_r$ *for each r with* $1 \leqslant r \leqslant n$.)

Remark
For $\alpha(x) \equiv x$ the above definition agrees with the Riemann definition when f is continuous.

17.11   The practical approach

Having rolled off the definition we need to ask the practical
question: how is this integral computed? We shall soon see that in
practice the computation boils down to two processes: some
ordinary integration plus a combination of 'point evaluations' i.e. a
contribution of the form

$$\sum_j f(c_j) \cdot h_j,$$

where the summation is taken over some of the places c_j where α, the
weighting function, has discontinuities. In a nutshell: integration plus
summation. We expand on this after looking at some special examples.

Example 1
(α a jump function). Let α be as illustrated, viz.,

$$\alpha(x) = \begin{cases} k_1 & a \leqslant x < c, \\ k_1 & x = c, \\ k_2 & c < x \leqslant b. \end{cases}$$

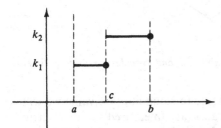

Fig. 17.15

with $k = k_2 - k_1 > 0$. Compute the integral of f with respect to α,
assuming that f is continuous at c.

Let $P = \{x_0, x_1, \ldots, x_n\}$ be any partition which includes the point c.
Suppose $c = x_s$. Then

$$S(P, \mathbf{t}) = f(t_1)\{\alpha(x_1) - \alpha(x_0)\} + f(t_2)\{\alpha(x_2) - \alpha(x_1)\} + \cdots$$
$$+ f(t_{s+1})\{\alpha(x_{s+1}) - \alpha(x_s)\} + \cdots + f(t_n)\{\alpha(x_n) - \alpha(x_{n-1})\}.$$

Since α is constant in all the subintervals $[x_{r-1}, x_r]$ *except* when $r = s + 1$
we obtain

$$S(P, \mathbf{t}) = f(t_{s+1})\{\alpha(x_{s+1}) - \alpha(x_s)\}$$
$$= f(t_{s+1}) \cdot k.$$

Now clearly the limiting value of this expression as P is progressively finer must be $f(c)$. Indeed, provided we can ensure that t_{s+1} is close enough to c, $f(t_{s+1})$ will be close to $f(c)$. For example, if we wanted $S(P, \mathbf{t})$ to be to within ε of the answer we would first need to know just how close t must be to c to obtain

$$|f(t) - f(c)| < \varepsilon/k.$$

Suppose that we need $|t - c| < h$. Then take $P_0 = \{a, c, c + h, b\}$ and observe that if P is finer than P_0 we shall have, for some s, $x_s = c$ and $x_{s+1} \leqslant c + h$. Consequently, the rather simple choice of P_0 controls the closeness of t_{s+1} to c. Under these circumstances, we have of course

$$|kf(t_{s+1}) - kf(c)| < \varepsilon,$$

i.e. $|S(P, \mathbf{t}) - kf(c)| < \varepsilon$. This proves that

$$\int_a^b f(x) d\alpha(x) = f(c) \cdot k.$$

Remark 1

Our result is easy to understand in the light of the story in section 17.9. Suppose the function $\alpha(x)$ introduced in the story gives the amount of pulp turned into cardboard sheet of length x and happens to have a graph like that in Example 1. Let us see what implications this holds. Clearly no pulp passes through the rollers for a while and the segment of sheet between $x = a$ and $x = c$ on the conveyor belt is actually void. When we reach $x = c$ a lump k of pulp suddenly passes onto the belt and thereafter no pulp follows. So Alice's sheet is mostly void ('infinitely thin' we might say) except for a line of cardboard at $x = c$ of mass k. So the cut-out has mass $f(x) \cdot k$. See Figure 17.16.

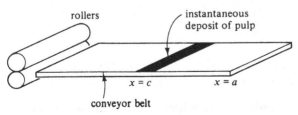

Fig. 17.16

Remark 2

Clearly the calculation generalizes to a weighting function $\alpha(x)$ with a finite number of jumps in $[a, b]$ say at positions $c_1 < c < \cdots < c_n$ where $a < c_1$ and $c_n < b$. We then have for $f(x)$ continuous that

$$\int_a^b f(x)\,d\alpha(x) = \sum_{i=1}^n k_i f(c_i),$$

where k_i is the jump at $x = c_i$.

Remark 3

The calculation above continues to hold when the left subinterval $[a, c]$ degenerates into the point (so that α jumps immediately after a). The same result also holds if we alter the value of $f(c)$ to k_2. Assuming this latter alteration is made we can allow the right subinterval $[c, b]$ to degenerate to a single point (this means that in this case α jumps only at the end of the interval).

Example 1 and the modification just mentioned should be compared closely with the next example.

Example 2   (behaviour exterior to $[a, b]$ irrelevant)

Compute the integral when α is 0 outside $[a, b]$ and 1 inside, as in Figure 17.17.

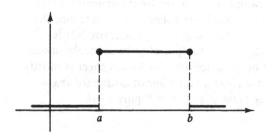

Fig. 17.17

We observe that

$$S(P, \mathbf{t}) = f(t_1)\{\alpha(x_1) - \alpha(x_0)\} + f(t_2)\{\alpha(x_2) - \alpha(x_1)\} + \cdots$$
$$+ f(t_n)\{\alpha(x_n) - \alpha(x_{n-1})\} = 0.$$

One might say that the only discontinuities which contribute to an integral $\int_a^b$ are those which are visible in $[a, b]$.

Example 3   (the sample distribution function)

Suppose $a_1, \ldots, a_n$ are points on the real line arranged in some definite order. (What the order is does not matter.) Define a function $F_n(x)$ by

$$F_n(x) = \frac{1}{n} \cdot \#\{i : a_i \leqslant x\}.$$

This function is known in statistics under the name '*sample distribution function*'. Its graph resembles a staircase (see Figure 17.18). (See also Section 17.13.)

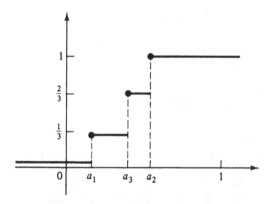

Fig. 17.18. Sample distribution function for a sample with $n = 3$

Referring to the probability theory discussion of Section 17.9, suppose $a_1, \ldots, a_n$ were the observed incomes of a sample of n individuals in the population. If the function $\alpha(x)$ giving the proportion of people with income $\leqslant x$ was not available, the function $F_n(x)$ based on our observation could be considered a natural estimate for $\alpha(x)$.

Let us use it to compute the 'expected' income. Recall that in our example $0 \leqslant x \leqslant 1$; assume further that $0 < a_i < 1$ for all i. We obtain from Example 1

$$\int_0^1 x dF_n(x) = \sum_{i=1}^n a_i \frac{1}{n},$$

which is just the *sample mean* (or average). Let us denote it by $\bar{a}$.

The quantity

$$\int_a^b x^k d\alpha(x)$$

is known as the *kth order moment* of the weighting function α, and the related quantity

$$\int_a^b (x - d)^k \, d\alpha(x)$$

is known as the *kth order moment about d*. Let us compute in our example the second order moment about the mean. We have

$$\int_a^b (x - \bar{a})^2 dF_n(x) = \sum_{i=1}^n (a_i - \bar{a})^2 \cdot \frac{1}{n}$$

$$= \frac{1}{n} \sum_{i=1}^n (a_i^2 - 2a_i\bar{a} + \bar{a}^2)$$

$$= \frac{1}{n}\left\{ \left[\sum_1^n a_i^2 \right] - 2\bar{a}\left[\sum_1^n a_i \right] + n\bar{a}^2 \right\}$$

$$= \frac{1}{n}\{[\sum a_i^2] - 2n\bar{a}^2 + n\bar{a}^2\}$$

$$= \left[\frac{1}{n}\sum a_i^2 \right] - \bar{a}^2.$$

The square root of this quantity is known as the *sample standard deviation*, and, as the name implies, is a measure of how much the observations deviate from the average. See also Chapter 3 for a discussion of topics in statistics.

17.12 Practical issues – part II

Given a function α the procedure to follow in practice is this. First decompose the interval $[a, b]$ into subintervals, whose endpoints are the positions where the *weighting* function α has discontinuities, as illustrated in Figure 17.19.

Our aim is to make use of the following basic property of the integral.

$$\int_a^b f(x)d\alpha(x) = \int_a^c f(x)d\alpha(x) + \int_c^b f(x)d\alpha(x).$$

(This is proved by messing about with partitions.) The implication is that we may, if we wish, compute separately the contributions from various subintervals. Next, we note the result

$$\int_c^d f(x)d\alpha(x) = \int_c^d f(x)\alpha'(x)dx \quad \text{provided } \alpha \text{ is smooth in } [c, d].$$

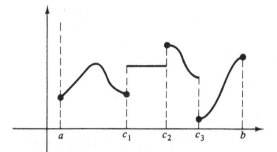

Fig. 17.19

By *smooth* we mean here that the function α is continuous in $[c, d]$ and differentiable in (c, d). Our result, intimated earlier, says that under smoothness the Riemann–Stieltjes integral reduces to an ordinary integral. A third basic fact which we need is

$$\int_a^b f(x)d\{\beta(x) + \gamma(x)\} = \int_a^b f(x)d\beta(x) + \int_a^b f(x)d\gamma(x)$$

and this holds provided any two of the integrals in the formula exist.

Before proceeding to the illustrated example let us consider the integral of $f(x)$ in the interval $[c_2, c_3]$. This would be easy if only the value of α at c_3 had been at the circled position of Figure 17.20, because then we

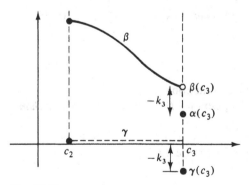

Fig. 17.20

should have had smoothness. We arrange for smoothness as follows. Let the limiting value of α from the left be $\alpha(c_3) + k_3$ (k_3 being the positive size of the jump). Define a function β on the interval $[c_2, c_3]$ thus:

$$\beta(x) = \begin{cases} \alpha(x), & x \neq c_3 \\ \alpha(c_3) + k_3, & x = c_3. \end{cases}$$

Also let

$$\gamma(x) = \begin{cases} 0, & x \neq c_3 \\ -k_3, & x = c_3. \end{cases}$$

Thus $\alpha(x) = \beta(x) + \gamma(x)$. Since β is smooth and γ is a jump function we immediately have

$$\int_{c_2}^{c_3} f(x)d\{\beta(x) + \gamma(x)\} = \int_{c_2}^{c_3} f(x)d\beta(x) + \int_{c_2}^{c_3} f(x)d\gamma(x).$$

$$= \int_{c_2}^{c_3} f(x)\beta'(x)dx - f(c_3)k_3$$

$$= \int_{c_2}^{c_3} f(x)\alpha'(x)dx - f(c_3)k_3,$$

where the last formula uses α' as a convenient abuse of notation.

Applying these results and Example 1 of Section 17.11, we see that for the case illustrated at the beginning of the section, we have

$$\int_a^b f(x)d\alpha(x) = \int_a^{c_1} f(x)\alpha'(x)dx + k_1 f(c_1) + k_2 f(c_2)$$

$$+ \int_{c_2}^{c_3} f(x)\alpha'(x)dx - k_3 f(c_3) + \int_{c_3}^b f(x)\alpha'(x)dx.$$

17.13 Some examples

Example 1   (the staircase function)

Let $\alpha(x) = [x]$, where $[x]$ denotes *the greatest integer less than or equal to* x.

Thus $[1/2] = 0$ and $[\pi] = 3$, but note that $[-1/2] = -1$ and $[-\pi] = -4$. See the illustration in Figure 17.21.

We claim that

$$\int_a^b f(x)d[x] = \sum_{a < n \leqslant b} f(n).$$

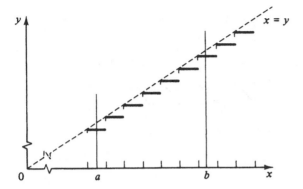

Fig. 17.21

Notice that even if a is an integer $f(a)$ does not appear
in the summation. This is because $[x]$ does not have a discontinuity at
a *within* the interval $[a, b]$; compare this with Example 2 in Section 17.11.
To prove the formula observe that the discontinuities occur at the integer
points and that $[x]$ is constant in each of the intervals $(n, n + 1)$. The
jumps are all positive and equal to unity.

Example 2
Find

$$\int_{1}^{10} x\,d(x - [x]).$$

Of course $x - [x]$ is the fractional part of x. Thus for example
$\pi - [\pi] = 0.14159\ldots$. Note that $-\pi - [-\pi] = 0.85840\ldots$. See the
graph in Figure 17.22.

$$\int_{1}^{10} x\,d(x - [x]) = \int_{1}^{10} x\,dx - \int_{1}^{10} x\,d[x]$$

$$= \left[\frac{x^2}{2}\right]_{1}^{10} - \sum_{2}^{10} n$$

$$= \frac{99}{2} - \{\tfrac{1}{2} \cdot 10 \cdot 11 - 1\}$$

$$= -4.5.$$

Note that we expect a negative answer since all the jumps in $x - [x]$ are
negative.

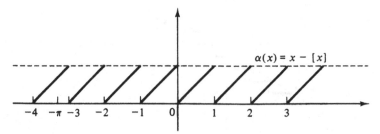

Fig. 17.22

17.4   Basic properties

In this section we list the linearity properties mentioned earlier and two very important results which carry over from ordinary integration, viz. change of variable and integration by parts.

1.

$$\int_a^b \{c_1 f(x) + c_2 g(x)\}\, d\alpha(x) = c_1 \int_a^b f(x)\, d\alpha(x) + c_2 \int_a^b g(x)\, d\alpha(x),$$

where c_1 and c_2 are constants and f and g are integrable with respect to α.

2.

$$\int_a^b f(x)\, d\{c_1 \beta(x) + c_2 \gamma(x)\} = c_1 \int_a^b f(x)\, d\beta(x) + c_2 \int_a^b f(x)\, d\gamma(x),$$

where c_1 and c_2 are constants and f and g are integrable with respect to α.

3.

$$\int_a^b f(x)\, d\alpha(x) = \int_a^c f(x)\, d\alpha(x) + \int_c^b f(x)\, d\alpha(x),$$

provided any two of the three integrals exist.

4. *Change of variable.* Let $\phi(t)$ be a strictly increasing and continuous function from $[c, d]$ onto $[a, b]$, then

$$\int_a^b f(x)\, d\alpha(x) = \int_c^d f(\phi(t))\, d\alpha(\phi(t)),$$

where it is assumed that one or other side of the equation exists.

Example
Find

$$I = \int_0^a x\, d[e^x].$$

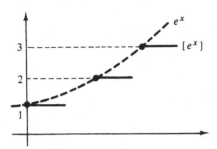

Fig. 17.23

The weighting function is graphed in Figure 17.23. Put $t = e^x$, i.e. $x = \log t$. Then, substituting in the usual way,

$$I = \int_1^{e^a} \log t \, d[t]$$

$$= \sum_2^{[e^a]} \log n = \log 2 + \log 3 + \cdots + \log [e^a]$$

$$= \log([e^a]!).$$

Note that we are making use of the function $\phi(t) \equiv \log t$ which maps $[1, e^a]$ onto $[0, a]$ and is strictly increasing and continuous in the specified range (cf. Figure 17.24).

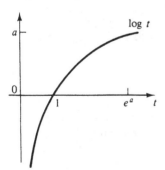

Fig. 17.24

5. *Integration by parts*

$$\int_a^b u(x) dv(x) = [u(x)v(x)]_a^b - \int_a^b v(x) du(x)$$

provided either side exists.

Example
Find

$$I = \int_0^a [e^x]\,dx.$$

This may be done directly by computing areas as illustrated in

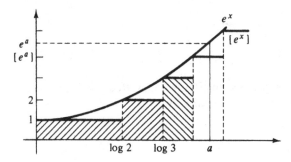

Fig. 17.25

Figure 17.25. But it is quicker to write

$$\int_0^a [e^x]\,dx = [x[e^x]]_0^b - \int_0^a x\,d[e^x].$$

$$= a[e^a] - \log([e^a]!).$$

Note how so much like ordinary integration these manipulations are, despite the wider range of functions to which the manipulations are applicable.

17.15 Example (sum of the integers)

We use integration by parts to obtain formulas for the sum of the integers from 1 to n and for their squares in 17.18.2. Observe that

$$L = \int_0^s x\,d[x] = [x[x]]_0^s - \int_0^s [x]\,dx.$$

But,

$$\int_0^s [x]\,dx = \int_0^1 [x]\,dx + \int_1^2 [x]\,dx + \cdots + \int_{[s]}^s [x]\,dx$$

$$= 1 + 2 + \cdots + ([s]-1) + (s-[s])[s]$$

$$= (L - [s]) + (s - [s])[s].$$

Hence,

$$L = s[s] - 0 - \{L - [s] + (s - [s])[s]\},$$

or

$$2L = s[s] + [s] - (s - [s])[s],$$

giving

$$L = \tfrac{1}{2}(1 + [s])[s].$$

17.16 When does the integral exist?

So far we have not troubled ourselves over this question, relying
in practical situations on the reduction to the Riemann integral.
The key to the problem is the family of functions which may be
represented as a difference of two increasing functions, i.e. as

$$\alpha(x) - \beta(x)$$

where α and β are increasing. The nice thing about such functions
is that a linear combination of them yields another function of the
same type (proof: exercise). Also, the product of two such functions
is again of the same type. It turns out that these are precisely the
functions whose graphs have finite length (it is tempting to say that
the graphs 'don't waggle about too much', in a sense that will be
made clear in the next section). In view of this very important
property they are known under the name '*functions of bounded
variation*'. We omit the proof of the fundamental result that: if $f(x)$
is a continuous function and $\alpha(x)$ is an *increasing* function and both
are defined on $[a, b]$ then the integral

$$\int_a^b f(x)d\alpha(x)$$

exists. See Appendix A.

It follows that the integral exists when f is continuous and α is
of bounded variation. And now we can make a very astute
inference using integration by parts: the integral also exists if f is of
bounded variation and α is continuous. The proof is easy. We know that

$$\int_a^b \alpha(x)df(x)$$

exists. Integrate this by parts to obtain the required integral.

We close this section with an example of an integral which does
not exist.

Warning example   (same-sided discontinuity)
The integral

$$\int_0^1 [x]\,d[x]$$

does not exist.

If $P = \{x_0, x_1, \ldots, x_n\}$ is a partition of $[0, 1]$ and t_r satisfies $x_{r-1} \leqslant t_r \leqslant x_r$ we have since $[x]$ is constant in $[0, 1]$

$$S(P, t) = [t_n] = \begin{cases} 1 & \text{if } t_n = 1, \\ 0 & \text{if } t_n < 1. \end{cases}$$

Thus no matter how close x_{n-1} is to x_n the sum $S(P)$ can take the values 0 and 1, depending on the choice of **t** (see Figure 17.26).

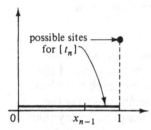

possible sites for $[t_n]$

x_{n-1}

Fig. 17.26

There is thus no limiting value.

This example shows that if $f(x)$ and $\alpha(x)$ share a same-sided discontinuity in $[a, b]$ the integral

$$\int_b^a f(x)\,d\alpha(x)$$

will not exist.

17.17 Curve length and functions of bounded variation

We have already seen how well the continuous functions behave in respect of integration. The 'intuitive' notion of a reasonably behaved function no doubt sees it as a function that may easily be drawn and consists alternately of stretches where it is increasing followed by stretches where it is decreasing. The key issue here appears to be whether the function does not waggle about too much and whether it can be drawn at all.

Now there is a rather obvious way of measuring how much the graph of a function 'waggles about' over an interval $[a,b]$ – it is to measure its curve length. To do this we would, as a first step, approximate the curve by a 'broken line' or 'polygonal arc' with

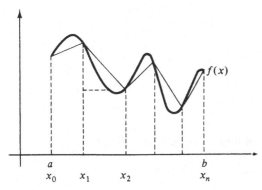

Fig. 17.27

vertices along the curve (cf. Figure 17.27). Let us take vertices at the points:

$$(x_0, f(x_0)), \quad (x_1, f(x_1)), \quad (x_2, f(x_2)), \dots, (x_n, f(x_n)).$$

Here $a = x_0 < x_1 < x_2 < \cdots < x_n = b$ is (inevitably) a partition, call it P, of $[a,b]$. Given this choice of P we arrive at the estimate for the curve length of

$$l(P) = \sum_{r=1}^{n} \sqrt{\{(x_r - x_{r-1})^2 + (f(x_r) - f(x_{r-1}))^2\}}.$$

It is not difficult to see that if Q is finer than P (i.e. Q includes the points of P) then $l(P) \leqslant l(Q)$. (If we put in one vertex B between A_{r-1} and A_r then, referring to Figure 17.28, by the triangle inequality $A_{r-1}A_r \leqslant A_{r-1}B + BA_r$.) Since the inclusion of more vertices seems intuitively to give a better estimate we take as our definition of curve length

$$\sup_{P} l(P).$$

Here we have again used the supremum as in Section 17.3. The next example shows that not all continuous functions have finite curve length.

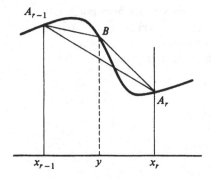

Fig. 17.28

Example
Consider the function illustrated in Figure 17.29

$$f(x) = x \cos \frac{\pi}{x} \quad 0 < x \leqslant 1$$

where we take $f(0) = 0$ (for reasons of continuity). This function has infinite curve length (see below). But note that the function

$$f(x) = x^2 \cos \frac{\pi}{x} \quad 0 < x \leqslant 1$$

has finite curve length.

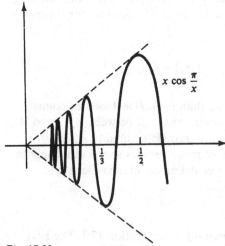

Fig. 17.29

The quantity $l(P)$ is a little unmanageable if all we want to test is the finiteness or otherwise of the curve length. We make the easy observation that

$$\sum_{r=1}^{n} |f(x_r) - f(x_{r-1})| \leqslant l(P)$$

$$\leqslant \sum_{r=1}^{n} \{|x_r - x_{r-1}| + |f(x_r) - f(x_{r-1})|\}$$

$$= \left\{ \sum_{r=1}^{n} |f(x_r) - f(x_{r-1})| \right\} + (b - a).$$

Definition
The quantity

$$\sup_{P} \sum_{r=1}^{n} |f(x_r) - f(x_{r-1})|,$$

where the supremum taken over all partitions P of $[a, b]$ is called the **variation of f on** $[a, b]$ *and is denoted* $V_f[a, b]$.

The point of this definition is that, by the earlier inequality, if the variation is finite, so is the curve length and vice-versa. Note that the formula now computes in absolute terms the amount of rise and fall in the function f.

Returning to the examples above, note that if

$$P_n = \left\{ 0, \frac{1}{n}, \frac{1}{n-1}, \frac{1}{n-2}, \dots, \frac{1}{2}, 1 \right\},$$

then in the case of the first function

$$\sum_{r=1}^{n} |f(x_r) - f(x_{r-1})| \geqslant \sum_{r=2}^{n} \frac{1}{r} \to \infty$$

as n tends to infinity. In the case of the second function note that for the same partition P_n

$$\sum_{r=1}^{n} |f(x_r) - f(x_{r-1})| \leqslant \sum_{r=1}^{n-1} \frac{1}{r^2} < \sum_{r=1}^{\infty} \frac{1}{r^2} < \infty.$$

Thus the curve length can be finite even though a function has infinitely many 'humps' (it is just because the humps rapidly decrease in size that the curve is of finite length).

Definition
   A function f is said to be of bounded variation if $V_f[a,b] < \infty$.
   The surprising fact, mentioned in Section 17.17, is that a function of bounded variation is the difference of two increasing functions. We mention that a natural alternative notation for the variation is

$$\int_a^b |df(x)|,$$

the symbol implying that we sum absolute values of the increments δf.

Example 1
If f is increasing $V_f[a,b] = f(b) - f(a)$.

$$\sum_{r=1}^{n} |f(x_r) - f(x_{r-1})| = \sum_{r=1}^{n} f(x_r) - f(x_{r-1}) = f(b) - f(a).$$

(This is illustrated in Figure 17.30.)

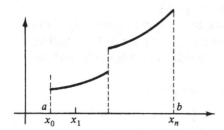

Fig. 17.30

Example 2
   If $f = \alpha - \beta$ with α and β both increasing then $V_f \leqslant V_\alpha + V_\beta$.
This time

$$|f(x_r) - f(x_{r-1})| = |\alpha(x_r) - \beta(x_r) - (\alpha(x_{r-1}) - \beta(x_{r-1}))|$$
$$= |\alpha(x_r) - \alpha(x_{r-1}) - (\beta(x_r) - \beta(x_{r-1}))|$$
$$\leqslant |\alpha(x_r) - \alpha(x_{r-1})| + |\beta(x_r) - \beta(x_{r-1})|$$
$$= \alpha(x_r) - \alpha(x_{r-1}) + (\beta(x_r) - \beta(x_{r-1})).$$

Summing we obtain (as in the previous example)

$$V_f[a,b] \leqslant \alpha(b) - \alpha(a) + \beta(b) - \beta(a).$$

Example 3

In the case illustrated in Figure 17.31

$$V_f[a,b] = v_1 + v_2 + v_3.$$

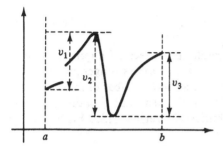

Fig. 17.31

This may be verified by considering partitions which include the points c_1 and c_2. Notice how this measures *cumulatively* (i.e. without cancellations) the rise and fall of the function f.

Example 4

If f is continuous on $[a,b]$ and $f'(x)$ *exists and is bounded in* (a,b) then f is of bounded variation.

Let $P = \{x_0, x_1, x_2, \ldots, x_n\}$ be a partition. Then we notice that

$$f(x_r) - f(x_{r-1}) = f'(z_r)(x_r - x_{r-1}).$$

We are told this by the mean value theorem of differential calculus (which asserts that some intermediate tangent is parallel to the chord illustrated, cf. Figure 17.32). Hence

$$|f(x_r) - f(x_{r-1})| = |f'(z_r)|(x_r - x_{r-1}) \leqslant M \cdot (x_r - x_{r-1}),$$

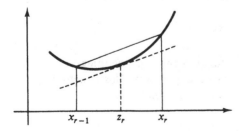

Fig. 17.32

where M is the bound mentioned in the assumptions. Now summing we obtain

$$\sum |f(x_r) - f(x_{r-1})| \leqslant M \sum (x_r - x_{r-1}),$$

so

$$V_f[a, b] \leqslant M \cdot (b - a).$$

This is an important example. It enables us to recognise quite quickly that a function is of bounded variation if it enjoys the stronger property that its domain $[a, b]$ may be split into a finite number of intervals in the interior of each of which the derivative exists and is bounded. Such functions are sometimes called 'subinterval – regular'. (See Figure 17.33.)

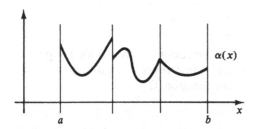

Fig. 17.33

17.18 Further properties of the integral

We start off with a deceptively simple result.

1.   First mean value theorem

If α is increasing and f is continuous on $[a, b]$ then for some x_0 in $[a, b]$

$$\int_a^b f(x) d\alpha(x) = f(x_0) \int_a^b d\alpha(x).$$

In other words we may pull the integrand through the integral sign by evaluating at an appropriate intermediate position. This is helpful in estimation procedures. The proof is almost trivial. Let

$$m = \min_{[a,b]} f(x), \quad M = \max_{[a,b]} f(x).$$

Now $f(x)$ is continuous so it takes all intermediate values between m

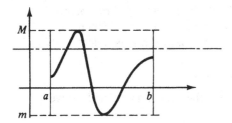

Fig. 17.34

and M (cf. Figure 17.34). But notice that

$$\sum m(\alpha(x_r) - \alpha(x_{r-1})) \leqslant S(P, t) \leqslant \sum M(\alpha(x_r) - \alpha(x_{r-1}))$$

i.e.

$$m(\alpha(b) - \alpha(a)) \leqslant S(P, t) \leqslant M(\alpha(b) - \alpha(a)),$$

so

$$m(\alpha(b) - \alpha(a)) \leqslant \int_b^a f(x)\, d\alpha(x) \leqslant M(\alpha(b) - \alpha(a)),$$

or

$$m \leqslant \frac{\displaystyle\int_a^b f(x)\, d\alpha(x)}{\alpha(b) - \alpha(a)} \leqslant M.$$

Thus, at some x_0, f takes the intermediate value indicated below (we have just shown that it lies between m and M), viz.

$$f(x_0) = \frac{\displaystyle\int_a^b f(x)\, d\alpha(x)}{\alpha(b) - \alpha(a)}.$$

This gives the required result. See below for an application.

2. Differentiability and other properties

Let $\alpha(t)$ be of bounded variation of $[a, b]$, let $f(t)$ be bounded and suppose $\int_a^b f(t)\, d\alpha(t)$ exists. Then

(i) $\beta(x) = \displaystyle\int_a^x f(t)\, d\alpha(t),$

exists for each x with $a \leqslant x \leqslant b$ and is itself of bounded variation on $[a,b]$;

    (ii) if $\alpha(t)$ is continuous at a point ξ, then so is the β defined above;

    (iii) if $f(t)$ is continuous at a point ξ and if $\alpha(t)$ is differentiable at ξ, then

$$\beta'(\xi) = f(\xi)\alpha'(\xi) \quad \text{for} \quad a < \xi < b;$$

    (iv) if $g(t)$ is continuous on $[a,b]$, then for β as defined above

$$\int_a^b g(x)\,d\beta(x) = \int_a^b g(t)f(t)\,d\alpha(t).$$

Example (sum of the squares)

We return to example 17.15 and apply (iv) above to the more meaty problem of the squares: We have

$$S = \int_0^n x^2 d[x] = \left[x \cdot \int_0^x t\,d[t] \right]_0^n - \int_0^n \int_0^x t\,d[t]\,dx$$

$$= n \cdot \frac{1}{2} n \cdot (n+1) - \int_0^n \frac{1}{2}[x]([x]+1)\,dx.$$

So,

$$2S = n^2(n+1) - \left\{ \int_0^n [x]^2\,dx + \int_0^n [x]\,dx \right\}$$

$$= n^2(n+1) - \{S - n^2 + (1+2+\cdots+(n-1))\}$$

$$= n^2(n+1) - S + n^2 - \tfrac{1}{2}(n-1)n$$

$$3S = \tfrac{1}{2}n\{2n(n+1) + 2n - (n-1)\}$$

$$S = \tfrac{1}{6}n\{2n^2 + 3n + 1\} = \tfrac{1}{6}n(n+1)(2n+1),$$

the well-known formula. This technique may be used to obtain higher powers as well.

3.   The second mean value theorem

If $\alpha(t)$ is continuous and $f(t)$ is increasing in $[a,b]$ then there is a point x_0 in $[a,b]$ such that

$$\int_a^b f(x)\,d\alpha(x) = f(a)\int_a^{x_0} d\alpha(x) + f(b)\int_{x_0}^b d\alpha(x).$$

Thus the integral is a weighted average of the two extreme values $f(a)$ and $f(b)$. This is proved by integration by parts. First notice that

$$\int_a^b f(x)\,d\alpha(x) = [f(x)\alpha(x)]_a^b - \int_a^b \alpha(x)\,df(x). \tag{1}$$

But by the first mean value theorem we have for some x_0

$$\int_a^b \alpha(x)\,df(x) = \alpha(x_0)\int_a^b df(x) = \alpha(x_0)\{f(b) - f(a)\}.$$

The right-hand side of (1) is thus equal to

$$\{\alpha(b)f(b) - \alpha(x_0)f(b)\} + \{\alpha(x_0)f(a) - \alpha(a)f(a)\}$$

and this is the required result.

Example
Estimate, as $t \to \infty$,

$$I_t = \int_\delta^b e^{-tx}\frac{\sin x}{x}\,dx$$

where $\delta > 0$. We take

$$\beta(x) = \int_\delta^x \sin t\,dt,$$

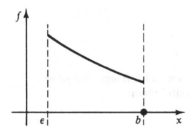

Fig. 17.35

and (see Figure 17.35)

$$f(x) = \begin{cases} \dfrac{1}{x}e^{-tx} & \delta \leqslant x < b, \\ 0 & x \doteq b. \end{cases}$$

Then, since β is *not* discontinuous at b and $-f$ is increasing we have

$$I_t = \int_\delta^b f(x)\,d\beta(x) = f(\delta)\int_\delta^{x_0} d\beta(x) + f(b)\int_{x_0}^b d\beta(x)$$

$$= \frac{e^{-t\delta}}{\delta}\int_\delta^{x_0}\sin t\,dt.$$

So since

$$\left|\int_\delta^{x_0}\sin t\,dt\right| < 1,$$

we have

$$|I_t| \leqslant \frac{e^{-t\delta}}{\delta}.$$

17.19   Exercises

1. Show directly from the definition of the integral that

   (i) $\displaystyle\int_0^1 x^3 \, d[x] = 1,$

   (ii) $\displaystyle\int_0^1 [2-x] \, dx = \int_0^1 [2-x] \, d[x] \neq \int_0^1 (2-[x]) \, d[x].$

2. Using only the definition of the integral show that

$$\int_1^2 f(x) \, d(x[x]) = \int_1^2 f(x) \, dx + 2f(2).$$

3. The functions $\alpha(t)$, $\beta(t)$ and $\gamma(t)$ are defined as follows.

    $\alpha(t) = \beta(t)\gamma(t),$

$$\beta(t) = \begin{cases} 1, & t < 1/2 \\ 2, & t \geqslant 1/2 \end{cases} \qquad \gamma(t) = \begin{cases} |t|e^{-t} & t \leqslant 1 \\ -t/2, & t > 1. \end{cases}$$

If $f(t)$ is defined and continuous on $[-1, 2]$ express the integral below as a Riemann integral together with some jump contributions.

$$\int_{-1}^2 f(x) \, d\alpha(x).$$

4. Calculate the following integrals

$$\int_0^2 x \, d([2^x]),$$

$$\int_0^3 x \, d(x[x^2]),$$

$$\int_2^5 [5-x] \, d(\log[x]),$$

$$\int_1^2 2x \, d(x[2^x]).$$

5. Using only the definition of the integral prove that

$$\int_a^b d\alpha(x) = \alpha(b) - \alpha(a).$$

6. Prove that

$$\int_0^n 2x \, d(x[x]) = n^3 + \sum_1^n r^2.$$

17.20 Exercises

1. The function $f(x)$ is defined as in Figure 17.36 so that

$$f(x) = 2^{-n} \quad \text{if} \quad 2^{-n-1} < x \leqslant 2^{-n}, \quad f(0) = 0.$$

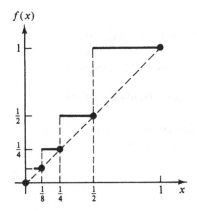

Fig. 17.36

Show that

(a) $\displaystyle\int_0^1 f(x) \, dx = 2/3,$

(b) $f(x) = 2^{-[-\log x/\log 2]},$

(c) $\displaystyle\int_0^t f(x) \, dx = tf(t) - \tfrac{1}{3}f(t)^2.$

2. Find

$$\int_0^n f(x) \, d(2x(1 + [x])) \quad \text{where} \quad f(x) = \int_x^\infty \exp(-t^2) \, dt.$$

What is the limit when $n \to \infty$?

3. Prove that

(a) $\displaystyle\sum_1^n \frac{1}{r^s} = \frac{1}{n^{s-1}} + s \int_1^n \frac{[x]}{x^{s+1}} \, dx,$

(b) $\displaystyle\sum_1^n \frac{(-1)^r}{r^s} = s \int_1^{2n} \frac{2[x/2]-[x]}{x^{s+1}}\, dx,$

(c) $\displaystyle\sum_1^n (-1)^{r+1} f(r) = \int_1^{2n} f'(x)(2[x/2]-[x])\, dx.$

4. If $\{a_n\}$ is a sequence of real numbers defined for $n \geqslant 1$, let

$$A(x) = \sum_{n \leqslant x} a_n.$$

Show, using integration by parts, that

$$\sum_{n \leqslant a} a_n f(n) = \left(\sum A(n)\{f(n) - f(n+1)\}\right) + A([a])f([a]+1).$$

5. Let $\langle x \rangle = x - [x]$. Assuming f' is continuous on $[a,b]$, show that

$$\sum_{a < n \leqslant b} f(n) = \int_a^b f(x)\, dx + \int_a^b f'(x)\langle x \rangle\, dx + f(a)\langle a \rangle - f(b)\langle b \rangle.$$

6. Amend the function in question 1 above so that $f(0) = 1$. Show that $f(x)$ does not have a minimum in $[0,1]$, but that its *infimum* is zero.

17.21   Exercises

1. Explain why

$$\int_0^x f(t)\, d[t] = \int_0^{[x]} f(t)\, d[t]$$

and also why

$$\int_0^n [t]^3\, dt = \int_0^{n-1} t^3\, d[t].$$

Integrate by parts

$$\int_0^n t^3\, d[t]$$

to obtain a formula for the 'sum of the cubes'

$$\sum_1^n r^3.$$

(Use the formulas earlier derived for the sum of the squares.)

2. Verify that

$$\int_0^y e^x\, d[x] = \frac{e^{[y]+1} - e}{e-1}$$

and find

$$\int_0^y xe^x\,d[x].$$

3. Show that for continuous f and for α and β of bounded variation

$$\int_a^b f\,d(\alpha\beta) = \int_a^b f\alpha\,d\beta + \int_a^b f\beta\,d\alpha.$$

4. Comment on the validity of the formula

$$\int_a^b 2\alpha(x)\,d\alpha(x) = \int_a^b d(\alpha^2(x)).$$

5. Find

$$\int_0^n x\,d(x^2[x^3]),$$

$$\int_0^n x\,d\left(\frac{[x]}{x}\right),$$

$$\int_0^3 x\,d[2^x].$$

6. Show that for a function f with continuous derivative

$$\int_0^n f'\,d(x[x]) = nf(n) + \int_0^n (xf' - f)\,d[x].$$

7. Verify the following identity when m and n are integers with $m < n$ and f is a differentiable function.

$$\sum_{r=m}^{n} f(r) = \int_m^n f(x)\,dx + \int_m^n f'(x)(x - [x] - \tfrac{1}{2})\,dx + \tfrac{1}{2}(f(m) + f(n)).$$

17.22 Exercises

1. Sketch two increasing functions α and β such that $\alpha - \beta$ has a graph like that in Section 17.17 Example 3.

2. Prove that the function $x^2 \cos(\pi/x)$ of Section 17.17 is of bounded variation on $[0, 1]$.

3. If f is of bounded variation on $[a, b]$ and $a \leqslant c \leqslant b$ show that f is of bounded variation on $[a, c]$ and on $[c, b]$ and that

$$V_f[a, b] = V_f[a, c] + V_f[c, b].$$

Deduce that $\alpha(x) = V_f[a, x]$ is increasing. Show that $V_f[x, y] \geqslant |f(y) - f(x)|$ and conclude that f is bounded and that $\beta(x) = f(x) - \alpha(x)$ is also increasing.

4. If α is increasing and f is continuous show that

$$\left| \int_1^2 f(x)\,d\alpha(x) \right| \le \int_1^2 |f(x)|\,d\alpha(x).$$

Extend this result appropriately to α of bounded variation. [Hint: use the increasing function V_α.]

5. If α and β are of bounded variation show that so are $\alpha + \beta$ and $\alpha \cdot \beta$. If moreover $|\beta|$ is bounded away from 0, show that β^{-1} is of bounded variation.

18

Manipulation of integrals

18.1 Overview

Here is a simple argument which will be justified in this chapter and which is typical of the trickery that we develop in the sections that follow.

Instead of computing by parts the integral

$$\int_1^2 x^n \log x \, dx,$$

we indulge in the following lateral thinking. Evidently for $t \neq -1$

$$\int_1^2 x^t \, dx = \frac{2^{t+1} - 1}{t+1}. \tag{1}$$

Now we differentiate both sides of the equation with respect to the parameter t and, remembering that $x^t = e^{t \log x}$, obtain

$$\frac{d}{dt} \int_1^2 x^t \, dx = \int_1^2 \frac{\partial}{\partial t}(x^t) \, dx = \int_1^2 x^t \log x \, dx.$$

But,

$$\frac{d}{dt} \frac{2^{t+1} - 1}{t+1} = \frac{(t+1)2^{t+1} \log 2 - 1(2^{t+1} - 1)}{(t+1)^2}.$$

Thus

$$\int_1^2 x^n \log x \, dx = \frac{(n+1)2^{n+1} \log 2 - 2^{n+1} + 1}{(n+1)^2}.$$

The integral in (1) is of the form

$$\int_a^b K(x, t) \, dx.$$

We say that the integrand $K(x, t)$ involves a parameter t. We shall see that if t varies over a *finite range*, say $c \leqslant t \leqslant d$ and that if K, as a function of the two variables x and t, is in an appropriate sense continuous in the domain $[a, b] \times [c, d]$ (cf. Figure 18.1) then the above argument will be valid.

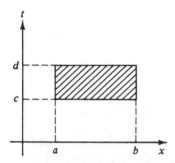

Fig. 18.1. *Finite range*

18.2   Joint continuity in two variables

Consider the function H illustrated in Figure 18.2. The function is given by the formula:

$$H(x, t) = \begin{cases} \dfrac{2xt}{x^2 + t^2} & \text{for } x \neq 0, \\ 0 & \text{if } x = 0. \end{cases}$$

If we take $t = mx$ with $x \neq 0$ we obtain

$$H(x, mx) = \frac{2m}{1 + m^2},$$

i.e. the value is constant on the straight line l_m in the $z = 0$ plane given by $t = mx$. Hence the surface may be represented as a set of line segments:

$$\left(x, mx, \frac{2m}{1 + m^2} \right) \quad (0 \leqslant x < \infty)$$

winding round the z-axis *plus* the point $(0, 0, 0)$. Notice that apart from the origin the z-axis has no points in common with the surface.

Observe that the limiting value of H as we approach $(0, 0)$ along the line l_m in the $z = 0$ plane equals $H(0, 0)$ only when $m = 0$ or $m = \infty$. Thus different ways of approaching the origin lead to different limiting values of H.

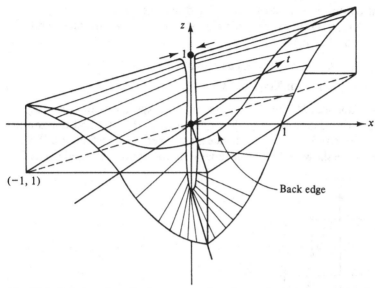

Fig. 18.2. Ruled surface. Surface approaches the z-axis without touching it.

We say that K is *jointly continuous* at the point (x_0, t_0) if

$$\lim_{n \to \infty} K(x_n, t_n) = K(x_0, t_0)$$

for all sequences $x_n \to x_0$ and $t_n \to t_0$ (cf. Figure 18.3). Thus *no matter how*
we approach (x_0, t_0) the value of the function K approaches $K(x_0, t_0)$.
As may be expected, K is said to be jointly continuous on $[a, b] \times [c, d]$
if it is jointly continuous at every point (x_0, t_0) in $[a, b] \times [c, d]$.

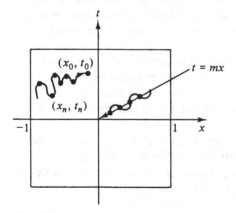

Fig. 18.3

Joint continuity should be contrasted with separate continuity. A function $K(x, t)$ is said to be *separately continuous* in the variables x and t at the point (x_0, t_0) if

(i) $K(x_0, t_0) = \lim K(x_0, t_n)$ as $t_n \to t_0$

(ii) $K(x_0, t_0) = \lim K(x_n, t_0)$ as $x_n \to x_0$

for any sequences $\{t_n\}$ and $\{x_n\}$ converging respectively to t_0 and x_0. Note that this definition considers only vertical $(t_n \to t_0)$ and horizontal $(x_n \to x_0)$ approaches to (x_0, t_0), hence is a weaker notion (cf. Figure 18.4). As an exercise show that $H(x, t)$ is separately continuous in x and t at $(0, 0)$.

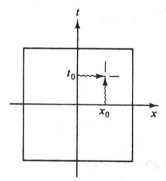

Fig. 18.4

18.3   Recognition of joint continuity

We usually recognise that a function $K(x, t)$ is jointly continuous on $[a, b] \times [c, d]$ by using the following facts. Henceforth we shall omit to use the word 'jointly' and so 'continuous' will always mean 'jointly continuous'.

1. The sum and product of two continuous function $G(x, t)$ and $K(x, t)$ viz. $G(x, t) + K(x, t)$ and $G(x, t)K(x, t)$ are again continuous;

2. The quotient of two continuous functions $G(x, t)$ and $K(x, t)$ viz. $G(x, t)/K(x, t)$ is again continuous *provided* $K(x, t) \neq 0$ throughout the domain of interest;

3. A function of a single variable, say $g(x)$, which is continuous under our (earlier) definition of continuity (applicable, evidently, only to functions of one variable) may be regarded as a function of two variables (albeit one of the two is absent); it is then also (jointly) continuous;

4. If $g : \mathbb{R} \to \mathbb{R}$ is continuous and $K(x, t)$ is continuous then so is $g(K(x, t))$.

Examples
(i) $2xt$ and $x^2 + t^2$ are continuous at the origin but their quotient is not.
(ii) Any polynomial in x and t is continuous.
(iii) For x positive, x^t is continuous since $x^t \equiv e^{t \log x}$ and e^z and $t \log x$ are continuous.

18.4 Manipulations involving finite ranges

For the moment let $K(x, t)$ be any function defined on $[a, b] \times [c, d]$. Let $\alpha(x)$ be a function of bounded variation on $[a, b]$. We consider three manipulations on the parameter t which arise from considering

$$f(t) = \int_a^b K(x, t)\,d\alpha(x).$$

1. Limits through the integral sign

Provided $K(x, t)$ is jointly continuous on $[a, b] \times [c, d]$ the function $f(t)$ is continuous on $[c, d]$; that is

$$f(t_0) = \lim f(t_n) \quad \text{as } t_n \to t_0$$

or,

$$\lim_{t \to t_0} \int_a^b K(x, t)\,d\alpha(x) = \int_a^b K(x, t_0)\,d\alpha(x).$$

Example
Take $K(x, t) = e^{-xt}$ and let

$$\alpha(x) = \int_a^x g(u)\,du,$$

where g is Riemann integrable on $[a, b]$. Then, even though g need not be continuous, we have

$$\lim_{t \to t_0} \int_a^b e^{-xt} g(x)\,dx = \int_a^b \exp(-xt_0) g(x)\,dx.$$

2. Differentiation under the integral sign

Provided the partial derivative

$$\frac{\partial}{\partial t} K(x, t)$$

is continuous on $[a,b] \times [c,d]$, then for any t with $c < t < d$, we have:

$$\frac{\partial}{\partial t} \int_a^b K(x,t)d\alpha(x) = \int_a^b \frac{\partial}{\partial t} K(x,t_0)d\alpha(x).$$

Here the range $[c,d]$ is *any* (proper) interval with t_0 in its interior, i.e. $c < t_0 < d$. This will be of use to us later (and is to be remembered quite generally in Chapter 21).

Example

$$\frac{\partial}{\partial t} \int_1^2 x^t dx = \int_1^2 x^t \log x\, dx \quad \text{for any } t.$$

Indeed, if t_0 is given, choose any two numbers c and d with $c < t_0 < d$. Evidently $a = 1$ and $b = 2$. We saw that $K(x,t) \equiv x^t = e^{t \log x}$ is continuous on $[1,2] \times [c,d]$; but this time we require the continuity of

$$\frac{\partial}{\partial t} K(x,t) = x^t \log x$$

and this is assured by Section 18.3. Note that the restriction $t \neq -1$ applies only to formula (1). We have now justified the example at the beginning of the chapter.

3.   Integration with respect to the parameter

Given a function of bounded variation $\beta(t)$ defined on $[c,d]$ we may wish to perform the integration

$$\int_c^d f(t)d\beta(t).$$

Here again, *provided $K(x,t)$ is continuous on $[a,b] \times [c,d]$* we can integrate with respect to t before integrating with respect to x

$$\int_c^d f(t)d\beta(t) = \int_a^b \left\{ \int_c^d K(x,t)d\beta(t) \right\} d\alpha(x).$$

Example
Evaluate

$$I_t = \int_0^\pi \frac{\log(1 + t\cos x)}{\cos x} dx \quad \text{for } |t| < 1.$$

We notice that

$$\frac{d}{dt}\int_0^\pi \frac{\log(1+t\cos x)}{\cos x}dx = \int_0^\pi \frac{dx}{1+t\cos x}.$$

This is true for any $|t| < 1$. Indeed if $|t_0| < 1$ choose c and d with $-1 < c < t_0 < d < 1$. Then since $1+t\cos x \neq 0$ in $[0,\pi] \times [c,d]$ the integrand on the right-hand side is continuous and part (2) applies. The trick involving the choice of c and d so that $[c,d]$ includes t_0 is designed merely to produce a rectangle $[0,\pi] \times [c,d]$ in which continuity is assured. Certainly it would have been inappropriate to take the rectangle $[0,\pi] \times [-1,1]$ since $1 + \cos(\pi) = 0$ and part (2) can no longer be applied. Returning to the calculation we have (cf. 18.16 Postscript)

$$\int_0^\pi \frac{dx}{1+t\cos x} = \frac{\pi}{\sqrt{(1-t^2)}}.$$

Integrating now with respect to t from 0 to s where $|s| < 1$ we have

$$I = \int_0^s\int_0^\pi \frac{dx}{1+t\cos x}dt = \pi\int_0^s \frac{dt}{\sqrt{(1-t^2)}} = \pi\arcsin(s).$$

18.5 Limits of integration depending on the parameter

We consider a more complicated dependence on the parameter given by

$$f(t) = \int_{p(t)}^{q(t)} K(x,t)dx,$$

for $c \leqslant t \leqslant d$. For simplicity we have considered only Riemann integration this time. We assume that $p(t)$ and $q(t)$ are differentiable and attempt to compute $f'(t)$. This can be done quite easily using the chain rule. But first recall that

$$\frac{d}{dt}\int_c^t \phi(x)dx = \phi(t)$$

and

$$\frac{d}{dt}\int_t^d \phi(x)dx = \frac{d}{dt}\left\{-\int_d^t \phi(x)dx\right\} = -\phi(t).$$

To make things a little easier to understand we first define

$$F(x_1,x_2,x_3) = \int_{x_1}^{x_2} K(x,x_3)dx,$$

where we assume that K is defined on a rectangle $[a,b] \times [c,d]$ which includes the graphs of $p(t)$ and $q(t)$, compare Figure 18.5. Now by the

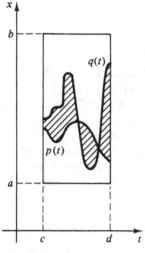

Fig. 18.5

remark made just before the definition of F we have

$$\frac{\partial F}{\partial x_2} = K(x_2, x_3) \quad \text{and} \quad \frac{\partial F}{\partial x_1} = -K(x_1, x_3),$$

while

$$\frac{\partial}{\partial x_3} F(x_1, x_2, x_3) = \int_{x_1}^{x_2} \frac{\partial}{\partial x_3} K(x, x_3) \, dx.$$

Now substituting $x_1 = p(t), x_2 = q(t)$ and $x_3 = t$ we have by the chain rule

$$f'(t) = \frac{d}{dt} F(p(t), q(t), t)$$

$$= \frac{\partial F}{\partial x_1} \cdot \frac{dx_1}{dt} + \frac{\partial F}{\partial x_2} \cdot \frac{dx_2}{dt} + \frac{\partial F}{\partial x_3} \cdot \frac{dx_3}{dt}$$

$$f'(t) = -K(p(t), t)p'(t) + K(q(t), t)q'(t) + \int_{p(t)}^{q(t)} \frac{\partial K}{\partial t}(x, t) \, dx.$$

Example
We indulge in some make-believe. We pretend that we do not know that the value of

$$f(t) = \int_1^t \frac{dx}{x}$$

is log t and deduce a basic property about the logarithm function straight from the formula for $f(t)$. Fix some $s > 0$ and observe that

$$f(st) - f(t) = \int_t^{st} \frac{dx}{x}.$$

Differentiating the right-hand side with respect to t (treating s as constant) gives

$$\frac{\partial}{\partial t} \int_t^{st} \frac{dx}{x} = -\frac{1}{t} + \frac{1}{st} \cdot s = 0.$$

Thus the value of $f(st) - f(t)$ is independent of t and is thus a constant (evidently a constant dependent on s). To find the value of the constant put $t = 1$ and note that $f(1) = 0$, thus

$$f(st) - f(t) = f(s) - f(1) = f(s),$$

i.e.

$$f(st) = f(s) + f(t),$$

which is the fundamental property of the logarithm function.

18.6 Infinite ranges: improper integrals

Until now we have been manipulating integrals over finite ranges. It is therefore time to leave this sheltered situation and consider infinite ranges such as are encountered, say, in the well-known expression

$$N(t) = \frac{2}{\sqrt{\pi}} \int_{-\infty}^{t} \exp(-x^2) \, dx.$$

We must first reconsider our definition of an integral so as to cover infinite ranges of integration.

Definition
Suppose $\alpha(x)$ is defined on $[a, \infty)$ and that for each $b \geqslant a$ the integral

$$\int_a^b f(t) \, d\alpha(t)$$

exists. If also the limit

$$\lim_{b \to \infty} \int_a^b f(t) \, d\alpha(t)$$

*exists, we say that the **improper** integral of f exists over $[a, \infty)$, and* *denote its value either by*

$$\int_a^{\to\infty} f(t)\,d\alpha(t)$$

(*when we wish to draw attention to the limit process involved*), *or just*

$$\int_a^{\infty} f(t)\,d\alpha(t).$$

To distinguish this limit process from a closely related one (which we introduce a little later) the integral is more fully described as **improper of the first kind**.

Examples

$$\int_1^b \frac{dx}{x^2} = \left[-\frac{1}{x}\right]_1^b = 1 - \frac{1}{b} \to 1 \quad \text{as} \quad b \to \infty,$$

$$\int_1^b \frac{dx}{x} = [\log x]_1^b = \log b \to \infty \quad \text{as} \quad b \to \infty,$$

$$\int_1^b e^{-x}dx = [-e^{-x}]_1^b = e^{-1} - e^{-b} \to e^{-1}, \quad \text{as} \quad b \to \infty,$$

$$\int_1^b \frac{d[x]}{x} = \sum_{1 < n \leqslant [b]} \frac{1}{n} \to \infty \quad \text{as} \quad b \to \infty \quad \text{(cf. Figure 18.6)},$$

$$\int_1^b \frac{dx}{1+x^2} = [\arctan x]_1^b \to \frac{\pi}{2} \quad \text{as } b \to \infty.$$

The integral over the range $(-\infty, a]$ is defined analogously (taking limits

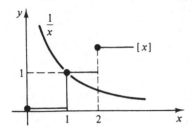

Fig. 18.6

as b tends to $-\infty$). *It is very important* to note that the integral

$$\int_{-\infty}^{+\infty} f(t)\,d\alpha(t)$$

is defined *only when for some a both the integrals*

$$\int_{a}^{\infty} f(t)\,d\alpha(t), \quad \int_{-\infty}^{a} f(t)\,d\alpha(t)$$

exist. In other words the definition for the integral over $(-\infty, +\infty)$ is

$$\lim_{b\to\infty} \int_{a}^{b} f(t)\,d\alpha(t) + \lim_{c\to-\infty} \int_{c}^{a} f(t)\,d\alpha(t),$$

where the two limits are taken independently. The following example clarifies the point of the definition.

Example
Observe that

$$\int_{-b}^{b} \frac{x\,dx}{1+x^2} = \left[\tfrac{1}{2}\log(1+x^2)\right]^{b}_{-b} = 0.$$

Nevertheless,

$$\lim_{b\to\infty} \int_{0}^{b} \frac{x\,dx}{1+x^2} = \lim_{b\to\infty} \left[\tfrac{1}{2}\log(1+x^2)\right]^{b}_{0} = \infty.$$

Hence the integral

$$\int_{-\infty}^{\infty} \frac{x\,dx}{1+x^2}$$

does not exist. A quick glance at the graph (Figure 18.7) will show that the area under the curve on the positive side of the y-axis is infinite and

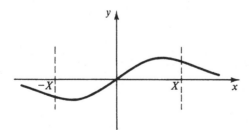

Fig. 18.7

is a symmetric image (reflection through the origin) of the area under the curve on the negative side of the y-axis. Since these two counterbalance, the earlier calculation had to come to zero. It is the infinite areas cropping up in the calculation which give rise to non-existence.

Remark 1

The example gives us a perfectly acceptable probability distribution function with non-existent 'expectation'. Take

$$\alpha(x) = \frac{1}{\pi} \int_{-\infty}^{x} \frac{dt}{1+t^2}.$$

This is an increasing function with $\lim \alpha(x) = 1$ (as $x \to \infty$). Now the expected value of the random variable X when it has $\alpha(x)$ for its probability distribution function is (see page 226 and Section 19.16)

$$\int_{-\infty}^{+\infty} x \, d\alpha(x),$$

but this does not exist.

Remark 2

We do have some use for the limit

$$\lim_{b \to \infty} \int_{-b}^{+b} f(t) \, d\alpha(t)$$

and when it exists it is known as the *Cauchy principal value*. For an application see Question 6 Exercises 22.11 in the Fourier Series section of Chapter 6.

Improper integrals of the second kind arise when, for example, the integrand is continuous in $(a, b]$ but 'blows up' at a, meaning that the limit of $f(t)$ as t tends to a is infinite (cf. Figure 18.8). Suppose under

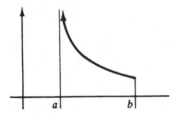

Fig. 18.8

such circumstances the limit

$$\lim_{\delta \to 0+} \int_{a+\delta}^{b} f(t)\,d\alpha(t)$$

exists (where the plus sign is used to show that in this limit process δ ranges through *positive values*). Then we say that the improper integral exists over $[a, b]$ and write it in either of the forms

$$\int_{\to a}^{b} f(t)\,d\alpha(t), \quad \int_{a}^{b} f(t)\,d\alpha(t).$$

Example
The following improper integral of the second kind exists.

$$\int_{\delta}^{1} \frac{dx}{\sqrt{x}} = [2\sqrt{x}]_{\delta}^{1} = 2 - 2\sqrt{\delta} \to 2 \quad \text{as} \quad \delta \to 0+.$$

Our next task is to learn to recognise in more complicated cases whether an improper integral exists without attempting to compute the integral in question.

18.7 Tests for convergence – part 1 (positive integrands)

Until further notice we assume that we are investigating only integrals in which the integrand is positive. This means that the area under its curve is either a finite (positive) number or just infinity. We will want to know whether the first case arises.

(i) Simple comparison

The typical argument proceeds as follows. We wish to know whether

$$\int_{1}^{\to \infty} \frac{dx}{x + x^2}$$

exists. We replace the given integrand by a function which is both larger and easier to compute. The reason for increasing the integrand is that, since the integrand is positive, we are merely increasing the area under the curve (see Figure 18.9). If that is finite then the smaller area was *a fortiori* also finite. For the case in point we note

$$x + x^2 \geqslant x^2.$$

Hence

$$\frac{1}{x + x^2} \leqslant \frac{1}{x^2}.$$

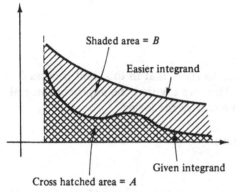

Shaded area = B

Easier integrand

Given integrand

Cross hatched area = A

Fig. 18.9. $0 < A < B$ so $B < \infty \Rightarrow A < \infty$

Thus

$$\int_1^b \frac{dx}{x+x^2} \leqslant \int_1^b \frac{dx}{x^2} = \left[\frac{1}{x}\right]_1^b = 1 - \frac{1}{b} \rightarrow 1 \quad \text{as} \quad b \rightarrow \infty.$$

So the larger integral is finite, and hence so also is the original integral. Thus the integral in question exists. Simple though the argument is, it has the drawback that occasionally in choosing an easier and bigger integrand more ingenuity is called for than is worth the effort. A better form of test is given in the next subsection. To see its power observe the reasoning in the next example.

Example
Test if

$$\int_1^{\rightarrow \infty} \frac{dx}{x^2 - x + 1}$$

exists. It will not do here to ignore the terms $-x+1$. We need to estimate them against the leading term. Note that for $x \geqslant 2$

$$x \leqslant \tfrac{1}{2}x^2.$$

Hence for $x \geqslant 2$

$$x^2 - x + 1 \geqslant x^2 - \tfrac{1}{2}x^2 + 1 \geqslant \tfrac{1}{2}x^2,$$

so for $x \geqslant 2$

$$\frac{1}{x^2 - x + 1} \leqslant \frac{2}{x^2}.$$

Thus

$$\int_2^b \frac{dx}{x^2 - x + 1} \leqslant \int_2^b \frac{2dx}{x^2} = \left[-\frac{2}{x} \right]_2^b = 1 - \frac{2}{b} < 1.$$

Perhaps it is already becoming clear from both examples that, in fact, the crucial observation underlying the two arguments for the convergence of the integral lies in the behaviour of the 'leading term'. The next test concentrates on this idea; interestingly enough the justification for the second test is in the simple comparison technique mentioned at the outset.

(ii)   Limit comparison

It is best to state the test and then to see it in action.

Theorem
Suppose $\alpha(x)$ is increasing and that $g(x) \geqslant 0$ and $f(x) \geqslant 0$ for all large enough x. Suppose also that the following two integrals exist for all $b \geqslant a$:

$$\int_a^b f(x)d\alpha(x), \quad \int_a^b g(x)d\alpha(x).$$

Suppose moreover that

$$\lim_{x \to \infty} \frac{f(x)}{g(x)} = c \quad \text{and} \quad 0 < c < \infty.$$

Then

$$\int_a^\infty f(x)d\alpha(x) \quad \text{exists if and only if} \quad \int_a^\infty g(x)d\alpha(x) \text{ exists.}$$

Proof. We reduce this situation to a direct comparison test. Since $c > 0$ we have $c/2 < c < c + 1$ (it is important that the inequalities are strict, implying that we have moved back from c to $c/2$ whilst staying to the right of 0, and have also moved forward to $c + 1$). Hence for all large enough x, say for all $x > X$, we have

$$\frac{c}{2} < \frac{f(x)}{g(x)} < c + 1,$$

so

$$\frac{c}{2}g(x) \leqslant f(x) \leqslant (c + 1)g(x),$$

and thus

$$\frac{c}{2}\int_X^b g\,d\alpha \leqslant \int_X^b f\,d\alpha \leqslant (c+1)\int_X^b g\,d\alpha.$$

Now if

$$\int_X^{\to\infty} g\,d\alpha$$

exists and we define

$$I_b = \int_X^b f\,d\alpha,$$

then for all b

$$I_b \leqslant (c+1)\int_X^b g\,d\alpha,$$

hence

$$\lim_{b\to\infty} I_b \leqslant (c+1)\int_X^{\to\infty} g\,d\alpha.$$

Since the bound on the right is finite and the limit of I_b can be only either infinity or some finite positive quantity, we conclude that it is *not* infinity and so

$$\int_X^{\to\infty} f\,d\alpha$$

exists. Conversely, if

$$\int_X^{\to\infty} g\,d\alpha = \infty,$$

then since

$$\frac{c}{2}\int_X^b g\,d\alpha \leqslant \int_X^b f\,d\alpha$$

it follows that I_b must increase with b unboundedly and so

$$\int_X^{\to\infty} f\,d\alpha = \infty,$$

that is, the integral above does not exist (since it diverges to infinity).

Remark

The way to use this test given an integrand f is to decide first the approximate behaviour of f for large x. We do this by picking a function g which we can easily integrate and which we suspect approximates to f for large x. The goodness of the approximation is tested by the limit

$$\lim_{x \to \infty} \frac{f(x)}{g(x)}.$$

If this is *finite and positive* (strictly positive, that is) the approximation is good. The choice of g inevitably comes from the stock of 'everyone's favourites' – well-known cases which might therefore be regarded in this connection as a 'tool kit' for convergence testing. The examples which follow will elucidate this point.

Example 1

Show that

$$\int_1^{\to \infty} \frac{dx}{x\{\sqrt{(x^2 + x)} - x\}}$$

does not exist.

At first one thinks to approximate $\sqrt{(x^2 + x)}$ by $\sqrt{(x^2)} = x$. This unfortunately makes the denominator of the integrand zero which is not a valid expression. To get rid of the square root in the denominator we rationalise as follows.

$$\frac{1}{x\{\sqrt{(x^2 + x)} - x\}} = \frac{\sqrt{(x^2 + x)} + x}{x\{(x^2 + x) - x^2\}}.$$

We now proceed to approximate the numerator crudely as $x + x$. We thus consider the integrand to be approximately

$$\frac{2x}{x\{x\}}$$

and we therefore take for our g the function $2/x$. We then have

$$\frac{f(x)}{g(x)} = \frac{x}{2} \cdot \frac{\sqrt{(x^2 + x)} + x}{x^2} = \frac{1}{2}\{1 + \sqrt{(1 + 1/x)}\},$$

and this tends to unity as x tends to ∞. But, evidently,

$$\int_1^{\to \infty} \frac{2dx}{x}$$

does not exist. Hence the integral in question does not exist.

Example 2

Test whether

$$\int_{1}^{\to\infty} \left(\frac{1}{x^{x+1}}\right)^{1/x} dx$$

exists.

The first question to ask about the integrand is what power is x being raised to, or more properly, how does the integrand compare with x^{-1}. The point is, of course, that integration of x^{-1} in this range returns a divergent answer, whereas for (fixed) powers $x^{-(1+\alpha)}$ with α positive the integral over this same range converges. Now the integrand here is

$$f(x) = \frac{1}{x^{(1+1/x)}}.$$

For large values of x the exponent $(1 + 1/x)$ is extremely close to 1, so we should half expect divergence. We calculate the ratio of f to g where $g(x) = x^{-1}$ to be $x^{1/x}$ and now need to know the limit of this ratio as x tends to infinity. To calculate this limit we examine the behaviour of the function

$$\phi(x) = x^{1/x} \quad \left(= \exp\left(\frac{1}{x}\log x\right)\right).$$

First we test whether ϕ is increasing or decreasing. The derivative is

$$\phi'(x) = x^{1/x}\left\{\frac{1}{x^2} - \frac{\log x}{x^2}\right\} = x^{1/x-2}\{1 - \log x\},$$

so this is negative when $\log x > 1$, i.e. for $x > e$. Since ϕ is always positive and is decreasing the limiting value of ϕ as x tends to infinity must itself be positive or zero. Let us put

$$a = \lim_{x\to\infty} x^{1/x}.$$

To evaluate a we attempt to relate a to the expression on the right of the limit sign. We have for all $x > e$, since ϕ is decreasing,

$$a \leqslant x^{1/x},$$

so $a^x \leqslant x$ (for $x \geqslant e$). It follows (why?) that $a \leqslant 1$. Suppose that $a < 1$. Choose b with $a < b < 1$. Then since b is greater than a, the function ϕ must eventually be below b. So for large enough x we have that

$$x^{1/x} \leqslant b,$$

so $x \leqslant b^x$. But since $b < 1$ the limit of b^x as x tends to infinity is 0 and this contradicts our last inequality. Hence, after all $a = 1$. By the ratio test the integral diverges.

Example 3
Show that the following integral converges for any t.

$$\int_{-\infty}^{t} \exp(-x^2/2)dx.$$

It suffices to show that this integral exists for $t = -1$. (Evidently it exists over the finite range $[-1, t]$.) By a change of variable this is equivalent to considering

$$\int_{1}^{\infty} \exp(-x^2/2)dx.$$

Now for $x > 1$ we have $x^2 > x$, so

$$\int_{1}^{\infty} \exp(-x^2/2)dx \leqslant \int_{1}^{\infty} \exp(-x/2)dx < \infty.$$

and existence is assured by simple comparison.

18.8 L'Hôpital's rule

Example 2 in the last section required careful work to establish the value of a ratio occurring in the ratio test. Ratio calculations can also arise in regard to the integrand itself. For example, if we needed to test for convergence the integral

$$\int_{\to 0}^{1} \frac{\sin x}{x} dx,$$

we would need to know that, as x tends to zero, the integrand tends to 1. This is common knowledge, but in many more awkward cases the calculation required can be much simplified by reference to a theorem named after L'Hôpital.

Theorem
Suppose that $f(x), g(x)$ have continuous derivatives in the neighbourhood to the left of $x = a$ and that their limits from the left are

$$\lim_{x \to a-} f(x) = \lim_{x \to a-} g(x) \quad \text{and the common value is 0 or } \infty.$$

Suppose also that

$$\lim_{x \to a-} \frac{f'(x)}{g'(x)}$$

exists, then

$$\lim_{x \to a-} \frac{f(x)}{g(x)} \quad \textit{exists and is equal to} \quad \lim_{x \to a-} \frac{f'(x)}{g'(x)}.$$

The theorem also holds when 'to the right' replaces 'to the left' provided $a-$ is replaced by $a+$. The values $\pm \infty$ for a are permitted.

Examples

(a)

$$\lim_{x \to 0} \frac{\sin x}{x} = \lim_{x \to 0} \frac{\cos x}{1} = 1.$$

(b)

$$\lim_{x \to 1} \frac{\log x}{1-x} = \lim_{x \to 1} \frac{1/x}{-1} = -1.$$

(c) Find for $p > 0$ the limit of $x^p \log x$ as x tends to 0.
We rearrange this expression as a ratio and calculate that:

$$\lim_{x \to \infty} \frac{\log x}{x^{-p}} = \lim_{x \to \infty} \frac{x^{-1}}{-px^{-p-1}} = 0.$$

18.9   Tests for convergence – part 2 (positive integrands)

We note that the blanket assumption that all the integrands considered are positive continues to hold until further notice. We start off with the remark that part of the Ratio Test can be salvaged even when the limiting ratio is zero or infinity. One should be very careful in applications when these values are involved. The results are as follows:
(i) if $c = 0$ where

$$c = \lim_{x \to \infty} \frac{f(x)}{g(x)}$$

and the integral of g is finite then so is the integral of f;
(ii) if $c = \infty$ and the integral of g diverges then the integral of f diverges.
   The next example illustrates this point.

Example 1

Test if the following integral converges for $s > 0$:

$$\int_1^{\to \infty} x^s e^{-sx} dx.$$

In view of the rapid decay of the exponential factor we expect convergence and test using $g(x) = x^{-2}$. The choice of g is merely that of a convenient function whose integral converges in this range. The ratio comes to

$$\lim_{x \to \infty} \frac{x^s e^{-sx}}{1/x^2} = \lim_{x \to \infty} x^{s+2} e^{-sx} = 0.$$

Since the integral of g converges, so does the integral in question.

Note however that the choice $g(x) = x^{-1}$ also leads to a zero limiting ratio. Nevertheless the integral in question converges despite the divergence of the test function x^{-1}. This is not a contradiction since the ratio test makes no claim in regard to divergence when the limit ratio $c = 0$.

We now continue with some further tests.

(iii)   Integral test

For this test we need to assume not only that $f(x)$ is positive but also that it is continuous and *decreasing* in the range $[a, \infty)$. We take $a = 1$ for convenience of notation. Under these assumptions we can relate the integral

$$\int_1^{\to \infty} f(x) dx$$

to the sum

$$\sum_{n=1}^{\infty} f(n)$$

and vice versa. This is because (cf. Figure 18.10) for any i

$$f(i+1) \leqslant \int_i^{i+1} f(x) dx \leqslant f(i).$$

Hence

$$\sum_2^n f(i) = \sum_1^{n-1} f(i+1) \leqslant \int_1^n f(x) dx \leqslant \sum_1^{n-1} f(i).$$

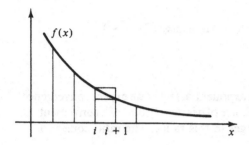

Fig. 18.10

Thus the sum and the integral

$$\sum_1^\infty f(i), \quad \int_1^\infty f(x)dx$$

either both converge or both diverge.

The above result may also be put to powerful use in conjunction with the following test.

(iv)   Condensation test

If $\{f(n)\}$ is a decreasing sequence of positive terms, then

$$\sum_1^\infty f(n) \quad \text{converges} \quad \Leftrightarrow \quad \sum_1^\infty 2^n f(2^n) \text{ converges.}$$

Proof We have for each n and $i = 1, 2, \ldots, 2^n$

$$f(2^{n+1}) \leqslant f(2^n + i) \leqslant f(2^n)$$

so

$$2^n f(2^{n+1}) \leqslant f(2^n + 1) + f(2^n + 2) + \cdots + f(2^n + 2^n) \leqslant 2^n f(2^n).$$

Thus

$$\frac{1}{2}\sum_0^\infty 2^{n+1} f(2^{n+1}) \leqslant \sum_2^\infty f(i) \leqslant \sum_0^\infty 2^n f(2^n).$$

Example 2

(a)

$$\sum_1^\infty \frac{1}{n} \quad \text{diverges, since } \sum 2^n \cdot \frac{1}{2^n} = \infty.$$

(b)

$$\sum \frac{1}{(\log n)^2} \quad \text{diverges, since} \quad \sum \frac{2^n}{(\log 2^n)^2} = \sum \frac{2^n}{n^2 (\log 2)^2} = \infty.$$

This last series diverges since its nth term tends to infinity.
(c) Test for convergence the integral

$$\int_2^{\to \infty} \frac{dx}{(\log x)^2}.$$

By example (*b*) this integral diverges.

Remark
   The use of the condensation test may be viewed as the analogue
of an exponential change of variable. In the last example put $x = e^y$ and
the integral becomes

$$\int_{\log 2}^{\to \infty} \frac{e^y dy}{y^2}.$$

This diverges because the integrand itself tends to infinity as $y \to \infty$. The
integral formula evidently resembles the series formula of example (b).

18.10   Tests for convergence – part 3 (integrand of variable sign)

In this section we shall lift the blanket assumption of positivity and turn
our attention to integrands which are of variable sign. Our first test
originates in the study of series. To apply it we must label the areas
under the curve lying successively above and below the x-axis as
$a_0, a_1, a_2, a_3, \ldots, a_n, \ldots$, these being the absolute values of the areas; see
Figure 18.11.
   With this in mind we may formulate our next test.

(v)   Alternating sign test

If $a_0 > a_1 > a_2 > \cdots \geqslant 0$ and $\lim_{n \to \infty} a_n = 0$, then

$$\sum_0^{\infty} (-1)^n a_n = a_0 - a_1 + a_2 - a_3 + \cdots + (-1)^n a_n + \cdots \text{converges}.$$

The explanation is quite simple. Consider s_n the partial sum up to the
nth term. Notice that in any of these partial sums each time that a
positive term is added on, it is less in magnitude than the preceding term

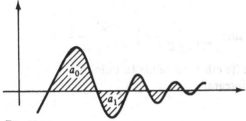

Fig. 18.11

which has been subtracted. As a result the partial sums with n odd form an increasing sequence:

$$s_1 < s_3 < s_5 < s_7 < \cdots.$$

Similarly, each time a term is subtracted the following term which is added, being smaller in magnitude, does not entirely manage to offset its predecessor. As a result the partial sums with n even this time form a decreasing sequence: $s_0 > s_2 > s_4 > s_6 > \cdots$. Both sequences are bounded, so they both have a limit; the two limits are identical since the discrepancy between an 'odd' term, say s_{2n+1}, and its 'even' counterpart s_{2n} is a_{2n+1}, but that tends to zero.

Example 1

$$I = \int_1^{\to \infty} \frac{(-1)^{[x]} dx}{x} \qquad \text{converges.}$$

Let

$$a_n = \int_{n+1}^{n+2} \frac{dx}{x} \qquad (n = 0, 1, 2, 3, \ldots)$$

(cf. Figure 18.12) then $I = -\sum (-1)^n a_n$ and, of course,

$$a_0 > a_1 > a_2 > a_3 > \cdots \to 0.$$

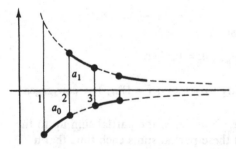

Fig. 18.12

Example 2

Show that $I = \displaystyle\int_{1}^{\to\infty} \dfrac{\sin x \, dx}{x}$ converges.

We give an alternative explanation in the exercises. We define

$$a_n = \left| \int_{(n+1)\pi}^{(n+2)\pi} \frac{\sin x \, dx}{x} \right| \qquad (n = 0, 1, 2, 3, \ldots)$$

(cf. Figure 18.13) then, as before, $I = -\sum (-1)^n a_n$ and to see that $a_0 > a_1 > a_2 > a_3 > \cdots \to 0$, we note that $|\sin(x + \pi)| = |\sin x|$ and

$$\frac{|\sin x|}{x} > \frac{|\sin(x + \pi)|}{x + \pi}.$$

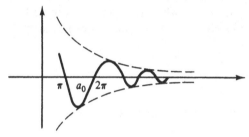

Fig. 18.13

(vi) Unconditional convergence

A relatively straightforward technique, much less powerful than the last test, though certainly easier, consists in replacing the integral in question

$$\int_{1}^{\to\infty} f(x) dx$$

by

$$\int_{1}^{\to\infty} |f(x)| dx.$$

If the latter converges we say that the original integral *converges unconditionally*. The procedure amounts to 'flipping over' the negative areas between the curve and x-axis so that they become positive (see Figure 18.14). To say that an integral is unconditionally convergent is to say that, after the flipping-over operation, all the areas (i.e. including also those previously negative) add up to a finite quantity A. It follows

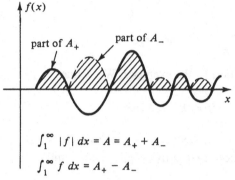

$$\int_1^\infty |f|\, dx = A = A_+ + A_-$$

$$\int_1^\infty f\, dx = A_+ - A_-$$

Fig. 18.14

that the areas which were originally positive must add up to a smaller and therefore also a finite quantity, say A_+, whilst those that were initially negative and were flipped over must likewise total to something less than A say to A_-. Evidently

$$A = A_+ + A_- = \int_1^{\to \infty} |f(x)|\, dx,$$

so that, restoring the flipped-over areas to their original sign,

$$I = A_+ - A_- = \int_1^{\to \infty} f(x)\, dx,$$

and I is finite. A convergent integral need not be unconditionally convergent, and we might therefore say in such a case that the integral, *converges conditionally*. In general, it need not be the case that the negative areas as well as the positive areas both add up to finite quantities. The alternating sign test allows for situations where piece by piece summation of alternately positive and negative areas gives cancellations between terms and this may yield a finite answer for the limiting sum. The next two examples elucidate this point.

Example 3

$$I = \int_1^{\to \infty} \frac{\sin x\, dx}{x^2} \qquad \text{converges unconditionally.}$$

If we take the modulus of the integrand we have

$$\frac{|\sin x|}{x^2} \leqslant \frac{1}{x^2}$$

and by the comparison test we obtain convergence.

Example 4 (warning example)

$$\int_1^{\to\infty} \frac{\sin x\,dx}{x} \qquad \text{converges conditionally only.}$$

We are saying here that

$$I = \int_1^{\to\infty} \frac{|\sin x|}{x}\,dx \qquad \text{does } not \text{ converge.}$$

The reason for this situation is the non-existence of

$$I = \int_1^{\to\infty} \frac{dx}{x}.$$

In order to apply the comparison test we note that in $[\pi/4, 3\pi/4]$ we have the inequality

$$|\sin x| \geqslant 1/\sqrt{2}.$$

Thus

$$I \geqslant \sum_1^\infty \int_{2n\pi+\pi/4}^{2n\pi+3\pi/4} \frac{|\sin x|\,dx}{x}$$

$$\geqslant \sum_1^\infty \frac{1}{\sqrt{2}} \cdot \frac{1}{2n\pi+3\pi/4} \cdot \frac{\pi}{2} = \sum_1^\infty \frac{\sqrt{2}}{8n+3} = \infty,$$

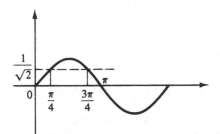

Fig. 18.15

where the individual integrals have been estimated by reference to the least value of the integrand in the relevant interval and to the length of the interval of integration.

18.11   Functions defined by infinite integrals

We now consider extending the range of validity for the manipulations
introduced in Section 18.4 to functions of the form

$$\int_a^{\to\infty} K(x,t)d\alpha(x) \qquad (c \leqslant t \leqslant d).$$

The domain of K is illustrated in Figure 18.16. Suppose for example that

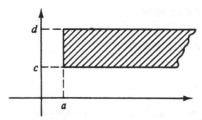

Fig. 18.16

the integral

$$\int_a^{\to\infty} g(x)dx$$

exists and we let

$$f(t) = \int_a^{\to\infty} e^{-tx}g(x)dx.$$

We ask whether it is true that

$$\lim_{t\to 0} \int_a^{\to\infty} e^{-tx}g(x)dx = \int_a^{\to\infty} g(x)dx.$$

The natural approach to answering such a question is to replace the
infinite range in the integral by a finite range which will allow
approximation of the given integral to within a specified accuracy.
Evidently, the manipulations could then be carried out in the finite
range. Suppose, for example, that we desire an accuracy of δ. Then for
some b (note that the choice may depend on t) we have

$$\int_a^{\to\infty} e^{-tx}g(x)dx = \int_a^b e^{-tx}g(x)dx + error,$$

where $|error| < \delta$. Then for this value of b we can safely assert that

$$\lim_{t \to 0} \int_a^b e^{-tx} g(x)\,dx = \int_a^b g(x)\,dx.$$

However, error may 'blow up' if the b above has to depend on t.

Example
Let

$$K(x,t) = 2xt(1-t)^x + x^2 t(1-t)^x \log(1-t) \qquad \text{for } 0 \leqslant t < 1.$$

The function is defined so that for fixed t we have the result:

$$\int_0^b K(x,t)\,dx = b^2 t(1-t)^b,$$

and this tends to 0 as b tends to infinity. Illustrated in Figure 18.17 are

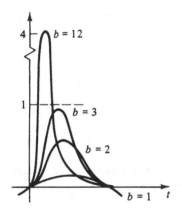

Fig. 18.17

the graphs of

$$y = f_b(t) = b^2 t(1-t)^b$$

for $b = 1, 2, \ldots$. Notice that for fixed b the graph peaks at

$$t = \frac{1}{1+b}$$

(which is close to the origin for large b) and the peak value is

$$\frac{b^2}{1+b}\left(\frac{b}{1+b}\right)^b.$$

This peak value tends to infinity as b tends to infinity.

Now notice that if for a given t we wish to approximate to

$$\int_0^{\to\infty} K(x,t)dx \qquad (=0)$$

by means of

$$\int_0^b K(x,t)dx \qquad (=b^2t(1-t)^b)$$

to an accuracy of, say, δ then the choice of b may be investigated by reference to the graphs indicated in Figure 18.18 as follows. Draw the

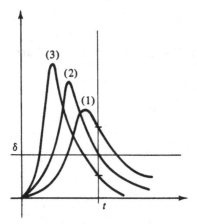

Fig. 18.18 *Graphs for (1) a small value of* b, *(2) the 'critical' value of* b, *(3) values of* b *beyond the critical all of which have intercept below* δ.

(vertical) ordinate line for the given t. Mark off $y = \delta$ on this vertical. For some values of b the graph of f_b cuts the vertical above δ. But for all large enough b beyond a critical starting value the peak of f_b is to the left of this vertical and the tail part of the curve is below δ. Now observe what happens when the value of t is taken closer and closer to 0. The appropriate critical value of b rockets away simply because of the behaviour of the peak values. In fact the smaller the value of t the larger the critical value of b must be. Certainly if

$$t = \frac{1}{1+n} \qquad (n = 1, 2, 3, \ldots)$$

then the critical value of b is greater than n. In other words for the approximation to be accurate when t is small we need b quite large, and there is a definite dependence between the size of t and the appropriate size of b.

We say that the convergence of the integral

$$\int_a^{\to\infty} K(x,t)d\alpha(x)$$

is *uniform* on $[c,d]$ if for a given $\varepsilon > 0$ there is a value of b dependent on ε, but not on t, such that whenever $c \leqslant t \leqslant d$ we have

$$\left| \int_a^{\to\infty} K(x,t)d\alpha(x) - \int_a^b K(x,t)d\alpha(x) \right| \leqslant \varepsilon.$$

In practice we look for a non-negative function $\phi(x)$ satisfying

$$|K(x,t)| \leqslant \phi(x) \quad \text{for} \quad x \geqslant a \quad \text{and} \quad c \leqslant t \leqslant d,$$

such that

$$\int_a^{\to\infty} \phi(x)d\alpha(x) < \infty.$$

If this happens, we say that ϕ *dominates* K and that K has *dominated convergence*.

Under the above circumstances, and *assuming α is increasing*, for given $\varepsilon > 0$ there will be a b such that

$$\int_b^{\to\infty} \phi(x)d\alpha(x) < \varepsilon.$$

Hence,

$$\left| \int_b^{\to\infty} K(x,t)d\alpha(x) \right| \leqslant \int_b^{\to\infty} |K(x,t)|d\alpha(x)$$

$$\leqslant \int_b^{\to\infty} \phi(x)d\alpha(x)$$

$$\leqslant \varepsilon,$$

as required by our definition.

We then have, for α an increasing function, the following.

(i) Assuming dominated convergence and continuity of $K(x,t)$ on $[a,\infty) \times [c,d]$ we have for t_0 in (c,d) that

$$\lim_{t\to t_0} \int_a^{\to\infty} K(x,t)d\alpha(x) = \int_a^{\infty} K(x,t_0)d\alpha(x).$$

(ii)   Assuming dominated convergence and continuity of

$$\frac{\partial K(x,t)}{\partial t}$$

on $[a, \infty) \times [c, d]$ we have for t in (c, d) that

$$\frac{\partial}{\partial t} \int_a^{\to \infty} K(x,t)\,d\alpha(x) = \int_a^\infty \frac{\partial}{\partial t} K(x,t)\,d\alpha(x).$$

(iii)   Assuming dominated convergence and continuity of $K(x,t)$ on $[a, \infty) \times [c, d]$ we have for any function $\beta(t)$ of bounded variation on $[c, d]$

$$\int_c^d \int_a^{\to \infty} K(x,t)\,d\alpha(x)\,d\beta(t) = \int_a^\infty \int_c^d K(x,t)\,d\beta(t)\,d\alpha(x).$$

Example 1
If

$$\int_0^{\to \infty} |g(x)|\,dx$$

exists, then $K(x,t) = g(x)e^{-tx}$ is dominated by $|g(x)|$ on $[0, \infty) \times [0,1]$ since $|K(x,t)| \leqslant |g(x)|$. Hence,

$$\lim_{t \to 0} \int_0^{\to \infty} g(x)e^{-tx}\,dx = \int_0^{\to \infty} g(x)\,dx.$$

Example 2
We amend the last example taking

$$g(x) = \begin{cases} \dfrac{\sin x}{x} & \text{for } x \neq 0, \\ 1 & \text{for } x = 0, \end{cases} \quad \text{i.e.} \quad K(x,t) = \begin{cases} e^{-xt}\dfrac{\sin x}{x}, \\ 1. \end{cases}$$

Here the partial derivative of K is (cf. p. 296 Question 12)

$$\frac{\partial}{\partial t} K(x,t) = -e^{-xt}\sin x.$$

If $c > 0$ and $d > c$ we shall have for $c < t < d$ that

$$|e^{-xt}\sin x| < e^{-cx}.$$

Since $\partial K/\partial t$ is continuous and dominated we have for $t > 0$:

$$\frac{\partial}{\partial t} \int_0^{\to \infty} e^{-xt} \frac{\sin x}{x} dx = \int_0^\infty -e^{-xt} \sin x \, dx = \frac{1}{1+t^2}.$$

Example 3
Let

$$f(t) = \int_0^{\to \infty} \cos(xt) \exp(-x^2/2) dx.$$

Then

$$\frac{d}{dt} f(t) = \int_0^{\to \infty} x \sin(xt) \exp(-x^2/2) dx.$$

This is justified since

$$|x \sin(xt) \exp(-x^2/2)| \leqslant x \exp(-x^2/2)$$

and, of course,

$$\int_0^b x \exp(-x^2/2) dx = [-\exp(-x^2/2)]_0^b$$

$$= 1 - \exp(-b^2/2) \to 1 \quad \text{as} \quad b \to \infty.$$

18.12 Power series

All the above manipulations on integrals may be very easily taken over to corresponding manipulations on power series, provided we remember that a power series may be represented as a Riemann–Stieltjes *integral*. We can turn

$$\sum_{n=1}^\infty a_n t^n \qquad \text{(we have ignored the term } a_0)$$

into the integral

$$\int_0^{\to \infty} a_x t^x d[x],$$

provided we interpret a_x as illustrated in Figure 18.19. Thus

$$K(x,t) = a_x t^x$$

is continuous (at any rate for $t \geqslant 0$).

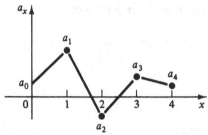

Fig. 18.19

We recall that if, for some $\xi > 0$, $\sum a_n \xi^n$ converges then so does $\sum a_n t^n$ for any t with $|t| < \xi$. The reason is as follows. First note that since $|a_n \xi^n| \to 0$ the sequence $a_n \xi^n$ is bounded, say by M. Secondly,

$$\sum_{n=0}^{\infty} |a_n t^n| = \sum_{0}^{\infty} |a_n \xi^n| \left|\frac{t}{\xi}\right|^n \leqslant M \sum_{0}^{\infty} \left|\frac{t}{\xi}\right|^n$$

and the series on the extreme right is a geometric progression with common ratio less than unity. In consequence, there is a largest number R, known as the *radius of convergence*, for which it is true that $\sum a_n x^n$ converges whenever x satisfies $|x| < R$.

The above argument also provides us with a proof of the dominated convergence of the Riemann–Stieltjes representation of the series. Let $d < R$ and let $d < s < R$, then for $x > 1$ and $0 \leqslant t \leqslant d$ we have

$$|a_x t^x| \leqslant |a_x d^x| \leqslant K \left(\frac{d}{s}\right)^x,$$

where K is a bound for $\{a_n s^n\}$. Thus

$$\int_1^{\to \infty} K \left(\frac{d}{s}\right)^x d[x] \leqslant K \sum_{n=2}^{\infty} \left(\frac{d}{s}\right)^n < \infty.$$

Hence by dominated convergence

$$\int_0^d \int_1^{\to \infty} a_x t^x d[x] \, dt = \int_1^{\to \infty} \int_0^d a_x t^x \, dt \, d[x],$$

$$\int_0^d \sum_{n=2}^{\infty} a_n t^n \, dt = \sum_{n=2}^{\infty} \int_0^d a_n t^n \, dt.$$

Thus term by term integration is justified:

$$\int_0^d \sum_{n=0}^{\infty} a_n t^n \, dt = \sum_{n=0}^{\infty} \int_0^d a_n t^n \, dt.$$

A similar argument can be given to show that a power series may be differentiated term by term. Here one needs to notice that

$$\sum_{n=1}^{\infty} |na_n t^n| = \sum_{1}^{\infty} n|a_n \zeta^n| \left|\frac{t}{\zeta}\right|^n \leqslant K \sum_{1}^{\infty} n \left|\frac{t}{\zeta}\right|^n < \infty.$$

18.13 Parameter range infinite

Suppose we know that, as in the last section,

$$\int_0^d \sum_{n=0}^{\infty} f_n(t)dt = \sum_{n=0}^{\infty} \int_0^d f_n(t)dt$$

and that this formula is valid for all $d < R$ (we allow the possibility that $R = \infty$). Suppose furthermore that $f_n(t) \geqslant 0$ for all n and all t. Then we may deduce that

$$\int_0^{\to R} \sum_{n=0}^{\infty} f_n(t)dt = \sum_{n=0}^{\infty} \int_0^{\to R} f_n(t)dt,$$

provided either side exists. For example, if the right-hand side exists and $R = \infty$, then, for each d

$$\int_0^d \sum_{n=0}^{\infty} f_n(t)dt = \sum_{n=0}^{\infty} \int_0^d f_n(t)dt \leqslant \sum_{n=0}^{\infty} \int_0^{\to \infty} f_n(t)dt,$$

since

$$\int_0^d f_n(t)dt \leqslant \int_0^{\infty} f_n(t)dt.$$

Hence

$$\int_0^{\to \infty} \sum_0^{\infty} f_n(t)dt \leqslant \sum_0^{\infty} \int_0^{\to \infty} f_n(t)dt. \qquad (1)$$

Now for each m

$$\int_0^{\to \infty} \sum_0^{\infty} f_n(t)dt \geqslant \int_0^{\infty} \sum_0^m f_n(t)dt = \sum_0^m \int_0^{\to \infty} f_n(t)dt,$$

i.e.

$$\int_0^{\to \infty} \sum_0^{\infty} f_n(t)dt \geqslant \sum_0^m \int_0^{\infty} f_n(t)dt.$$

Letting m tend to infinity we obtain the reverse inequality to (1).

Example

Observe that

$$-\log(1-x) = \sum_1^{\infty} \frac{x^n}{n} \quad \text{for} \quad |x| < 1.$$

Thus

$$\int_0^{1-\delta} -\log(1-x)\,dx = \int_0^{1-\delta} \sum_1^{\infty} \frac{x^n}{n}\,dx$$

$$= \sum_1^{\infty} \int_0^{1-\delta} \frac{x^n}{n}\,dx$$

$$= \sum_1^{\infty} \frac{(1-\delta)^{n+1}}{n(n+1)}$$

$$= \int_0^{\infty} \frac{(1-\delta)^{x+1}}{x(x+1)}\,d[x].$$

We shall now take the limit of the last expression as $\delta \to 0$. We observe that taking limits as $\delta \to 0$ through this Riemann–Stieltjes integral is permissible provided the integrand is dominated. But for $0 \leqslant \delta \leqslant 1$:

$$\left| \frac{(1-\delta)^{x+1}}{x(x+1)} \right| \leqslant \frac{1}{x(x+1)}$$

and

$$\sum_2^{\infty} \frac{1}{n(n+1)} \leqslant \sum_2^{\infty} \frac{1}{n^2} \leqslant \int_1^{\infty} \frac{dx}{x^2} < 1.$$

Now $(1-\delta)^n \to 1$ as δ tends to zero. Moreover

$$\sum_1^{m} \frac{1}{n(n+1)} = \sum_1^{m} \left(\frac{1}{n} - \frac{1}{n+1} \right) = \left(1 - \frac{1}{2} \right) + \left(\frac{1}{2} - \frac{1}{3} \right) + \cdots$$

$$= 1 - \frac{1}{m+1} \to 1 \quad \text{as} \quad m \to \infty.$$

It now follows by the argument at the beginning of this section|that

$$\int_0^1 \log(1-x)\,dx = -1.$$

The argument at the beginning of the section proves the following

Theorem

    Suppose α and β are increasing functions defined on $[a, \infty)$ and $[c, \infty)$ respectively. Suppose moreover that $K(x, t) \geqslant 0$ for all x and t in

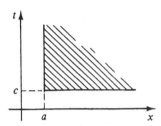

Fig. 18.20

the domain of K (Figure 18.20) and that it is known that

$$\int_c^d \int_a^\infty K(x,t)\,d\alpha(x)\,d\beta(t) = \int_a^\infty \int_c^d K(x,t)\,d\beta(t)\,d\alpha(x) \quad \text{for } d \geqslant c.$$

and

$$\int_a^b \int_c^\infty K(x,t)\,d\alpha(x)\,d\beta(t) = \int_c^\infty \int_a^b K(x,t)\,d\beta(t)\,d\alpha(x) \quad \text{for } b \geqslant a.$$

Then, provided either side exists:

$$\int_c^\infty \int_a^\infty K(x,t)\,d\alpha(x)\,d\beta(t) = \int_a^\infty \int_c^\infty K(x,t)\,d\beta(t)\,d\alpha(x).$$

18.14   Exercises

1. Using

$$\arctan x = \int_0^x \frac{dt}{1+t^2},$$

show that

$$\arctan(x) + \arctan(y) = \arctan \frac{x+y}{1-xy}.$$

2. Evaluate

$$\int_0^{\pi/2} \frac{\sin^2 x \, dx}{(1 + t \sin^2 x)^2}.$$

3. Test whether the following improper integrals exist: $(p > 1)$.

(i) $\displaystyle\int_{2}^{\to \infty} x e^{-x^2}\,dx$

(ii) $\displaystyle\int_{0}^{\to \infty} x^2 e^{-x^2}\,dx$

(iii) $\displaystyle\int_{1}^{\to \infty} \frac{\log x\,dx}{x^p}$

(iv) $\displaystyle\int_{2}^{\to \infty} \frac{dx}{x(\log x)^p}$

(v) $\displaystyle\int_{1}^{\infty} \frac{dx}{x(\sqrt{(x+1)} - \sqrt{x})}$

(vi) $\displaystyle\int_{0}^{\infty} (\sqrt[3]{(x^3 + 4)} - x)\,dx$

(vii) $\displaystyle\int_{0}^{1} \frac{\sin x}{x}\,dx$

(viii) $\displaystyle\int_{1}^{\to \infty} \frac{\sin x}{x}\,dx$ [Hint: Integration by parts.]

4. For what values of p, q do the following exist.

(i) $\displaystyle\int_{0}^{1} x^p (1 - x^2)^q\,dx$

(ii) $\displaystyle\int_{0}^{\to \infty} x^x \exp(-x^p)\,dx$

(iii) $\displaystyle\int_{0}^{\to \infty} \frac{dx}{x^p}$.

5. (i) Find
$$\sum_{r=1}^{\infty} r x^r.$$

(ii) Does
$$\sum^n \sqrt{\{(\tfrac{2}{3})^n + (\tfrac{3}{4})^n\}}$$
converge?

6. How do the results of Section 18.4 specialise when
$$\alpha(x) = \int_{a}^{x} g(t)\,dt?$$

7. Let

$$P(x) = \sum_{n \leqslant x} \frac{\lambda^n e^{-\lambda}}{n!}.$$

Show

$$\int_{-1}^{\to \infty} x \, dP(x) = \lambda,$$

$$\int_{-1}^{\to \infty} x^2 \, dP(x) = \lambda(1 + \lambda).$$

8. Show that

$$\int_{1}^{\to \infty} \frac{\log x \, dx}{x^{n+1}} = \frac{1}{n^2}.$$

9. For the function $K(x, t)$ illustrated in Figure 18.21, show

$$\lim_{t \to 0} \int_{0}^{\infty} K(x, t) dx \neq \int_{0}^{\infty} \lim_{t \to 0} K(x, t) dx.$$

10. For the example on p. 285, if $t = 1/6$, how large must b be to ensure

$$\int_{0}^{\to \infty} K(x, t) dt < 1/2?$$

11. Referring to Example 2 of Section 18.11, let

$$f(t) = \int_{0}^{\infty} e^{-xt} \frac{\sin x}{x} dx.$$

Show that $f(t) = c - \arctan t$ (for $t > 0$). Show also that $c = \pi/2$ by considering the limit of each side as $t \to \infty$. Finally justify the validity of this latter formula for f when $t = 0$.
[Hint: Write the integral so that e^{ix} replaces $\sin x$ and then integrate by parts and take limits as $t \to 0$.]

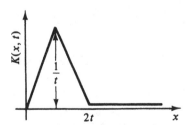

Fig. 18.21

12. If $f(0) = 1$ and $f(x) = \sin x/x$ for $x \neq 0$, use L'Hôpital's rule to show that

$$\lim_{h \to 0} \frac{f(0) - f(h)}{h} = 0 \quad \text{and} \quad \lim_{h \to 0} f'(h) = 0.$$

18.15 Further exercises

1.[†] Check which of the following integrals converge

$$\int_0^1 \frac{dx}{\sqrt{-\log x}} \qquad \int_2^\infty \frac{dx}{(\log x)^3} \qquad \int_0^\pi \frac{1 - \cos x}{x^2} dx$$

$$\int_1^\infty \frac{x^b}{(1 + x)^a} dx \qquad \int_0^\infty e^{-x} \log(1 + e^x) dx$$

$$\int_0^1 \frac{dx}{x^{4/3} + x^{1/3}} \qquad \int_{-1}^0 \frac{dx}{x^{4/3} + x^{1/3}}$$

$$\int_0^\infty \frac{x \sin x \, dx}{\sqrt{(x^4 + 1)}} \qquad \int_{10}^\infty \frac{dx}{x \log(\log x)}$$

$$\int_0^1 \frac{\sqrt{x} \, dx}{e^{\sin x} - 1} \qquad \int_0^1 \frac{\log x \, dx}{\sqrt{x}}$$

$$\int_0^1 (\log x)^2 dx \qquad \int_0^1 \frac{(\log x)^2}{1 + x^2} dx$$

$$\int_0^{\pi/2} \frac{x^p \, dx}{\sqrt{(1 - \cos x)}}, \qquad \int_0^\infty \sin(x^p) dx$$

$$\int_0^\infty \frac{x^p \sin x}{1 + x^2} dx \qquad \int_1^\infty \frac{dx}{x\sqrt{(x^2 - 1)}},$$

$$\int_0^1 \frac{dx}{\sqrt{(x \log(1 + x))}}, \qquad \int_1^\infty \frac{\log x \, dx}{x + e^{-x}}.$$

2. For what values of p and q do the following converge

(a) $\int_0^1 x^p(1 - x)^q dx$ (b) $\int_1^\infty x^p \frac{x + \sin x}{x - \sin x} dx$

(c) $\int_0^1 \frac{x^p - x^q}{\log x} dx$ (d) $\int_0^1 \frac{x^{p-1} \log x}{1 - x} dx$

(e) $\int_0^1 \frac{x^{p-1} dx}{1 - x}$

[†] In (1) and (2) simplify the integrand $f(x)$ by judicious use of a substitution or an approximation (strike out insignificant terms, refer to Taylor series e.g. for $\cos x, \ldots$) arriving at $g(x)$. Check $\lim f/g$ (cf. p. 271 and p. 276).

3. Show that when the integral in Question 2(c) exists its value is

$$\log\left(\frac{1+p}{1+q}\right).$$

4. Using a power series expansion for $(1-x)^{-1}$ calculate

$$\int_{\to 0}^{1-\varepsilon}\frac{x^{p-1}dx}{1-x}$$

and hence deduce that

$$\int_{\to 0}^{\to 1}\frac{x^{p-1}\log x}{1-x}dx = -\sum_{n=0}^{\infty}\frac{1}{(n+p)^2}.$$

Justify any limit manipulations performed.

5. Using the fact that

$$\int_0^1 x^n dx = \frac{1}{n+1}(n>-1)$$

obtain the value of the infinite series

$$\sum_1^{\infty}\frac{(-1)^n}{3n-2}.$$

6. Verify the dominated convergence of

$$\int_1^{\infty} t^{-x}d[x]$$

for $2\leqslant t<\infty$. Deduce that the series

$$\sum_2^{\infty}\frac{1}{y^n}$$

may be integrated term by term in $[2, R]$ for any $R\geqslant 2$. Hence evaluate the series

$$\sum_1^{\infty}\frac{1}{n2^n}.$$

7. By expanding $\exp(-x\log x)$ as a series show that

$$\int_0^1 x^{-x}dx = \sum_{m=1}^{\infty}\frac{1}{m^m}.$$

[Hint: Consider the maximum of $|x\log x|$.]

8. By first calculating

$$\int_0^{\pi/2}\frac{\cos x dx}{\sin(x+\beta)}$$

show that for $t > 0$

$$\int_0^{\pi/2} \frac{\cos x\, dx}{t\cos x + \sin x} = \frac{\pi}{2}\frac{t}{1+t^2} - \frac{\log t}{1+t^2}.$$

Hence evaluate

$$\int_0^{\pi/2} \frac{\cos^2 x\, dx}{(2\cos x + \sin x)^2}.$$

Justify any manipulations performed.

9. Show that

$$\int_0^{\pi} \frac{\log(1 + \cos\alpha\cos x)}{\cos x}\, dx = \pi\left[\frac{\pi}{2} - \alpha\right]$$

for $0 < \alpha \leqslant \pi/2$. Does the result extend to $\alpha = 0$?

10. For $t > 0, h > 0$ let

$$I(t) = \int_0^\infty \exp\{-(x - t/x)^2\}\, dx,$$

$$I_h(t) = \int_h^\infty \exp\{-(x - t/x)^2\}\, dx.$$

What is the limiting value of $I_h(t)$ as $t \to 0$? Justify your answer. Deduce the limiting value of $I(t)$ as $t \to 0$.

Show that for any $t > 0$ the integral $I(t)$ may be differentiated under the integral sign and that $I'(t) = 0$. (Hint: Substitute $z = t/x$). Deduce that

$$I(t) = \frac{\sqrt{\pi}}{2}.$$

11. Prove by differentiating with respect to x that for a continuous function f,

$$\int_0^x \int_0^t f(u)\, du\, dt = \int_0^x (x - u) f(u)\, du.$$

12. Check that for $\alpha > 0$

$$I(\alpha) = \int_0^\infty \exp(-\alpha x^2)\, dx = \frac{\sqrt{\pi}}{2\sqrt{\alpha}}$$

and by differentiating this function with respect to α find a formula for

$$\int_0^\infty x^{2n} \exp(-\alpha x^2)\, dx,$$

where n is a natural number.

13. Show that $I(t) = \int_0^\infty e^{-ax}\cos tx\, dx = a/(a^2 + t^2)$ (for $a > 0$). By integrating

twice with respect to the parameter t deduce that

$$\int_0^\infty e^{-ax}\frac{1-\cos tx}{x^2}dx = t\tan^{-1}\left[\frac{t}{a}\right] - \frac{a}{2}\log\left[1+\frac{t^2}{a^2}\right].$$

Putting $t=1$ and letting $a \to 0$ deduce that

$$\int_0^\infty \frac{1-\cos x}{x^2}dx = \frac{\pi}{2}.$$

You should justify all your manipulations.
(Remark: $\int \tan^{-1}\theta\, d\theta = \theta\tan^{-1}\theta - \frac{1}{2}\log(1+\theta^2)$)

18.16 Postscript

Several of the examples in earlier sections required the evaluation of an integral by standard trigonometric substitutions. For reference purposes we recall these substitutions.

Suppose $f(u,v)$ is a function of two variables and we are to evaluate an integral of the form

$$\int_a^b f(\sin x, \cos x)dx.$$

The appropriate substitutions are as follows:

(a) If $f(u,v) = -f(-u,v)$, i.e. the integrand is odd with respect to the sine, put

$$w = \cos x.$$

(b) If $f(u,v) = -f(u,-v)$, i.e. the integrand is odd with respect to the cosine, put

$$w = \sin x.$$

(c) If $f(u,v) = f(-u,-v)$, i.e. the integrand is even with respect to cos-and-sine, put

$$w = \tan x.$$

(d) In all other cases put

$$w = \tan x/2.$$

19

Multiple integrals

This chapter considers the higher dimensional analogues of the integral studied in Chapter 17. There we were essentially calculating 'area under the graph'. Here we will calculate items like 'volume under a surface', but we now have an added source of variety since the volume need not stand simply over a rectangle as in illustration (i) of Figure 19.1 but may stand over a more general region like D in illustration (ii). Case (i) amounts to performing *two* integration processes corresponding to the two axes. Case (ii) is more awkward; it is handled by a reduction to case (i) and the techniques for this reduction form the main material of this chapter.

Our first task is to define the double integral

$$\iint_D f(x, y)\, dx\, dy$$

and to see how it reduces to a repeated integral in the case when D is a rectangle.

19.1 Definition

We begin with the definition in the simpler case when D is the rectangle $[a, b] \times [c, d]$, which we denote by R. Let $z = f(x, y)$ be a continuous function which is defined and positive on R. The definition goes much as before. We take two partitions, one of $[a, b]$ and one of $[c, d]$, say P_1 and P_2, with

$$P_1 = \{x_0, x_1, \ldots, x_n\},$$
$$P_2 = \{y_0, y_1, \ldots, y_n\}.$$

These can be used to divide up R into small rectangles (cf. Figure 19.2)

$$R_{ij} = [x_{i-1}, x_i] \times [y_{j-1}, y_j]$$

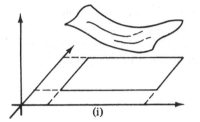 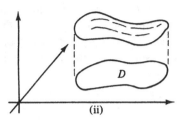

Fig. 19.1

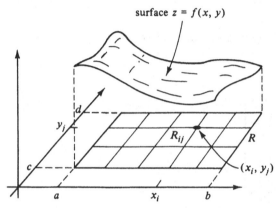

Fig. 19.2

whose area will be σ_{ij}. A lower estimate for the area beneath the surface is

$$L(P_1, P_2) = \Sigma m_{ij}\sigma_{ij},$$

where

$$m_{ij} = \inf\{f(x, y):(x, y)\in R_{ij}\}.$$

The supremum of all these lower estimates over all relevant partitions P_1, P_2 is then what we mean by the integral

$$\iint_R f(x, y)\, dx\, dy.$$

We note that just as in the single variable case we can formulate an alternative definition by taking the infimum over all the upper estimates:

$$U(P_1, P_2) = \Sigma M_{ij}\sigma_{ij},$$

where

$$M_{ij} = \sup\{f(x, y):(x, y)\in R_{ij}\}.$$

Although this alternative definition coincides with our earlier one when the function f is continuous, the two might lead to distinct values for a general function f. In such cases one talks of the lower and upper integral (respectively) and these are then denoted by the symbols:

$$\underline{\iint_R} f(x, y)\, dx\, dy$$

and

$$\overline{\iint_R} f(x, y)\, dx\, dy.$$

If for a given function f the two values coincide we say that f is integrable.

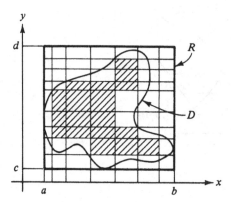

Fig. 19.3

Finally we consider a more general *bounded* region D (cf. Figure 19.3). We can still 'tile' with small rectangles, but we must first fit D into a rectangle R. If $R = [a, b] \times [c, d]$ and P_1, P_2 are as before we can form a lower estimate by summing

$$\sum m_{ij}\sigma_{ij}$$

over rectangles R_{ij} contained in D. Clearly, the finer the tiling the more of D will be included.

Upper estimates may also be obtained by summing

$$\sum M_{ij}\sigma_{ij}$$

over rectangles R_{ij} meeting D, it being understood now that

$$M_{ij} = \sup \{f(x, y) : (x, y) \in R_{ij} \cap D\}.$$

When f is continuous and D is an open domain (or the closure of an open domain) the supremum of lower estimates will coincide with the infimum of upper estimates. Their common value is precisely:

$$\iint_D f(x, y)\,dx\,dy.$$

19.2 Repeated integrals

By reorganizing the sums

$$\sum m_{ij}\sigma_{ij},$$

we can make the lower estimates 'look' like

$$\int_c^d \left\{ \int_a^b f(x, y)\,dx \right\} dy.$$

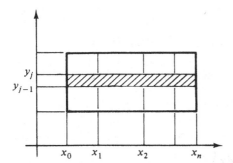

Fig. 19.4

Indeed if we fix a value of j and sum over i (cf. Figure 19.4) we obtain a typical Riemann sum approximating to

$$\int_a^b f(x, y_j)\,dx.$$

A more precise calculation justifies this. Notice

$$m_{ij} \leqslant f(x, y) \quad \text{for } (x, y)\in R_{ij},$$

so for any y with $y_{j-1} \leqslant y \leqslant y_j$,

$$m_{ij} \leqslant \min\{f(x, y) : x_{i-1} \leqslant x \leqslant x_i\}$$
$$= m_i(y), \quad \text{(say)}$$

hence

$$m_{ij}\sigma_{ij} \leqslant m_i(y)\cdot(x_i - x_{i-1})\cdot(y_j - y_{j-1})$$

and summing over i (that is over the shaded row)

$$\sum_i m_{ij}\sigma_{ij} \leqslant (y_j - y_{j-1})\cdot\sum_i m_i(y)(x_i - x_{i-1})$$

$$\leqslant (y_j - y_{j-1})\cdot\int_a^b f(x, y)dx.$$

Let us write

$$F(y) = \int_a^b f(x, y)\,dx.$$

Thus for any y in $[y_{j-1}, y_j]$

$$\sum_i m_{ij}\sigma_{ij} \leqslant F(y)(y_j - y_{j-1}),$$

so

$$\sum_i m_{ij}\sigma_{ij} \leqslant \min\{F(y): y \in [y_{j-1}, y_j]\}\cdot(y_j - y_{j-1})$$

and the expression on the right summed over j is less than

$$\int_a^b F(y)\,dy.$$

Thus

$$\sum_{ij} m_{ij}\sigma_{ij} \leqslant \int_c^d \left\{\int_a^b f(x, y)dx\right\}dy.$$

A similar calculation shows that

$$\int_c^d \left\{\int_a^b f(x, y)dx\right\}dy \leqslant \sum M_{ij}\sigma_{ij}.$$

Hence

$$\iint_R f(x, y)\,dxdy = \int_c^d \int_a^b f(x, y)\,dxdy.$$

Similarly, by summing over i first, we arrive at

$$\iint_R f(x, y)\,dxdy = \int_a^b \int_c^d f(x, y)\,dy\,dx.$$

For the more discerning reader we remark that if $f(x, y)$ is not continuous in the region of integration the equations above may fail. An example is given in the exercises.

19.3 General regions: direct method

Our first approach to the evaluation of multiple integrals over a general region extends the idea of the last section. It is easiest explained in the case of a double integral

$$\iint_D f(x, y)\, dx\, dy$$

when the boundary of D conveniently splits into a pair of continuous curves as illustrated in Figure 19.5.

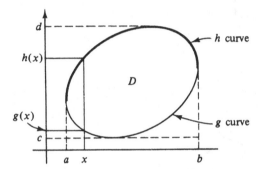

Fig. 19.5

Thus

$$D = \{(x, y): a \leqslant x \leqslant b \text{ and } g(x) \leqslant y \leqslant h(x)\}.$$

The argument of the last section can easily be extended to show that

$$\iint_D f(x, y)\, dx\, dy = \int_a^b dx \int_{g(x)}^{h(x)} f(x, y)\, dy.$$

The more general result that this anticipates is *Fubini's theorem* which asserts that for a general region D, say, lying in the strip $a \leqslant x \leqslant b$,

$$\iint_D f(x, y)\, dx\, dy = \int_a^b dx \int_{D(x)} f(x, y)\, dy, \tag{*}$$

where $D(x)$ is the appropriate vertical section of D (compare Figure 19.6). Unfortunately to make sense of this formula we would need to define the symbol

$$\int_S f(t)\, dt$$

when the region of integration is more general than an interval. To avoid
a more advanced treatment of integration we hastily advocate that for
a region like the one illustrated in Figure 19.6, it is best to split D into
several regions as in Figure 19.7 each of which is separately covered by
formula (∗).

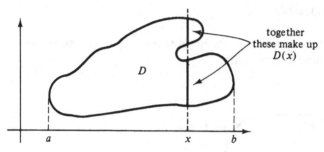

Fig. 19.6

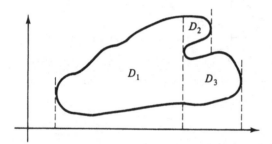

Fig. 19.7

Here,

$$\iint_D = \iint_{D_1} + \iint_{D_2} + \iint_{D_3}.$$

The interesting point about the Fubini formula we quoted is that it is
an instance of a general rule for multiple integrals. Thus in three
dimensions we similarly have

$$\iiint_V f(x, y, z)\,dx\,dy\,dz = \int_a^b dx \iint_{V(x)} f(x, y, z)\,dy\,dz$$

where $V(t)$ for $a \leqslant t \leqslant b$ is the intersection of V with the plane $x = t$.
   All the points raised so far are illustrated in the next example.

19.4 Example

Calculate

$$I = \iiint_B x^n \, dx\, dy\, dz,$$

where B is the unit ball in $\mathbb{R}^3$ and n is an integer.

Clearly for n odd the answer is zero (negative and positive contributions balance). So suppose n is even. In order to simplify matters (see comment at end) we let V be the positive orthant of B. See Figure 19.8.

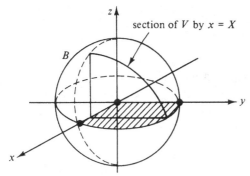

Fig. 19.8. See also Figure 19.9

Then, for reasons of symmetry,

$$I = 8 \iiint_V x^n \, dx\, dy\, dz.$$

Thus,

$$\tfrac{1}{8} I = \int_0^1 dx \iint_{V(x)} x^n \, dy\, dz.$$

But,

$$\iint_{V(x)} x^n \, dy\, dz = \int_0^{\sqrt{(1-x^2)}} dy \int_0^{\sqrt{(1-x^2-y^2)}} x^n \, dz.$$

The idea is that as x takes different values in $[0, 1]$ so V is first *dissected* into slices $V(x)$, and then these in turn are dissected into line segments $D(y)$ where $D = V(x)$; see Figure 19.9.

But now we have, performing the innermost integration

$$\iint_{V(x)} x^n \, dy\, dz = \int_0^{\sqrt{(1-x^2)}} x^n \sqrt{(1-x^2-y^2)} \, dy$$

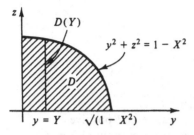

Fig. 19.9. A slice of $D = V(X)$

(substituting $y = \sqrt{(1 - x^2)} \sin \theta$)

$$= \int_0^{\pi/2} x^n(1 - x^2) \cos^2 \theta \, d\theta$$

$$= \frac{\pi}{4} x^n(1 - x^2).$$

Thus,

$$I = 8 \int_0^1 \frac{\pi}{4} x^n(1 - x^2) \, dx$$

$$= 2\pi \left(\frac{1}{n+1} - \frac{1}{n+3} \right) = \frac{4\pi}{(n+1)(n+3)}.$$

Remark
The argument from symmetry simplifies some tedious calculations associated with the use of the complete sections $B(x)$; for these the argument would need to start off as follows.

$$\iint_{B(x)} x^n \, dy \, dz = \int_{-\sqrt{(1-x^2)}}^{+\sqrt{(1-x^2)}} dy \int_{-\sqrt{(1-x^2-y^2)}}^{+\sqrt{(1-x^2-y^2)}} x^n \, dz.$$

Note that, for each x, the circle is decomposed into the two curves:

$$z = \pm \sqrt{(1 - x^2 - y^2)}.$$

19.5   General regions: change of variable

A second approach to the evaluation of a multiple integral over a general region is to transform the region into a rectangle or at least into a region with a more manageable boundary. This approach is often combined with an attempt to simplify the integrand by the same change of variable. A wealth of examples abounds. We start with an easy one.

Example
Calculate

$$\iint_D \frac{dx\,dy}{xy},$$

where D is the region (cf. Figure 19.10) bounded by the lines

$$y = x, \qquad y = 2x,$$
$$x + y = 1, \qquad x + y = 2.$$

We can 'straighten out' D into a rectangle by using the change of variables (defined over D):

$$u = x + y \qquad v = y/x.$$

The lines $u = $ constant and $v = $ constant for $1 \leqslant u \leqslant 2$ and $1 \leqslant v \leqslant 2$ 'sweep out' D in the obvious sense (see Figure 19.11). Thus the region D transforms into

$$\Delta = \{(u, v): 1 \leqslant u \leqslant 2 \quad \text{and} \quad 1 \leqslant v \leqslant 2\}$$

which is a rectangle. We now have to invoke a theorem which tells us how to calculate the given integral in the new variables. We state the result only for double integrals; but, of course, the result is valid mutatis mutandis in higher dimensions. For another example see Section 19.15.

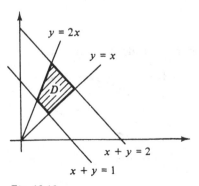

Fig. 19.10

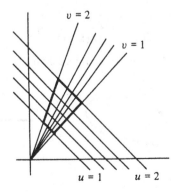

Fig. 19.11

Theorem (change of variable)
Suppose the following transformation

$$x = \phi(u,v), \quad y = \psi(u,v)$$

*takes a subset Δ of the (u,v)-plane to the set D in the (x,y)-plane (cf.
Figure 19.12). Suppose further that:*

(i)   *the transformation is one-to-one,*
(ii) *ϕ and ψ have continuous partial derivatives,*
(iii) *the Jacobian*

$$\frac{\partial(x,y)}{\partial(u,v)} = \begin{vmatrix} \dfrac{\partial\phi}{\partial u} & \dfrac{\partial\phi}{\partial v} \\[2ex] \dfrac{\partial\psi}{\partial u} & \dfrac{\partial\psi}{\partial v} \end{vmatrix}$$

*exists throughout Δ and is **never** zero in Δ.*
    Then,

$$\iint_D f(x,y)\,dx\,dy = \iint_\Delta f(\phi(u,v), \psi(u,v)) \left| \frac{\partial(x,y)}{\partial(u,v)} \right| du\,dv.$$

where the vertical bars denote absolute value.
We comment in detail on this theorem in the next section. For the
moment notice that the theorem speaks of a transformation from (u,v)
to (x,y) whilst *usually* (see the example above) we arrive at a change
of variables by introducing (u,v) in terms of (x,y), quite the opposite
order to that required in the theorem.

It would therefore seem necessary to obtain a formula for the inverse
transformation for an application of the theorem. In practice this may
often be avoided, particularly since

$$\frac{\partial(x,y)}{\partial(u,v)} = \left\{ \frac{\partial(u,v)}{\partial(x,y)} \right\}^{-1},$$

provided this is non-zero and (ii) above holds. Indeed as far as the

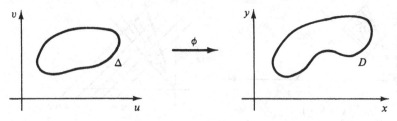

Fig. 19.12

theorem is concerned only the existence of the transformation is needed, not the formula specifying it.

Example (continued)
We have

$$\frac{\partial(u,v)}{\partial(x,y)} = \begin{vmatrix} 1 & 1 \\ -y/x^2 & 1/x \end{vmatrix} = \frac{1}{x} + \frac{y}{x^2} = \frac{x+y}{x^2}.$$

Thus, since this is positive throughout D, we have, for Δ as in Figure 19.13,

$$I = \iint_\Delta \frac{1}{xy} \cdot \frac{x^2}{x+y} \, du \, dv$$

$$= \iint_\Delta \frac{x}{y(x+y)} \, du \, dv$$

$$= \iint_\Delta \frac{du \, dv}{uv}$$

$$= \int_1^2 \frac{du}{u} \int_1^2 \frac{dv}{v} = (\log 2)^2.$$

We did not need to know x, y in terms of u, v because of the convenient cancellations. A good exercise here is to calculate

$$\iint_D xy \, dx \, dy.$$

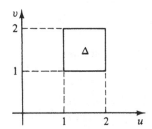

Fig. 19.13

19.6 Change of variables: comments

In this section we comment on some of the details of the theorem of the last section. Of course the appearance of the Jacobian is perhaps the first item to explain. We prefer to postpone to Section 19.11 a revision of the argument that it represents a local magnification factor. The reason in brief is this. The matrix (see Part I, Chapter 14)

$$\frac{d(x, y)}{d(u, v)}$$

gives a local linear approximation to the given transformation and an invertible matrix transforms area/volume magnifying it by a factor equal to its own determinant.

19.6.1 Significance of the modulus bars

It is easier to see what is happening, by reference to the one-dimensional case. Consider therefore

$$I = \int_a^b f(x)\, dx$$

and the transformation $x = g(u)$. Suppose g transforms the interval $[c, d]$ into $[a, b]$ as in Figure 19.14. Our illustrative argument splits according to whether g is increasing or decreasing.

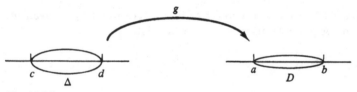

Fig. 19.14

(Actually, since we assume that g' is continuous and that g is one-to-one, g can only be strictly increasing or strictly decreasing. The point about the formulas (∗) and (∗∗) which we develop below is that they both describe the region of integration as $[c, d]$ with c and d appearing in the natural order; the description altogether ignores the fact that g may turn an interval back to front. Similar considerations apply in higher dimensions and are the reason for the modulus bars round the Jacobian. For example, in two dimensions the corresponding question is whether the image of a point tracing the boundary of Δ describes the boundary of D in the same sense (clockwise or anti-clockwise) as the point tracing out the boundary of Δ.)

If g is *increasing* then $g(c) = a$ and $g(b) = d$ while $g'(u) > 0$ throughout.

We thus have

$$I = \int_c^d f(g(u)) \cdot g'(u)\, du. \tag{*}$$

But if g is *decreasing* then in fact $g(c) = b$ and $g(d) = a$, while $g'(u) < 0$. Hence,

$$I = \int_d^c f(g(u)) \cdot g'(u)\, du$$

$$= \int_c^d f(g(u)) \cdot (-g'(u))\, du$$

$$= \int_c^d f(g(u)) \cdot |g'(u)|\, du. \tag{**}$$

19.6.2 One-oneness and non-zero Jacobian

The assumption that the transformation should be one-to-one is rather obvious: the object of transforming is to re-state the problem in a different coordinate system and in the absence of one-onenness we may run the risk of loosing the equivalence of the two problems. There is also a more tangible reason. It could be that the re-stated problem halves or doubles the required answer (for example if the mapping is one-to-two or two-to-one).

Fortunately, the Jacobian also provides a safety check because of the following connections:

(1) If the transformation is one-to-one and the region is 'connected' (i.e. any two points can be linked by a continuous path/curve lying wholly in the region), then the Jacobian does not change sign in the region.

(2) If the Jacobian of the transformation is non-zero at some point (x_0, y_0), then close to that point (i.e. in some disc round the point (x_0, y_0)) the transformation is one-to-one. Thus a formula for the inverse transformation exists in a 'patch' round (x_0, y_0). We say that a *local inverse* is guaranteed. This implies that any *bounded* (and closed) region of integration can be split into a finite number of regions of integration inside each of which the transformation is one-to-one (cf. Figure 19.15).

Warning example
Consider

$$\left.\begin{array}{l} u = e^x \cos y \\ v = e^x \sin y \end{array}\right\}.$$

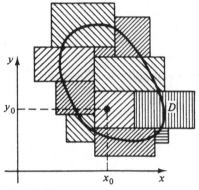

Fig. 19.15

Then

$$\frac{\partial(u,v)}{\partial(x,y)} = \begin{vmatrix} e^x \cos y & -e^x \sin y \\ e^x \sin y & e^x \cos y \end{vmatrix} = e^{2x} > 0.$$

The transformation is not one-to-one in the whole plane; (x, y) and $(x, y + 2\pi)$ transform to the same point (u, v). Nevertheless an open horizontal strip of width no more than 2π gives a region in which the transformation is one-to-one.

Thus if we were to use the change of variable in an integration problem involving the region D illustrated in Figure 19.16 we would need to split D into non-overlapping sections each of width at most 2π (e.g. as shown). We would then need to calculate separately the integrals over D_1, D_2 and D_3 and add the three results.

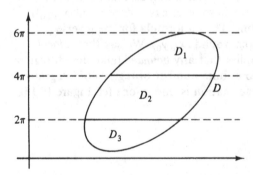

Fig. 19.16

19.6.3 Significance of a zero Jacobian

Let (x_0, y_0) be a point such that the Jacobian

$$\frac{\partial(u, v)}{\partial(x, y)}$$

vanishes. Let $u_0 = u(x_0, y_0)$ and $v_0 = v(x_0, y_0)$. Now consider the contours in the (x, y) plane given by

$$u(x, y) = u_0$$
$$v(x, y) = v_0.$$

We observe that the normal at (x, y) to the first of these curves is

$$\left(\frac{\partial u}{\partial x}, \frac{\partial u}{\partial y} \right).$$

Hence, if at (x_0, y_0) we have

$$\begin{vmatrix} \dfrac{\partial u}{\partial x}, & \dfrac{\partial u}{\partial y} \\ \dfrac{\partial v}{\partial x}, & \dfrac{\partial v}{\partial y} \end{vmatrix} = 0,$$

then not only do the u_0 and v_0-contours cross at (x_0, y_0), but they are also tangential, since their normal vectors are parallel (provided they are non-zero). There is therefore a strong chance (see Figure 19.17) that other v-contours which are close enough, cut the u_0-contour *twice*; so, unless one of the two intersections is excluded from the region of integration in the (x, y) plane, the transformation is not one-to-one. See Example 19.7. If this does happen it would be appropriate as a first step to split up the region of integration into two parts – one with the Jacobian

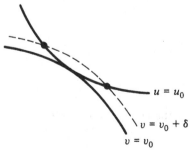

$u = u_0$

$v = v_0 + \delta$

$v = v_0$

Fig. 19.17

positive, the other with the Jacobian negative. Note that a zero Jacobian need not *necessarily* imply absence of one-oneness. Consider for example, the following transformation (whose *u*- and *v*-contours are sketched in Figure 19.18).

$$u = x^3 - y,\}$$
$$v = y. \quad\}$$

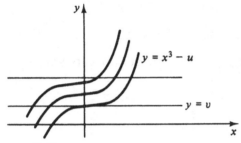

Fig. 19.18

Conclusions

In order to apply the theorem to a change of variable such as is given by the formula

$$u = u(x, y),$$
$$v = v(x, y),$$

it may be necessary to split the region of integration into subregions in each of which
(1)   the Jacobian is of fixed sign, *and*
(2)   the transformation has an inverse (in the subregion).
The calculations are then performed separately for each subregion and summed.

19.7   Example

Show how to transform the integral

$$I = \iint_D f(x, y)\exp(-x^2 y^2)\,dx\,dy,$$

where *D* is the half-plane $x + y \geq 1$ using the change of variable

$$u = x + y\}$$
$$v = xy \quad\}.$$

This is a classic example. Note that the transformation is appropriate to the problem since D is straightforwardly dissected into parallel lines $u = $ constant. We disregard the problems associated with infinite domains, returning to them in a later Section (19.13).

We begin by calculating the Jacobian:

$$J = \frac{\partial(u, v)}{\partial(x, y)} = \begin{vmatrix} 1 & 1 \\ y & x \end{vmatrix} = x - y.$$

The Jacobian vanishes along $y = x$ ergo we split D (cf. Figure 19.19) into two subregions D_+ and D_- where $J > 0$ and where $J < 0$ (respectively).

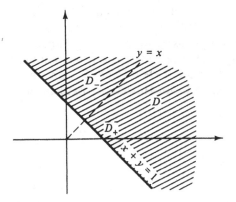

Fig. 19.19

We notice that both of the points (x, y), (y, x) are transformed to the one point (u, v) but that they are fortunately in different subregions.

Now we look for an inverse transformation. Clearly x, y are the roots of the quadratic

$$t^2 - (x + y)t + xy = 0.$$

Hence

$$x, y = \frac{u \pm \sqrt{(u^2 - 4v)}}{2},$$

so that in D the given transformation is two-to-one. The choice of sign determining x, y now depends on the subregion (so in particular no further splitting of the subregions is required). For example, in D_+ we have

$$x = \frac{u + \sqrt{(u^2 - 4v)}}{2} \qquad y = \frac{u - \sqrt{(u^2 - 4v)}}{2}.$$

Our next task is to find Δ_+ and Δ_-. One approach is analytical: transform the inequalities defining D_+ (and D_-). These are

$$\left.\begin{array}{r} x + y \geqslant 1 \\ x \geqslant y \end{array}\right\}.$$

The *fallacious answer* would be to say $u \geqslant 1$ (since the second inequality is automatic in D_+). In fact we need to *verify that x, y are real*. This requires that

$$u^2 - 4v \geqslant 0.$$

A second approach is to trace the u, v contours. This will in fact immediately throw light on the vanishing of the Jacobian and on the inequality just noted above.

Consider a contour $v = v_0$ i.e. $xy = v_0$. This is a rectangular hyperbola. The contour $u = u_0$ (i.e. $x + y = u_0$) is in D provided $u_0 \geqslant 1$, but will in general cut $v = v_0$ in two points on either side of $x = y$. Hence D_+ contains only one of these intersections (see Figure 19.20). However, it may happen

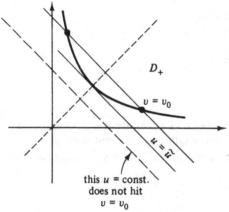

Fig. 19.20. *NB. The critical position has* $x = y = \frac{1}{2}\tilde{u}$ *so* $v_0 = xy = \frac{1}{4}\tilde{u}^2$.

that $u = u_0$ cuts $v = v_0$ *once only*, tangentially, (and this is on $x = y$), say at $u = \tilde{u}$ or, alternatively, *not at all*, when

$$-\tilde{u} < u < \tilde{u}$$

(this also takes into account the other branch of the hyperbola). Evidently the condition here corresponds to the inequality arising in the earlier argument, so

$$\tilde{u} = \sqrt{(4v_0)}.$$

Finally,

$$I = \iint_{D_+} + \iint_{D_-}$$

and noting that

$$(x - y)^2 = (x + y)^2 - 4xy = u^2 - 4v$$

we have

$$\iint_{D_+} = \iint_{\Delta_-} \frac{e^{-v}}{\sqrt{(u^2 - 4v)}} f\left(\frac{u + \sqrt{(u^2 - 4v)}}{2}, \frac{u - \sqrt{(u^2 - 4v)}}{2}\right) du\, dv.$$

Remark

If for example $f(x, y) = x - y$, then clearly the contributions from D_+ and D_- cancel each other out, whilst if $f(x, y) = |x - y|$ they are equal. This illustrates a point made earlier in the section.

19.8 Ambiguities of the inverse transformation

Example 19.7 contained a search for the local inverse transformations leading to the ambiguity

$$x, y = \frac{u \pm \sqrt{(u^2 - 4v)}}{2}.$$

The ambiguity arises because in fact the transformation

$$\left.\begin{array}{l} u = x + y \\ v = xy \end{array}\right\}$$

is two-to-one (both (x, y) and (y, x) yield the same point in the (u, v)-plane). Fortunately in D_+ the ambiguity is resolved. The next example presents a different type of ambiguity.

19.9 Example

Evaluate

$$I = \iint_D \frac{dx\, dy}{x^3 y^2},$$

where D is the shaded region bounded by the curves

$$y = x^2, \qquad y = 2x^2,$$
$$x^2 y = 1, \qquad x^2 y = 2.$$

The obvious choice of new variables is

$$u = y/x^2 \quad v = x^2 y \quad (+).$$

This is because the contours $u = u_0$ for $1 \leqslant u_0 \leqslant 2$ and $v = v_0$ for $1 \leqslant v_0 \leqslant 2$ trace out intervening curves that fill out D (see Figure 19.21) and 'straighten out' the curvilinear rectangle D into Δ (as in example 19.5.1).

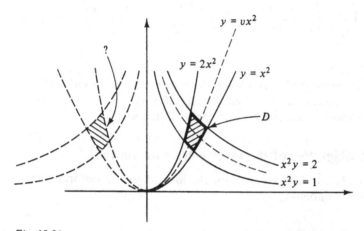

Fig. 19.21

We now calculate

$$\frac{\partial(u, v)}{\partial(x, y)} = \begin{vmatrix} -2yx^{-3} & x^{-2} \\ 2xy & x^2 \end{vmatrix} = -\frac{4y}{x}.$$

So the Jacobian is negative in D and we seem set to go straight to the calculation.

However, upon reflection one might doubt that the transformation from Δ to D is *one-to-one*. Indeed the u and v contours intersect twice: once in D and once in a mirror image (because of the other branch of the curve $x^2 y = v$). To settle this doubt we calculate that

$$y^2 = uv \quad \text{and} \quad x^4 = v/u$$

and there is now the ambiguity

$$y = \pm \sqrt{(uv)} \quad x = \pm \sqrt[4]{(v/u)}.$$

In other words the formulas for u and v above do not in fact as yet specify a transformation from the u, v-plane to the x, y-plane. However, we can choose to define

$$\phi(u, v) = \sqrt[4]{(v/u)} \quad \psi(u, v) = \sqrt{(uv)}.$$

Now that we have actually defined a transformation from (u, v) to (x, y) which satisfies $(+)$ we may proceed to the calculation. The ambiguity was only apparent and our doubts were founded on the absence of a (u, v) to (x, y) transformation.

$$I = \iint_{\Delta} \frac{1}{y^2 x^3} \cdot \frac{x}{4y} \, du \, dv = \iint_{\Delta} \frac{1}{uv} \cdot \frac{1}{4v} \, du \, dv = \tfrac{1}{8} \log 2.$$

19.10 Plotting the transform of the region of integration

A frequent source of difficulty in multiple integration is the indentification of the transform Δ of the region of integration D once a change of variables is selected. It is best to think of the new variables given by

$$u = u(x, y)$$
$$v = v(x, y)$$

as providing a 'curvilinear' grid (or co-ordinate system) over an area in the (x, y)-plane which includes at least the region of integration (cf. Figure 19.22). It is then necessary to calculate for each fixed value of, say, u the range of values of v for which the v contours meet the fixed u contour at points within the region of integration. Compare Figure 19.23. This can be done either by plotting the contours and inspecting them, or by re-writing the relations defining D in terms of u and v. The latter usually requires the transformation and manipulation of inequalities; care must be exercised so that all manipulations return equivalent statements (in chains of 'if and only if' statements). The next

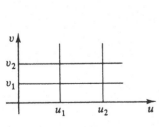
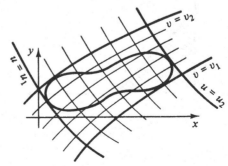

Fig. 19.22

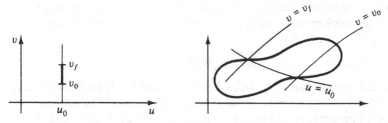

Fig. 19.23. *For each value u_0 of u find the starting value v_0 and the finishing value v_f for v.*

two examples are devoted to just such routines (see also 19.7) and a third more awkward example follows. See also Question 9 page 346.

Example

Let D be the triangle bounded by $x + y = 1$, $x = 0$, $y = 0$. What is the image of D under the following transformation?

$$u = x + y, v = x - y.$$

(i) Solution by contour plotting. Consider a contour $u = u_0$ (cf. Figure 19.24). This cuts D only if $0 \leqslant u_0 \leqslant 1$. Fix such a u_0. The contour $v = v_0$ is a straight line, viz. $y = x - v_0$. The latter will cut $u = u_0$ in D only if $-u_0 \leqslant v_0 \leqslant u_0$. Hence Δ is as shown in Figure 19.25.

(ii) Solution by transformation of the defining relations. D is defined by the inequalities

$$x + y \leqslant 1, x \geqslant 0, y \geqslant 0.$$

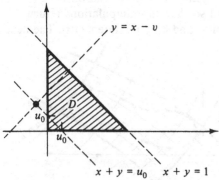

Fig. 19.24

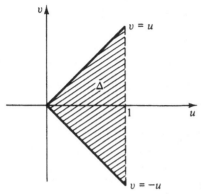

Fig. 19.25

Now
$$x = \tfrac{1}{2}(u + v), \quad y = \tfrac{1}{2}(u - v).$$
Thus
(a) $x + y \leqslant 1 \Leftrightarrow u \leqslant 1,$
(b) $x \geqslant 0 \Leftrightarrow \tfrac{1}{2}(u + v) \geqslant 0 \Leftrightarrow v \geqslant -u,$
(c) $y \geqslant 0 \Leftrightarrow \tfrac{1}{2}(u - v) \geqslant 0 \Leftrightarrow u \geqslant v.$

Example
For the transformation
$$u = y^2 - x^2$$
$$v = x + y$$
find the image of the positive quadrant.

(i) Solution by contour plotting. The u-contours are rectangular hyperbolas with asymptotes $x - y = 0$ and $x + y = 0$ (Figure 19.26). The v-contours are thus all parallel to one of the asymptotes and hence they cut the u-contours once only. To find the transformation of the positive quadrant we note that $v > 0$ and that we are obliged to consider both negative and positive values of u. Fix a value of v, say $v_0 > 0$. It is clear from the position of the intercepts of the u and v contours on the x and y axes that the pair (u, v_0) gives a point in the positive quadrant precisely when

(i) $\sqrt{u} \qquad < v_0$ for $u > 0,$
     and
(ii) $\sqrt{(-u)} < v_0$ for $u < 0.$

We thus reach the conclusion that the image is given by
$$v > 0 \quad \& \quad -v^2 < u < v^2.$$

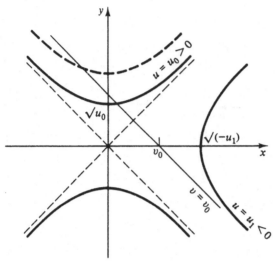

Fig. 19.26

The above, of course, required us to have some accurate knowledge of the intercepts. The advantage of the analytical approach is that the necessary information comes automatically.

A more complex situation occurs in Question 4 of Section 19.17.

(ii) Solution by transformation of the defining relations. Consider that

$$u = (y - x)(y + x).$$

Thus

$$u/v = y - x$$

and

$$v = y + x.$$

Thus

$$x > 0 \leftrightarrow v - u/v > 0$$

$$\leftrightarrow \frac{v^2 - u}{v} > 0$$

$$\leftrightarrow either \ (v > 0 \quad \& \quad v^2 - u > 0)$$

$$or \quad (v < 0 \quad \& \quad v^2 - u < 0).$$

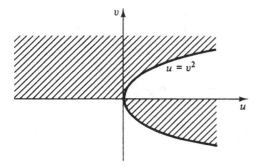

Fig. 19.27

This may be identified as in Figure 19.27. Similarly,

$$y > 0 \leftrightarrow v + u/v > 0$$

$$\leftrightarrow \frac{v^2 + u}{v} > 0$$

$$\leftrightarrow either \ (v > 0 \quad \& \quad v^2 > -u)$$
$$or \quad (v < 0 \quad \& \quad v^2 < -u).$$

Combining the two results we obtain (see Figure 19.28)

$$\{(u, v): v > 0 \quad \& \quad -v^2 < u < v^2\}.$$

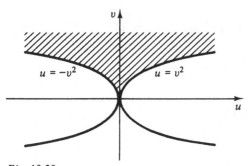

Fig. 19.28

19.11 Presence of the Jacobian

We begin by considering the effect of linear transformations on area and volume.

Let M be a 2×2 non-singular matrix with entries m_{ij} and consider the linear function $L: \mathbb{R}^2 \to \mathbb{R}^2$ defined by $y = L(x) = Mx$.

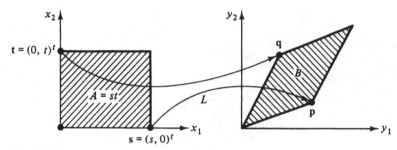

Fig. 19.29

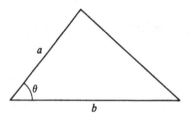

Fig. 19.30

What is the relation between the areas A and B in Figure 19.29? We begin by calculating B using the fact that the area of the triangle in Figure 19.30 is equal to $\frac{1}{2}ab\sin\theta$.
Thus

$$B^2 = \|\mathbf{p}\|^2\|\mathbf{q}\|^2\sin^2\theta = \|\mathbf{p}\|^2\|\mathbf{q}\|^2(1-\cos^2\theta)$$

and so

$$B^2 = \|\mathbf{p}\|^2\|\mathbf{q}\|^2 - \langle\mathbf{p},\mathbf{q}\rangle^2.$$

But $\mathbf{p} = M\mathbf{s}$ and $\mathbf{q} = M\mathbf{t}$. Therefore

$$\left.\begin{array}{l} p_1 = m_{11}s \\ p_2 = m_{21}s \end{array}\right\} \quad \left.\begin{array}{l} q_1 = m_{12}t \\ q_2 = m_{22}t \end{array}\right\}.$$

It follows that

$$\|\mathbf{p}\|^2 = (m_{11}^2 + m_{21}^2)s^2; \quad \|\mathbf{q}\|^2 = (m_{12}^2 + m_{22}^2)t^2$$
$$\langle\mathbf{p},\mathbf{q}\rangle = (m_{11}m_{12} + m_{21}m_{22})st.$$

Thus

$$B^2 = \{(m_{11}^2 + m_{21}^2)(m_{12}^2 + m_{22}^2) - (m_{11}m_{12} + m_{21}m_{22})^2\}s^2t^2$$
$$= (\det M)^2 A^2.$$

(To check this last step, expand $(\det M)^2 = (m_{11}m_{22} - m_{12}m_{21})^2$.)

We have shown that $B^2 = (\det M)^2 A^2$ and so either $B = (\det M)A$ or else $B = -(\det M)A$. Since $B > 0$ we require the alternative with a positive right-hand side. Since $A > 0$, this means that

$$B = |\det M| A.$$

This useful geometric interpretation of a determinant extends to the 3×3 case provided that A and B are interpreted as volumes. It also extends to the $n \times n$ case with A and B interpreted as 'hypervolumes'.

When applying the result in a calculation below, we shall be using affine functions rather than linear functions. If M is a 2×2 non-singular matrix and the affine function $\alpha : \mathbb{R}^2 \to \mathbb{R}^2$ is defined by

$$\mathbf{y} = \alpha(\mathbf{x}) = M(\mathbf{x} - \boldsymbol{\xi}) + \boldsymbol{\eta},$$

then the equation $B = |\det M| A$ still holds with A and B equal to the areas indicated in Figure 19.31.

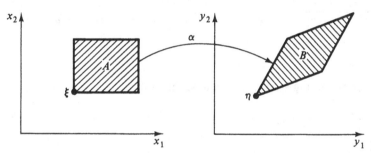

Fig. 19.31

To see this, simply take $\mathbf{X} = \mathbf{x} - \boldsymbol{\xi}$ and $\mathbf{Y} = \mathbf{y} - \boldsymbol{\eta}$ and consider the equation $\mathbf{Y} = M\mathbf{X}$. A similar result holds, of course, in the case when M is an $n \times n$ matrix.

We can now explain the presence of the Jacobian in the transformed integral. Suppose given a transformation as in the theorem of a subset Δ in the u, v plane to the set D in the x, y plane with

$$x = \phi(u, v) \quad y = \psi(u, v) \tag{1}$$

and suppose that as in the theorem the inverse transformation exists and is given by

$$u = U(x, y) \quad v = V(x, y).$$

We divide the region Δ into a large number of small rectangles (see Section 19.1). The lines $u = u_i$ and $v = v_i$ in the u, v plane correspond to the curves $U(x, y) = u_i$ and $V(x, y) = v_i$ in the x, y plane. Our subdivision of Δ therefore induces a corresponding subdivision of D.

We shall write $\mathbf{z} = (x, y)^t$ and $\mathbf{w} = (u, v)^t$. The shaded area with a corner at $\mathbf{z}_i = (x_i, y_i)^t$ will be denoted by B_i and the corresponding shaded area with a corner at $\mathbf{w}_i = (u_i, v_i)^t$ will be denoted by A_i. See Figure 19.32.

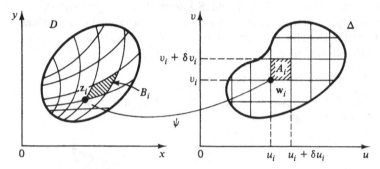

Fig. 19.32

An approximating sum to the double integral

$$\iint_D f(x, y)\,dx\,dy \tag{2}$$

is

$$\Sigma f(x_i, y_i)B_i = \Sigma F(u_i, v_i)B_i,$$

where

$$F(u, v) = f(\varphi(u, v), \psi(u, v))),$$

i.e. $F(u, v)$ is obtained by substituting for x and y in $f(x, y)$ using equation (1).

The next step is to express the area B_i in terms of the area $A_i = \delta u_i\, \delta v_i$. We begin by observing that

$$\alpha(\mathbf{w}) = \frac{d(u_i, v_i)}{d(x, y)}(\mathbf{w} - \mathbf{w}_i) + (\phi(\mathbf{w}_i), \psi(\mathbf{w}_i))^t$$

is a good approximation to $(\phi(\mathbf{w}), \psi(\mathbf{w}))^t$ for values of $\mathbf{w}$ close to $\mathbf{w}_i$. The reason is that $\mathbf{z} = \alpha(\mathbf{w})$ is tangent to $\mathbf{z} = (\phi(\mathbf{w}), \psi(\mathbf{w}))^t$ at the point $\mathbf{w}_i$ (see Chapter 14). It follows that C_i in Figure 19.33 is a good approximation to B_i provided that δu_i and δv_i are small.

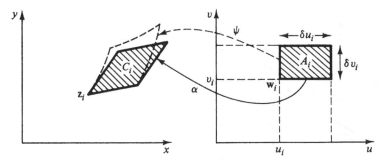

Fig. 19.33

But, since $\alpha(\mathbf{w})$ is an affine transformation, it follows that

$$B_i \approx C_i = \left| \det \frac{d(u_i, v_i)}{d(x, y)} \right| A_i$$

$$= \left| \det \frac{d(u_i, v_i)}{d(x, y)} \right| \delta u_i \, \delta v_i.$$

Returning now to our approximating sum, we obtain that

$$\Sigma f(x_i, y_i) B_i \approx \Sigma F(u_i, v_i) \left| \det \frac{d(u_i, v_i)}{d(x, y)} \right| \delta u_i \, \delta v_i.$$

But this final sum is an approximating sum for the double integral

$$\iint_\Delta F(u, v) \left| \det \frac{d(u, v)}{d(x, y)} \right| du \, dv. \tag{3}$$

We conclude that (2) is equal to (3).

19.12 Example

For the spherical polar transformation (cf. Figure 19.34)

$$x = r \sin \theta \cos \phi$$
$$y = r \sin \theta \sin \phi$$
$$z = r \cos \theta$$

find the image of the ball V:

$$x^2 + y^2 + (z + k)^2 \leqslant a^2$$

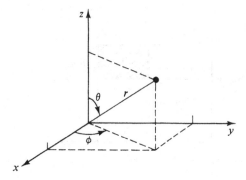

Fig. 19.34

and hence evaluate

$$I = \iiint_B \frac{dx\,dy\,dz}{\sqrt{\{x^2 + y^2 + (z - t)^2\}}},$$

where B is the sphere

$$x^2 + y^2 + z^2 \leqslant a^2 \quad \text{and} \quad 0 < a < t.$$

Solution by inspection. This type of problem – based firmly on geometric considerations (compare Question 6 in 19.20) can really only best be done by inspection. We consider only the more awkward case $k > a$. We refer to Figure 19.35. Clearly the range of θ is constrained between the values

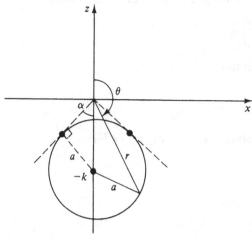

Fig. 19.35

$\pi - \alpha$ and $\pi + \alpha$ where α is as on the diagram. To find the r range corresponding to a given value of θ note that by the cosine rule

$$a^2 = r^2 + k^2 - 2kr\cos(\pi - \theta),$$

so

$$a^2 = r^2 + k^2 + 2kr\cos\theta,$$

or

$$r^2 + 2k\cos\theta \cdot r + (k^2 - a^2) = 0,$$

with roots r_+ and r_- given by

$$-k\cos\theta \pm \sqrt{\{k^2\cos^2\theta - (k^2 - a^2)\}},$$

or

$$-k\cos\theta \pm \sqrt{\{a^2 - k^2\sin^2\theta\}}.$$

Clearly, for a given θ, r runs from r_- to r_+ (note also that $\cos\theta < 0$). We have thus found the image set of V.

Evidently from the right-angled triangle with vertex at the tangency point we have

$$\sin\alpha = a/k.$$

To evaluate the integral we first find that

$$J = \frac{\partial(x,y,z)}{\partial(r,\theta,\phi)} = \begin{vmatrix} \sin\theta\cos\phi & r\cos\theta\cos\phi & -r\sin\theta\sin\phi \\ \sin\theta\sin\phi & r\cos\theta\sin\phi & r\sin\theta\cos\phi \\ \cos\theta & -r\sin\theta & 0 \end{vmatrix}$$

$$= r^2\cos^2\theta\sin\theta\{\cos^2\phi + \sin^2\phi\} + r^2\sin^3\theta\{\cos^2\phi + \sin^2\phi\}$$

$$= r^2\sin\theta.$$

Next we change co-ordinates to:

$$X = x \quad Y = y \quad Z = z - t$$

and obtain

$$I = \iiint_V \frac{dX\,dY\,dZ}{\sqrt{\{X^2 + Y^2 + Z^2\}}},$$

where this time V is the ball

$$X^2 + Y^2 + (Z + t)^2 \leq a^2.$$

We can now apply the previous work taking $k = t$. Dropping the upper case letters to lower, we pass to spherical polar co-ordinates. But the

Jacobian changes sign in $[\pi - \alpha, \pi + \alpha]$ and so we need to compute separately in $[\pi - \alpha, \pi]$ where $J > 0$ and in $[\pi, \pi + \alpha]$ where $J < 0$. Let the corresponding contributions to I be I_1 and I_2. We have

$$I_1 = \int_0^{2\pi} d\phi \int_{\pi-\alpha}^{\pi} d\theta \int_{r_-}^{r_+} r^2 \frac{\sin\theta \, dr}{r}$$

$$= 2\pi \int_{\pi-\alpha}^{\pi} \left[\frac{r^2}{2} \right]_{r_-}^{r_+} \sin\theta \, d\theta$$

$$= \pi \int_{\pi-\alpha}^{\pi} (r_+ + r_-)(r_+ - r_-) \sin\theta \, d\theta$$

$$= \pi \int_{\pi-\alpha}^{\pi} -2t\cos\theta \sqrt{\{a^2 - t^2 \sin^2\theta\}} \sin\theta \, d\theta$$

$$= \pi \left[\frac{2}{3}(a^2 - t^2 \sin^2\theta)^{3/2} \cdot \frac{1}{t} \right]_{\pi-\alpha}^{\pi}$$

$$= \frac{2\pi}{3t} a^3 \qquad \text{(since } a^2 - t^2 \sin^2\alpha = 0\text{)}.$$

Clearly

$$I = 2I_1 = \frac{4\pi a^3}{3t}.$$

Note that we would have received a zero answer if we had not tested the sign of the Jacobian (i.e. I_1 and I_2 should not cancel each other).

19.13 Improper integrals: infinite range

Recall the traditional computation of

$$I = \int_0^{\infty} \exp(-x^2/2) \, dx.$$

The usual argument is

$$I^2 = \int_0^{\infty} \exp(-x^2/2) \int_0^{\infty} \exp(-y^2/2) \, dy$$

$$= \iint_D \exp\{-(x^2 + y^2)/2\} \, dx \, dy,$$

where D is the positive quadrant. The step from first to second line is based on a reduction of a repeated integral to a double integral. We

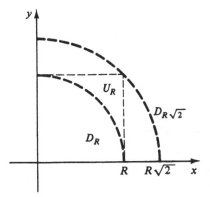

Fig. 19.36

shall comment on this in a moment when we have completed the calculation.

We change variables to polar co-ordinates

$$x = r \cos \theta$$
$$y = r \sin \theta$$

so that Δ is described by $0 \leqslant r < \infty$, $0 \leqslant \theta \leqslant \pi/2$.
Thus

$$\frac{\partial(x, y)}{\partial(r, \theta)} = \begin{vmatrix} \cos \theta & -r \sin \theta \\ \sin \theta & r \cos \theta \end{vmatrix} = r\{\cos^2 \theta + \sin^2 \theta\} = r.$$

Finally,

$$I^2 = \iint_\Delta \exp(-r^2/2) r \, dr \, d\theta = \int_0^{\pi/2} d\theta \int_0^\infty \exp(-r^2/2) r \, dr$$

$$= \frac{\pi}{2} [\exp(-r^2/2)]_0^\infty = \frac{\pi}{2}.$$

So $I = \sqrt{(\pi/2)}$.

To justify the second line we simply note that with reference to Figure 19.36

$$\iint_D \exp\{-(x^2 + y^2)/2\} \, dx \, dy$$

$$= \lim_{R \to \infty} \iint_{U_R} \exp\{-(x^2 + y^2)/2\} \, dx \, dy,$$

and we are entitled to write

$$\iint_{U_R} \exp\{-(x^2+y^2)/2\} \, dx \, dy$$

$$= \int_0^R dx \int_0^R \exp(-x^2/2)\exp(-y^2/2) dy$$

$$= \int_0^R \exp(-x^2/2) dx \cdot \int_0^R \exp(-y^2/2) dy$$

$$\to I \cdot I \quad \text{as} \quad R \to \infty.$$

Now

$$\lim_{R \to \infty} \iint_{D_R} \exp\{-(x^2+y^2)/2\} \, dx \, dy$$

$$= \lim_{R \to \infty} \iint_{U_R} \exp\{-(x^2+y^2)/2\} \, dx \, dy,$$

because, since the integrand is positive, we have

$$\iint_{D_R} \leqslant \iint_{U_R} \leqslant \iint_{D_{R\sqrt{2}}}.$$

So

$$\iint_{D_R} \exp\{-(x^2+y^2)/2\} \, dx \, dy = \iint_{\Delta_R} \exp(-r^2/2) r \, dr \, d\theta$$

where Δ_R is described by $0 \leqslant r \leqslant R$, $0 \leqslant \theta \leqslant \pi/2$. Thus

$$\iint_{\Delta_R} \exp(-r^2/2) r \, dr \, d\theta = \int_0^{\pi/2} d\theta \int_0^R \exp(-r^2/2) r \, dr \to \pi/2.$$

19.14 Improper integrals: integrand discontinuous

Consider the problem of evaluating

$$I = \iint_D \cos\left(\frac{x-y}{x+y}\right) dx \, dy,$$

where D is the triangle bounded by $x+y=1$, $x=0$ and $y=0$.

Notice that the integrand is discontinuous at the origin, so we are dealing here with the improper integral

$$\lim_{\delta \to 0} \iint_{\substack{D \\ x+y \geqslant \delta}} \cos\left(\frac{x-y}{x+y}\right) dx \, dy.$$

We note that the limit exists because the integrand remains bounded at the origin (integration over $x + y \leqslant \delta$ will contribute less than $\frac{1}{2}\delta^2$, since $|\cos\{(x - y)/(x + y)\}| \leqslant 1$).

We should obviously take $u = x + y$ and $v = x - y$ so that,

$$x = \tfrac{1}{2}(u + v) \quad y = \tfrac{1}{2}(u - v).$$

To find the transform of D observe that, e.g. when $u = \delta$ only the lines $v = \text{const.}$ for $-\delta \leqslant \text{const.} \leqslant \delta$ intersect $u = \delta$. Hence the transform Δ of D is bounded by $v = \pm u$ and $u = 1$ (cf. p. 322–3). Now

$$\frac{\partial(x, y)}{\partial(u, v)} = \begin{vmatrix} 1/2 & 1/2 \\ 1/2 & -1/2 \end{vmatrix} = -\frac{1}{2}.$$

Thus

$$I = \int_0^1 du \int_{-u}^{u} \cos\left(\frac{v}{u}\right) \cdot \tfrac{1}{2} dv$$

$$= \frac{1}{2} \cdot \int_0^1 \left[\sin\left(\frac{v}{u}\right) \right]_{-u}^{u} u \cdot du = \tfrac{1}{2} \cdot \sin 1.$$

19.15 Further examples

Example 1 (on dissection)

Find the volume of the pyramid D

$$x + y + z \leqslant h,$$
$$x, y, z \geqslant 0.$$

The volume is just

$$V = \iiint_D dx \, dy \, dz.$$

We follow the argument of 19.4

$$u = x + y + z.$$

Thus $u = u_0$ for $0 \leqslant u_0 \leqslant h$ gives a section of D parallel to the shaded face of Figure 19.37. Now the section $D(u_0)$ may similarly be dissected into line segments parallel to the edge marked in bold in Figure 19.38. Its equation is

$$y = 0 \quad \& \quad x + z = u_0.$$

Parallel segments will thus satisfy

$$x + z = \text{const.} \quad \& \quad y = \text{const.}$$

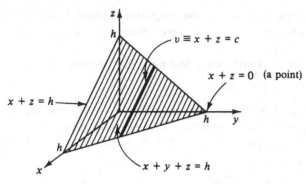

Fig. 19.37

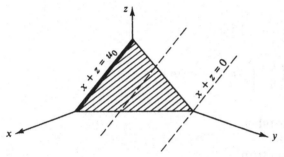

Fig. 19.38. *The section $D(u_0)$ is dissected into parallel lines $x + z = $ constant.*

So put

$$v = x + z$$

and then

$$0 \leqslant v \leqslant u_0.$$

Finally to describe the points along a line segment $v = v_0$ use x or z as a parameter, say x. Thus $0 \leqslant x \leqslant v_0$. In conclusion

$$u = x + y + z,$$
$$v = x + z,$$
$$w = x.$$

Thus

$$\frac{\partial(u, v, w)}{\partial(x, y, z)} = \begin{vmatrix} 1 & 1 & 1 \\ 1 & 0 & 1 \\ 1 & 0 & 0 \end{vmatrix} = 1$$

and so

$$V = \int_0^h du \int_0^u dv \int_0^v dw = \int_0^h du \int_0^u v\,dv = \int_0^h u^2/2\,du = h^3/6.$$

What we did was to select new variables in such a way as to describe the points of D using a dissection into planes, then into line segments and, finally, points, in the manner required by Fubini's formula (Section 19.3). Note that the domain of integration used in the (u, v, w)-spaced is also pyramid.

Example 2 (Jacobian with variable sign)

Evaluate

$$I = \iint_D \exp\left\{-\left(\frac{x}{x+y}\right)^2 - (x+y)^2\right\} dx\,dy$$

over the region shown ($y \leqslant x$). It is assumed that the integrand is zero whenever $x + y = 0$. We let

$$u = \frac{x}{x+y}, \quad v = x + y.$$

Hence

$$u = \text{const.} \quad \left(\text{along which } y = \frac{1-u}{u}x\right)$$

describe a straight line (see Figure 19.39). It is easy to see that $u = 0$

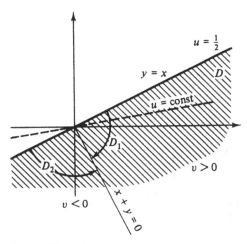

Fig. 19.39

corresponds to $x = 0$ while $u = 1$ corresponds to $y = 0$. Thus D_1 is traced out when

$$\tfrac{1}{2} \leqslant u < \infty \quad \& \quad v > 0.$$

On the other hand D_2 is traced out when

$$-\infty < u \leqslant \tfrac{1}{2} \quad \& \quad v < 0.$$

Now

$$x = uv,$$
$$y = v - uv,$$

so

$$J = \frac{\partial(x, y)}{\partial(u, v)} = \begin{vmatrix} v & u \\ -v & 1 - u \end{vmatrix} = v.$$

Thus referring to Figure 19.40 $|J| = v$ in Δ_1 but $|J| = -v$ in Δ_2.

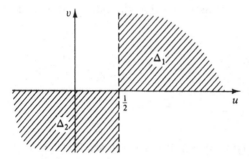

Fig. 19.40

Consequently,

$$
\begin{aligned}
I &= \int_{-\infty}^{1/2} du \int_{-\infty}^{0} (-v) \exp(-u^2 - v^2) dv \\
&\quad + \int_{1/2}^{\infty} du \int_{0}^{\infty} v \exp(-u^2 - v^2) dv \\
&= \int_{-\infty}^{1/2} \exp(-u^2) du \left[\frac{\exp(-v^2)}{2} \right]_{-\infty}^{0} \\
&\quad + \int_{1/2}^{\infty} \exp(-u^2) du \cdot \left[-\frac{\exp(-v^2)}{2} \right]_{0}^{\infty} \\
&= \frac{1}{2} \int_{-\infty}^{\infty} \exp(-u^2) du = \frac{\sqrt{\pi}}{2}.
\end{aligned}
$$

Remark

The integrand is discontinuous only at the origin (e.g. approach the origin along $x = 0$ to obtain a limiting value 1 for the integrand, and approach along $x + y = 0$ to obtain a limiting value of 0). However, since

$$\exp\left\{-\left(\frac{x}{x+y}\right)^2\right\} \leqslant e^0,$$

the integrand is bounded near the origin. So just as in the example of 19.14 the existence of the improper integral over any bounded part of D is assured.

19.16   Problems arising in statistics

Statisticians interest themselves in the probability that a randomly selected individual from a given population has some numerical property or other. For example, the throw of a die randomly chooses one face from a population of six faces and the numerical property of interest could be the number of dots on the face.

In general we consider a population denoted Ω and a numerical property X. The numerical value of X corresponding to the individual ω in Ω is written $X(\omega)$. Thus, mathematically speaking, X is a function from Ω to $\mathbb{R}$. Statisticians refer to such functions as random variables (since these are the variable quantities whose value is to be randomly selected).

Examples

1. Labelling the six faces of a die as $1, 2, \ldots, 6$ (according to the number of dots) we obtain $\Omega = \{1, 2, \ldots, 6\}$ and the function which assigns the number of dots to the face is then quite straightforwardly,

$$X(i) = i.$$

2. In Monopoly the score is determined by throwing two dice and taking the sum of the dots on the faces turned up. This amounts to a random selection of two faces one from each of two dice. Here the population consists of the ordered pairs of faces conveniently labelled (m, n). The score is given by the function

$$X(m, n) = m + n.$$

Note also the other two numerical properties/random variables

$$X_1(m, n) = m \quad X_2(m, n) = n,$$

viz. the number of dots on the first and the second face.

The function defined by

$$F(x) = \text{Prob}\,(X \leqslant x)$$

is known as the *probability distribution function* of X.

In Example 1, assuming an unbiased die

$$F(x) = \frac{\#\{n : n \leqslant x\}}{\#\{n : n = 1, 2, \ldots, 6\}} = \frac{[x]}{6}.$$

In Example 2, the distribution function of X_1 is

$$F_1(x) = \frac{\#\{(m, n) : m \leqslant x\}}{\#\{(m, n) : m = 1, \ldots, 6, n = 1, \ldots, 6\}} = \frac{[x] \cdot 6}{6 \cdot 6} = \frac{[x]}{6},$$

where $[x]$ denotes the greatest integer less than or equal to x.

If there is a function $f(t)$ such that

$$F(x) = \int_{-\infty}^{x} f(t)\,dt$$

then $f(t)$ is called the *density function* of X. Quite often we are given a *joint distribution function* of several random variables $X_1, X_2, \ldots, X_n$, which is defined to be:

$$F(x_1, x_2, \ldots, x_n) = \text{Prob}\,(X_1 \leqslant x_1 \,\&\, X_2 \leqslant x_2 \,\&\, \ldots \,\&\, X_n \leqslant x_n).$$

This might well be available in the form

$$F(x_1, x_2, \ldots, x_n) = \int_{-\infty}^{x_1} dt_1 \int_{-\infty}^{x_2} dt_2 \int_{-\infty}^{x_n} f(t_1, t_2, \ldots, t_n)\,dt_n$$

where $f(t_1, t_2, \ldots, t_n)$ is the *joint density* function. We are then required to find the probability density of combinations of the random variables. Let us illustrate this.

Example
The random variables X, Y have joint density function

$$f(x, y) = \frac{1}{2\pi} \exp\{-(x^2 + y^2)/2\}.$$

Find the density function of Y/X. By definition we have to consider

$$F(t) = \text{Prob}\,(Y/X \leqslant t)$$

$$= \iint_D f(x, y)\,dx\,dy$$

where

$$D = \{(x,y) : y/x \leqslant t\}.$$

This last equation is based on the intuition that $f(x,y)$ is the approximate probability that

$$x \leqslant X \leqslant x + \delta x \quad \& \quad y \leqslant Y \leqslant y + \delta y.$$

Now $y/x \leqslant t$ for $x > 0$ implies $y \leqslant tx$; while for $x < 0$ it implies $y \geqslant tx$. Hence $D = D_1 \cup D_2$ as shown in Figure 19.41.

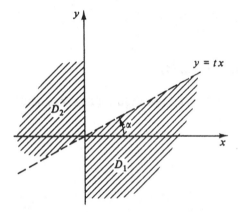

Fig. 19.41

Changing to polar co-ordinates

$$F(t) = \frac{1}{2\pi} \int_{-\pi/2}^{\alpha} d\theta \int_0^{\infty} \exp(-r^2/2) r \, dr$$

$$+ \frac{1}{2\pi} \int_{\pi/2}^{\pi+\alpha} d\theta \int_0^{\infty} \exp(-r^2/2) r \, dr$$

$$= \frac{1}{2\pi}\left(\alpha + \frac{\pi}{2}\right) + \frac{1}{2\pi}\left(\alpha + \frac{\pi}{2}\right) = \frac{\tan^{-1}(t)}{\pi} + \frac{1}{2},$$

since $\tan \alpha = t$. We see (by differentiation) that Y/X has the probability density function

$$\frac{1}{\pi(1 + t^2)}.$$

In anticipation of a later chapter we note the following.

Example (convolutions)

Suppose X_1 and X_2 are random variables with density functions $f_1(x)$, $f_2(x)$. To say that X_1 and X_2 are independent means that the joint distribution of X_1 and X_2 is

$$f(x_1, x_2) = f_1(x_1) \cdot f_2(x_2).$$

We now obtain the density function of $X_1 + X_2$ assuming independence. We have

$$F(t) = \text{Prob}(X_1 + X_2 \leqslant t)$$

$$= \iint_D f_1(x_1) \cdot f_2(x_2) \, dx_1 dx_2$$

$$= \int_{-\infty}^{\infty} dx_1 \int_{-\infty}^{t-x_1} f_1(x_1) \cdot f_2(x_2) dx_2,$$

where D is the region $\{(x_1, x_2) : x_1 + x_2 \leqslant t\}$. See Figure 19.42. Assuming

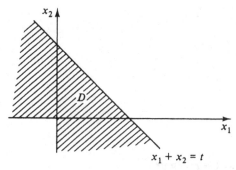

Fig. 19.42

f_1, f_2 are continuous (and supposing that for each i

$$\int_{-\infty}^{\infty} |f_i(x_i)| dx_i$$

exists), we may differentiate with respect to t to obtain

$$f(t) = \int_{-\infty}^{\infty} f_1(x_1) \cdot f_2(t - x_1) dx_1.$$

This is known as the convolution of f_1 and f_2. For a finite population such as $\Omega = \{1, 2, \ldots, n\}$ where we have $f(m) = \text{Prob}(X_1 + X_2 = m)$, we see that the appropriate formula would read

$$\sum_1^n p_i q_{m-i},$$

where $p_i = \text{Prob}(X_1 = i)$ and $q_j = \text{Prob}(X_2 = j)$.

19.17 Exercises. Problems on transformations

For each of the transformations given below obtain, where relevant, the transform of the region indicated.

1. $u = \frac{1}{2}(x^2 - y^2)$

    $v = xy$

The positive quadrant.

2. $u = \dfrac{x}{x^2 + y^2}$

    $v = \dfrac{y}{x^2 + y^2}$

Find the inverse transformation and resolve any ambiguities.

3. $u = xy$

    $v = x + y$

The unit disc.

4. $u = \dfrac{x(x + y)}{y}$

    $v = x + y$

The positive quadrant.

5. $u = x^2 + y^2$

    $v = xy$

The sector $0 < y < x$.

6. $u = y - x^2$

    $v = x^2 - y^2$

Show that the plane splits into four regions where the Jacobian is of constant sign. Find the inverse transformation in each region and the image of the half strip $x > 0$ & $0 < y < 1/2$.

7. $u = x + y$

    $v = y/x$ (for $x \neq 0$)

The strip $0 \leqslant y \leqslant 1$.

8. $u = x^2 - yx$

    $v = y^2 - x^2$.

The lines $x = $ constant.

19.18   Exercises. Evaluation of integrals

1. Find

$$\iint_D \exp\{-(x+y)\}\,dx\,dy$$

where D is bounded by $x=0$, $y=0$ and $x-y\leqslant k$.

2. Find the volume common to the cylinders $x^2+y^2\leqslant a^2$ and $x^2+z^2\leqslant a^2$.

3. Find the area enclosed between $xy=1$, $xy=2$, $x^2-y^2=1$, $x^2-y^2=2$ in the positive quadrant.

4. Obtain the volume of the sphere $x^2+y^2+z^2\leqslant R^2$ using the transformation

$$u=\sqrt{(x^2+y^2+z^2)},\quad v=y,\quad w=z.$$

5. Evaluate

$$I=\int_0^\infty\int_0^\infty \frac{x^2+y^2}{1+(x^2-y^2)^2}e^{-2xy}\,dx\,dy.$$

Use $u=x^2-y^2$, $v=2xy$.

6. Express the following integral as an improper double integral. Show that the integral exists and evaluate it using the substitution $u=x+y, v=x-y$

$$\int_0^{1/2} dx \int_x^{1-x}\left(\frac{x-y}{x+y}\right)^2 dy.$$

7. Show that

$$\int_0^1\left\{\int_0^1\frac{x-y}{(x+y)^3}\,dy\right\}dx=\frac{1}{2}\neq\int_0^1\left\{\int_0^1\frac{x-y}{(x+y)^3}\,dx\right\}dy.$$

19.19   Exercises. Problems from statistics

1. The joint probability density function of the three random variables X, Y, Z is given below. Find $\text{Prob}(X<Y<Z)$.

$$f(x,y,z)=\begin{cases}8xyz & \text{for } 0\leqslant x,y,z\leqslant 1,\\ 0 & \text{otherwise.}\end{cases}$$

2. The joint probability density function of the two random variables X, Y is given below. A third random variable Z is also defined below. Find its density.

$$f(x,y)=\begin{cases}1 & \text{for } 0\leqslant x,y\leqslant 1\\ 0 & \text{otherwise.}\end{cases}\qquad Z(\omega)=\begin{cases}X(\omega)+Y(\omega) & \text{if less than }1\\ X(\omega)+Y(\omega)-1, & \text{otherwise.}\end{cases}$$

3. The random variables Z_1,Z_2,Z_3 are independent and have the identical probability density function

$$\frac{1}{\sqrt{(2\pi)}}\exp(-t^2/2).$$

By first finding the region where the inequalities below are satisfied,
- (i) $\frac{1}{2}(z_1^2 + z_2^2) \leqslant k$
- (ii) $kz_1 \leqslant \sqrt{(z_1^2 + z_2^2)}$ $(k > 1)$
- (iii) $|z_1| \leqslant k\sqrt{(z_2^2 + z_3^2)}$

and integrating over these regions show that
- (i) $\text{Prob}(\frac{1}{2}(Z_1^2 + Z_2^2) \leqslant k) = 1 - e^{-k}$
- (ii) $\text{Prob}(kZ_1 \leqslant \sqrt{(Z_1^2 + Z_2^2)}) = -\dfrac{\arctan\sqrt{(k^2 - 1)}}{\pi} + 1$

- (iii) $\text{Prob}(|Z_1| \leqslant k\sqrt{(Z_2^2 + Z_3^2)}) = 1 - \dfrac{k}{\sqrt{(1 + k^2)}}$

Discuss the integration problem involved in solving for k

$$\text{Prob}(Z_1 + kZ_2 \leqslant 3) = x.$$

19.20   Exercises. Miscellaneous

1. A transformation is given for $x \neq 0$ by:

$$u = x + y, \quad v = y/x.$$

Find the image of the strip $0 \leqslant y \leqslant 1$.

2. A transformation of $\mathbb{R}^2$ into $\mathbb{R}^2$ is given by:

$$\left.\begin{aligned} u &= x^2 - xy \\ v &= y^2 - xy \end{aligned}\right\}.$$

Show that $u = \text{const.}$ and $v = \text{const.}$ contours are hyperbolas. What are the asymptotes? Show that in appropriate regions the inverse transformations are:

$$x = \frac{\pm u}{\sqrt{(u + v)}}, \quad y = \frac{\mp v}{\sqrt{(u + v)}}.$$

3. A transformation of the positive quadrant is given by

$$u = \tfrac{1}{2}(x + y), \quad v = \sqrt{(xy)}.$$

Find the image of the positive half-lines along $x = \text{const.}$ and $y = \text{const.}$ for positive constants. Find the image of the rectangle

$$\left.\begin{aligned} a &\leqslant y \leqslant b \\ b &\leqslant x \leqslant c \end{aligned}\right\},$$

where $0 < a < b < c$.

4. Evaluate

$$\int_0^{a\sin\beta} dy \int_y^{\sqrt{(a^2 - y^2)}} \log(x^2 + y^2)\, dx,$$

for $0 < \beta < \pi/4, 0 < a$.
[Recall that $\int \operatorname{cosec}^2 \theta\, d\theta = -\cot\theta + c$.]

5. Evaluate

$$\iint_D (\sqrt{x} - \sqrt{y}) x^n y^n \, dx \, dy,$$

where D is the finite region bounded by the x and y axes and the parabola $\sqrt{x} + \sqrt{y} = 1$.

6. Let u and v be the distances of the point (x, y, z) to $(0, 0, 1)$ and $(0, 0, -1)$ respectively. Let θ be the angle between the plane through the three points just mentioned and the plane $x = 0$. Define the variables ξ, η by

$$\xi = \tfrac{1}{2}(u + v), \quad \eta = \tfrac{1}{2}(u - v).$$

Find the Jacobian $\partial(\xi, \eta, \theta)/\partial(x, y, z)$ and show that

$$\iiint_{R^3} \frac{1}{uv} \exp\{-(u + v)/2\} \, dx \, dy \, dz = \frac{4\pi}{e}.$$

7. Show that for $0 < \alpha < \pi$

$$\int_0^\infty \int_0^\infty \exp\{-x^2 - 2xy \cos \alpha - y^2\} \, dx \, dy = \frac{\alpha}{2 \sin \alpha}.$$

[Hint: Complete the square, apply a linear transformation and then use polar co-ordinates.]

8. $u = x^2 + y^2$

$$v = y - x^2$$

As in Question 2 find the inverse transformations appropriate to different parts of the plane. Also find the image of the positive quadrant.

9. For the transformation

$$u = x^2 - xy$$
$$v = x^2 + y^2$$

calculate the Jacobian and show that it vanishes on two perpendicular lines. What is their relationship to the hyperbolas $u = $ constant? Show that the eight local inverse transformations are given by

$$x = \pm \tfrac{1}{2}\sqrt{\{(2u + v) \pm \sqrt{\{(2u + v)^2 - 8u^2\}}\}},$$

$$y = \pm \tfrac{1}{2}\sqrt{\{(3v - 2u) \mp \sqrt{\{(2u + v)^2 - 8u^2\}}\}},$$

and identify the appropriate signs by reference to the sign of the Jacobian.

10. Find the volume of the four-dimensional sphere S_4 given by

$$x^2 + y^2 + z^2 + t^2 \leqslant a^2.$$

It may be useful to observe that

$$\iiiint_{S_4} dx \, dy \, dz \, dt = 16 \iiiint_{S_4 \cap \Omega} dx \, dy \, dz \, dt,$$

where Ω is the positive orthant. You may assume that the volume of a three-dimensional sphere of radius a is $\tfrac{4}{3}\pi a^3$.

20

Differential and difference equations (revision)

This is a somewhat brisk chapter intended as background material to the following topics: Laplace transforms, the upper triangular form, series solutions of differential equations, calculus of variations. For a slower paced account see, for example, *Calculus* by K. G. Binmore.

Recall that a differential equation is said to be *partial* when the derivatives of the unknown function occurring in the equation are themselves partial; otherwise the differential equation is *ordinary*. The *order* of a differential equation is the order of the highest-ordered derivative appearing in the equation; its *degree* is the algebraic degree of the highest ordered derivative (i.e. the power it is raised to). It is important to note that by a *solution* of a differential equation we mean any relation between the variables which is free from derivatives and is consistent with the given differential equation. We begin by considering *ordinary* differential equations.

20.1 Elementary methods for order 1 degree 1

(a) *Separation of variables* is said to occur when the differential equation may be written in the form

$$P(y)\frac{dy}{dx} = Q(x) \tag{1}$$

where $P(y)$ and $Q(x)$ are respectively functions of y only and x only. The equation (1) may in principle be solved by integrating the relation

$$\int P(y)dy = \int Q(x)dx.$$

Example
Solve

$$\frac{dy}{dx} = \frac{y^3}{x+3}.$$

We separate the variables by writing

$$\frac{1}{y^3}\frac{dy}{dx} = \frac{1}{x+3}.$$

Next we integrate the equation

$$\int \frac{dy}{y^3} = \int \frac{dx}{x+3}$$

to obtain the solution

$$-\frac{1}{2y^2} = \log(x+3) + c,$$

where c is a constant of integration. (Note that in the present context we are not concerned to find y as a function of x.)

(b)   *Use of an integrating factor* facilitates the solution of an equation that can be written in the form

$$\frac{dy}{dx} + Q(x)y = R(x), \tag{2}$$

where $Q(x)$ and $R(x)$ are functions of x only. The method is based on observing that if $g(x)$ is any function of x, then

$$\frac{d}{dx}(g(x) \cdot y) = g(x)\frac{dy}{dx} + g'(x)y. \tag{3}$$

We compare the right-hand side of (3) with the left-hand side of

$$g(x)\frac{dy}{dx} + g(x)Q(x)y = g(x)R(x), \tag{4}$$

where (4) is obtained from (2) by multiplying by the factor $g(x)$.
    We try to arrange matters so that

$$g(x)Q(x) = g'(x), \tag{5}$$

or,

$$Q(x) = \frac{g'(x)}{g(x)} = \frac{d}{dx}(\log\{g(x)\}).$$

But this gives

$$g(x) = \exp\left\{\int Q(x)\,dx\right\}. \tag{6}$$

The function in (6), known as an *integrating factor* for equation (2), enables us to integrate up equation (4) using equation (3). We obtain for $g(x)$ as in (6) that:

$$g(x)y = \int g(x)R(x)dx.$$

Example
Solve

$$\frac{dy}{dx} = x^2 + \frac{2y}{x}.$$

We have

$$\frac{dy}{dx} - \frac{2}{x}y = x^2. \tag{7}$$

So

$$g(x) = \exp\left\{-\int \frac{2}{x}dx\right\}$$
$$= \exp\{-2\log x\}$$
$$= x^{-2}.$$

Multiplying (7) by x^{-2} on both sides gives

$$\frac{1}{x^2}\frac{dy}{dx} - \frac{2}{x^3}y = 1$$

or

$$\frac{d}{dx}\left(\frac{1}{x^2}y\right) = 1,$$

so

$$\frac{1}{x^2}y = x + c.$$

Thus

$$y = x^3 + cx^2,$$

where c is a constant. (Note that earlier on we had no need for a constant of integration when computing the integrating factor.)

(c) *Change of variables* often simplifies a differential equation, though occasionally some inspired guesswork is needed. We mention three examples.

(i)   The *homogeneous equation* of the form

$$N(x, y)\frac{dy}{dx} = M(x, y),$$

where both $M(x, y)$ and $N(x, y)$ are homogeneous functions of x and y both of the same order (cf. page 183). The latter condition means that for any number $\lambda \neq 0$

$$\frac{N(\lambda x, \lambda y)}{M(\lambda x, \lambda y)} = \frac{N(x, y)}{M(x, y)}. \tag{8}$$

In this case use is made of the new variable

$$v = y/x.$$

Now since

$$y = vx$$

we have

$$\frac{dy}{dx} = x\frac{dv}{dx} + v. \tag{9}$$

It is easy to see that the differential equation may then be solved by separation of variables.

Example
Solve

$$\frac{dy}{dx} = \frac{x^2 - xy + y^2}{xy}.$$

Note that here

$$N(\lambda x, \lambda y) = \lambda^2 xy = \lambda^2 N(x, y),$$
$$M(\lambda x, \lambda y) = \lambda^2(x^2 - xy + y^2) = \lambda^2 M(x, y),$$

so M and N are both homogeneous of degree 2 and thus (8) is satisfied. We have, by (9),

$$x\frac{dv}{dx} + v = \frac{x^2 - x^2 v + x^2 v^2}{x^2 v}$$
$$= \frac{v^2 - v + 1}{v},$$

so

$$x\frac{dv}{dx} = \frac{1-v}{v}.$$

Separating the variables

$$\frac{v}{1-v}\frac{dv}{dx} = \frac{1}{x}$$

and integrating

$$\int\frac{vdv}{v-1} \equiv \int\frac{v-1+1}{v-1}dv = -\int\frac{dx}{x},$$

we obtain

$$v + \log(v-1) = -\log x - \log c,$$

where $\log c$ is a constant. Thus

$$e^{-v} = cx(v-1)$$

or

$$e^{-y/x} = c(y-x).$$

   (ii)   Sometimes *interchanging the variables* helps. In the example below we consider y as the independent variable and x as the dependent variable. We are of course assuming then that

$$\frac{dx}{dy} = \left(\frac{dy}{dx}\right)^{-1}$$

which will be true if the derivative of y is non-zero.

Example
Solve

$$\frac{dy}{dx} = \frac{y}{2x+y^3}.$$

Rewriting this as

$$\frac{dx}{dy} = \frac{2x+y^3}{y} = \frac{2x}{y} + y^2,$$

we are now to solve

$$\frac{dx}{dy} - \frac{2x}{y} = y^2.$$

This is the same problem as in (7) with x and y interchanged.

(iii)   *Changes* inspired by context are worth investigating.

Example

Solve

$$\frac{dy}{dx} = \frac{1}{x-y} + 1.$$

(10)

Put

$$z = x - y,$$

so that

$$\frac{dz}{dx} = 1 - \frac{dy}{dx}.$$

Equation (10) now simplifies to

$$1 - \frac{dz}{dx} = \frac{1}{z} + 1,$$

or

$$\frac{dz}{dx} = -\frac{1}{z}.$$

Hence

$$\int z \, dz = - \int dx,$$

or

$$\tfrac{1}{2} z^2 = -x + c.$$

Thus

$$x - y = \sqrt{\{2(c-x)\}}$$

and

$$y = x - \sqrt{\{2(c-x)\}}.$$

(d)   *Exactness.* This concept is useful in looking for a solution to the general first-order differential equation of the form

$$M(x, y) + N(x, y)\frac{dy}{dx} = 0,$$

(11)

where $M(x, y)$ and $N(x, y)$ are functions of x and y only. We are interested to know whether for some function $f(x, y)$ the solution to (11) may be written in the form

$$f(x, y) \equiv 0.$$

(12)

Assuming (12) holds we differentiate with respect to x and obtain from the chain rule that

$$\frac{\partial f}{\partial x} + \frac{\partial f}{\partial y}\frac{dy}{dx} = 0. \tag{13}$$

We now ask if (11) as it stands is in the form (13). If that is the case we will have

$$\frac{\partial M(x, y)}{\partial y} = \frac{\partial N(x, y)}{\partial x}, \tag{14}$$

since both would equal

$$\frac{\partial^2 f}{\partial x \partial y}.$$

If the equation (11) satisfies (14) we say that it is *exact*, and it is then true that we may find f by solving the simultaneous partial differential equations

$$\left.\begin{array}{l} \dfrac{\partial f}{\partial x} = M(x, y) \\[2mm] \dfrac{\partial f}{\partial y} = N(x, y) \end{array}\right\}. \tag{15}$$

The reason for the solubility of (15) is indicated in the course of the example below. Equation (11) might not be exact and it is then an open problem to find a function $\phi(x, y)$ such that

$$\phi(x, y)M(x, y) + \phi(x, y)N(x, y)\frac{dy}{dx} = 0$$

is exact. Such a function ϕ, if found, is known as an *integrating factor*.

Example
Solve

$$(x + y + 1) + (x - y^2 + 3)\frac{dy}{dx} = 0.$$

We have here that

$$\left.\begin{array}{l} \dfrac{\partial}{\partial y}(x + y + 1) = 1 \\[2mm] \dfrac{\partial}{\partial x}(x - y^3 + 3) = 1 \end{array}\right\}. \tag{16}$$

Since the differential equation is exact we are assured that a function $f(x, y)$ may be found such that

$$\left.\begin{aligned} \frac{\partial f}{\partial x} &= x + y + 1 \\[2mm] \frac{\partial f}{\partial y} &= x - y^2 + 3 \end{aligned}\right\}. \qquad (17)$$

Integrating the first of these equations with respect to x whilst holding y fixed, we obtain

$$f(x, y) = \tfrac{1}{2}x^2 + (y + 1)x + C(y).$$

The constant of the integration with respect to x may depend on y so we have included here a function $C(y)$. Repeating the procedure on the other equation in (17) we obtain

$$f(x, y) = -\frac{y^3}{3} + (x + 3)y + D(x).$$

Here the constant of the integration with respect to y may depend on x, so a term $D(x)$ is included. Equating the two results we have

$$-\frac{y^3}{3} + xy + 3y + D(x) = \frac{1}{2}x^2 + xy + x + C(y).$$

Note that the term in xy on both sides is the same; the reason for it is the exactness condition (14) i.e. (16). Because of this we are able to rewrite the last equation in such a way that one side depends only on x and the other only on y, thus:

$$-\frac{y^3}{3} + 3y - C(y) = \frac{1}{2}x^2 + x - D(x).$$

It follows that both sides must be constant in value (why?). Say this value is k. Thus we have:

$$\left.\begin{aligned} D(x) &= k - \tfrac{1}{2}x^2 - x \\[2mm] C(y) &= -\frac{y^3}{3} + 3y - k \end{aligned}\right\}.$$

Hence, finally

$$f(x, y) = \frac{1}{2}x^2 + x + xy - \frac{y^3}{3} + 3y - k,$$

where k is a constant.

20.2 Partial differential equations – (a glance)

The theory of partial differential equations is vast. Classical methods, concerned with solving those equations that arise in physics, often make use of series expansions; modern methods avail themselves of the geometric insights afforded by the tools of (infinite-dimensional) vector spaces. The present section is more an appendix to Section 20.1(d) and to chapter 14 of Part I than an apology for an introduction.

We have already seen the simplest of arguments in the example of Section 20.1(d). We amplify the scope of that approach with two further examples.

Example (1)
Solve

$$\frac{\partial f}{\partial x} = (ax + by)\frac{\partial f}{\partial y}. \tag{18}$$

Rewrite this as

$$\frac{\partial f}{\partial x} - (ax + by)\frac{\partial f}{\partial y} = 0. \tag{19}$$

Suppose $z = f(x, y)$ solves (19) and consider a contour

$$f(x, y) = c. \tag{20}$$

If the contour is expressed in the form $y = y(x)$ we have from (20) that

$$\frac{\partial f}{\partial x} + \frac{\partial f}{\partial y}\frac{dy}{dx} = 0,$$

hence, comparing with (18), we have on the contour that

$$\frac{dy}{dx} = -(ax + by).$$

Assume $b \neq 0$ and make the obvious change of variable

$$u = ax + by,$$

so that

$$\frac{du}{dx} = a + b\frac{dy}{dx}.$$

Thus

$$\frac{1}{b}\left[\frac{du}{dx} - a\right] = -u,$$

or

$$\frac{du}{dx} = a - bu.$$

So separating the variables and integrating

$$\int \frac{du}{a - bu} = \int dx,$$

we obtain

$$-\frac{1}{b} \log (a - bu) = x + c_1,$$

where c_1 is a constant. Thus

$$(a - bu)^{-1/b} = c_2 e^x$$

where $c_2 = e^{c_1}$, or, finally

$$k = e^x (a - bu)^{1/b}, \tag{21}$$

where k is a constant. Now let ψ be the function which maps the constant c of equation (20) to the constant k of (21). Thus the

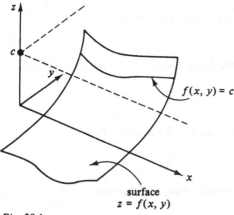

Fig. 20.1

contour (see Figure 20.1)

$$f(x, y) = c$$

may be re-written as

$$(a - bu)^{1/b} e^x = \psi(c).$$

The function $\psi(c)$ is one-to-one, for if $k = \psi(c_1) = \psi(c_2)$, then for any point $(\tilde{x}, \tilde{y})$ with

$$(a - b\tilde{u})^{1/b} e^{\tilde{x}} = k$$

where $\tilde{u} = a\tilde{x} + b\tilde{y}$ we have since $k = \psi(c_1)$ and $k = \psi(c_2)$ that

$$\left. \begin{array}{l} f(\tilde{x}, \tilde{y}) = c_1 \\ f(\tilde{x}, \tilde{y}) = c_2 \end{array} \right\}.$$

So $c_1 = c_2$. Let ϕ be the inverse function of ψ, then

$$\begin{aligned} f(x, y) = c &= \phi(\psi(c)) = \phi(k) \\ &= \phi((a - bu)^{1/b} e^x). \end{aligned}$$

Thus the general solution of (18) is for $b \neq 0$

$$f(x, y) = \phi((a - abx - b^2 y)^{1/b} e^x),$$

where $\phi(t)$ is an arbitrary (differentiable) function.

Example 2 (using characteristic curves)
We attempt to solve the equation

$$P(x, y, z) \frac{\partial z}{\partial x} + Q(x, y, z) \frac{\partial z}{\partial y} = R(x, y, z), \tag{22}$$

which generalises the previous example, by looking for a solution in the form of a surface

$$\phi(x, y, z) = 0.$$

Assuming that we can solve this last equation for z in the form $z = f(x, y)$, we obtain by differentiating the identity

$$\phi(x, y, f(x, y)) \equiv 0$$

that

$$\left. \begin{array}{l} \dfrac{\partial \phi}{\partial x} + \dfrac{\partial \phi}{\partial z} \dfrac{\partial f}{\partial x} = 0, \\[3mm] \\ \dfrac{\partial \phi}{\partial y} + \dfrac{\partial \phi}{\partial z} \dfrac{\partial f}{\partial y} = 0. \end{array} \right\} \tag{23}$$

and

Suppose now that the function ϕ satisfies

$$\frac{\partial \phi}{\partial x} P(x, y, z) + \frac{\partial \phi}{\partial y} Q(x, y, z) + \frac{\partial \phi}{\partial z} R(x, y, z) = 0, \tag{24}$$

then by (23) we shall have

$$P(x, y, z)\frac{\partial f}{\partial x} + Q(x, y, z)\frac{\partial f}{\partial y} = R(x, y, z).$$

So our original equation may be solved by solving the equation (24) for ϕ. But the equation (24) asserts that the vector

$$\nabla\phi = \left[\frac{\partial\phi}{\partial x}, \frac{\partial\phi}{\partial y}, \frac{\partial\phi}{\partial z}\right]^t$$

is orthogonal to the vector $(P, Q, R)^t$.

We now examine curves along $\phi(x, y, z) = 0$ parametrised in the form

$$\left.\begin{array}{l} x = x(t) \\ y = y(t) \\ z = z(t) \end{array}\right\}.$$

(See Figure 20.2.)

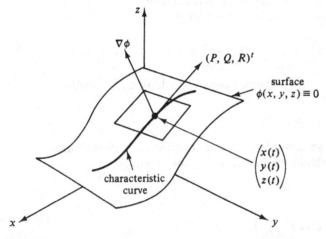

Fig. 20.2

Evidently we may differentiate the identity

$$\phi(x(t), y(t), z(t)) \equiv 0$$

to obtain

$$\frac{\partial\phi}{\partial x}\frac{dx}{dt} + \frac{\partial\phi}{\partial y}\frac{dy}{dt} + \frac{\partial\phi}{\partial z}\frac{dz}{dt} = 0. \tag{25}$$

This equation says that the vector $(dx/dt, dy/dt, dz/dt)^t$ is tangential to the surface $\phi(x, y, z) = 0$. We therefore now seek curves whose tangent vector is proportional to

$$(P(x(t), y(t), z(t)), Q(x(t), y(t), z(t)), R(x(t), y(t), z(t)))^t$$

since by (25) we will have satisfied (24). We thus have for some function $\lambda(t)$ (which establishes the desired proportionality) that

$$\left.\begin{aligned}\dot{x} &= \lambda(t)P(x, y, z)\\ \dot{y} &= \lambda(t)Q(x, y, z)\\ \dot{z} &= \lambda(t)R(x, y, z)\end{aligned}\right\} \tag{26}$$

where the dotted variables denote derivatives of the respective undotted variables with respect to the parameter t.

If P, Q, R are non-zero, we can eliminate λ and write (26) in the more memorable form:

$$\frac{\dot{x}}{P(x, y, z)} = \frac{\dot{y}}{Q(x, y, z)} = \frac{\dot{z}}{R(x, y, z)}.$$

Curves satisfying this equation are known as '*characteristic curves*' of the differential equation.

Once we have found the characteristic curves we are left with the problem of fitting them together to form a surface. Let us choose x as our parameter. We thus have to solve

$$\frac{dy}{dx} = \frac{Q(x, y, z)}{P(x, y, z)} \qquad\qquad \frac{dz}{dx} = \frac{R(x, y, z)}{P(x, y, z)}.$$

In principle we might expect to write the solution curves in the form

$$\left.\begin{aligned}y &= h_1(x, c_1)\\ z &= h_2(x, c_2)\end{aligned}\right\}$$

where c_1 and c_2 are constants of integration. For later convenience we suppose this curve to be presented in the form

$$\left.\begin{aligned}\psi_1(x, y, z) &= c_1\\ \psi_2(x, y, z) &= c_2\end{aligned}\right\}.$$

Now let $\gamma(t)$ be any (differentiable) function, we claim that the relationship

$$\psi_1(x, y, z) - \gamma(\psi_2(x, y, z)) \equiv 0 \tag{27}$$

is the most general form of the solution to our differential equation.

We check only that this is a solution. We have by (27)

$$\frac{\partial \psi_1}{\partial x} = \gamma' \frac{\partial \psi_2}{\partial x}, \frac{\partial \psi_1}{\partial y} = \gamma' \frac{\partial \psi_2}{\partial x}, \frac{\partial \psi_1}{\partial z} = \gamma' \frac{\partial \psi_2}{\partial z},$$

then since evidently (by construction)

$$\frac{\partial \psi_i}{\partial x} \dot{x} + \frac{\partial \psi_i}{\partial y} \dot{y} + \frac{\partial \psi_i}{\partial z} \dot{z} = 0 \quad (i = 1, 2)$$

we may add the equation for $i = 1$ to a multiple of the equation for $i = 2$ by γ' to see that $\psi_1 - \gamma(\psi_2)$ satisfies the same form of equation. A worked example follows.

Example 3
Solve

$$x \frac{\partial z}{\partial x} + (x + y) \frac{\partial z}{\partial y} = \frac{1}{z}.$$

The characteristic curves satisfy for some $\lambda = \lambda(t)$:

$$\dot{x} = \lambda x \quad \dot{y} = \lambda \cdot (x + y) \quad \dot{z} = \frac{\lambda}{z}.$$

Choosing x as our parameter we obtain

$$\frac{dy}{dx} = \frac{x + y}{x}, \quad \frac{dz}{dx} = \frac{1}{zx}.$$

Thus

$$\frac{dy}{dx} - \frac{1}{x} y = 1,$$

so (since the integrating factor is $1/x$)

$$\frac{d}{dx} \left(\frac{1}{x} y \right) = \frac{1}{x}, \quad \text{or} \quad \frac{1}{x} y = \log x + c_1.$$

Also

$$\int z \, dz = \int \frac{dx}{x} \quad \text{so} \quad \frac{1}{2} z^2 = \log x + c_2$$

and the general solution is therefore

$$\frac{y}{x} - \log x - \gamma \left(\frac{1}{2} z^2 - \log x \right) = 0.$$

Example 4

Find a solution of the last example passing through the
parabola $y = x^2, z = 0$.

We have when $y = x^2$ and $z = 0$ that for all x

$$x - \log x - y(-\log x) = 0.$$

Putting $t = -\log x$ we have thus

$$y(t) = t - e^{-t}.$$

20.3 Linear differential equations with constant coefficients

This section is devoted to a revision of a standard technique for
solving equations of the form

$$\frac{d^n y}{dx^n} + a_{n-1}\frac{d^{n-1}y}{dx^{n-1}} + \cdots + a_1 \frac{dy}{dx} + a_0 y = f(x). \tag{1}$$

An alternative approach is provided by Laplace Transforms, see
Chapter 21. The coefficients $a_0, a_1, \ldots$ are constant.

If y^1 and y^2 are two solutions of (1) then $y = y^1 - y^2$ satisfies the
homogeneous equation

$$\frac{d^n y}{dx^n} + a_{n-1}\frac{d^{n-1}_y}{dx^{n-1}} + \cdots + a_1 \frac{dy}{dx} + a_0 y = 0. \tag{2}$$

Two conclusions are drawn from this: any two solutions of (1)
differ by a function which solves (2); the general solution y^2 to (1)
may be expressed as the sum $y^1 + y$ of two functions where y^1 is
any one *particular* solution of (1) and y is the general solution to
(2). We therefore concentrate our attention first on the
homogeneous equation (2).

20.3.1 The homogeneous case

The substitution $y = e^{\lambda x}$ reduces the equation (2) to

$$e^{\lambda x}\{\lambda^n + a_{n-1}\lambda^{n-1} + \cdots + a_1\lambda + a_0\} = 0.$$

Hence if λ is a root of the *auxiliary equation*

$$\lambda^n + a_{n-1}\lambda^{n-1} + \cdots + a_1\lambda + a_0 = 0 \tag{3}$$

then $y = e^{\lambda x}$ solves (2). When the roots of (3) are all *distinct*, say
they are $\lambda_1, \ldots, \lambda_n$, it turns out that any solution of (2) may be

written in the form

$$A_1 e^{\lambda_1 x} + A_2 e^{\lambda_2 x} + \cdots + A_n e^{\lambda_n x}, \tag{4}$$

where $A_1, \ldots, A_n$ are constants. Thus the solutions form a vector subspace of the space $\mathbb{R}^{\mathbb{R}}$ spanned by the 'vectors'

$$f_1, f_2, \ldots, f_n \quad \text{where } f_i(x) = e^{\lambda_i x}$$

(see Part I Chapter 1 Section 1.4). When the roots are *not distinct* complications arise in the formula (4). If, for instance, λ_1 has multiplicity k, say $\lambda_1 = \lambda_2 = \cdots = \lambda_k$, then it is necessary to use instead

$$
\begin{aligned}
f_1 &= e^{\lambda_1 x} \\
f_2 &= x e^{\lambda_1 x}, \ldots, \\
f_k &= x^{k-1} e^{\lambda_1 x}
\end{aligned}
\tag{5}
$$

with similar treatment of any other repeated roots. The fact that (4), or (4) modified as indicated, gives *all* possible solutions of (2) of course needs proof. This we do in the next section.

20.3.2 The general solution of the homogeneous case

The simplest equation of the form (2) is

$$\frac{dy}{dx} = 0$$

the solution of which is known to be $y = \text{const.}$ (See, for example, K. G. Binmore *Mathematical Analysis*.) Our whole job is to reduce the more general problem to an application of this simple fact. The appropriate language for the task is that of linear algebra.

Let C^∞ be the set of all functions $f(x)$ from $\mathbb{R}$ into $\mathbb{R}$ which have derivatives of all orders. This is obviously a vector subspace of $\mathbb{R}^{\mathbb{R}}$ and contains all polynomials. We can define a transformation D: $C^\infty \to C^\infty$ by the formula

$$D f(x) = \frac{d f(x)}{dx}$$

e.g. if $f(x) = x^2$ then Df is the function g whose value at x is $2x$; thus $g(x) = Df(x) = 2x$. Of course D is a linear transformation; that is, for any scalars α, β and any functions f, g

$$D(\alpha f + \beta g) = \alpha Df + \beta Dg.$$

Given two linear transformations S, T from C^∞ to C^∞ we can define the transformations $\alpha S + \beta T$ (for scalars α, β) and ST by the rule:

$$\left.\begin{array}{c} (\alpha S + \beta T)f = \alpha(Sf) + \beta(Tf) \\ STf = S(Tf) \end{array}\right\}. \tag{6}$$

In particular then $D^2 f = D(Df)$, so

$$D^2 f = \frac{d^2 f(x)}{dx^2} \quad \text{and generally } D^n f(x) = \frac{d^n f(x)}{dx^n}.$$

We may also put

$$S = D^n + a_{n-1}D^{n-1} + \cdots + a_1 D + a_0 I,$$

where I is the identity transformation which satisfies $If = f$. We are thus interested in solving for f:

$$Sf \equiv D^n f + a_{n-1}D^{n-1} + \cdots + a_1 Df + a_0 f = 0.$$

Now we note that the rules (6) allow us to factorise S. By this we mean that if $\lambda_1, \ldots, \lambda_n$ are the roots of the auxiliary equation then just as we may write

$$\lambda^n + a_{n-1}\lambda^{n-1} + \cdots + a_1\lambda + a_0 = (\lambda - \lambda_1)(\lambda - \lambda_2)\cdots(\lambda - \lambda_n)$$

so too we have $S = (D - \lambda_1 I)(D - \lambda_2 I)\cdots(D - \lambda_n I)$.
In this formula the order of the factors does not matter. However it is *not* generally true that $ST = TS$.

Example
The transformation S with $Sf(x) = xf(x)$ is linear (e.g. $S(f + g)(x) = xf(x) + xg(x)$). Observe however that

$$SDf(x) = x\frac{df(x)}{dx}, \quad \text{but } DSf(x) = f(x) + x\frac{df(x)}{dx}.$$

What is nevertheless true is that if S and T are transformations which are polynomial expressions built up from one transformation W then $ST = TS$.

Example
When $S = 3I + W$ and $T = I + W^2$ we have:

$$\begin{aligned} ST &= (3I + W)(I + W^2) = 3I(I + W^2) + W(I + W^2) \\ &= 3I + 3W^2 + W + W^3 \\ &= W^3 + 3W^2 + W + 3I. \end{aligned}$$

$$\begin{aligned} TS &= (I + W^2)(3I + W) = I(3I + W) + W^2(3I + W) \\ &= 3I + W + 3W^2 + W^3. \end{aligned}$$

The relevant fact here is that $W \cdot W^2 = W^2 \cdot W$ and generally

$$W^l W^k = \underbrace{(W \cdots W)}_{l \text{ factors}} \cdot \underbrace{(W \cdots W)}_{k \text{ factors}} = W^{k+l} = W^k W^l.$$

Now we write

$$S = (D - \alpha)^{l_1}(D - \beta)^{l_2} \cdots (D - \gamma)^{l_m}$$

where $\alpha, \beta, \dots \gamma$ are the m distinct roots of the auxiliary equation. Note that for example $(D - \alpha)$ here stands for $(D - \alpha I)$. We stop to consider the equation

$$(D - \alpha)f = 0.$$

Write

$$y = e^{-\alpha x} f.$$

Then

$$0 = (D - \alpha I)f = (D - \alpha I)e^{\alpha x} y$$
$$= \alpha e^{\alpha x} y + e^{\alpha x} Dy - \alpha e^{\alpha x} y$$

i.e.

$$0 = Dy.$$

Hence $y = \text{constant} = A$, so $f(x) = A e^{\alpha x}$.

What about $(D - \alpha)^2 f = 0$? Put $z = (D - \alpha)f$ then $(D - \alpha)z = 0$ so by the last result $z = A e^{\alpha x}$ and so $(D - \alpha)f = A e^{\alpha x}$.

Hence

$$e^{-\alpha x} Df - \alpha e^{-\alpha x} f = A$$

or

$$D(e^{-\alpha x} f) = A,$$

so

$$e^{-\alpha x} f = Ax + B,$$

thus

$$f = Ax e^{\alpha x} + B e^{\alpha x}.$$

The technique is now clear. To solve $(D - \alpha)^3 f = 0$ we put $y_1 = (D - \alpha)^2 f$ then $(D - \alpha)y_1 = 0$, so $y_1 = A e^{\alpha x}$. Next put $y_2 = (D - \alpha)f$. Then $(D - \alpha)y_2 = y_1 = A e^{\alpha x}$, hence

$$e^{-\alpha x} Dy_2 - \alpha e^{-\alpha x} y_2 = A,$$

so, as before,

$$e^{-\alpha x} y_2 = Ax + B,$$

or

$$y_2 = (Ax + B)e^{\alpha x}$$

and

$$(D - \alpha)f = (Ax + B)e^{\alpha x}.$$

Thus
$$e^{-\alpha x}Df - \alpha e^{-\alpha x}f = Ax + B,$$

so
$$e^{-\alpha x}f = \tfrac{1}{2}Ax^2 + Bx + C,$$

and finally
$$f = (A'x^2 + Bx + C)e^{\alpha x}.$$

Thus
$$\{f:(D-\alpha)^k f = 0\} = N((D-\alpha)^k)$$
$$= \mathrm{Lin}(e^{\alpha x}, xe^{\alpha x}, \ldots, x^{k-1}e^{\alpha x}).$$

Let us denote by $\mathbb{V}_k(\alpha)$ the space $\mathrm{Lin}(e^{\alpha x}, \ldots, x^{k-1}e^{\alpha x})$.

Now we consider $(D-\alpha)^k(D-\beta)^l f = 0$.

Rather than pursue the argument above using integrating factors, we prefer an algebraic argument. Consider the action of $(D-\beta)$ on $\mathbb{V}_k(\alpha)$. We have

$$(D-\beta)e^{\alpha x} = (\alpha - \beta)e^{\alpha x}$$
$$(D-\beta)x^m e^{\alpha x} = mx^{m-1}e^{\alpha x} + \alpha x^m e^{\alpha x} - \beta x^m e^{\alpha x}$$
$$= mx^{m-1}e^{\alpha x} + (\alpha - \beta)x^m e^{\alpha x}.$$

Thus $(D-\beta)$ transforms $\mathbb{V}_k(\alpha)$ into $\mathbb{V}_k(\alpha)$.

Also observe that

$$(D-\beta)\sum_1^k \gamma_i x^i e^{\alpha x} = [\gamma_0(\alpha - \beta) + \gamma_1]e^{\alpha x}$$
$$+ [\gamma_1(\alpha - \beta) + 2\gamma_2]xe^{\alpha x}$$
$$+ \cdots + [\gamma_k(\alpha - \beta)]x^k e^{\alpha x}.$$

If the right-hand side is identically zero it must be that all the coefficients are zero. Thus $\gamma_k = 0$ since $\alpha \neq \beta$. Working backwards through the terms on the right we deduce successively that $\gamma_{k-1} = 0$, then that $\gamma_{k-2} = 0$ and so on. Thus $N(D-\beta) \cap \mathbb{V}_k(\alpha) = \{0\}$ i.e. $(D-\beta)$ is invertible on $\mathbb{V}_k(\alpha)$. Hence all its powers are invertible on $\mathbb{V}_k(\alpha)$. Now we use a simple theorem in algebra to complete the argument.

For the notion of direct sum used below refer to Part I Chapter 4.

20.3.3 Sum theorem

Let S, T be linear transformations of a vector space $\mathbb{V}$ into $\mathbb{V}$. Suppose that $ST = TS$, that $N(S) \cap N(T) = \{0\}$ and that T maps the set $N(S)$ onto $N(S)$. Then

$$N(ST) = N(S) \oplus N(T).$$

Proof   Let $STx = 0$ then $u = Tx \in N(S)$. This does not mean that $x \in N(S)$; however, by the hypothesis of the theorem, for some $y \in N(S)$, $u = Ty$. (T is invertible on $N(S)$ by hypothesis.) Thus $Ty = u = Tx$ or $T(x - y) = 0$ so $z = x - y \in N(T)$, that is $x = y + z$ and $y \in N(S)$, $z \in N(T)$.

Conversely, if $x = y + z$ and $y \in N(S)$, $z \in N(T)$ then $STx = STy + STz = TSy + STz = 0$ as required.

We apply the theorem by taking $T = (D - \alpha)^k$ and $S = (D - \beta)^l$ to obtain

$$N((D - \alpha)^k(D - \beta)^l) = \mathrm{Lin}\,\{e^{\alpha x}, \ldots, x^{k-1}e^{\alpha x}, e^{\beta x}, \ldots, x^{l-1}e^{\beta x}\}.$$

The argument easily extends to three factors and more. For example if $\gamma \notin \{\alpha, \beta\}$ then $(D - \gamma)$ is easily seen to be invertible both on $\mathbb{V}_k(\alpha)$ and $\mathbb{V}_l(\beta)$ and hence on their direct sum (because $T^{-1}(x + y) = T^{-1}(x) + T^{-1}(y)$). Thus we have:

Corollary
The set of functions f satisfying

$$(D - \alpha)^k(D - \beta)^l \cdots (D - \gamma)^m f = 0,$$

is

$$\mathbb{V}_k(\alpha) \oplus \mathbb{V}_l(\beta) \oplus \cdots \oplus \mathbb{V}_m(\gamma).$$

20.3.4 Examples

1. Find all the solutions of $(D^2 - 1)y = 0$.
We have $z^2 - 1 = (z - 1)(z + 1)$, so $D^2 - 1 = (D - 1)(D + 1)$.
Any solution takes the form $y = Ae^x + Be^{-x}$.
2. Find all the solutions of $(D^2 + 1)y = 0$.
We have $z^2 + 1 = (z + i)(z - i)$, so $D^2 + 1 = (D + i)(D - i)$.
Any solution takes the form $y = Ae^{ix} + Be^{-ix}$.
We write this in an alternative way. Using the identities

$$e^{ix} = \cos x + i \sin x, \ e^{-ix} = \cos x - i \sin x$$

we have $y = (A + B)\cos x + (iA - iB)\sin x = G\cos x + H\sin x$, which is a more useful form when working with real solutions (complex A and B can give rise to real G and H).
3. Find all the solutions of

$$(D^5 + 3D^4 + 7D^3 + 13D^2 + 12D + 4)y = 0.$$

We need to factorise the auxiliary polynomial

$$P(z) \equiv z^5 + 3z^4 + 7z^3 + 13z^2 + 12z + 4.$$

Since the product of the roots is 4 we check if ± 1, ± 2, ± 4 are roots. Indeed $z = -1$ is a root, i.e. $(z + 1)$ is a factor. We perform a long division

$$
\begin{array}{r}
z^4 + 2z^3 + 5z^2 + 8z + 4 \\
\hline
(z+1)\sqrt{z^5 + 3z^4 + 7z^3 + 13z^2 + 12z + 4} \\
z^5 + z^4 \\
\hline
2z^4 + 7z^3 \\
2z^4 + 2z^3 \\
\hline
5z^3 + 13z^2 \\
5z^3 + 5z^2 \\
\hline
8z^2 + 12z \\
8z^2 + 8z \\
\hline
4z + 4 \\
4z + 4 \\
\hline
\end{array}
$$

Oddly enough $z = -1$ also satisfies $z^4 + 2z^3 + 5z^2 + 8z + 4 = 0$. Continuing the process yields

$$P(z) \equiv (z+1)^3(z^2+4) \equiv (z+1)^3(z+2i)(z-2i).$$

Any solution therefore takes the form

$$y = e^{-x}(A + Bx + Cx^2) + (G\cos 2x + H\sin 2x).$$

4. Solve $(D^2 + D + 1)y = 0$.

The roots of $z^2 + z + 1 = 0$ are

$$z = \frac{-1 \pm \sqrt{-3}}{2}.$$

The solutions are $A\exp[-\tfrac{1}{2} + i\sqrt{3/2}]x + B\exp[-\tfrac{1}{2} - i\sqrt{3/2}]x$

$$= e^{-1/2x}[A\exp(i\sqrt{3/2})x + B\exp(-i\sqrt{3/2})x]$$
$$= e^{-x/2}[G\cos(\sqrt{3/2})x + H\sin(\sqrt{3/2})x] \text{ in } real \text{ form.}$$

20.3.5 General solution of the homogeneous difference equation

The techniques of the last section may be applied to the solution of the equation

$$y_{x+n} + a_{n-1}y_{x+n-1} + \cdots + a_1 y_{x+1} + a_0 y_x = 0. \tag{7}$$

We must think of the sequences $\{y_x : x = 1, 2, \ldots\}$ as vectors with infinitely many components $(y_1, y_2, \ldots)$ analogously to n-tuples. These form a vector space (just like the n-tuples) which we shall denote by S. It is also useful to think of a sequence $(y_1, y_2, \ldots)$ as being a function $y(x)$ defined only for positive integers x. We define a linear transformation $E : S \to S$ by the formula

$$Ey_x = y_{x+1},$$

i.e.

$$E(y_1, y_2, y_3, \ldots) = (y_2, y_3, y_4, \ldots).$$

E is a *shift* operator. Let

$$S = E^n + a_{n-1}E^{n-1} + \cdots + a_0 I,$$

where I is the operator $Iy = y$. Again we may write

$$S = (E - \lambda_1)(E - \lambda_2)\cdots(E - \lambda_n)$$

where $\lambda_1, \ldots, \lambda_n$ are the roots of the auxiliary equation

$$\lambda^n + a_{n-1}\lambda^{n-1} + \cdots + a_1\lambda + a_0 = 0.$$

Let us solve the equation $(E - \alpha)y_x = 0$.

We have $y_{x+1} = \alpha y_x = \alpha^2 y_{x-1} = \cdots = \alpha^x y_1$, that is $y_x = A\alpha^x$ where $A = (y_1/\alpha)$ for $\alpha \neq 0$. Of course if $\alpha = 0$ then $y_{x+1} = 0^x y_1 = 0$ for $x = 1$, $2, 3, \ldots$ that is y_1 is arbitrary and $y_{1+1} = y_{2+1} = y_{3+1} = \cdots = 0$. Let us define $\delta_1(x)$ by

$$\delta_1(x) = \begin{cases} 1 & x = 1 \\ 0 & x \neq 1 \end{cases}$$

so that the sequence δ_1 is $(1, 0, 0, \ldots)$.

Thus the solution of $Ey_x = 0$ is $y = A\delta_1$ with A arbitrary.

It will now come as no surprise that just as with differential equations, if the roots $\lambda_1, \ldots, \lambda_n$ of the auxiliary equation are all *distinct* the general solution of (7) takes the form

$$A_1\lambda_1^x + A_2\lambda_2^x + \cdots + A_n\lambda_n^x \quad (x = 1, 2, 3, \ldots)$$

provided of course we remember to replace the function λ^x by $\delta_1(x)$, when $\lambda = 0$. Thus the solution space is spanned by $f_i(x) \equiv \lambda_i^x$ (except when $\lambda_i = 0$).

When the roots are *not distinct* a more complicated formula is required. If for instance λ_1 has multiplicity k and say $\lambda_1 = \cdots = \lambda_k$ then it is necessary to use for f_i the functions (sequences):

$$\left. \begin{aligned} f_1(x) &= \lambda_1^x \\ f_2(x) &= x\lambda_1^x, \ldots, \\ f_k(x) &= x^{k-1}\lambda_1^x \end{aligned} \right\}$$

and again we must make a special case for $\lambda_1 = 0$; here we would need

$$\left.\begin{aligned} f_1(x) &= \delta_1(x) \\ f_2(x) &= \delta_2(x) \\ &\cdots \\ f_k(x) &= \delta_k(x) \end{aligned}\right\},$$

where

$$\delta_i(x) = \begin{cases} 1 & x = i, \\ 0 & x \neq i. \end{cases}$$

Again a proof is required for these assertions and that is provided by the next section.

20.3.6 Justification

We begin by noting that the equation $(E - \alpha)^2 y_x = 0$ can be dealt with as follows. Put $z_x = (E - \alpha)y_x$. Then $(E - \alpha)z_x = 0$, so $z_x = A\alpha^x$.
To solve

$$(E - \alpha)y_x = A_0\alpha^x,$$

observe

$$(E - \alpha)x\alpha^x = (x + 1)\alpha^{x+1} - \alpha x\alpha^x = \alpha \cdot \alpha^x,$$

whence

$$(E - \alpha)\left\{ y_x - \frac{A_0}{\alpha}\alpha^x x \right\} = 0,$$

so for some B

$$y_x - \frac{A_0}{\alpha}x\alpha^x = B\alpha^x,$$

thus for some A and B

$$y_x = (Ax + B)\alpha^x.$$

To solve $(E - \alpha)^3 y_x = 0$, let $z_x = (E - \alpha)y_x$ then

$$(E - \alpha)^2 z_x = 0,$$

so for some A_0 and B_0

$$(E - \alpha)y_x = z_x = (A_0x + B_0)\alpha^x.$$

Now observe that

$$(E - \alpha)x^2\alpha^x = (x^2 + 2x + 1)\alpha^{x+1} - \alpha x^2\alpha^x$$
$$= \alpha(2x + 1)\alpha^x.$$

So

$$(E - \alpha)\left\{ y_x - \frac{A_0}{2\alpha}x^2\alpha^x - \left(\frac{B_0}{\alpha} - \frac{A_0}{2\alpha}\right)x\alpha^x \right\} = 0.$$

Hence for some C_0

$$y_x = \frac{A_0}{2\alpha}x^2\alpha^x - \left(\frac{A_0}{2\alpha} - \frac{B_0}{\alpha}\right)x\alpha^x + C_0\alpha^x.$$

Thus in general for $\alpha \neq 0$ and $k \neq 0$

$$N((E - \alpha)^k) = \text{Lin } \{\alpha^x, x\alpha^x, \ldots, x^{k-1}\alpha^x\}.$$

For $\alpha \neq 0$ let us denote by $\mathbb{V}_k(\alpha)$ the space Lin $\{\alpha^x, \ldots, x^{k-1}\alpha^x\}$. When $\alpha = 0$ the analysis is different.

$$E^k y_k = 0 \qquad (x = 1, 2, \ldots)$$

or equivalently

$$y_{x+k} = 0 \quad \text{for} \quad x = 1, 2, \ldots,$$

implies that

$$y_x = 0 \quad \text{for} \quad x = k + 1, k + 2, \ldots \text{ and}$$
$$y_1, y_2, \ldots, y_k \text{ are arbitrary.}$$

Recall that $\delta_i = (0, \ldots, 0, 1, 0, 0 \ldots)$, with the 1 entry in position i.

Thus for $k \neq 0$, $N(E^k) = \text{Lin } \{\delta_1, \delta_2, \ldots, \delta_k\} = \mathbb{V}_k(0)$. We finally note that if $\alpha \neq \beta$ then $E - \beta$ maps $\mathbb{V}_k(\alpha)$ onto $\mathbb{V}_k(\alpha)$ because

$$(E - \beta)x^t \alpha^x = (x + 1)^t \alpha \cdot \alpha^x - \beta x^t \alpha^x$$

$$= (\alpha - \beta)x^t \alpha^x + \binom{t}{1} x^{t-1} \alpha^x + \cdots.$$

Also if $\alpha = 0$, $(E - \beta)\delta_i = \delta_{i-1} - \beta\delta_i$. (Note $E(y_1, y_2, \ldots) = (y_2, y_3, \ldots)$ so $E(0, 1, 0, \ldots) = (1, 0, 0, \ldots)$.) It follows that the mapping is onto and by the Sum theorem (20.3.3) that

Corollary

The solutions of $Sf = 0$ i.e. of

$$E^k(E - \alpha)^l \cdots (E - \gamma)^m y = 0$$

are precisely the elements of

$$\mathbb{V}_k(0) \oplus \mathbb{V}_l(\alpha) \cdots \oplus \mathbb{V}_m(\gamma).$$

20.3.7 Examples

1. Solve $(E^2 - 1)y_x = 0$.

$$E^2 - 1 \equiv (E - 1)(E + 1).$$

So the general solution is

$$y_x = A(1)^x + B(-1)^x$$
$$= A + B(-1)^x.$$

2. Solve $(E^3 + E)y_x = 0$.

$$E^3 + E = E(E^2 + 1) = E(E + i)(E - i).$$

The general solution is

$$y_x = A\delta_1 + Bi^x + C(-i)^x.$$

To convert to real form, note that $i = e^{i\pi/2}$ whence, as $i^x = e^{i\pi x/2}$,
$y_x = A\delta_1 + (B + C)\cos(\pi x/2) + i(B - C)\sin(\pi x/2)$. Thus
$y_x = A\delta_1 + G\cos(\pi x/2) + H\sin(\pi x/2)$.
3. Solve $(E^2 + 4E + 5)y_x = 0$.
The roots of $z^2 + 4z + 5 = 0$ are $x = -2 \pm \sqrt{-1}$, so the general
solution is

$$y_x = A(-2 + i)^x + B(-2 - i)^x.$$

To put this in real form, note $|-2 + i| = \sqrt{\{4 + 1\}} = \sqrt{5}$.
Thus

$$-2 + i = \sqrt{5}\left(-\frac{2}{\sqrt{5}} + i\frac{1}{\sqrt{5}}\right).$$

Now we look for an angle θ with $\cos\theta = -2/\sqrt{5}$ and $\sin\theta = 1/\sqrt{5}$.
(Clearly $\pi/2 < \theta < \pi$.) Thus $-2 + i = \sqrt{5}e^{i\theta}$ so
$(-2 + i)^x = \sqrt{5}^x(\cos\theta x + i\sin\theta x)$. Hence

$$y_x = A\sqrt{5}^x e^{i\theta x} + B\sqrt{5}^x e^{-i\theta x},$$

or in real form
$$y_x = 5^{x/2}(G\cos\theta x + H\sin\theta x).$$

Alternatively we may observe that

$$(-2 + i)^x = \mathrm{Re}\,(-2 + i)^x + i\,\mathrm{Im}\,(-2 + i)^x,$$
$$(-2 - i)^x = \mathrm{Re}\,(-2 + i)^x - i\,\mathrm{Im}\,(-2 + i)^x.$$

Hence $y_x = G\,\mathrm{Re}\,[(-2 + i)^x] + H\,\mathrm{Im}\,[(-2 + i)^x]$.
These real and imaginary parts may, for given x, be calculated by using
the Binomial Theorem.
4. Find all the solutions of

$$E^2(E^4 - 12E^3 + 56E^2 - 120E + 100)y_x = 0.$$

It turns out (sic!) that the factorised form is

$$E^2(E - 3 - i)^2(E - 3 + i)^2 y_x = 0.$$

So the general solution is

$$y_x = A\delta_1 + B\delta_2 + (Cx + D)(3 + i)^x + (Gx + H)(3 - i)^x$$
$$= A\delta_1 + B\delta_2 + (C'x + D')\,\mathrm{Re}\,((3 + i)^x) + (G'x + H')\,\mathrm{Im}\,(3 - i)^x.$$

20.3.8 The inhomogeneous (differential/difference) equation

We now consider the equation

$$(D^n + a_{n-1}D^{n-1} + \cdots + a_1 D + a_0 I)y = f(x)$$

and its analogue

$$(E^n + a_{n-1}E^{n-1} + \cdots + a_1 E + a_0 I)y_x = f_x.$$

We write either of these equations indiscriminately as

$$Sy = f$$

where $S = T^n + a_{n-1}T^{n-1} + \cdots + a_0 I$ and T stands either for D or E depending on context. Suppose we can find a polynomial in T, call it $Q(T)$ such that

$$Q(T)f = 0.$$

We term Q an *annihilator* of f. In other words we seek a homogeneous equation of which $f(x)$ itself is a solution. For example, if $f(x) = e^{2x} \sin x$, we recognise $e^{2x} \sin x$ as arising from a combination of $e^{2x}e^{ix}$ and $e^{2x}e^{-ix}$, which functions give the solution space of the homogeneous equation

$$(D - (2 + i))(D - (2 - i))y = 0,$$

or

$$((D - 2)^2 + 1)y = 0.$$

Thus $Q(D) \equiv D^2 - 4D + 5$ is an annihilator of $e^{2x} \sin x$. Now if $Q(T)$ annihilates f we see that any solution y of $Sy = f$ must satisfy

$$Q(T)Sy = Q(T)f = 0.$$

Hence a particular solution of $Sy = f$ lies in the kernel of $Q(T)S$, which we can in principle write down using the method of the previous sections as follows.

We factorise S and Q into their distinct factors

$$S = (T - \alpha)^k(T - \beta)^l \cdots (T - \gamma)^m$$
$$Q(T) = (T - a)^K(T - b)^L \cdots (T - c)^M.$$

Two situations may arise.

Case 1 The numbers $a, b, \ldots, c$ are distinct from $\alpha, \beta, \ldots, \gamma$
By the Corollary to Theorem 20.3.3

$$N(Q(T)S) = \mathbb{V}_k(\alpha) \oplus \cdots \oplus \mathbb{V}_m(\gamma) \oplus \mathbb{V}_K(a) \oplus \cdots \oplus \mathbb{V}_M(c), \qquad (8)$$

This equation tells us where to find a particular solution.

Example

Solve $(D^2 - 4)y = e^{2x} \sin x$.

A particular solution takes the form
$Ae^{2x} + Be^{-2x} + Ge^{2x+ix} + He^{2x-ix}$. We may rewrite this as

$$Ae^{2x} + Be^{-2x} + G'e^{2x} \cos x + H'e^{2x} \sin x.$$

Now substitute in the given equation in order to determine the constants.

$$(D^2 - 4)\{Ae^{2x} + Be^{-2x} + G'e^{2x} \cos x + H'e^{2x} \sin x\}$$
$$= (D^2 - 4)\{G'e^{2x} \cos x + H'e^{2x} \sin x\} \ (since\ (D^2 - 4)e^{\pm 2x} = 0)$$
$$= G'\{4e^{2x} \cos x + 2 \cdot 2e^{2x}(-\sin x) + e^{2x}(-\cos x)\}$$
$$\quad - 4G'e^{2x} \cos x + H'\{4e^{2x} \sin x + 2 \cdot 2e^{2x}(\cos x)$$
$$\quad + e^{2x}(-\sin x)\} - 4H'e^{2x} \sin x$$
$$= e^{2x} \cos x\{4H' - G'\} - e^{2x} \sin x\{4G' - H'\}$$

If this is to equal $e^{2x} \sin x$ we have $4H' - G' = 0$ and $4G' - H' = -1$ so
$4G' - G'/4 = -1$ or $G' = -\frac{4}{15}$ and $H' = -\frac{1}{15}$. Thus a particular solution
is $-\frac{4}{15}e^{2x} \cos x - \frac{1}{15}e^{2x} \sin x$.

Case 2 $Q(T)$ and S have common roots

Suppose for example that

$$S = (T - \alpha)^k (T - \beta)^l \cdots (T - \sigma)^p \cdot (T - \tau)^r \cdots (T - \gamma)^m$$

$$Q(T) = \underbrace{(T - \alpha)^K (T - \beta)^L \cdots (T - \sigma)^P}_{\text{same for } S \text{ and } Q} \cdot \underbrace{(T - t)^R \cdots (T - c)^M}_{\text{all distinct}}$$

Then $Q(T)S = (T - \alpha)^{k+K} \cdots (T - \sigma)^{p+P}(T - \tau)^r \cdots (T - \gamma)^m (T - t)^R \cdots$
$(T - c)^M$. Hence

$$N(Q(T)S) = \mathbb{V}_{k+K}(\alpha) \oplus \cdots \oplus \mathbb{V}_{p+P}(\sigma) \oplus \mathbb{V}_r(\tau) \cdots$$
$$\oplus \mathbb{V}_m(\gamma) \oplus \mathbb{V}_R(t) \cdots \oplus \mathbb{V}_M(c). \tag{9}$$

But

$$S[N(Q(T)S)] = \mathbb{V}_K(\alpha) \oplus \cdots \oplus \mathbb{V}_P(\sigma) \oplus \mathbb{V}_R(t) \oplus \cdots \oplus \mathbb{V}_M(c). \tag{10}$$

This is because for example

$$(D - \alpha)x^n e^{\alpha x} = nx^{n-1}e^{\alpha x},$$

whence

$$(D - \alpha)^k x^n e^{\alpha x} = n(n-1)\cdots(n-k+1)x^{n-k}e^{\alpha x},$$

whereas

$(D - \beta)$ transforms $\mathbb{V}_{k+K}(\alpha)$ into $\mathbb{V}_{k+K}(\alpha)$.

The equation (10) confirms that $S[N(Q(T)S)] = N(Q(T))$ and so there is a vector y in $N(Q(T)S)$ satisfying $Sy = f$. The equation (9) tells us that it is not sufficient to look for particular solutions in the same spaces $\mathbb{V}_k(\alpha)$ in whose span f lies, but in higher order ones of type $\mathbb{V}_{k+\kappa}(\alpha)$.

Example
Solve $(D^2 - 1)y = xe^x$.

We recognise that $xe^x \in \mathbb{V}_2(1)$ i.e. that $(D-1)^2 xe^x = 0$. Thus to find a particular solution we solve

$$(D-1)^2(D^2-1)y = (D-1)^2 xe^x = 0,$$

or

$$(D-1)^2(D-1)(D+1)y = 0,$$

that is

$$(D-1)^3(D+1)y = 0.$$

So a particular solution takes the form

$$y = Ae^{-x} + (a + bx + cx^2)e^x.$$

Substituting in the original equation we have

$$
\begin{aligned}
(D^2-1)y &= (D^2-1)Ae^{-x} + (D^2-1)\{ae^x + bxe^x + cx^2 e^x\} \\
&= (D^2-1)\{bx + cx^2\}e^x \text{ (because } (D^2-1)e^{\pm x} = 0). \\
&= e^x\{bx + cx^2\} + 2\{b + 2cx\}e^x \\
&\quad + \{2c\}e^x - \{bx + cx^2\}e^x \\
&= 2(b + c)e^x + 4cxe^x \\
&\equiv xe^x.
\end{aligned}
$$

So $4c = 1$ and $b + c = 0$. Thus a particular solution is

$$y = (\tfrac{1}{4}x^2 - \tfrac{1}{4}x)e^x.$$

20.3.9 Remark on the general solution of $Sy = f$

If y_p is a particular solution of $Sy = f$, and y^1 is any other solution then of course

$$S(y^1 - y_p) = Sy^1 - Sy_p = f - f = 0.$$

So, if $z = y^1 - y_p$, then $Sz = 0$ and $y^1 = z + y_p$.

Thus the general solution of $Sy = f$ may be expressed in the form of a particular solution plus the general solution of $Sy = 0$.

20.3.10   Some useful annihilators

---

1. $(D - \alpha)^l x^k e^{\alpha x} = 0$ provided $l > k$

This is obviously true if $k = 0$. Suppose this is true for $k = j$. We deduce that it is true for $k = j + 1$. Now if $l > j + 1$, then $l - 1 > j$, thus

$$(D - \alpha)^l x^{j+1} = (D - \alpha)^{l-1}\{(D - \alpha)x^{j+1}e^{\alpha x}\}$$
$$= (D - \alpha)^{l-1}\{(j+1)x^j e^{\alpha x} + \alpha x^{j+1}e^{\alpha x} - \alpha x^{j+1}e^{\alpha x}\}$$
$$= (j+1)(D - \alpha)^{l-1}x^j e^{\alpha x} = 0.$$

The result follows by induction.

Note in particular that $x^k e^{\gamma x} \sin \omega x$ falls into this category since $e^{(\gamma + i\omega)x} = e^x(\cos \omega x + \sin \omega x)$ provided we use $(D - \alpha)^l(D - \bar{\alpha})^l$ as the annihilator.

---

2. $(E - \alpha)^l x^k \alpha^x = 0$ provided $l > k$

The proof is similar.

Here we note that $x^k r^x \sin \theta x$ 'falls' under this category, since $(re^{i\theta})^x = r^x(\cos \theta x + i \sin \theta x)$, i.e. the relevant annihilator is $(E - re^{i\theta})^l(E - re^{-i\theta})^l$.

20.3.11   An example

Solve $(E - 1)(E - 2)^2 y = 2^x + x$.

The homogeneous equation $(E - 1)(E - 2)^2 y = 0$ has general solution $A + B2^x + Cx2^x$.

The quantity $2^x + x$ satisfies the equation $(E - 2)(E - 1)^2 y = 0$ and hence any particular solution of $(E - 1)(E - 2)^2 y = 2^x + x$ must satisfy

$$(E - 1)^3(E - 2)^3 y = 0$$

and so takes the form

$$(A + B2^x + Cx2^x) + Dx + Fx^2 + Gx^2 2^x.$$

We forget the $A + B2^x + Cx2^x$ part since we already know that this satisfies $(E-1)(E-2)^2 y = 0$ and substitute the rest into $(E-1)(E-2)^2 y = 2^x + x$.

Now

$$(E-1)(E-2)^2 = E^3 - 5E^2 + 8E - 4$$

and

$$(E^3 - 5E^2 + 8E - 4)x = (x+3) - 5(x+2) + 8(x+1) - 4x = 1$$
$$(E^3 - 5E^2 + 8E - 4)x^2 = (x+3)^2 - 5(x+2)^2 + 8(x+1)^2 - 4x^2$$
$$= x^2 + 6x + 9 - 5x^2 - 20x - 20$$
$$+ 8x^2 + 16x + 8 - 4x^2$$
$$= 2x - 3.$$
$$(E^3 - 5E^2 + 8E - 4)x^2 2^x = (x+3)^2 2^{x+3} - 5(x+2)^2 2^{x+2}$$
$$+ 8(x+1)^2 2^{x+1} - 4x^2 2^x$$
$$= 2^x \{8x^2 + 48x + 72 - 20x^2 - 80x$$
$$- 80 + 16x^2 + 32x + 16 - 4x^2\}$$
$$= 2^x \{8\}.$$

Substituting $Dx + Fx^2 + Gx^2 2^x$ into $(E-1)(E-2)^2 = 2^x + x$ therefore yields

$$D + F(2x - 3) + G(8 \cdot 2^x) = x + 2^x.$$

From this we obtain that

$$\left.\begin{array}{ll} D - 3F = 0 & 8G = 1 \\ 2F = 1 & \end{array}\right\}.$$

This finally yields $D = \frac{3}{2}$, $F = \frac{1}{2}$, $G = \frac{1}{8}$.
Hence the general solution is

$$y = A + B2^x + Cx2^x + \frac{3x}{2} + \frac{1}{2}x^2 + \frac{1}{8}x^2 2^x.$$

All this is tiresome but quite straightforward (the author does not guarantee the accuracy of any of these calculations!).

20.4   Exercises

1. Find the general solutions of the following differential equations.

$$y = x\frac{dy}{dx} + \frac{1}{y}$$

$$x\frac{dy}{dx} = y + \sqrt{\{x^2 + y^2\}}$$

$$4x^2 + 2y^2 - xy\frac{dy}{dx} = 0$$

$$x\frac{dy}{dx} + 2 + 3y = xy$$

$$\frac{dy}{dx} = (x + y + 1)^2$$

$$3x^2y^2 + 2xy + (2x^3y + x^2)\frac{dy}{dx} = 0$$

2. Show that the substitution $v = y/x$ allows a first-order homogeneous differential equation to be solved by separation of variables.

3. Solve

$$(1 - x^2)\frac{dy}{dx} - xy = 3xy^2.$$

4. Solve

(a) $$3x\frac{d^2y}{dx^2} + 2\frac{dy}{dx} = 0$$

(b) $$\frac{d^2y}{dx^2} + 3\frac{dy}{dx}y = 0$$

5. Solve the equation

$$\frac{\partial f(x, y)}{\partial x} = \frac{\partial f(x, y)}{\partial y}$$

using first the contour method and then the method of characteristic curves.

6. Solve as in question 5 by both methods the equation

$$(x + y)\frac{\partial f}{\partial x} + (x - y)\frac{\partial f}{\partial y} = 0.$$

7. Solve by the method of characteristic curves the equations:

(a) $$x\frac{\partial f}{\partial x} + y\frac{\partial f}{\partial y} = rf.$$

(b) $$\frac{1}{x}\frac{\partial f}{\partial x} + \frac{1}{y}\frac{\partial f}{\partial y} = x^2 - y^2.$$

(c) $$x \sin y\frac{\partial f}{\partial x} + \frac{1}{y}\frac{\partial f}{\partial y} = f.$$

(d) $(x + y)\dfrac{\partial f}{\partial x} + (x - y)\dfrac{\partial f}{\partial y} = 1.$

Remark. Part (a) is Euler's equation. Deduce from your solution that f is homogeneous of degree r.

8. Using the method of characteristic curves to solve the equation

$$\frac{\partial f}{\partial x} = (ax + by)\frac{\partial f}{\partial y}.$$

9. Find the equation of the surface all of whose normals meet the z-axis and which contains the line $x = a, y = z$.

10. Find *all* solutions of the following homogeneous differential equations.

(i) $3\dfrac{d^3 y}{dx^3} + 5\dfrac{d^2 y}{dx^2} - 2\dfrac{dy}{dx} = 0$

(ii) $\dfrac{d^2 y}{dx^2} + 2\dfrac{dy}{dx} - 3y = 0$

(iii) $\dfrac{d^3 y}{dx^3} - 4\dfrac{d^2 y}{dx^2} + \dfrac{dy}{dx} + 6y = 0$

(iv) $\dfrac{d^4 y}{dx^4} + 2\dfrac{d^3 y}{dx^3} + \dfrac{d^2 y}{dx^2} = 0$

(v) $\dfrac{d^4 y}{dx^4} - 2\dfrac{d^3 y}{dx^3} - 3\dfrac{d^2 y}{dx^2} + 4\dfrac{dy}{dx} + 4y = 0$

(vi) $\dfrac{d^2 y}{dx^2} + \dfrac{dy}{dx} + y = 0$

(vii) $\dfrac{d^3 y}{dx^3} - \dfrac{d^2 y}{dx^2} + 9\dfrac{dy}{dx} + 13y = 0$

11. Find *all* solutions of the following homogeneous difference equations.
 (i) $y_{x+2} + 2y_{x+1} - 3y_x = 0$
 (ii) $y_{x+3} - 4y_{x+2} + y_{x+1} + 6y_x = 0$
 (iii) $y_{x+4} - 2y_{x+3} - 3y_{x+2} + 4y_{x+1} + 4y_x = 0$
 (iv) $y_{x+2} + y_{x+1} + y_x = 0$
 (v) $y_{x+3} - 3y_{x+2} + 9y_{x+1} + 13y_x = 0$
 (vi) $3y_{x+3} + 5y_{x+2} - 2y_{x+1} = 0$
 (vii) $y_{x+4} + 2y_{x+3} + y_{x+2} = 0$

12. Find *all* solutions of the differential equation

$$3\frac{d^3 y}{dx^3} + 5\frac{d^2 y}{dx^2} - 2\frac{dy}{dx} = f(x)$$

in the following cases.

(i) $f(x) = e^x$
(ii) $f(x) = \sin x$
(iii) $f(x) = x^2$
(iv) $f(x) = e^{-2x}$

13. Find *all* solutions of the difference equation

$$y_{x+3} - 3y_{x+2} + 9y_{x+1} + 13y_x = f_x$$

in the following cases.

(i) $f_x = 1$
(ii) $f_x = x$
(iii) $f_x = (-1)^x$
(iv) $f_x = 1 + x + (-1)^x.$

21

Laplace transforms

In this chapter we develop a tool for solving differential equations by transforming the problem from one in calculus to one in algebra. The method works particularly well when the differential equations are linear (i.e. are linear expressions involving the independent variable and its derivatives, see below) and especially if all the coefficients are constant. It remains attractive also as a tool in more general contexts.

21.1 Definition and existence

Let $f(t)$ be a real-valued (or complex-valued) function defined in the interval $(0, \infty]$. Its Laplace transform, written variously as $L\{f\}$ or $\tilde{f}(s)$ is the function defined by the formula

$$\tilde{f}(s) = \int_0^\infty e^{-st} f(t) dt$$

provided this exists. The variable s may be real or complex.

The transform often exists only for a specified range of values of s but usually the range is of the form $(k, \infty]$ when s is real. The class of functions f for which the transform is available is quite wide as we soon see.

Let us say that the function $f(t)$ defined on $[0, \infty]$ is 'of exponential growth at most' if for some positive constants M and γ it is the case that

$$|f(t)| \leqslant M e^{\gamma t}.$$

If this happens note the following inequality for $s > \gamma$:

$$\int_0^\infty |e^{-st} f(t)| dt \leqslant \int_0^\infty M e^{-st} e^{\gamma t} dt$$

$$= M \int_0^\infty \exp\{(\gamma - s)t\}\, dt$$

$$= \frac{M}{s - \gamma}$$

implies the unconditional convergence of $L\{f\}$ (see Chapter 18 Section 11).

We know already from Chapter 20 that the solutions of homogeneous ordinary linear differential equations are of exponential growth at most, hence the Laplace transform will be applicable to them.

21.2   Example

Calculate $L\{e^{at}\}$.

We have

$$L\{e^{at}\} = \lim_{T \to \infty} \int_0^T e^{-st} e^{at}\, dt = \lim_{T \to \infty} \left[\frac{\exp\{-(s-a)t\}}{s-a} \right]_0^T$$

$$= \lim_{T \to \infty} \left\{ \frac{1}{s-a} - \frac{\exp\{-(s-a)T\}}{s-a} \right\} = \frac{1}{s-a}$$

which is valid only for $s > a$ (or $\mathrm{Re}(s) > a$ if s is complex).

21.3   Why and wherefore

On the face of it, the formula above seems to replace a given function by a more complicated expression. Actually, however, the use of the transform often simplifies matters quite considerably. We shall illustrate this point by tabulating the transforms of commonly occurring functions and then using them to solve some differential equations. The technique is rather like using logarithm tables to simplify the process of multiplication. Instead of multiplying the raw numbers, their logarithms are added. Thus a multiplication problem is *transformed* into an addition problem.

The essential idea here is contained in the following calculation.

$$\int_0^T e^{-st} \frac{df}{dt}\, dt = [e^{-st} f(t)]_0^T + s \int_0^T e^{-st} f(t)\, dt$$

so if $\lim_{T \to \infty} f(T) e^{-sT} = 0$ (which will be the case for large enough s when f is of exponential growth at most), then

$$\boxed{L\{f'\} = sL\{f\} - f(0).}$$

Similarly

$$L\{f''\} = s^2 L\{f\} - \{sf(0) + f'(0)\},$$

since

$$L\{f''\} = sL\{f'\} - f'(0)$$
$$= s^2 L\{f\} - sf(0) - f'(0).$$

Generally,

$$\boxed{L\{f^{(n)}\} = s^n L\{f\} - \{s^{n-1}f(0) + s^{n-2}f'(0) + \cdots + f^{(n-1)}(0)\}}$$

21.4   Example

Solve $\dfrac{d^2 f}{dt^2} - 3\dfrac{df}{dt} + 2f = e^{3t}$ subject to $f(0) = f'(0) = 1$.

We take transforms of both sides of the equation to obtain

$$s^2 \tilde{f} - \{s + 1\} - 3s\tilde{f} + 3 + 2\tilde{f} = \int_0^\infty e^{-st} e^{3t}\, dt$$

$$= \left[\frac{\exp\{(3 - s)t\}}{3 - s} \right]_0^\infty,$$

so

$$(s^2 - 3s + 2)\tilde{f} + 2 - s = \frac{1}{s - 3} \quad (s > 3).$$

(Note that the Laplace transform of e^{3t} does not exist for $s \leqslant 3$.) Thus

$$\tilde{f} = \frac{1}{(s^2 - 3s + 2)} \left\{ \frac{1}{s - 3} + s - 2 \right\}.$$

We now need to ask whether this formula *uniquely* determines the function f. The answer is: yes. If f and g are different then so are $\tilde{f}$ and $\tilde{g}$. Sometimes f is called the anti-transform (or inverse transform) of $\tilde{f}$. An inversion formula for obtaining f from $\tilde{f}$ does indeed exist but, since it involves complex contour-integration, we prefer to omit it. For our purposes it will be sufficient to remember the anti-transforms of a few standard functions (just as one remembers the indefinite integrals of a few standard functions). For the problem in hand, we observe that

$$s^2 - 3s + 2 = (s - 1)(s - 2).$$

Hence, .

$$\tilde{f} = \frac{1}{(s-1)(s-2)(s-3)} + \frac{1}{s-1}$$

$$= \frac{1}{(s-1)(1-2)(1-3)} + \frac{1}{(2-1)(s-2)(2-3)}$$

$$+ \frac{1}{(3-1)(3-2)(s-3)} + \frac{1}{(s-1)}$$

$$= \frac{1}{2(s-1)} - \frac{1}{s-2} + \frac{1}{2(s-3)} + \frac{1}{s-1}$$

$$= \frac{3}{2(s-1)} - \frac{1}{s-2} + \frac{1}{2(s-3)}$$

or

$$\tilde{f} = \tfrac{3}{2}L\{e^{t}\} - L\{e^{2t}\} + \tfrac{1}{2}L\{e^{3t}\}.$$

Thus

$$f(t) = \tfrac{3}{2}e^{t} - e^{2t} + \tfrac{1}{2}e^{3t}.$$

The point to take note of is the simplicity of the method. No messing about with particular solutions is required. Just factorise, use partial fractions and turn the handle!

21.5   A historical note

An appreciation of the elegance of the technique can be obtained by looking at its predecessor, the operational technique. The latter still finds use and is itself worthy of interest although deeper mathematical arguments are needed (from operator theory) to justify it. A cavalier exposition runs as follows.

First introduce the differential operator

$$D = \frac{d}{dt} \quad \text{so that} \quad D^{-1} = \int_{0}^{t} \cdots dt.$$

Then, since

$$(D-1)(D-2)f(t) = e^{3t},$$

surely it ought to be correct that

$$f(t) = (D-2)^{-1}(D-1)^{-1}e^{3t}.$$

Continuing to treat D as an algebraic variable, we write

$$(D-1)^{-1} = -(1-D)^{-1} = -(1 - D + D^{2} - \cdots)$$

which, if valid, would yield

$$(1 - D)^{-1}e^{3t} = (1 - D + D^2 - \cdots)e^{3t}$$
$$= (1 - 3 + 9 - \cdots)e^{3t}$$

which unfortunately diverges. Some legerdemain is therefore required. Rewriting, we have

$$(D - 1)^{-1} = D^{-1}(1 - D^{-1})^{-1} = D^{-1}(1 - D^{-1} + D^{-2} - \cdots)$$

$$(D - 1)^{-1}e^{3t} = e^{3t}\{\tfrac{1}{3} - \tfrac{1}{9} + \tfrac{1}{27} - \cdots\} = e^{3t}\frac{1/3}{1 - 1/3} = \frac{1}{2}e^{3t},$$

where we have been very remiss in omitting the constants of integration (they add up to $-1/2$). Continuing in the same vein, we obtain

$$(D - 2)^{-1} = D^{-1}(1 - 2D^{-1})^{-1} = D^{-1}(1 - 2D^{-1} + 4D^{-2} - \cdots)$$

$$(D - 2)^{-1}\tfrac{1}{2}e^{3t} = e^{3t}\{\tfrac{1}{2} - \tfrac{1}{3} + \tfrac{2}{9} - \tfrac{4}{27}\} = \tfrac{1}{2}e^{3t}\frac{1/3}{1 - 2/3}.$$

Thus we appear to obtain the solution (which is actually only a 'particular solution' of the differential equation)

$$f(t) = \tfrac{1}{2}e^{3t}.$$

Of course, if we had included the various constants of integration we should have received the full solution, but with some effort. The reader should try for himself!

21.6   Some elementary transformations

$$L\{1\} = \int_0^\infty e^{-st}\,dt = \left[\frac{e^{-st}}{-s}\right]_0^\infty = \frac{1}{s}$$

$$L\{t\} = \int_0^\infty e^{-st}t\,dt = \left[t\frac{e^{-st}}{-s}\right]_0^\infty - \int_0^\infty 1\frac{e^{-st}}{-s}\,dt = \frac{1}{s^2}$$

$$L\{t^2\} = \int_0^\infty e^{-st}t^2\,dt = \left[t^2\frac{e^{-st}}{-s}\right]_0^\infty - \int_0^\infty 2t\frac{e^{-st}}{-s}\,dt = \frac{2}{s^2}\cdot\frac{1}{s} = \frac{2}{s^3}$$

and, by induction,

$$L\{t^n\} = \int_0^\infty e^{-st}t^n\,dt = \left[t^n\frac{e^{-st}}{-s}\right]_0^\infty - \int_0^\infty nt^{n-1}\frac{e^{-st}}{-s}\,dt = \frac{n!}{s^{n+1}}.$$

This last formula is in fact valid for n ranging through all real numbers greater than − 1, provided we reinterpret the symbol n!. More on this later.

To find $L\{\cos at\}$, we observe that

$$L\{\cos at + i \sin at\} = \int_0^\infty \exp(iat - st)\,dt$$

$$= \left[\frac{\exp\{(-s+ia)t\}}{-s+ia} \right]_0^\infty = \frac{1}{s-ia} = \frac{s+ia}{s^2+a^2}.$$

So

$$L\{\cos at\} = \frac{s}{s^2+a^2} \qquad\qquad L\{\sin at\} = \frac{a}{s^2+a^2}.$$

(Note that the implied limit in the formula above exists, since

$$|\exp\{(-s+ia)T\}| = e^{-sT} \to 0$$

as $T \to \infty$ provided $s > 0$). We state without proof the result that

$$L\{\log t\} = \frac{1}{s}\{-\log s - \gamma\}$$

where $\gamma \approx 0.577\,22$ is Euler's constant and is defined by

$$\gamma = \lim_{n \to \infty}\left\{1 + \frac{1}{2} + \cdots + \frac{1}{n} - \log n\right\}.$$

To make the result look plausible we offer the following argument. Begin by differentiating with respect to a the identity

$$\int_0^\infty t^a e^{-st}\,dt = \frac{a!}{s^{a+1}}.$$

Recall that $t^a = \exp(a \log t)$, $s^{-a} = \exp(-a \log s)$. Hence for fixed s

$$\int_0^\infty \frac{\partial}{\partial a} t^a e^{-st}\,dt = \frac{d}{da}\left(\frac{a!}{s^{a+1}}\right)$$

$$\int_0^\infty t^a e^{-st} \log t\,dt = \frac{d}{da}(a!)\frac{1}{s^{a+1}} + \frac{(a!)}{s}\frac{d}{da}(s^{-a})$$

$$= \frac{d}{da}(a!)\frac{1}{s^{a+1}} + \frac{(a!)}{s}(s^{-a})\cdot(-\log s).$$

Now put $a = 0$, to obtain

$$\int_0^\infty e^{-st} \log t\,dt = \frac{\text{const.}}{s^1} + \frac{1}{s}(-\log s).$$

We may now tabulate our preliminary results.

$f(t)$	$\tilde{f}(s)$
1	$\dfrac{1}{s}$
t^n	$\dfrac{n!}{s^{n+1}}$
e^{at}	$\dfrac{1}{s-a}$
$\cos at$	$\dfrac{s}{s^2+a^2}$
$\sin at$	$\dfrac{a}{s^2+a^2}$
$\log t$	$-\dfrac{1}{s}\{\log s + \gamma\}$

21.7   Three basic facts

It is obvious that $L\{f+g\} = L\{f\} + L\{g\}$ and that $L\{\alpha f\} = \alpha L\{f\}$ when α is a constant. Thus, using the language of linear algebra, L is a linear transformation acting on the space of real functions f for which $L\{f\}$ exists. Implicit use of this fact was needed for the calculation in Section 21.4. Two other facts are not quite so obvious. They are called the shift property and the scale change property.

The shift property asserts that

$$L\{f(t)e^{at}\} = \tilde{f}(s-a).$$

Clearly,

$$\tilde{f}(s-a) = \int_0^\infty \exp\{-(s-a)t\}f(t)\,dt = \int_0^\infty e^{-st} \cdot e^{at} f(t)\,dt.$$

The scale change property asserts that for $a > 0$

$$L\{f(at)\} = (1/a)\tilde{f}(s/a)$$

which is justified by changing t to at thus

$$\frac{1}{a} \cdot \tilde{f}(s/a) = \int_0^\infty \exp\{-(s/a)t\}f(t)\frac{1}{a}\,dt$$

$$= \int_0^\infty \exp\{-(s/a)at\}f(at)\frac{1}{a}\,d(at) = \int_0^\infty e^{-st}f(at)\,dt.$$

21.8 Example (simultaneous differential equations)

Solve the simultaneous system:

$$\frac{dz_1}{dt} = z_1 + 2z_2,$$

$$\frac{dz_2}{dt} = z_2 - 2z_1.$$

Taking transforms gives

$$s\tilde{z}_1 - z_1(0) = \tilde{z}_1 + 2\tilde{z}_2 \quad \text{or} \quad (s-1)\tilde{z}_1 - 2\tilde{z}_2 = z_1(0),$$
$$s\tilde{z}_2 - z_2(0) = \tilde{z}_2 - 2\tilde{z}_1 \quad \text{or} \quad 2\tilde{z}_1 + (s-1)\tilde{z}_2 = z_2(0).$$

Eliminating the second variable gives

$$\{(s-1)(s-1) + 2\cdot 2\}\tilde{z}_1 = (s-1)z_1(0) + 2z_2(0)$$

$$\tilde{z}_1 = \frac{(s-1)z_1(0)}{(s-1)^2 + 4} + \frac{2z_2(0)}{(s-1)^2 + 4}.$$

Hence,

$$z_1 = z_1(0)e^t \cos 2t + z_2(0)e^t \sin 2t,$$
$$z_2 = -z_1(0)e^t \sin 2t + z_2(0)e^t \cos 2t.$$

21.9 Example

Solve

$$\frac{d^2 f}{dt^2} + 3\frac{df}{dt} + 2f = t.$$

Taking transforms we have

$$s^2\tilde{f} - \{f(0)s + f'(0)\} + 3s\tilde{f} - 3\{f(0)\} + 2\tilde{f} = \frac{1}{s^2},$$

$$(s^2 + 3s + 2)\tilde{f} = f(0)s + (f'(0) + 3f(0)) + \frac{1}{s^2}.$$

But

$$(s^2 + 3s + 2) \equiv (s+1)(s+2),$$

so

$$\tilde{f} = \frac{f(0)s}{(s+1)(s+2)} + \frac{(f'(0) + 3f(0))}{(s+1)(s+2)} + \frac{1}{s^2(s+1)(s+2)}$$

$$= \frac{f(0)}{(s+1)} + \frac{(f'(0) + f(0))}{(s+1)(s+2)} + \frac{1}{s^2(s+1)(s+2)}.$$

Partial fractions are now required. Since the last term is a little more complicated than usual, we pause for some revision of the 'cover-up' rule.

There are constants A, B, C, D such that

$$\frac{1}{s^2(s+1)(s+2)} = \frac{A}{s^2} + \frac{B}{s} + \frac{C}{(s+1)} + \frac{D}{(s+2)}. \qquad (*)$$

This can be checked by expressing the right-hand side as a single fraction with a common denominator and then finding the values of A, B, C and D which make the coefficients of s^2, s and 1 in the numerator equal to $0, 0$ and 1 respectively.

The 'cover-up' rule says that to find A, or C, or D (but alas not B) delete in the denominator of the left-hand side the factor corresponding to the letter (thus for A it would be s^2 and for C it would be $(s+1)$). Then evaluate the new expression substituting for s the value which makes the deleted factor vanish. Thus we have:

$$A = \frac{1}{(0+1)(0+2)} = \frac{1}{2}.$$

This may be justified by multiplying the equation $(*)$ through by s^2 and putting $s = 0$. All terms vanish on the right except that involving A. But note that this does not work with s itself; B must be obtained by an alternative method. To obtain B subtract A/s^2 from both sides of $(*)$, then

$$\text{Left-hand side} = \frac{1}{s^2(s+1)(s+2)} - \frac{1}{2s^2} = \frac{2 - (s+1)(s+2)}{2s^2(s+1)(s+2)}$$

$$= \frac{-s^2 - 3s}{2s^2(s+1)(s+2)} = \frac{-(s+3)}{2s(s+1)(s+2)}$$

$$= \frac{B}{s} + \frac{C}{(s+1)} + \frac{D}{(s+2)}.$$

And now the 'cover-up' rule applied to the last but one expression will give B. In fact

$$B = -\tfrac{3}{4}, \quad C = 1, \quad D = -\tfrac{1}{4}.$$

Hence,

$$\bar{f} = \frac{f(0)}{(s+1)} + \frac{(f'(0) + f(0))}{(s+1)(s+2)} + \frac{1}{s^2(s+1)(s+2)}$$

$$= \frac{f(0)}{(s+1)} + \frac{(f'(0)+f(0))}{(s+1)(-1+2)} + \frac{(f'(0)+f(0))}{(s+2)(-2+1)}$$

$$+ \frac{1}{2s^2} - \frac{3}{4s} + \frac{1}{(s+1)} - \frac{1}{4(s+2)}.$$

Consequently,

$$f(t) = Ae^{-t} + Be^{-2t} + \tfrac{1}{2}t - \tfrac{3}{4},$$

where now

$$A = 1 + f(0) + \{f'(0) + f(0)\} \quad \text{and} \quad B = -\tfrac{1}{4} + \{f'(0) + f(0)\}.$$

21.10 Initial and final values

In this section we show how to obtain information about $f(0)$ and about the limit of $f(t)$ as $t \to \infty$ from a knowledge of the transform $\tilde{f}$.

$$s\tilde{f}(s) - f(0) = \int_0^\infty e^{-st} f'(t)\, dt$$

$$\lim_{s \to 0} \{s\tilde{f}(s)\} - f(0) = \int_0^\infty f'(t)\, dt = \lim_{T \to \infty} [f(t)]_0^T = \lim_{T \to 0} f(T) - f(0)$$

where the limit may be taken through the integral sign in the presence of uniform convergence (see Section 18.11). Thus, we obtain the final value theorem

$$\lim_{t \to \infty} f(t) = \lim_{s \to 0} s\tilde{f}(s).$$

There is an analogous computation for $\lim_{t \to 0} f(t)$ which is particularly interesting when f has a discontinuity at $t = 0$. This is likely to be the case if, for example, $f(t)$ describes the behaviour of a dynamical system which suffered a violent shock at $t = 0$. Then the said limit gives information about the immediate response of the system. We have as earlier (by integration by parts):

$$s\tilde{f}(s) - f(0+) = \lim_{\substack{T \to \infty \\ \delta \to 0}} \left\{ [f(t)e^{-st}]_\delta^T + s \int_\delta^T f(t)e^{-st}\, dt \right\}$$

$$= \int_0^\infty e^{-st} f'(t)\, dt.$$

Assuming that $s\tilde{f}(s)$ has a limit as s tends to infinity and supposing that $f'(t)$ does not grow too fast one can show that $L(f')$ tends to zero as s

tends to infinity. That this is plausible can be seen from the behaviour of the exponential factor. We thus obtain the initial value theorem:

$$\lim_{t \to 0} f(t) = \lim_{s \to \infty} s\tilde{f}(s)$$

Let us see why this is true. We assume, as is usual in applications, that the function $f'(t)$ is at most of exponential growth, thus for some M and k positive

$$|f'(t)| < Me^{kt} \quad \text{for} \quad t > 0.$$

Hence, for $s > k$ and $\tau > 0$:

$$\left| \int_{\tau}^{\infty} e^{-st} f'(t) dt \right| < M \int_{\tau}^{\infty} \exp\{(k-s)t\} dt = \frac{M \exp(k-s)\tau}{s-k}.$$

Thus, this part of the integral can be made as small as required by taking s large enough. As regards the contribution from the range $[0, \tau]$, this can be estimated similarly, if, say, f' were continuous (and therefore bounded); however, a stronger result is available from Analysis, namely that if $|g(t)|$ is integrable then

$$\lim_{s \to \infty} \int_0^{\tau} e^{-st} g(t) dt = 0.$$

21.11 Transform of an integral

Provided we can justify the formal manipulation below, we have, changing the order of integration by reference to Figure 21.1

$$L\left\{ \int_0^t f(u) du \right\} = \int_0^{\infty} dt\, e^{-st} \int_0^t f(u) du$$

$$= \int_0^{\infty} du\, f(u) \int_u^{\infty} e^{-st} dt$$

$$= \int_0^{\infty} \frac{e^{-su}}{s} f(u) du$$

$$= \frac{1}{s} \tilde{f}(s),$$

so under the transformation the integration process becomes division by s (the expected inverse operation to the transform of differentiation). An alternative derivation of the formula above is also available (see Exercise 21.15 Question 14).

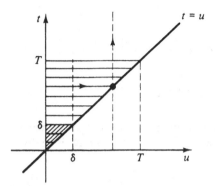

Fig. 21.1

To justify the manipulation, we consider a limiting process both at 0 and at ∞. We have, of course,

$$\int_{\delta}^{T} dt\, e^{-st} \int_{0}^{t} f(u)\,du = \int_{\delta}^{T} du\, f(u) \int_{u}^{T} e^{-st}\,dt$$

$$+ \int_{0}^{\delta} f(u)\,du \int_{\delta}^{T} e^{-st}\,dt.$$

We leave the rest as an exercise.

21.12 The convolution theorem

We now derive the extremely important result that with qualification:

$$L\left\{\int_{0}^{t} f(u)g(t-u)\,du\right\} = \bar{f}(s)\bar{g}(s).$$

We shall illustrate the significance of this result in the next section. The expression

$$\int_{0}^{t} f(u)g(t-u)\,du$$

is known as the convolution of f with g. See also Example 3, Section 19.16. To prove the result we shall need to assume that we are dealing with a value of s for which both the limits

$$\lim_{T\to\infty} \int_{0}^{T} e^{-st}|f(t)|\,dt \qquad \lim_{T\to\infty} \int_{0}^{T} e^{-st}|g(t)|\,dt$$

exist and are finite. Consider now

$$I_T = \int_0^T e^{-su} f(u)\, du \int_0^T e^{-sv} g(v)\, dv$$

$$= \int_0^T \int_0^T \exp\{-s(u+v)\} f(u)g(v)\, du\, dv.$$

This integral can be written as a sum of integrals over the four areas indicated in Figure 21.2. Interest focuses on the area Δ which is really

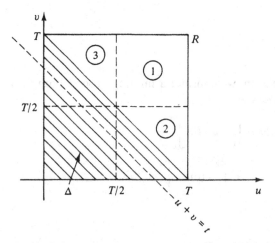

Fig. 21.2

the only significant contribution, as we shall soon see. We consider

$$I_T = \iint_\Delta \exp\{-s(u+v)\} f(u)g(v)\, du\, dv.$$

Let us put $t = u + v$, then on Δ, $0 \leqslant t \leqslant T$ and, since $0 \leqslant v \leqslant T$, we also have $0 \leqslant u = t - v \leqslant t$. Consequently, since the Jacobian is unity,

$$I_\Delta = \int_0^T e^{-ts} \int_0^t f(u)g(t-u)\, du\, dt.$$

We now show that $|I_T - I_\Delta| \to 0$ as $T \to \infty$. To this end consider an arbitrary small positive number δ. By the assumptions which we made at the outset, there is T so large that

$$\int_{T/2}^\infty e^{-st} |f(t)|\, dt < \delta \quad \text{and} \quad \int_{T/2}^\infty e^{-st} |g(t)|\, dt < \delta.$$

Thus in area 1

$$\left| \iint \exp\{-s(u+v)\} f(u)g(v)\,du\,dv \right|$$

$$\leqslant \iint \exp\{-s(u+v)\} |f(u)|\,|g(v)|\,du\,dv$$

$$< \int_{T/2}^{T} e^{-su}|f(u)|\,du \int_{T/2}^{T} e^{-sv}|g(v)|\,dv < \delta^2,$$

whilst in area 2

$$\left| \iint \exp\{-s(u+v)\} f(u)g(v)\,du\,dv \right|$$

$$\leqslant \iint \exp\{-s(u+v)\} |f(u)|\,|g(v)|\,du\,dv$$

$$< \int_{T/2}^{T} e^{-su}|f(u)|\,du \int_{0}^{T} e^{-sv}|g(v)|\,dv$$

$$< \delta \int_{0}^{\infty} e^{-sv}|g(v)|\,dv.$$

Finally, in area 3

$$\left| \iint \exp\{-s(u+v)\} f(u)g(v)\,du\,dv \right|$$

$$\leqslant \iint \exp\{-s(u+v)\} |f(u)|\,|g(v)|\,du\,dv$$

$$< \int_{0}^{T} e^{-su}|f(u)|\,du \int_{T/2}^{T} e^{-sv}|g(v)|\,dv$$

$$< \delta \int_{0}^{\infty} e^{-su}|f(u)|\,du.$$

Hence, since δ is arbitrary, all the contributions above can be made as small as we wish, so indeed $|I_\Delta - I_T| \to 0$ as $T \to \infty$ and the theorem is proved since the limit of I_Δ is the required transform.

21.13 Application of the convolution theorem

Example 1
We begin with a simple example. Suppose we wish to find the

function whose transform is

$$\frac{1}{s(s+2)}.$$

Then taking $f(t) = 1$, $g(t) = e^{-2t}$ we have $\tilde{f}(s) = 1/s$, $\tilde{g}(s) = 1/(s - (-2))$. By the convolution theorem the function we need is

$$\int_0^t g(u)f(t-u)du = \int_0^t e^{-2u} \cdot 1\, du = \tfrac{1}{2}\{1 - e^{-2t}\}$$

(notice the crafty transposition of f and g). Clearly this is simpler than partial fractions.

Example 2
Find the function whose transform is:

$$\frac{1}{s\sqrt{(s+1)}}.$$

Again we take $f(t) = 1$ and noting that the transform of $t^{-1/2}$ is $\sqrt{\pi}s^{-1/2}$ (a fact that remains to be proved, see Section 21.14 below) we take $g(t) = t^{-1/2}$; we have by the shift property

$$L\{g(t)e^{-t}\} = \frac{\sqrt{\pi}}{\sqrt{(s+1)}}.$$

Thus the required function is

$$\int_0^t \frac{1}{\sqrt{\pi}} e^{-u} g(u) f(t-u)\, du = \int_0^t \frac{1}{\sqrt{\pi}} e^{-u} u^{-1/2} \cdot 1\, du$$

$$= \frac{2}{\sqrt{\pi}} \int_0^{\sqrt{t}} e^{-v^2}\, dv,$$

where we have used the substitution $u = v^2$ (so that $du = 2u^{1/2}dv$). The function appearing rightmost is the error-function evaluated at $\sqrt{t}$.

Example 3 Find $L\{\sqrt{t}\}$.
Suppose that $\tilde{f} = L\{\sqrt{t}\}$ then $\tilde{f}^2 = L\{\int_0^t \sqrt{u}\sqrt{(t-u)}du\}$. Now we evaluate, using $u = t\sin^2\theta$ (so that $du = 2t\sin\theta\cos\theta\, d\theta$),

$$I = \int_0^t \sqrt{(ut - u^2)}\, du$$

$$= \int_0^{\pi/2} 2t\sin\theta\cos\theta \cdot t\sqrt{(\sin^2\theta - \sin^4\theta)}\, d\theta$$

$$= \int_0^{\pi/2} 2t^2 \sin^2 \theta \cos^2 \theta \, d\theta = \frac{t^2}{2} \int_0^{\pi/2} \sin^2 2\theta \, d\theta$$

$$= \frac{t^2}{4} \int_0^{\pi/2} (1 - \cos 4\theta) \, d\theta = \frac{\pi t^2}{8}.$$

Thus

$$\bar{f}^2 = L\left\{\frac{\pi t^2}{8}\right\} = \frac{\pi}{8} \cdot \frac{2!}{s^3}$$

and so

$$\bar{f}(s) = \frac{\sqrt{\pi}}{2} \cdot \frac{1}{s^{3/2}}.$$

Example 4 (input–output systems)

A large class of dynamical system can be reduced to the model which follows. The system receives a time dependent input $u(t)$ which causes an output $y(t)$ related to the input by the differential equation

$$D^n y + a_{n-1} D^{n-1} y + \cdots + a_0 y = D^m u + b_{m-1} D^{m-1} u + \cdots + b_0 u.$$

Taking transforms we obtain:

$$s^n \tilde{y} + a_{n-1} s^{n-1} \tilde{y} + \cdots + a_0 \tilde{y} + Q_{n-1}$$
$$= s^m \tilde{u} + b_{m-1} s^{m-1} \tilde{u} + \cdots + b_0 \tilde{u} + R_{m-1}.$$

where Q_{n-1} and R_{m-1} denote the polynomials in s of orders $n-1$ and $m-1$ arising from the initial condition of the output and of the input of the system (i.e. at time $t = 0$) in accordance with the formulas of Section 2.3. With the obvious notation we thus have

$$\tilde{y} = \frac{P_m(s)\tilde{u}}{P_n(s)} + \frac{I(s)}{P_n(s)},$$

where all the initial conditions have been lumped together in one polynomial $I(s)$. Now the inverse Laplace transform of the second term on the right may be found, say by partial fractions. Call it $h(t)$. Similarly find the inverse transform of

$$G(s) = \frac{P_m(s)}{P_n(s)}$$

and call it $g(t)$. Then we have by the convolution theorem

$$y(t) = \int_0^t u(t - \tau) g(\tau) \, d\tau + h(t).$$

$G(s)$ is known as the system's *transfer function*. Clearly $G(s) = \bar{y}(s)/\bar{u}(s)$ provided we have zero initial conditions. $g(t)$ is known as the *impulse response*, since $y(t) = g(t)$ if $\bar{u} = 1$. Note however that 1 is not the transform of any function; nevertheless it almost is. For if f_δ is the function illustrated

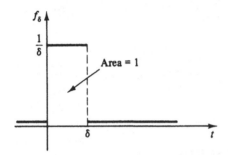

Fig. 21.3

in Figure 21.3, then

$$L\{f_\delta\} = \frac{1}{\delta}\int_0^\delta e^{-st}\,dt = \frac{-\{e^{-s\delta} - 1\}}{s\delta}$$

$$\to 1 \quad \text{as} \quad \delta \to 0.$$

Note that $e^{-s\delta} = 1 - s\delta + s^2\delta^2/2 - \cdots$. Thus $L^{-1}\{1\}$ may be regarded as an instantaneous impulse of zero duration and infinite size.

Example 5 (the renewal equation)

This example comes from statistics and before we present it we need to motivate the definition. Reference should also be made to the final section of Chapter 19.

A *stochastic process* on the set T assigns a random variable $X(t)$ to each t in the set T. In the current example t represents the time with $T = [0, \infty)$ and $X(t)$ will be a *counting process* – i.e. $X(t)$ counts the number of times something has happened between time 0 and time t. A much quoted example involves the number of V2 rockets which fell on London during the Second World War but we shall consider instead the much more mundane example in which $X(t)$ is the number of customers served at a particular supermarket till between time 0 and time t. A random variable X_n which records the time it takes for X to get from the value $n - 1$ to the value n is called an *inter-arrival time*. A *waiting time* S_n is the time it takes for X to get from 0 to n i.e.

$$S_n = \sum_1^n X_i.$$

A *counting process* $X(t)$ is called a *renewal process* if the random variables $X_1, X_2, \ldots$ are independent and identically distributed. We denote their common probability distribution function by F. This is a reasonable hypothesis because we are asserting that the process behaves from any point in time in exactly the same way (probabilistically) as it did originally, i.e. the process renews itself at any point in time. The expected number of times the event being counted occurs by time t is known as the *renewal function*. It is defined by

$$m(t) = E(X(t)).$$

Clearly, letting $f(x)$ be the probability density,

$$m(t) = \int_0^\infty E[X(t)|X_1 = x]f(x)dx,$$

where $E[X|Q]$ denotes the expected value of the random variable X conditionally on X having property Q.

Now,

$$E[X(t)|X_1 = x] = \begin{cases} 1 + m(t-x) & \text{if } x \leqslant t, \\ 0 & \text{if } t < x. \end{cases}$$

Hence

$$m(t) = \int_0^t \{1 + m(t-x)\} f(x)dx$$

$$= F(t) + \int_0^t m(t-x)f(x)dx.$$

This result is known as the *renewal equation*. We solve the equation by taking Laplace transforms using the fact that the integral above is a convolution. We have

$$\tilde{m}(s) = \tilde{F}(s) + \tilde{m}(s)\tilde{f}(s),$$

or

$$\tilde{m}(s)(1 - \tilde{f}(s)) = \tilde{F}(s),$$

giving

$$\tilde{m}(s) = \frac{\tilde{F}(s)}{1 - \tilde{f}(s)}.$$

Thus in principle we have found the renewal function.

21.14 The gamma and beta functions

We introduce these special functions in order both to calculate $L\{t^\alpha\}$ for $\alpha > -1$ and to illustrate the power of the convolution theorem.

Definition
The gamma and beta functions are:

$$\Gamma(\alpha) = \int_0^\infty t^{\alpha-1} e^{-t} dt \qquad (\alpha > 0)$$

$$B(p, q) = \int_0^1 t^{p-1} (1-t)^{q-1} dt \quad (p, q > 0).$$

Observe that

$$\Gamma(1) = \int_0^\infty e^{-t} dt = 1$$

and that

$$\Gamma(\alpha + 1) = \int_0^\infty t^\alpha e^{-t} dt \qquad\qquad (\alpha + 1 > 0)$$

$$= \left[t^\alpha \frac{e^{-t}}{-1} \right]_0^\infty + \int_0^\infty \alpha t^{\alpha-1} e^{-t} dt. \quad (\alpha \geqslant 0).$$

Note that the last integral exists for $\alpha \geqslant 0$. Thus for $\alpha > 0$, $\Gamma(\alpha + 1) = \alpha\Gamma(\alpha)$ and hence by induction $\Gamma(n + 1) = n!$. [Exercise: What happens when $\alpha = 0$?] Finally, note that the existence of $\Gamma(\alpha)$ is assured for $\alpha > 0$ since

$$\int_\delta^1 t^{\alpha-1} dt = \left[\frac{t^\alpha}{\alpha} \right]_\delta^1 \to \frac{1}{\alpha} \quad \text{as} \quad \delta \to 0.$$

A similar calculation will establish that for $\alpha \leqslant 0$ the gamma integral diverges. The same type of argument also establishes convergence for the beta integral for $p, q > 0$ precisely. Now let v be given with

$$v > -1$$

and let

$$\alpha = v + 1 > 0.$$

Then we have

$$L\{t^v\} = L\{t^{\alpha - 1}\} = \int_0^\infty t^{\alpha - 1} e^{-st} dt$$

$$= \int_0^\infty \left[\frac{\tau}{s}\right]^{\alpha - 1} e^{-\tau} \frac{1}{s} d\tau \quad \text{(writing } st = \tau\text{)}$$

$$= \frac{1}{s^\alpha} \int_0^\infty \tau^{\alpha - 1} e^{-\tau} d\tau$$

$$= \frac{\Gamma(\alpha)}{s^\alpha} = \frac{\Gamma(v + 1)}{s^{v + 1}}. \qquad (v > -1).$$

We infer from an example in the last section that

$$\Gamma\left(\frac{3}{2}\right) = \frac{\sqrt{\pi}}{2} \quad \text{and so} \quad \Gamma\left(\frac{1}{2}\right) = \sqrt{\pi}.$$

The relationship between the gamma and beta functions can be obtained by introducing the auxiliary function

$$B(p, q, t) = \int_0^t u^{p - 1} (t - u)^{q - 1} du, \quad (p, q > 0)$$

so that $B(p, q, 1) = B(p, q)$. Now by the convolution theorem

$$L\{B(p, q, t)\} = L\{t^{q - 1}\} L\{t^{p - 1}\}$$

$$= \frac{\Gamma(q)\Gamma(p)}{s^{p + q}}$$

$$= \frac{\Gamma(q)\Gamma(p)}{\Gamma(p + q)} \cdot \frac{\Gamma(p + q)}{s^{p + q}} = \frac{\Gamma(q)\Gamma(p)}{\Gamma(p + q)} L\{t^{p + q - 1}\}.$$

Hence,

$$B(p, q, t) = \frac{\Gamma(q)\Gamma(p)}{\Gamma(p + q)} \cdot t^{p + q - 1}.$$

Thus

$$B(p, q) = \frac{\Gamma(q)\Gamma(p)}{\Gamma(p + q)}.$$

Example 1

Suppose the random variable X has standard normal probability density function (see Chapter 19 Section 16):

$$\frac{1}{\sqrt{2\pi}}\exp(-x^2/2).$$

Find the expected value of X^{2k} for $k = 0, 1, 2, \ldots$. ($E(X^r)$ is known as the *rth order moment* of the distribution.) We have

$$E(X^{2k}) = \frac{1}{\sqrt{2\pi}}\int_{-\infty}^{\infty} x^{2k}\exp(-x^2/2)\,dx$$

$$= \frac{\sqrt{2}}{\sqrt{\pi}}\int_{0}^{\infty} x^{2k}\exp(-x^2/2)\,dx$$

(put $z = x^2/2$)

$$= \frac{\sqrt{2}}{\sqrt{\pi}}\int_{0}^{\infty} (2z)^k e^{-z}(2z)^{-1/2}\,dz$$

$$= \frac{1}{\sqrt{\pi}}2^k\int_{0}^{\infty} z^{k-1/2}e^{-z}\,dx$$

$$= \frac{1}{\sqrt{\pi}}2^k\Gamma(k + (1/2)).$$

Using $\Gamma(\frac{1}{2}) = \sqrt{\pi}$ and $\Gamma(\alpha + 1) = \alpha\Gamma(\alpha)$ the last term may be simplified down to

$$(2k - 3)(2k - 5)\cdots 5\cdot 3\cdot 1.$$

Example 2

A random variable X with values in $(0, 1)$ is said to have *beta distribution* if its density function is for some $p, q > 0$

$$f(x) = \frac{1}{B(p, q)}x^{p-1}(1-x)^{q-1} \quad 0 < x < 1.$$

Calculate its kth order moment.

We are asked to find

$$\frac{1}{B(p, q)}\int_{0}^{1} x^k x^{p-1}(1-x)^{q-1}\,dx$$

$$= \frac{B(p + k, q)}{B(p, q)}$$

For example if $k = 1$ we have

$$E(X) = \frac{B(p+1,q)}{B(p,q)} = \frac{\Gamma(p+1)\Gamma(q)}{\Gamma(p+1+q)} \cdot \frac{\Gamma(p+q)}{\Gamma(p)\Gamma(q)}$$

$$= \frac{p}{p+q}.$$

Remark

A random variable with values in $(0, \infty)$ is said to have a *gamma distribution function* if for some $\alpha > 0$, $\lambda > 0$ its density function is

$$f(x) = \frac{\lambda^\alpha}{\Gamma(\alpha)} x^{\alpha-1} e^{-\lambda x}$$

(cf. Question 7, Exercises 18.14).

21.15 Exercises

1. Find the Laplace transforms of the following functions, specifying in each case the values of s for which the transform exists.

$$2e^{4t}, \quad 3e^{-2t}, \quad 5t - 3, \quad t^2 - 2e^{-t}, \quad 3\cos 5t,$$
$$(t^2 + 1)^2, \quad t^3 e^{-3t}, \quad (t + 2)^2 e^t.$$

2. Determine the functions whose Laplace transforms are:

$$\frac{1}{s-a}, \quad \frac{1}{s^2+a^2}, \quad \frac{1}{s^n}, \quad \frac{5s+4}{s^3}, \quad \frac{6}{2s-3},$$

$$\frac{2s-18}{s^2+9} + \frac{24-30\sqrt{s}}{s^4}, \quad \frac{3s+7}{s^2-2s-3}.$$

3. Solve the following differential equations using transforms

(i) $f'' + 2f' + 5f = t - 2$

subject to $f(2) = f'(2) = 1$
[Hint: Define $g(t) = f(t+2)$.]

(ii) $g' + g + 2f = e^{2t}$
$\quad\ 2g' - g + f' = 0$

subject to $f(0) = g(0) = 0$.

4. Use the convolution theorem to find the functions whose transforms are:

$$\frac{1}{(s+3)(s-1)}, \quad \frac{1}{(s+1)(s^2+1)}.$$

5. Prove that

$$\int_0^{\pi/2} \sin^{2p-1}\theta \cos^{2q-1}\theta\, d\theta = \tfrac{1}{2}B(p,q).$$

Hence evaluate

$$\int_0^{\pi/2} \sin^4\theta \cos^6\theta\, d\theta.$$

6. Evaluate $\Gamma(5/2)$ and $L\{t^{1/2} + t^{-1/2}\}$.

7. Solve using Laplace transforms

$$f(t) = t + 2\int_0^t \cos(t-u)f(u)\, du.$$

8. Use the convolution theorem to solve the problem

$$f''(t) + 4f(t) = g(t),$$

where g is a given function and $f(0) = a$, $f'(0) = b$.

9. Solve using the convolution theorem the equation

$$f(t) = 1 + t + \int_0^t (t-u)f(u)\, du.$$

10. Show that

$$L\{t^n f(t)\} = (-1)^n \frac{d^n}{ds^n}\,\bar{f}(s).$$

What limitations does this impose on simplifying by transforms a linear ordinary differential equation with non-constant coefficients?

11. Use Laplace transforms to solve the equation

$$tf''(t) + (2-t)f'(t) + 2f(t) = 0.$$

12. Show that

$$L\left\{\int_0^t \frac{f(u)}{u}\, du\right\} = \frac{1}{s}\int_s^\infty \bar{f}(v)\, dv.$$

Hence find

$$L\left\{\int_0^t \frac{\sin u}{u}\, du\right\}.$$

Also find

$$L^{-1}\left\{\log\frac{s+a}{s+b}\right\}.$$

13. Find

$$L\left\{\frac{\sin u}{u}\right\}.$$

[Hint: Differentiate $\int_0^\infty e^{-st}(\sin at)/t\,dt$ with respect to a.]

14. Obtain the result of Section 21.11 by taking the transform of f' where

$$f(t) = \int_0^t g(u)du$$

and g is a continuous function.

15. Use the relationship between the gamma and beta functions to find $\Gamma(3/2)$ and hence show that

$$\int_0^1 \sqrt{(-\log x)}\,dx = \frac{\sqrt{\pi}}{2}.$$

21.16 Further exercises

1. Use a series expansion of the exponential function to verify the result

$$L\left\{\frac{2}{\sqrt{\pi}}\int_0^{\sqrt{t}} \exp(-u^2)du\right\} = \frac{1}{s\sqrt{(s+1)}}.$$

Remark. This computes the transform of $\text{erf}(\sqrt{t})$; compare Example 2 of Section 21.13.

2. Show that the function defined for $t \geqslant 0$ by

$$f(t) = \int_0^\infty \frac{\exp(-tx^2)}{1+x^2}dx$$

satisfies the equation

$$\frac{df}{dt} - f = -\frac{\sqrt{\pi}}{2\sqrt{t}}, \quad f(0) = \pi/2$$

and hence find $f(t)$ in terms of the error function. (Hint: Refer to the last question.)

3. Following on from Question 5 above (Section 21.15) prove that

$$\int_0^{\pi/2} \sin^{2p} 2\theta\, d\theta = \int_0^{\pi/2} \sin^{2p} \theta\, d\theta$$

and from this deduce that

$$\Gamma(2p)\Gamma(1/2) = 2^{2p-1}\Gamma(p)\Gamma(p+1/2).$$

[Hint: put $\psi = 2\theta$.]

4. A function $f(x)$ is defined for $x > 0$ by

$$f(x) = \int_0^\infty \cos(xu^2)du.$$

404   Laplace transforms

Show by first putting $w = xu^2$ and then integrating by parts that the integral in the definition is convergent. Show also that

$$f(x) = \frac{1}{\sqrt{x}} f(1).$$

By reversing the order of integration and making a change of variable obtain the result

$$L\{f(t)\} = \frac{1}{4\sqrt{s}} \int_0^1 v^{-1/4} (1 - v)^{-3/4} dv.$$

Assuming that the formula

$$\int_0^\infty \exp(-ax^2) dx = \frac{\sqrt{\pi}}{2\sqrt{\alpha}}$$

is valid for $\sqrt{\alpha} = \exp(i\pi/4)$ find

$$\int_0^\infty \exp(-iv^2) dv$$

and hence deduce from the Question 2 (in this section) that

$$\Gamma(1/2)\Gamma(3/4) = \pi\sqrt{2}.$$

5. A continuous function $f(t)$ defined for $t \geqslant 0$ is known to have as its transform

$$\bar{f}(s) = e^{-\sqrt{s}}.$$

Differentiate this equation twice and show that $\bar{f}$ satisfies a second order differential equation with coefficients that are constants or are linear in s. From this deduce a differential equation for $f(t)$ and hence find f. [Use Question 10 of Section 21.15.]

6. Use the convolution theorem to show that the equation

$$g(t) = \int_0^t (t - u)^{-\alpha} f(u) du \quad (0 < \alpha < 1)$$

has the solution

$$f(t) = C \frac{d}{dt} \left\{ \int_0^t (t - u)^{\alpha - 1} g(u) du \right\}$$

where C is a constant depending on α.

7. Show that if $f(t) = 0$ for $t \leqslant 0$ then the transform of $f(t - n)$ is

$$e^{-sn} \bar{f}(s).$$

If it is further known that $f(t)$ satisfies the lag equation:

$$\frac{d}{dt} f(t) + f(t - 1) = t \quad \text{for} \quad t \geqslant 0$$

find f. [Hint: Expand $(1 + e^{-s}/s)^{-1}$ as a geometric series in e^{-s}/s. Be careful to check that in fact your answer is only apparently an infinite series. This problem is related to z-transforms, a form of Laplace transform appropriate to the solution of difference equations.]

8. Define

$$L_\delta\{f\} = \int_\delta^\infty f(t)e^{-st}\,dt \quad \text{and} \quad L_0 = L.$$

Find $L_\delta\{f'\}$ in terms of $L_\delta\{f\}$. A function $g(t)$ is defined for $t \geq 0$ by

$$g(t) = \int_0^{\sqrt{t}} \exp(u^2)\,du.$$

Differentiate g twice and show that g satisfies a homogeneous second order differential equation involving coefficients that are linear in t. Why is it not possible to take a Laplace transform of this equation in the usual way? Applying the transform L_δ to the equation obtain a formula for $L_\delta(g)$ and passing to the limit as $\delta \to 0$ find $L\{g\}$. [Use the Final-Value Theorem to determine the constant of integration.]

9. For $x > -1$ define

$$f(n, t) = \begin{cases} (1 - t/n)^n t^{x-1} & \text{for } 0 \leq t \leq n, \\ 0 & \text{for } t \geq n. \end{cases}$$

Find $\lim f(n, t)$ as $n \to \infty$ and use this result to prove Gauss' formula

$$\Gamma(x) = \lim_{n \to \infty} \frac{n^x \cdot n!}{x(x+1)\cdots(x+n)}$$

[Hint: show that $f(n, t)$ has dominated convergence. For this refer to e^{-t}.]

10. Use Gauss' formula to prove that, for $0 < x < 1$:

$$\Gamma(x)\cdot\Gamma(1-x) = \pi/\sin\pi x$$

[Hint: You will need to know the result

$$\sin x = x \prod_{n=1}^\infty (1 - x^2/(n^2\pi^2))$$

which is proved in Chapter 22.]

11. Show that

(i) $L\{[t]\} = \dfrac{1}{s(e^s - 1)}$,

where $[t]$ is the staircase function,

(ii) $L\left\{\int_t^\infty \dfrac{1}{u}e^{-u}\,du\right\} = \dfrac{1}{s}\log(1 + s)$

[Hint: Change the order of integration and use the series expansion for e^{-su}.]

22

Series solution of linear ordinary differential equations

22.1 General considerations

Consider the problem of solving a homogeneous differential equation with non-constant coefficients of the form

$$a_2(x)\frac{d^2y}{dx^2} - a_1(x)\frac{dy}{dx} + a_0(x)y = 0.$$

Guided by Taylor's theorem we are tempted to express the solution function $y(x)$ in the neighbourhood of some point x_0 of interest as an infinite power series (see Section 18.12)

$$y(x) = c_0 + c_1(x - x_0) + c_2(x - x_0)^2 + \cdots.$$

Assuming $a_0(x)$, $a_1(x)$, $a_2(x)$ can also be expressed as power series we could substitute the above expression for $y(x)$ into the differential equation, perform all the relevant differentiations and obtain a relationship governing the constants c_i. This turns out to be a feasible programme provided $a_2(x_0) \neq 0$. If however $a_2(x_0) = 0$, the series for y needs to be replaced by the generalised power series

$$y(x) = (x - x_0)^{\sigma}\{c_0 + c_1(x - x_0) + c_2(x - x_0)^2 + \cdots\},$$

where σ might need to be fractional. Even then we are not guaranteed success; the method works provided the 'order' of the zero of $a_0(x)$ at x_0 is appropriately related to its order for $a_1(x)$ and for $a_2(x)$. More precisely, if we can write

$$a_0(x) = (x - x_0)^r\{a_0^0 + a_1^0(x - x_0) + a_2^0(x - x_0)^2 + \cdots\},$$
$$a_1(x) = (x - x_0)^s\{a_0^1 + a_1^1(x - x_0) + a_2^1(x - x_0)^2 + \cdots\},$$
$$a_2(x) = (x - x_0)^t\{a_0^2 + a_1^2(x - x_0) + a_2^2(x - x_0)^2 + \cdots\},$$

where r, s, t are positive integers with $r > 1$ and $s > 2$, we require

$$s \geqslant r - 1 \quad \text{and} \quad t \geqslant r - 2.$$

See Exercise 22.11 Question 8 for an example where these conditions are not met.

The drawback of this method is that, although it gives us *one* solution at least, there is nevertheless no guarantee of obtaining a *second* solution which is linearly independent of the first, whereas of course we would expect to have two linearly independent solutions in the case of a second order equation (see Section 22.3). Fortunately there is an alternative method for finding the second solution. See Exercise 22.10 Question 10.

Henceforth we shall assume, for convenience, that $x_0 = 0$.

22.2 An example

Solve

$$\frac{d^2y}{dx^2} - x\frac{dy}{dx} + 2y = 0.$$

Suppose

$$y = \sum_0^\infty c_n x^{n+\sigma} \quad (c_0 \neq 0)$$

is a solution converging for $|x| < R$, where R is the radius of convergence for this series expansion. We may differentiate (twice) term by term within the radius of convergence, thus

$$Dy = \sum (n+\sigma)c_n x^{n+\sigma-1},$$
$$D^2 y = \sum (n+\sigma)(n+\sigma-1)c_n x^{n+\sigma-2}.$$

Thus, on substitution into the equation, we have

$$\left\{\sum (n+\sigma)(n+\sigma-1)c_n x^{n+\sigma-2}\right\} - \left\{\sum (n+\sigma)c_n x^{n+\sigma}\right\}$$
$$+ 2\left\{\sum c_n x^{n+\sigma}\right\} = 0.$$

We collect like terms, and, starting with the lowest power (here $x^{\sigma-2}$), we set the coefficients all equal to zero. (Why?) Thus:

$x^{\sigma-2}$: $c_0\sigma(\sigma-1) = 0$ (this is known as the *indicial equation*),
$x^{\sigma-1}$: $c_1(\sigma+1)\sigma = 0,$
$x^{\sigma+n}$: $c_{n+2}(n+2+\sigma)(n+2+\sigma-1) - c_n(n+\sigma) + 2c_n = 0 \ (n \geqslant 0).$

The coefficient of the general term gives us the recurrence relation:

$$c_{n+2}(n+2+\sigma)(n+1+\sigma) = c_n(n-2+\sigma).$$

The indicial equation gives us two choices for σ: 0 or 1. Taking $\sigma = 1$ leads to $c_1 = 0$ and hence $0 = c_1 = c_3 = c_5 = \cdots = c_{2k+1}$, i.e. all the odd coefficients are zero. The even coefficients are governed by the equation

$$c_{n+2}(n+3)(n+2) = c_n(n-1),$$

so that

$$c_{2k} = \frac{(2k-3)}{(2k+1)(2k)} c_{2k-2},$$

and

$$c_{2k} = \frac{(2k-3)(2k-5)}{(2k+1)2(k)(2k-1)2(k-1)} c_{2k-4}$$

$$= \frac{(2k-3)(2k-5)(2k-7)}{(2k+1)2(k)(2k-1)2(k-1)(2k-3)2(k-2)} c_{2k-6} = \cdots$$

$$= \frac{(2k-3)(2k-5)(2k-7)\cdots(1)(-1)}{(2k+1)2(k)(2k-1)2(k-1)(2k-3)2(k-2)\cdots3\cdot2} c_0$$

$$= \frac{-1}{2^k k!(2k+1)(2k-1)} c_0 = -2^{-k}\frac{1}{k!(4k^2-1)} c_0.$$

Thus

$$y = -\sum_0^\infty c_0 \frac{2^{-k}x^{2k+1}}{k!(4k^2-1)}.$$

This is seen to converge for all x. We have thus found a solution.

The case $\sigma = 0$ warrants attention. We now have c_1 arbitrary. Also

$$c_{n+2} = \frac{(n-2)}{(n+2)(n+1)} c_n,$$

so that putting $n = 2$ we have $c_2 = 0$, whence $0 = c_4 = c_6 = c_8 = \cdots = c_{2k}$. Taking $c_1 = 0$, we also have $0 = c_1 = c_3 = c_5 = \cdots$. Thus we obtain a second solution

$$y = c_0 - c_0 x^2 = c_0(1 - x^2).$$

22.3   A second example

In the following example we obtain just one solution by the series method:

$$x^2D^2y + xDy + (x^2 - k^2)y = 0.$$

This is known as *Bessel's equation* of order k.

Again we assume that there is a solution y of the form

$$y = \sum_0^\infty c_n x^{n+\sigma} \quad (c_0 \neq 0),$$

where the convergence is assumed within some radius R. On substitution we obtain

$$\sum (n+\sigma)(n+\sigma-1)c_n x^{n+\sigma} + \sum (n+\sigma)c_n x^{n+\sigma} + \sum c_n x^{n+\sigma+2}$$
$$- k^2 \sum c_n x^{n+\sigma} = 0.$$

The lowest power is x^σ and the *indicial equation* reads

$$c_0(\sigma(\sigma-1)+\sigma-k^2) = 0 \quad \text{or} \quad c_0(\sigma^2-k^2) = 0.$$

Hence $\sigma = \pm k$. We take $\sigma = k \geqslant 0$. The coefficient at x^{k+1} is then

$$c_1 k(k+1) + c_1(k+1) - k^2 c_1 = 0.$$

Thus

$$c_1(2k+1) = 0 \quad \text{so that} \quad c_1 = 0.$$

The next highest power has a coefficient taking contributions from each of the terms of the equation and is thus already an instance of the general case. We have

$$c_n(k+n)(k+n-1) + c_n(k+n) - k^2 c_n + c_{n-2} = 0.$$

Thus

$$c_n\{(k+n)^2 - k^2\} = -c_{n-2}$$

or

$$c_n = \frac{-c_{n-2}}{n(2k+n)}.$$

But $c_1 = 0$, so $0 = c_1 = c_3 = c_5 = \cdots$ and

$$c_{2m} = \frac{-c_{2(m-1)}}{2m(2k+2m)} = \frac{-c_{2(m-1)}}{4m(k+m)},$$

giving

$$y = c_0 x^k \sum_{m=0}^\infty \frac{k!}{m!(m+k)!} \left(-\frac{x^2}{4}\right)^m.$$

The intended interpretation of the factorial signs when k is not an integer is for example $\{(k+m)(k+m-1)(k+m-2)\cdots k\}^{-1}$. The solution above is known as *Bessel's function of the first kind*, denoted $J_k(x)$.

When one repeats the calculation for $\sigma = -k$ ($k \geqslant 0$) all is well provided k is *not* an integer, and a second solution $J_{-k}(x)$ is obtained. However, if k is an integer the equation

$$-c_{n-2} = n(-2k+n)c_n$$

tells us first that $c_{2k-2} = 0$ and then by backward application of the formula that $c_0 = 0$. This is a contradiction.

A second (i.e. linearly independent) solution is known to be

$$Y_k(x) = \lim_{\eta \to k} \frac{J_\eta(x)\cos\eta\pi - J_{-\eta}(x)}{\sin\eta\pi}$$

and goes by the name of Bessel's function of the second kind. See also Exercises 22.10 Question 10.

22.4   Trigonometric series expansion

Consider the differential equation

$$D^2 y + \omega^2 y = f(x).$$

The solution of the homogeneous version of the equation is of course

$$y = A\cos\omega x + B\sin\omega x.$$

Observe that

$$\cos\omega(x + 2\pi/\omega) = \cos\omega x$$

and

$$\sin\omega(x + 2\pi/\omega) = \sin\omega x.$$

Let

$$l = \frac{2\pi}{\omega}.$$

Thus $y(x + l) = y(x)$ and the function is said to be *periodic* with period l. Now suppose that the function $f(x)$ is itself periodic. In this case one hopes to write y as a sum of sines and cosines rather than as a Taylor series, since the former are periodic functions. Thus the aim is to write y in the form of the *Fourier series*

$$f(x) = \tfrac{1}{2}a_0 + \sum_1^\infty [a_n\cos nx + b_n\sin nx]$$

(the $\tfrac{1}{2}$ appears for convenience only). Before considering for what functions f such an expansion might be vaild, we examine some elementary examples. The idea behind them is that of summing a geometric progression with common ratio $e^{i\theta} = \cos\theta + i\sin\theta$.

Example 1

We compute

$$f_r(x) = \frac{1}{2} + \sum_1^\infty r^n \cos nx \quad \text{for} \quad |r| < 1.$$

$$= \frac{1}{2} + \sum_1^\infty \mathrm{Re}\,\{r^n e^{inx}\}$$

$$= \mathrm{Re}\left\{\frac{1}{2} + \frac{re^{ix}}{1 - re^{ix}}\right\}$$

$$= \mathrm{Re}\left\{\frac{re^{ix}(1 - re^{-ix})}{(1 - re^{ix})(1 - re^{-ix})} + \frac{1}{2}\right\}$$

$$= \mathrm{Re}\left\{\frac{re^{ix} - r^2}{1 - r(e^{ix} + e^{-ix}) + r^2} + \frac{1}{2}\right\}$$

$$= \frac{2(r\cos x - r^2)}{2(1 - 2r\cos x + r^2)} + \frac{1 - 2r\cos x + r^2}{2(1 - 2r\cos x + r^2)}$$

$$= \frac{1 - r^2}{2(1 - 2r\cos x + r^2)}.$$

Example 2

$$\tfrac{1}{2} + \cos x + \cos 2x + \cos 3x + \cdots + \cos nx = \frac{\sin(n + \tfrac{1}{2})x}{\sin \tfrac{1}{2}x}.$$

Proof

$$1 + e^{ix} + e^{2ix} + \cdots + e^{inx} = \frac{e^{i(n+1)x} - 1}{e^{ix} - 1}$$

$$= \frac{\exp\{i(n + \tfrac{1}{2})x\} - \exp(-\tfrac{1}{2}ix)}{\exp(\tfrac{1}{2}ix) - \exp(-\tfrac{1}{2}ix)}$$

$$= \frac{\{\cos(n + \tfrac{1}{2})x - \cos \tfrac{1}{2}x\} + i\{\sin(n + \tfrac{1}{2})x + \sin \tfrac{1}{2}x\}}{2i \sin \tfrac{1}{2}x}$$

and taking real parts of each side we obtain the stated result.

Example 3

Show that $\tfrac{1}{2}(\pi - x) = \sum_{n=1}^\infty \dfrac{\sin nx}{n}$ (for $0 < x < 2\pi$).

Proof. Integrating the result in Example 2 between the limits π and x:

$$\frac{1}{2}\int_{\pi}^{x} d\theta + \sum_{1}^{N}\int_{\pi}^{x}\cos n\theta \, d\theta = \int_{\pi}^{x}\frac{\sin(N+\frac{1}{2})\theta}{\sin\frac{1}{2}\theta}\,d\theta$$

$$\tfrac{1}{2}(\pi - x) + \sum_{1}^{N}\frac{\sin nx}{n} = \frac{1}{N+\frac{1}{2}}\left\{\left[-\frac{\cos(N+\frac{1}{2})\theta}{\sin\frac{1}{2}\theta}\right]_{\pi}^{x}\right.$$
$$\left. - \int_{\pi}^{x}\frac{\cos(N+\frac{1}{2})\theta\cos\frac{1}{2}\theta}{2\sin^{2}\frac{1}{2}\theta}\,d\theta\right\}.$$

Observe that for fixed x with $0 < x < 2\pi$, we have $0 < \frac{1}{2}x < \pi$ and hence in $[\pi, x]$ or $[x, \pi]$ the function $\sin^{2}\frac{1}{2}\theta$ is bounded from below by $\sin^{2}\frac{1}{2}x$. Thus the integrand is bounded and so letting $N \to \infty$ we obtain the required result.

Since the sum below is periodic with period 2π the graph of the function defined by

$$f(x) = \sum_{n=1}^{\infty}\frac{\sin nx}{n}$$

is as shown in Figure 22.1.

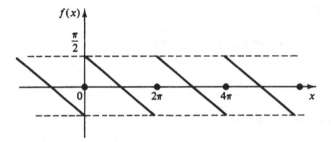

Fig. 22.1

Notice that f is discontinuous at the points $x = 0, \pm 2\pi, \pm 4\pi$, $\pm 6\pi,\ldots$ where its value is obviously zero and equals the average of the limiting value on the left of x and of the limiting value to the right of x, i.e.

$$\tfrac{1}{2}\{f(x-) + f(x+)\}.$$

We shall soon see that this is 'typical' behaviour for discontinuities.

22.5 Calculation of Fourier expansions

The key to all calculations is the following theorem.

Theorem
Suppose that $f(x)$ is integrable in $[-\pi, \pi]$ and that

$$f(x) = \tfrac{1}{2}a_0 + \sum_1^\infty a_n \cos nx + b_n \sin nx$$

the convergence being assumed uniform. Then

$$a_n = \frac{1}{\pi} \int_{-\pi}^\pi f(x) \cos nx \, dx \quad b_n = \frac{1}{\pi} \int_{-\pi}^\pi f(x) \sin nx \, dx.$$

Warning The theorem does not say that an integrable function can be expanded into a Fourier series. It claims only: *assuming that it can be expanded*, the coefficients are as stated. We shall have more on this question when we come to state **Dirichlet's theorem.**

Definition
*Observe that the formulas above are meaningful whenever f is integrable. We therefore define the **Fourier series of f** to be the formal expression*

$$\tfrac{1}{2}a_0 + \sum_1^\infty a_n \cos nx + b_n \sin nx,$$

where the coefficients are given by the formulas in the theorem. This leaves open the two question which we delay answering:
1. *Does the Fourier series converge?*
2. *What does the Fourier series converge to?*

Proof. We first note that

$$\int_{-\pi}^\pi \cos rx \cos nx \, dx = \begin{cases} 0 & \text{if} \quad r \neq n, \\ \pi & \text{if} \quad r = n. \end{cases}$$

Indeed

$$\cos rx \cos nx = \tfrac{1}{2}\{\cos(r+n)x + \cos(r-n)x\},$$

so that for $r \neq n$ we have

$$\int_{-\pi}^\pi \cos rx \cos nx \, dx = \frac{1}{2}\left[\frac{1}{r+n}\sin(r+n)x + \frac{1}{r-n}\sin(r-n)x\right]_{-\pi}^\pi = 0$$

whereas for $r = n$ we have

$$\int_{-\pi}^{\pi} \cos^2 nx \, dx = \frac{1}{2} \int_{-\pi}^{\pi} (1 + \cos 2nx) \, dx = \frac{1}{2} \left[x + \frac{1}{2n} \sin 2nx \right]_{-\pi}^{\pi}$$

$$= \pi \quad \text{(for } n \geq 1\text{)}.$$

Similarly it may be shown that

$$\int_{-\pi}^{\pi} \sin rx \sin nx = \begin{cases} 0 & \text{if } r \neq n, \\ \pi & \text{if } r = n \geq 1, \end{cases}$$

and

$$\int_{-\pi}^{\pi} \sin rx \cos nx = 0.$$

The results we have just derived may be referred to as the *orthogonality relations* for the trigonometric functions. They correspond to the inner product on the vector space of functions (see Part I, Chapter 2) given by the formula

$$\langle f, g \rangle = \int f(x) g(x) \, dx.$$

Now if $n \geq 1$ is given, by the definition of uniform convergence we have for any $\varepsilon > 0$ that for all m large enough (say for $m \geq N_\varepsilon$) and for any x in $[-\pi, \pi]$:

$$\left| f(x) - \tfrac{1}{2} a_0 - \sum_1^m a_r \cos rx + b_r \sin rx \right| < \varepsilon.$$

Hence, also

$$\left| f(x) \cos nx - \tfrac{1}{2} a_0 \cos nx - \sum_1^m a_r \cos rx \cos nx + b_r \sin rx \cos nx \right| < \varepsilon,$$

or, on integrating between $-\pi$ and π, and using the orthogonality relations

$$\left| \int_{-\pi}^{\pi} f(x) \cos nx \, dx - a_n \pi \right| < 2\pi\varepsilon.$$

Since ε was arbitrary the required result follows. A similar argument proves the formula for b_n. Finally note that for $n = 0$ we may deduce

$$\left| \int_{-\pi}^{\pi} f(x) \, dx - a_0 \pi \right| < 2\pi\varepsilon,$$

so

$$a_0 = \frac{1}{\pi} \int_{-\pi}^{\pi} f(x)\,dx$$

and the latter formula agrees with the general case when $n = 0$. This last calculation explains the convention of writing the constant as $\frac{1}{2}a_0$. Note also that $\frac{1}{2}a_0$ is in fact the average value of f over $[-\pi, \pi]$.

22.6 Example

Calculate the Fourier series of f when

$$f(x) = |x| \quad \text{for} \quad -\pi \leqslant x \leqslant \pi$$

and f is defined periodically elsewhere (see Figure 22.2).

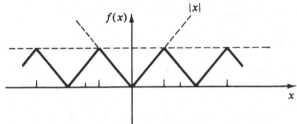

Fig. 22.2

We have

$$a_0 = \frac{1}{\pi} \int_{-\pi}^{\pi} |x|\,dx = \frac{2}{\pi} \int_0^{\pi} x\,dx = \pi$$

and since $|-x| = |x|$ and $\cos(-nx) = \cos nx$, we also have

$$a_n = \frac{1}{\pi} \int_{-\pi}^{\pi} |x| \cos nx\,dx = \frac{2}{\pi} \int_0^{\pi} x \cos nx\,dx$$

$$= \frac{2}{\pi}\left[x\frac{\sin nx}{n} \right]_0^{\pi} - \frac{2}{\pi} \int_0^{\pi} \frac{\sin nx}{n}\,dx$$

$$= \frac{2}{\pi}\left[\frac{\cos nx}{n^2} \right]_0^{\pi} = \begin{cases} 0 & \text{if } n \text{ is even,} \\ -4/(\pi n^2) & \text{if } n \text{ is odd.} \end{cases}$$

On the other hand, since $|-x| = |x|$ and $\sin(-nx) = -\sin nx$, we have

$$b_n = \frac{1}{\pi} \int_{-\pi}^{\pi} |x| \sin nx\,dx = \frac{1}{\pi} \int_0^{\pi} x \sin nx\,dx - \frac{1}{\pi} \int_{-\pi}^{0} x \sin nx\,dx = 0.$$

Thus the Fourier series for the given function is

$$\frac{\pi}{2} - \frac{4}{\pi} \sum_0^\infty \frac{1}{(2k+1)^2} \cos(2k+1)x$$

22.7   Dirichlet's theorem

This is the assertion that if f is integrable in $[-\pi, \pi]$ and is of *bounded variation* (see Section 17.17) then the Fourier series of f converges at each point x to the value

$$\tfrac{1}{2}\{f(x+) + f(x-)\},$$

where $f(x+)$ denotes the limit of $f(x+\delta)$ as δ tends to zero through positive values (this limit is known to exist for functions of bounded variation). The definition of $f(x-)$ is analogous. We recall that the variation of the function f on an interval $[a,b]$ is the number

$$V_f[a,b] = \sup_P \sum |f(x_i) - f(x_{i-1})|,$$

where the supremum is taken over all partitions $P = \{x_0, x_1, \ldots, x_n\}$ of $[a,b]$. The function f is of bounded variation on $[a,b]$ if the variation defined above is finite.

For a proof of the theorem see A. Zygmund, *Trigonometrical Series*.

22.8   Periodic solutions of differential equations

We may now solve the equation

$$D^2 y + \omega^2 y = f(x)$$

for functions f satisfying Dirichlet's theorem. First expand $f(x)$ as a Fourier series:

$$f(x) = \tfrac{1}{2} A_0 + \sum_1^\infty [A_n \cos nx + B_n \sin nx].$$

Next expand $y(x)$ as a Fourier series

$$y(x) = \tfrac{1}{2} a_0 + \sum_1^\infty [a_n \cos nx + a_n \sin nx].$$

In order to proceed we have to assume that the series for y converges uniformly in $[-\pi, \pi]$. This justifies term by term differentiation and we have

$$Dy(x) = \sum na_n(-\sin nx) + nb_n(\cos nx),$$
$$D^2y(x) = \sum -n^2 a_n \cos nx - n^2 b_n \sin nx,$$

provided both series converge uniformly. Finally we substitute into the differential equation and compare coefficients at like trigonometric functions. We thus obtain

$$\frac{\omega^2 a_0}{2} = \frac{A_0}{2} \quad \text{i.e.} \quad a_0 = \frac{A_0}{\omega^2},$$

$$-n^2 a_n + \omega^2 a_n = A_n \quad \text{i.e.} \quad a_n = \frac{A_n}{\omega^2 - n^2}$$

$$-n^2 b_n + \omega^2 b_n = B_n \quad \text{i.e.} \quad b_n = \frac{B_n}{\omega^2 - n^2}.$$

Thus

$$y(x) = \frac{A_0}{2\omega^2} + \sum_1^\infty \frac{A_n \cos nx + B_n \sin nx}{\omega^2 - n^2}.$$

As regards the convergence assumptions, it can be shown that the series for y, Dy and D^2y converge uniformly provided the series for f converges uniformly (see Exercise 22.10 Question 6).

Note that the above analysis breaks down if $\omega^2 = n^2$ for some integer n. However, in this case we can write

$$g(x) = f(x) - A_\omega \cos \omega x + B_\omega \sin \omega x$$

and then we can solve separately

$$D^2y + \omega^2 y = g(x)$$

and

$$D^2y + \omega^2 y = A_\omega \cos \omega x + B_\omega \sin \omega x.$$

The latter equation gives rise to the 'non-periodic' solutions $(Ax + B)\cos \omega x$ and $(A'x + B')\sin \omega x$. The sum of the solutions of the two equations gives a solution to the original equation. Also note that

$$D^2y + \omega^2 y = f(x)$$

has the solution (obtained by the use of an integrating factor):

$$y(x) = \exp(-i\omega x) \int_0^x \exp(2i\omega t)\left(\int_0^t e^{i\omega s} f(s)ds\right)dt.$$

22.9   The small parameter method

It is possible to extend the analysis of the previous sections to non-linear differential equations involving a *small* non-linear term. For example, consider

$$D^2 y + 2y = \sin x + \mu y^2,$$

where μ is 'small'. We regard μ for a moment as a further *variable* in the problem and seek a solution of the form

$$y(x, \mu) = y_0(x) + \mu y_1(x) + \mu^2 y_2(x) + \cdots,$$

so that we are using a power series in μ. We then consider an approximation to the original problem in which we elect to ignore terms of order μ^2 or higher. The non-linear term y^2 in the differential equation is approximated for each x about $y_0(x)$ by applying Taylor's Theorem to $F(y) = y^2$. Thus

$$F(y) = y_0^2 + 2y_0(y - y_0) + \text{Remainder}.$$

The differential equation then reads

$$D^2 y_0 + \mu D^2 y_1 + \mu^2 D^2 y_2 + \cdots + 2y_0 + 2\mu y_1 + 2\mu^2 y_2 + \cdots$$
$$= \sin x + \mu y_0^2 + \{\mu 2 y_0 (\mu y_1 + \mu^2 y_2 + \cdots)\}.$$

To solve this we consider the system of equations

$$\left.\begin{array}{l} D^2 y_0 + 2y_0 = \sin x \\ D^2 y_1 + 2y_1 = y_0^2 \\ \quad\text{etc.} \end{array}\right\}.$$

Solving this for y_0 gives $y_0 = \sin x$. The equation for y_1 then reads

$$D^2 y_1 + 2y_1 = \sin^2 x = \frac{1 - \cos 2x}{2}.$$

Solving for y_1 we obtain

$$y_1 = \tfrac{1}{4} + \tfrac{1}{4}\cos 2x.$$

Thus if we ignore terms of order μ^2 and higher we obtain the *approximate* solution

$$y = \sin x + \frac{\mu}{4}(1 + \cos 2x) \quad \text{(correct to order } \mu\text{)}.$$

We do not take the matter further, wishing only to illustrate the usefulness of the series technique.

22.10   Exercises

1. Solve $D^2y - xy = 0$ by series substitution and evaluate the radius of convergence of the series.

2. Solve $x^2D^2y + xDy + (m^2x^2 - n^2)y = 0$. [Hint: Make the change of variable $t = mx$.]

3. Solve by series technique the following equation near $x = 0$

$$(1 - x^2)D^2y - 2xDy + k(k + 1)y = 0.$$

What is the radius of convergence of your series? What happens if k is an integer?

4. Find the Fourier series for the function f defined to be $f(x) = x$ in $[-\pi, \pi]$ and defined elsewhere by periodicity.

5. Show that if f is periodic with period 2π then for any l:

$$a_n = \frac{1}{\pi}\int_{l-\pi}^{l+\pi} f(x)\cos nx\, dx \quad b_n = \frac{1}{\pi}\int_{l-\pi}^{l+\pi} f(x)\sin nx\, dx$$

6. Let $\{u_n\}$ and $\{v_n\}$ be two sequences. Define

$$U_n = u_0 + u_1 + u_2 + \cdots + u_n.$$

Show that

$$\sum_{r=n}^{m} u_r v_r = \left\{\sum_{r=n}^{m-1} U_r(v_r - v_{r+1})\right\} - U_{n-1}v_n + U_m v_m.$$

Deduce the following.

(i)   If the convergence of

$$u_0(x) + u_1(x) + u_2(x) + \cdots$$

is uniform and $u_i(x)$ are all bounded, then $\sum v_i u_i(x)$ converges uniformly provided

$$\sum |v_i - v_{i-1}| < \infty.$$

[Hint: Assume v_i converges to v.]

(ii)   If the convergence of

$$\sum_1^\infty A_n \cos nx + B_n \sin nx$$

is uniform, then so is that of the following series:

$$\sum_1^\infty \frac{A_n \cos nx + B_n \sin nx}{\omega^2 - n^2},$$

$$\sum_1^\infty \frac{n(-A_n \sin nx + B_n \cos nx)}{\omega^2 - n^2},$$

$$\sum_1^\infty \frac{n^2(A_n \cos nx + B_n \sin nx)}{\omega^2 - n^2}.$$

7. If f is continuous in $[-\pi, \pi]$ show that if a_n, b_n are the Fourier coefficients of f then

$$\frac{1}{\pi} \int_{-\pi}^{\pi} \{f(x)\}^2 dx \geq \tfrac{1}{2} a_0^2 + \sum_{1}^{\infty} a_n^2 + b_n^2.$$

[Hint: Consider $\int (f - g)^2 dx$ over $[-\pi, \pi]$ with $g = $ sum to n terms of the Fourier series for f.]

8. Solve $D^2 y + 3y = \cos x + \mu y^2$ by the small parameter method.

9. Find periodic solutions of the equation

$$D^2 y + 2Dy + y = \sum_{1}^{\infty} \frac{\sin nt}{n^4}.$$

10. If $y(x) = u(x)$ is one solution of the general equation:

$$a_2(x)D^2 y + a_1(x)Dy + a_0(x)y = 0,$$

use the substitution below to reduce the equation to first order.

$$y(x) = u(x) \int z(x)dx.$$

Hence,

(i)   solve the following, noting that $u(x) = x$ is a solution

$x^3 D^2 y - xDy + y = 0,$

(ii) find a second solution to Bessel's equation.

22.11   Further exercises

1. Form the Fourier series for the periodic function $f(x)$ with period 2π defined by

$$f(x) = x \sin x \quad -\pi \leqslant x \leqslant \pi.$$

2. Prove by forming a Fourier series expansion that

$$\sum_{n=1}^{\infty} \frac{\sin^2 nx}{n^2} = \frac{1}{2}(\pi|x| - x^2).$$

3. Prove that for all x

$$|\sin x| = \frac{4}{\pi} \sum_{n=1}^{\infty} \frac{1 - \cos 2nx}{4n^2 - 1}.$$

4. Show that

$$f(x) \equiv \cos x - \frac{\cos 5x}{5} + \frac{\cos 7x}{7} - \frac{\cos 11x}{11} + \cdots$$

$$= \begin{cases} \dfrac{\pi}{2\sqrt{3}} & 0 < x < \pi/3 \\[2mm] 0 & \pi/3 < x < \dfrac{2\pi}{3} \\[2mm] -\dfrac{\pi}{2\sqrt{3}} & \dfrac{2\pi}{3} < x < \pi \end{cases}$$

What other values does this series take?
[Hint: Find the Fourier series expansion of $f(x)$ by reference to the values on the right hand side of the brace.]
5. Let $f(x)$ be the function defined for all $x \in \mathbb{R}$ with $f(x + 2\pi) = f(x)$ and $f(x) = \cos \lambda x$ for $-\pi \leqslant x \leqslant \pi$. Find the Fourier series of f (assuming that λ is not an integer) and deduce that

$$\pi\lambda \cot \pi\lambda = 1 + 2\lambda^2 \sum_{n=1}^{\infty} \frac{1}{\lambda^2 - n^2}.$$

Show further, by integration from ε to θ that for $0 < \theta < \pi$

$$\frac{\sin \pi\theta}{\pi\theta} = \prod_{n=1}^{\infty}\left[1 - \frac{\theta^2}{n^2}\right],$$

where the infinite product denotes

$$\lim_{N \to \infty} \prod_{n=1}^{N}\left[1 - \frac{\theta^2}{n^2}\right].$$

Comment briefly on the justifiability of the manipulations involving integrals.
6. Let

$$I_n = \int \frac{\sin nx}{\sin x}\,dx.$$

Show that

$$I_{n+2} = I_n + 2\int \cos(n+1)x\,dx.$$

Use this to obtain the Fourier coefficients a_n, b_n (for $n \geqslant 1$) of the function, with period 2π, defined by

$$f(x) = \log\left(\cot \frac{x}{2}\right), \quad 0 < x < \pi,$$

$$f(0) = 0,$$

$$f(x) = f(-x), \quad -\pi < x < 0.$$

Explain why the Cauchy principal value of the integral for a_0 is zero. Using this value for a_0, find a periodic solution of the differential equation

$$\frac{d^2y}{dx^2}+\frac{dy}{dx}+y=f(x),$$

valid in $0<x<\pi$.

Justify the assumption of $a_0=0$ for this solution procedure, by considering the function $f(x)$ modified so that its value is zero in $[0,\varepsilon)$ and equal to its original value in $(\varepsilon,\pi]$.

7. Let

$$I_n=\int\frac{(1-\cos x)\sin nx}{\sin x}\,dx.$$

Show that

$$I_{n+1}=-I_n+\int\{\cos nx-\cos(n+1)x\}\,dx.$$

Use this recurrence formula to obtain the Fourier coefficients of the even function $f(x)$ defined by

$$f(x)=\log\left(\cos\frac{x}{2}\right),\quad -\pi<x<\pi,$$

and defined elsewhere by periodicity. Hence find a periodic solution of the differential equation

$$\frac{d^2y}{dx^2}+\frac{dy}{dx}+2y=f(x).$$

8. Find by power series expansion the general solution to the equation

$$x\frac{d^2y}{dx^2}+(x+2)\frac{dy}{dx}+y=0$$

and express your answer in its simplest form using the exponential function. Explain why the same approach will not solve the equation

$$x^3\frac{d^2y}{dx^2}-x\frac{dy}{dx}+y=0.$$

Use the substitution $z=1/x$ to solve the equation.
[Hint: $dy/dx=-z^2\,dy/dz$.]

23

Calculus of variations

23.1 Path-finding problems

Here are three classic problems whose solution requires us to extend differential calculus beyond ordinary variables:

(a) Brachistochrone problem

Given two points A, B in space, A higher than B, but not vertically above B, what shape of wire connecting A to B will have the property that a bead sliding smoothly along it under gravity gets from A to B in shortest time. (See Figure 23.1.)

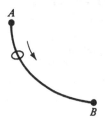

Fig. 23.1

(b) Geodesic problem

Given a surface in space, with equation $\phi(x, y, z) = 0$, and two points on it, A and B, find a path along the surface from A to B of shortest length (Figure 23.2).

(c) Isoperimetric problem

Among all plane curves of fixed length l find the one which encloses maximum area (Figure 23.3).

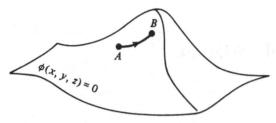

Fig. 23.2

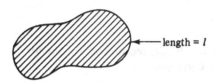

length = l

Fig. 23.3

The essential feature common to all three problems is that of finding some special curve Γ. In each case, too, a property $F(\Gamma)$ of curves is given: time of descent, length of curve, area enclosed by the curve; and we have to minimize/maximize $F(\Gamma)$ over all Γ. The variable of interest here is Γ and we need to know how F varies when Γ is varied.

It will be convenient to represent a curve Γ by an equation

$$x = x(t)$$

with

$$0 \leqslant t \leqslant T,$$

subject to $x(0) = A$

and

$$x(T) = B,$$

or perhaps by a vector equation, e.g.

$$x = x(t) = (x_1(t), x_2(t), x_3(t)) \quad (0 \leqslant t \leqslant T)$$

depending on context.

The property F of the 'curve x' will, more often than not, take the form

$$F(x) = \int_0^T f(x(t), \dot{x}(t), t)\,dt,$$

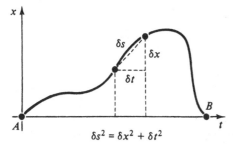

Fig. 23.4

where $\dot{x} \equiv dx/dt$. For example, curve length, in the case illustrated in Figure 23.4, is given by

$$\int_0^T \sqrt{1 + \dot{x}^2}\, dt$$

and area below the curve is

$$\int_0^T x(t)\, dt.$$

We have purposefully used the letter x to denote a function rather than a real number, since we shall be interested in varying x. Since F assigns to each function x a real number, it is itself a function. We call a real-valued function acting on functions, a *functional*.

23.2 Variation of a functional

Suppose F is a functional and the function $\xi(t)$ $(0 \leqslant t \leqslant T)$ maximizes F. This means that for any other curve $x(t)$

$$F(x) \leqslant F(\xi).$$

Just as in calculus, we can try to compare $F(x)$ with $F(\xi)$ when x is 'close to ξ'. We can think of a function

$$x(t) = \xi(t) + h(t),$$

where $h(t)$ is also a function, as arising from an 'increment' h added to ξ. See Figure 23.5. We call h a *variation* of ξ. If also we want

$$x(0) = \xi(0) = A,$$
$$x(T) = \xi(T) = B,$$

then we require that $h(0) = h(T) = 0$. Thus not all variations will do.

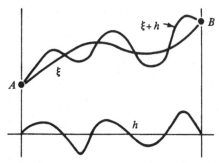

Fig. 23.5

Whenever we restrict our variations to fit in with the side conditions of a problem we will refer to them as *admissible* variations.

But now observe that in our problem sh is admissible for any real number s.

Thus the function

$$\psi(s) = F(\xi + sh)$$

has a maximum when $s = 0$. Hence, if ψ is differentiable in some interval round $s = 0$ we must have

$$\psi'(0) = 0,$$

i.e.

$$\frac{d}{ds} f(\xi + sh)\big|_{s=0} = 0.$$

Since this formula looks like a derivative in direction h we define

$$D_h F(x) = \lim_{s \to 0} \frac{F(x + sh) - F(x)}{s} = \frac{d}{ds} F(x + sh)\big|_{s=0}.$$

This is sometimes referred to as the weak derivative of F in direction h and sometimes as the *variation of F relative to h*.

Examples

(i) $F(x) = \displaystyle\int_0^T x(t)\,dt.$

Note for future purposes that

$$F(x + h) - F(x) = \int_0^T h(t)\,dt.$$

Also

$$D_h F(x) = \frac{d}{ds} \int_0^T (x(t) + sh(t))dt|_{s=0}$$

$$= \int_0^T \frac{d}{ds}(x(t) + sh(t))dt|_{s=0}$$

$$= \int_0^T h(t)dt.$$

(ii) $F(x) = \int_0^T \{x(t)\}^2 dt.$

Thus

$$F(x + sh) = \int_0^T \{x(t) + sh(t)\}^2 dt,$$

$$\frac{d}{ds}F(x + sh) = \int_0^T \frac{d}{ds}(x + sh)^2 dt = \int_0^T (2hx + 2sh^2)dt$$

so, setting $s = 0$

$$D_h F(x) = \int_0^T 2hx \, dt.$$

Note that

$$F(x + h) - F(x) = \int_0^T \{2hx + h^2 x\} dt.$$

(iii) $\qquad\qquad F(x) = \int_0^T \sqrt{1 + \dot{x}^2} \, dt.$

$$F(x + sh) = \int_0^T \sqrt{1 + (x + sh)^{\cdot 2}} \, dt = \int_0^T \sqrt{1 + (\dot{x} + s\dot{h})^2} \, dt.$$

$$\frac{d}{ds}F(x + sh) = \int_0^T \tfrac{1}{2}\{1 + (\dot{x} + s\dot{h})^2\}^{-1/2} \cdot 2(\dot{x} + s\dot{h})\dot{h} \, dt.$$

Thus

$$D_h F(x) = \int_0^T \frac{\dot{x}\dot{h}}{\{1 + \dot{x}^2\}^{1/2}} \, dt.$$

Notice that in all our examples $D_h F(x)$, as a function of h (with x

fixed), is *linear*. Particularly interesting is example (iii) which we follow up:

$$D_{\alpha h_1 + \beta h_2} F(x) = \int_0^T \frac{\dot{x}(\alpha h_1 + \beta h_2)}{\sqrt{1 + \dot{x}^2}} dt$$

$$= \int_0^T \frac{\dot{x}\alpha\dot{h}_1}{\sqrt{1 + \dot{x}^2}} + \int_0^T \frac{\dot{x}\beta\dot{h}_2}{\sqrt{1 + \dot{x}^2}}$$

$$= \alpha D_{h_1} F(x) + \beta D_{h_2} F(x).$$

It is often, though by no means always, also the case that $D_h F(x)$ is the *linear part* of $F(x + h) - F(x)$ in the sense that

$$F(x + h) - F(x) - D_h F(x)$$

is of higher order in h (compare Taylor's theorem).

23.3   The Euler–Lagrange equation

We now obtain a very useful equation that is necessarily satisfied by the function $\xi(t)$ which maximises/minimizes

$$F(x) \equiv \int_0^T f(x(t), \dot{x}(t), t) dt$$

subject to $x(0) = a$, $x(T) = b$.

We already know that the variation $D_h F(\xi)$ vanishes for all admissible h. We derive from this fact a differential equation to be satisfied by ξ. For this purpose we need to assume that $f(x, y, z)$ as a function of the *real* variables x, y, z has continuous partial derivatives $f_x(x, y, z)$ and $f_y(x, y, z)$. We have:

$$\frac{d}{ds} F(x + sh) = \int_0^T \left\{ f_x(x + sh, \dot{x} + s\dot{h}, t) \frac{d}{ds}(sh) \right.$$

$$\left. + f_y(x + sh, \dot{x} + s\dot{h}, t) \frac{d}{ds}(s\dot{h}) \right\} dt$$

(applying the chain rule!) and the latter equals

$$= \int_0^T \left\{ f_x(x + sh, \dot{x} + s\dot{h}, t)h + f_y(x + sh, \dot{x} + s\dot{h}, t)\dot{h} \right\} dt.$$

Thus writing this out in full

$$D_h F(x) = \int_0^T \left\{ f_x(x(t), \dot{x}(t), t)h(t) + f_y(x(t), \dot{x}(t), t)\dot{h}(t) \right\} dt.$$

Hence if $\xi(t)$ maximizes/minimizes F we have

$$D_h F(\xi) = 0,$$

or

$$\int_0^T \{f_x(\xi, \xi, t)h + f_y(\xi, \xi, t)\dot{h}\} \, dt = 0,$$

for all admissible h (i.e. for all h subject to $h(0) = h(T) = 0$).
Now we may integrate the last equation by parts to obtain

$$0 = \left[h(t) \int_0^t f_x(\xi(t), \xi(t), t) \, dt \right]_0^T + \int_0^T \left\{ f_y - \int_0^t f_x \right\} \dot{h} \, dt.$$

Thus for all h:

$$\int_0^T \left\{ f_y - \int_0^t f_x \right\} \dot{h} \, dt = 0.$$

We claim that the curly brackets are constant over $[0, T]$.

Lemma
Suppose g is continuous on $[0, T]$ and that

$$\int_0^T g(t)\dot{h}(t) \, dt = 0$$

for all (continous) functions h with $h(0) = h(T) = 0$ such that $\dot{h}(t)$ is continuous; then g is constant.

Remark
This result is motivated by the observation that if g had been known to have a continuous derivative $\dot{g}$, then integrating by parts:

$$0 = \int_0^T g\dot{h} \, dt = [gh]_0^T - \int_0^T \dot{g}h \, dt,$$

so

$$\int_0^T \dot{g}h \, dt = 0 \quad \text{for all admissable } h, \text{ hence } \dot{g} = 0.$$

(Observe that $\dot{g}(t) > 0$ for some t, implies $\dot{g} > 0$ in a small interval round t; now take h zero outside this interval and positive inside the interval; for such an h the integral is positive.) In the present context however, all we know is that g is continuous.

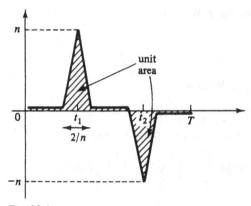

Fig. 23.6

Proof of the lemma. Let $0 < t_1 < t_2 < T$. We wish to prove $g(t_1) = g(t_2)$. For each n define the function $p_n(t)$ as illustrated in Figure 23.6. Take

$$h(t) = \int_0^t p_n(u)\,du.$$

Thus $h(0) = 0$ and $h(T) = 0$ (counterbalancing areas). By continuity of g at t_1 and t_2, g is approximately equal to $g(t_1)$ in the interval round t_1, and is equal approximately to $g(t_2)$ in the small interval round t_2. Hence

$$0 = \int_0^T g(t)\dot{h}(t)\,dt = \int_0^T g(t)p_n(t)\,dt$$

$$\approx g(t_1)\int_{t_1 - 1/n}^{t_1 + 1/n} p_n(t)\,dt + g(t_2)\int_{t_2 - 1/n}^{t_2 + 1/n} p_n(t)\,dt$$

$$= g(t_1)\cdot\{1\} + g(t_2)\cdot\{-1\}.$$

The error committed in the second line may be made as small as we please provided n is large enough. Hence, in the limit as n tends to infinity,

$$g(t_1) - g(t_2) = 0,$$

as required. Thus g is constant in value throughout $(0, T)$ and hence also throughout $[0, T]$ (by continuity).

We conclude that in our present context

$$f_y(\xi(t), \dot{\xi}(t), t) = \int_0^t f_x(\xi(t), \dot{\xi}(t), t)\,dt + \text{const.}$$

But the right-hand side is differentiable, consequently:

$$\frac{d}{dt}\{f_y(\xi(t), \dot{\xi}(t), t)\} = f_x(\xi(t), \dot{\xi}(t), t).$$

This is known as the Euler–Lagrange equation. It is sometimes written in the easily memorized forms

$$\frac{d}{dt}(f_{\dot{x}}) = f_x, \quad \text{or} \quad \frac{d}{dt}\left(\frac{df}{d\dot{x}}\right) = \frac{df}{dx}.$$

Before we attempt to solve some problems let us observe a special form of the Euler–Lagrange equation. Suppose we are maximizing/minimizing

$$F(x) \equiv \int_0^T f(x(t), \dot{x}(t))dt$$

with

$$x(0) = x_0, \quad x(T) = x_1.$$

Here the integrand f does not explicitly depend on t, that is f is a function of only *two* real variables $f(x, y)$ and there is *no third variable*. On the assumption that the optimal curve $\xi(t)$ possesses a second derivative $\xi''(t)$ the Euler–Lagrange equation is equivalent to

$$f(\xi, \xi') - \xi' f_y(\xi, \xi') = \text{const.}$$

Indeed by the chain rule

$$\frac{d}{dt}\{f - \xi' f_y\} = f_x(\xi, \xi')\xi' + f_y(\xi, \xi')\xi''$$

$$- \xi'' f_y - \xi' \frac{d}{dt} f_y$$

$$= \xi'\{f_x - \frac{d}{dt} f_y\} = 0.$$

Integrating this equation leads to the desired result.

23.4 Example. The brachistochrone problem

Choose axes through A as origin, measuring x downwards vertically and s horizontally, so that $B = (1, 1)$ (cf. Figure 23.7). The equation of motion

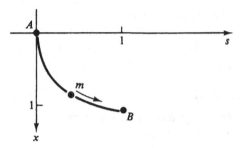

Fig. 23.7

(in the absence of friction) of a bead of mass m along the smooth wire is:

kinetic energy gained = potential energy lost

i.e.

$$\tfrac{1}{2}m(\dot{x}^2 + \dot{s}^2) = mgx,$$

or

$$\frac{ds}{dt}\left(\left(\frac{\dot{x}}{\dot{s}}\right)^{1/2} + 1\right) = \sqrt{(2gx)}.$$

Hence

$$\text{time taken} = T = \int_0^T dt = \int_0^1 k\left(\frac{1 + (\dot{x}/\dot{s})^2}{x}\right)^{1/2} ds$$

(where $k = 1/\sqrt{(2g)}$). But $\dot{x}/\dot{s} = dx/ds$, hence we have to find the curve $x = x(s)$ which minimizes

$$k \cdot \int_0^1 f\left(x(s), \frac{dx}{ds}\right) ds$$

with

$$x(0) = 0 \quad \text{and} \quad x(1) = 1,$$

where

$$f(x, y) = ((1 + y^2)/x)^{1/2}.$$

Hence the Euler–Lagrange equation in integrated form reads (with s for t)

$$\left(\frac{1 + \left(\dfrac{dx}{ds}\right)^2}{x}\right)^{1/2} - \frac{dx}{ds} \cdot \frac{\dfrac{dx}{ds}}{\sqrt{x\left(1 + \left(\dfrac{dx}{ds}\right)^2\right)^{1/2}}} = c = \text{const.},$$

since

$$f_y = \frac{1}{\sqrt{x}}\frac{1}{2}(1 + y^2)^{-1/2}\cdot 2y$$

$$= \frac{y}{\sqrt{\{x(1 + y^2)\}}}.$$

Thus

$$\left\{1 + \left(\frac{dx}{ds}\right)^2\right\} - \left(\frac{dx}{ds}\right)^2 = c\sqrt{x}\left(1 + \left(\frac{dx}{ds}\right)^2\right)^{1/2},$$

hence

$$1 = c^2 x\left\{1 + \left(\frac{dx}{ds}\right)^2\right\},$$

or

$$\left(\frac{dx}{ds}\right)^2 = \frac{1}{c^2 x} - 1 = \frac{1 - c^2 x}{c^2 x},$$

so

$$\int ds = \int\left(\frac{c^2 x}{1 - c^2 x}\right)^{1/2} dx.$$

Put $x = (1/c^2)\sin^2\theta$.
Integrating we obtain:

$$s - B = \int \frac{2}{c^2}\frac{\sin\theta}{\cos\theta}\sin\theta\cos\theta\,d\theta$$

$$= \frac{1}{c^2}\int(1 - \cos 2\theta)d\theta = A(\theta - \tfrac{1}{2}\sin 2\theta),$$

where $A = (1/c^2)$ and B is a constant.
   But we have the implications

$$\theta = 0 \Rightarrow s = 0.$$

So $B = 0$.
   Writing $2\theta = \phi$ and $a = A/2$ we obtain the parametric representation:

$$\left.\begin{aligned} s &= a(\phi - \sin\phi) \\ x &= a(1 - \cos\phi) \end{aligned}\right\}$$

of a curve known as a *cycloid*. Note the geometric interpretation: x is traced by a fixed point P on the rim of a wheel rolling along the s-axis. See Figure 23.8.

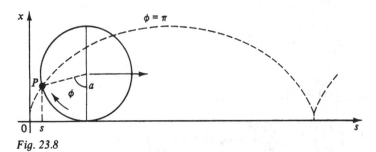

Fig. 23.8

23.5   Extension to vector-valued functions

We have considered so far only problems involving a curve

$$x = x(t) \quad 0 \leqslant t \leqslant T.$$

Problems involving curves in space will involve looking at functions

$$x(t) = (x_1(t), x_2(t), x_3(t)) \quad 0 \leqslant t \leqslant T,$$

or more generally

$$x(t) = (x_1(t), \dots x_n(t)).$$

Typically, we then deal with

$$F(x) = \int_0^T f(x_1(t), \dots x_n(t), \dot{x}_1(t), \dots \dot{x}_n(t), t) dt$$

subject to

$$x_1(0) = x_1^0, \dots, x_n(0) = x_n^0$$
$$x_1(T) = x_1^1, \dots, x_n(T) = x_n^1.$$

If $\xi(t) = (\xi_1(t), \dots, \xi_n(t))$ solves the problem, then clearly the function $\xi_1(t)$ solves the problem of maximising/minimising:

$$\int_0^T f(x_1(t), \xi_2(t), \dots, \xi_n(t), \dot{x}_1(t), \xi_2(t) \dots \xi_n(t), t) dt$$

subject to

$$x_1(0) = x_1^0, \quad x_1(T) = x_1^1.$$

Hence a necessary condition for ξ to be optimal, is that for each $i = 1, 2, \dots n$ the Euler–Lagrange equation

$$\frac{d}{dt}(f_{\dot{x}_i}) = f_{x_i}$$

must be satisfied.

23.6 Conditional maximum or minimum

Suppose, as in the geodesic problem, that we require to find
$x(t) = (x_1(t), \ldots x_n(t))$ which maximises or minimises a functional

$$F(x) = \int_0^T f(x_1, \ldots x_n, \dot{x}_1, \ldots \dot{x}_n, t)dt$$

subject to a number of constraints of the form

$$\left. \begin{array}{l} \phi_1(x_1(t), \ldots x_n(t), t) = 0 \\ \phi_2(x_1(t), \ldots x_n(t), t) = 0 \\ \vdots \\ \phi_m(x_1(t), \ldots x_n(t), t) = 0 \end{array} \right\},$$

where $m < n$.

   To apply our previous technique we do the same as in *ordinary*
calculus. We introduce Lagrange multipliers. The only difference is that
here Lagrange multipliers, not surprisingly, become functions
$\lambda_1(t), \ldots, \lambda_m(t)$ and we then find a stationary point of

$$\int_0^T \left\{ f(x, \dot{x}, t) + \sum \lambda_i(t)\phi_i(x(t)) \right\} dt.$$

Thus writing

$$l(x, \dot{x}, \lambda) \equiv f(x, \dot{x}, t) + \sum \lambda_i \phi_i(x)$$

we seek x, λ to satisfy

$$\frac{d}{dt}(l_{\dot{x}_i}(x, \dot{x}, \lambda)) = l_{x_i}(x, \dot{x}, \lambda)$$

together with

$$\phi_j(x(t), t) = 0 \quad (j = 1, \ldots m).$$

subject as usual to

$$x(0) = (x_1^0, \ldots x_n^0),$$
$$x(T) = (x_1^1, \ldots x_n^1).$$

Generally speaking the $m + n$ equations for $x_1, \ldots x_n, \lambda_1, \ldots, \lambda_m$ are
sufficient to determine these functions and the boundary conditions will
(if non-contradictory) determine the $2n$ arbitrary constants arising from
the differential equations. More precisely, it is necessary to assume that
the constraints $\phi_1 = 0, \ldots, \phi_m = 0$ are independent, that is that the

Jacobian:

$$\begin{bmatrix} \dfrac{\partial \phi_1}{\partial x_1}, \ldots, \dfrac{\partial \phi_1}{\partial x_n} \\ \ldots \\ \dfrac{\partial \phi_m}{\partial x_1}, \ldots, \dfrac{\partial \phi_m}{\partial x_n} \end{bmatrix}$$

should have rank n. We brush such niceties aside (leaving them to courses on functional analysis).

23.7   Examples

1   Geodesic problem

If $A = (x^0, y^0, z^0)$ and $B = (x^1, y^1, z^1)$ lie on the surface $\phi(x, y, z) = 0$, find the shortest path from A to B lying on the surface (cf. Figure 23.9). Thus we seek to minimise

$$\int_0^T \sqrt{\dot{x}^2 + \dot{y}^2 + \dot{z}^2}\, dt$$

subject to

$$\phi(x(t), y(t), z(t)) = 0 \quad \text{for } 0 \leqslant t \leqslant T,$$

with

$$(x(0), y(0), z(0)) = A \quad \text{and} \quad (x(T), y(T), z(T)) = B.$$

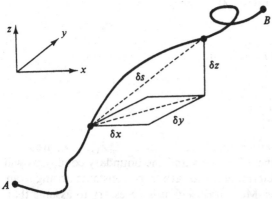

Fig. 23.9. $(\delta s)^2 = (\delta x)^2 + (\delta y)^2 + (\delta z)^2$

We form $l(x, \dot{x}, \lambda) = \sqrt{\dot{x}^2 + \dot{y}^2 + \dot{z}^2} + \lambda(t)\phi(x(t), y(t), z(t))$. The Euler Lagrange equations give

$$\left.\begin{aligned}
\frac{d}{dt}\left(\frac{\dot{x}(t)}{\sqrt{\{\dot{x}^2 + \dot{y}^2 + \dot{z}^2\}}}\right) &= \lambda(t)\phi_x(x(t), y(t), z(t)) \\
\frac{d}{dt}\left(\frac{\dot{y}(t)}{\sqrt{\{\dot{x}^2 + \dot{y}^2 + \dot{z}^2\}}}\right) &= \lambda(t)\phi_y(x(t), y(t), z(t)) \\
\frac{d}{dt}\left(\frac{\dot{z}(t)}{\sqrt{\{\dot{x}^2 + \dot{y}^2 + \dot{z}^2\}}}\right) &= \lambda(t)\phi_z(x(t), y(t), z(t))
\end{aligned}\right\} .$$

Sometimes, however, the problem simplifies down.

2   Geodesics on a cylinder

The two points A and B lie on $\Phi(x, y, z) \equiv x^2 + y^2 - R^2 = 0$. Introducing cylindrical polar co-ordinates (cf. Figure 23.10)

$$(x, y, z) = (r \cos \phi, r \sin \phi, z)$$

we have on the surface that $r = R$, hence

$$\dot{x}(t) = - R \sin \phi \cdot \dot{\phi}, \quad \dot{y} = + R \cos \phi \cdot \dot{\phi}.$$

Thus we are to minimize

$$\int \sqrt{\{R^2 \sin^2 \phi \cdot \dot{\phi}^2 + R \cos^2 \phi \cdot \dot{\phi}^2 + \dot{z}^2\}} \, dt = \int \sqrt{\{R^2 \dot{\phi}^2 + \dot{z}^2\}} \, dt.$$

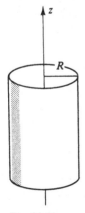

Fig. 23.10

The change of variables decreases the number of variables (the problem is now essentially two-dimensional); moreover ϕ and z are unconstrained. We are thus left with an unconstrained problem for which the Euler–Lagrange equations read

$$\frac{d}{dt}\left\{\frac{\dot{\phi}}{\sqrt{\{R^2\dot{\phi}^2 + \dot{z}^2\}}}\right\} = 0, \quad \frac{d}{dt}\left\{\frac{\dot{z}}{\sqrt{\{R^2\dot{\phi}^2 + \dot{x}^2\}}}\right\} = 0$$

hence for some constants A and B

$$\dot{\phi} = A\sqrt{\{R^2\dot{\phi}^2 + \dot{z}^2\}} \quad \text{and} \quad \dot{z} = B\sqrt{\{R^2\dot{\phi}^2 + \dot{z}^2\}},$$

so

$$\frac{dz}{d\phi} = \frac{\dot{z}}{\dot{\phi}} = \text{const} = a, \quad \text{say. Thus } x = a\phi + b \text{ a } \textit{spiral}.$$

3   Isoperimetric problem

We take the problem in the form (cf. Figure 23.11): find $x(t)$ for $-1 \leqslant t \leqslant 1$ with $x(-1) = x(1) = 0$ so as to maximise

$$\int_{-1}^{1} x(t)dt$$

subject to

$$\ell = \int_{-1}^{1} \sqrt{\{1 + \dot{x}^2\}}\, dt \qquad \text{(fixed arc-length)}.$$

To turn this problem into the kind considered above define

$$y(t) = \int_{-1}^{t} \sqrt{\{1 + \dot{x}^2\}}\, dt.$$

Thus

$$\dot{y}(t) = \sqrt{\{1 + \dot{x}^2\}}, \qquad y(-1) = 0 \quad \text{and} \quad y(1) = \ell.$$

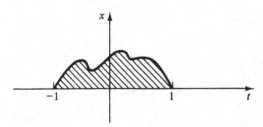

Fig. 23.11

We take $\phi(x, y, \dot{x}, \dot{y}) \equiv \dot{y} - \sqrt{\{1 + \dot{x}^2\}} = 0$ and introduce a Lagrange multiplier $\lambda(t)$.

Now we seek to maximise

$$\int_{-1}^{1} (x + \lambda(t)\{\dot{y} - \sqrt{[1 + \dot{x}^2]}\})dt.$$

The Euler–Lagrange equations are thus

$$\frac{d}{dt}\left\{\frac{-\lambda\dot{x}}{\sqrt{\{1 + \dot{x}^2\}}}\right\} = 1,$$

$$\frac{d}{dt}\{\lambda\} = 0.$$

Hence $\lambda(t)$ is in fact a constant. So we solve the first equation:

$$\frac{-\lambda\dot{x}}{\sqrt{(1 + \dot{x}^2)}} = t + c,$$

whence

$$\frac{\dot{x}^2}{1 + \dot{x}^2} = \frac{(t + c)^2}{\lambda^2}$$

and so

$$\dot{x}^2 = \frac{(t + c)^2}{\lambda^2 - (t + c)^2}.$$

Thus

$$\int dx = \int \frac{(t + c)\,dt}{\sqrt{\{\lambda^2 - (t + c)^2\}}}$$

and since λ is a constant

$$x(t) - a = \sqrt{\{\lambda^2 - (t + c)^2\}},$$

so

$$(x(t) - a)^2 + (t + c)^2 = \lambda^2,$$

that is the curve is part of a circle.

23.8   Exercises

1. Write down and solve the Euler–Lagrange equation corresponding to the problem of maximising/minimising

$$\int_0^T f(x(t), \dot{x}(t), t)dt$$

subject to boundary conditions when $f(x, y, z)$ is

(i) $4xz - y^2$,

(ii) $xy - 2y^2$,

(iii) $\dfrac{1}{x}\sqrt{\{1 - y^2\}}$,

(iv) $x^2 - 6xz$,

(v) $-y^2z^{-3}$.

2. Find the curve $x(t)$ with endpoints A, B so that the area of the surface of revolution (generated by rotating the curve round the t-axis) is minimised. See Figure 23.12.

3. At time $t = 0$ a man possesses £s. His total satisfaction over the time interval $[0, T]$ is assumed to be

$$\int_0^T e^{-\beta t}U(r(t))dt$$

where $r(t)$ is his rate of expenditure and $U(r) = \log(1 + r)$. Let $x(t)$ be his capital at time t thus

$$\dot{x}(t) = \alpha x(t) - r(t),$$

where the constant α is the interest rate. Find his optimal $x(t)$ if he seeks to maximise total satisfaction subject to

$$x(T) = 0$$

(no inheritors!).

4. A cable of *fixed* length l hangs between two supports in the shape of a curve $x(t)$ parametrised by t with $0 \leqslant t \leqslant T$. If the cable hangs so as to minimise the potential energy

$$V = mg \int_0^T x\sqrt{\{1 + \dot{x}^2\}}\,dt,$$

show that $x(t) = A \cosh((t + B)/A) + C$.

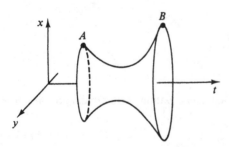

Fig. 23.12

5. Show by checking the equations in Section 23.1 that the line

$$x = 1 - t,$$
$$y = 1 + t,$$
$$z = t,$$

is a geodesic joining the points $(1, 1, 0)$ and $(0, 2, 1)$ on the surface $(x + z)(y - z) = 1$.

III SOLUTIONS TO SELECTED EXERCISES

Solutions to selected exercises on vector spaces

Exercises 1.9

6. We give two proofs.

Hard slog proof. The first step is to reduce the size of the $(n + 1) \times (n + 1)$ determinant by subtracting $(x_0 \times i$th row$)$ from the $i + 1$st, thus:

$$
\begin{vmatrix}
1 & 1 & 1 & \cdots & 1 \\
x_0 & x_1 & x_2 & & x_n \\
x_0^2 & x_1^2 & x_2^2 & & x_n^2 \\
\vdots & & & & \\
x_0^{n-1} & x_1^{n-1} & x_2^{n-1} & & x_n^{n-1} \\
x_0^n & x_1^n & x_2^n & & x_n^n
\end{vmatrix}
$$

$$
= \begin{vmatrix}
1 & 1 & 1 & \cdots & 1 \\
x_0 - x_0 & x_1 - x_0 & x_2 - x_0 & & x_n - x_0 \\
x_0^2 - x_0^2 & x_1^2 - x_0 x_1 & x_2^2 - x_0 x_2 & & x_n^2 - x_0 x_n \\
\vdots & & & & \\
x_0^{n-1} - x_0^{n-1} & x_1^{n-1} - x_0 x_1^{n-2} & x_2^{n-1} - x_0 x_2^{n-2} & \cdots & x_n^{n-1} - x_0 x_n^{n-2} \\
x_0^n - x_0^n & x_1^n - x_0 x_1^{n-1} & x_2^n - x_0 x_2^{n-1} & \cdots & x_n^n - x_0 x_n^{n-1}
\end{vmatrix}
$$

$$
= \begin{vmatrix}
1 & 1 & 1 & \cdots & 1 \\
0 & (x_1 - x_0) & (x_2 - x_0) & & (x_n - x_0) \\
0 & (x_1 - x_0)x_1 & (x_2 - x_0)x_2 & & (x_n - x_0)x_n \\
\vdots & & & & \\
0 & (x_1 - x_0)x_1^{n-2} & (x_2 - x_0)x_2^{n-2} & \cdots & (x_n - x_0)x_n^{n-2} \\
0 & (x_1 - x_0)x_1^{n-1} & (x_2 - x_0)x_2^{n-1} & \cdots & (x_n - x_0)x_n^{n-1}
\end{vmatrix}
$$

$$
= \begin{vmatrix}
1 & 1 & \cdots & 1 \\
\vdots & & & \\
x_1^{n-2} & x_2^{n-2} & & x_n^{n-2} \\
x_1^{n-1} & x_2^{n-1} & & x_n^{n-1}
\end{vmatrix} \cdot (x_1 - x_0)(x_2 - x_0) \cdots (x_n - x_0).
$$

At the end of this step we have arrived at a van Der Monde determinant of size $n \times n$. Repeating the argument several times we wind up with more factors and a van Der Monde determinant of size 2×2:

$$\begin{vmatrix} 1 & 1 \\ x_{n-1} & x_n \end{vmatrix} = (x_n - x_{n-1}).$$

The rest of the argument should now be clear.

Trick proof. Let us call the van Der Monde determinant with $x_0, x_1, \ldots, x_n$ as variables $D(x_0, x_1, \ldots, x_n)$. For each variable x_i this is a polynomial in x_i of degree n. For the moment think of x_0 as a variable and $x_1, \ldots, x_n$ as fixed numbers. Substituting $x_0 = x_1$ gives the determinant a value of 0 (since two rows become identical). The same is true for $x_0 = x_2$, etc. D must therefore factorise as, say, $K(x_1, \ldots, x_n) \cdot (x_1 - x_0) \cdot \cdots \cdot (x_n - x_0)$ where $K(x_1, \ldots, x_n)$ is a constant dependent on $x_1, \ldots, x_n$. Similar arguments are available with x_0 and x_1 interchanged. We conclude that

$$D(x_0, x_1, \ldots, x_n) = \text{const.} \times \prod_{i > j} (x_i - x_j).$$

Comparing the coefficient of the term $x_1^1 x_2^2 \cdots x_n^n$ on each side we see that the constant above is 1.

7. The Wronskian is

$$\begin{vmatrix} \exp(\alpha_1 x) & \cdots & \exp(\alpha_n x) \\ \alpha_1 \exp(\alpha_1 x) & \cdots & \alpha_n \exp(\alpha_n x) \\ \alpha_1^2 \exp(\alpha_1 x) & \cdots & \alpha_n^2 \exp(\alpha_n x) \\ \alpha_1^{n-1} \exp(\alpha_1 x) & \cdots & \alpha_n^{n-1} \exp(\alpha_n x) \end{vmatrix}$$

$$= \begin{vmatrix} 1 & 1 & \cdots & 1 \\ \alpha_1 & \alpha_2 & & \alpha_n \\ \vdots & & & \\ \alpha_1^{n-2} & \alpha_2^{n-2} & & \alpha_n^{n-2} \\ \alpha_1^{n-1} & \alpha_2^{n-1} & & \alpha_n^{n-1} \end{vmatrix} \cdot e^{(\alpha_1 + \alpha_2 + \cdots + \alpha_n)x}.$$

By the last question we know that the determinant will not vanish if the coefficients $\alpha_1, \ldots, \alpha_n$ are distinct.

Solutions to selected exercises on geometry in $\mathbb{R}^n$

Exercises 2.9

4.
$$\|x + y\|^2 + \|x - y\|^2 = \{\langle x + y, x + y \rangle + \langle x - y, x - y \rangle\}$$
$$= \{\langle x, x \rangle + \langle x, y \rangle + \langle y, x \rangle + \langle y, y \rangle$$
$$+ \langle x, x \rangle - \langle x, y \rangle - \langle y, x \rangle + \langle y, y \rangle\}$$
$$= 2\|x\|^2 + 2\|y\|^2.$$

This argument is valid in $\mathbb{R}^n$ and equally so in $\mathbb{C}^n$. The result says that the sum of the squares on the four sides of a parallelogram equals the sum of the squares on its two diagonals.

5. $x \neq 0 \Rightarrow 0 \neq \|x\| = \langle x, x \rangle$ so it is not true that $\langle x, y \rangle = 0$ for all y.

6. (i) As in Question 4, but working entirely in $\mathbb{R}^n$

$$\{\|x + y\|^2 - \|x - y\|^2\}$$
$$= \{\|x\|^2 + 2\langle x, y \rangle + \|y\|^2\} - \{\|x\|^2 - 2\langle x, y \rangle + \|y\|^2\}$$
$$= 4\langle x, y \rangle.$$

(ii) We repeat some of the calculation from the last question, working in $\mathbb{C}^n$

$$\|x + y\|^2 - \|x - y\|^2$$
$$= \{\|x\|^2 + \langle x, y \rangle + \langle y, x \rangle + \|y\|^2\}$$
$$\quad - \{\|x\|^2 - \langle x, y \rangle - \langle y, x \rangle + \|y\|^2\}$$
$$= 2\{\langle x, y \rangle + \langle y, x \rangle\} = 4 \operatorname{Re}(\langle x, y \rangle).$$

Similarly,

$$i\{\|x + iy\|^2 - \|x - iy\|^2\} = 4i \operatorname{Re}(\langle x, iy \rangle) = 4i \operatorname{Re}(-i\langle x, y \rangle)$$
$$= 4i \operatorname{Im}\langle x, y \rangle.$$

Adding these equations gives on the right-hand side

$$4 \operatorname{Re}\langle x, y \rangle + 4i \operatorname{Im}\langle x, y \rangle$$

which is the same as $4\langle x, y \rangle$, thus proving the required result.

8. The equation of the plane is $\langle x - (1, 2, 1)^t, (2, 1, 2)^t \rangle = 0$ or $2x_1 + x_2 + 2x_3 = 6$. Normalising $(2, 1, 2)^t$ to

$$(\tfrac{2}{3}, \tfrac{1}{3}, \tfrac{2}{3})^t$$

we have the equation in the standard form

$$\tfrac{2}{3}x_1 + \tfrac{1}{3}x_2 + \tfrac{2}{3}x_3 = 2.$$

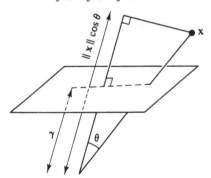

Hence the distance of this plane from the origin is 2. From the diagram we see that the distance of the point x from the plane is $\|x\| \cos \theta - 2$ or $\langle v, x \rangle - 2$. The distance is therefore $(\tfrac{1}{3})\langle (2, 1, 2)^t, (1, 2, 3)^t \rangle - 2 = \tfrac{4}{3}$.

9. If v is normal to the required plane then v is orthogonal to $(1, 0, -1)^t$ and

to $(1, 2, 3)^t$. Thus

$$v_1 - v_3 = 0 \quad \text{and} \quad v_1 + 2v_2 + 3v_3 = 0.$$

So $2v_2 = -4v_1$ and $\mathbf{v} = \lambda(1, -2, 1)^t$ for some scalar λ. The equation of the plane is thus $\langle \mathbf{x} - (3, 1, 2)^t, (1, -2, 1) \rangle = 0$ or $x_1 - 2x_2 + x_3 = 3$.

10. If your suspicions have been aroused, try the plane of the last question: yes, it works. Otherwise it is slog: assume the plane has equation $\langle (a, b, c)^t, (x, y, z)^t \rangle = p$ and solve the system

$$\left. \begin{array}{l} 3a + b + 2c = p \\ 2a + b + 3c = p \\ a + 2b + 3c = 0 \end{array} \right\}.$$

Solutions to selected exercises on matrices

Exercises 3.12

6. We have

$$\mathbf{x} = \begin{bmatrix} 1 & 3 & 2 \\ 2 & 1 & 3 \\ 3 & 2 & 1 \end{bmatrix} \mathbf{X} \quad \text{i.e.} \quad \mathbf{x} = P\mathbf{X},$$

where $\mathbf{x}$ is the vector of co-ordinates relative to the natural basis and $\mathbf{X}$ the vector of co-ordinates for the same point but relative to the new basis vectors. (The new basis vectors of course make up the columns of P.) So to find $\mathbf{X} = P^{-1}\mathbf{x}$ we compute the adjoint of P which is:

$$\begin{bmatrix} -5 & 1 & 7 \\ 7 & -5 & 1 \\ 1 & 7 & -5 \end{bmatrix}.$$

Since $\det P = -5 - 3 \cdot (-7) + 2 = 18$ we have $\mathbf{X} = \frac{1}{18}(\text{adj } P)\mathbf{x}$.

7. If $\mathbf{x} = P\mathbf{X}$ with the columns of P being the new basis vectors, we have

$$\mathbf{X} = P^{-1}\mathbf{x} = \begin{bmatrix} 1 & 2 & 1 \\ 2 & 1 & 0 \\ 1 & 1 & 1 \end{bmatrix} = Q\mathbf{x}.$$

We compute that $\det Q = 1 + (-3) = -2$. Thus $P = -\frac{1}{2}\text{adj } Q$ which is

$$-\frac{1}{2} \begin{bmatrix} 1 & -1 & -1 \\ -2 & 0 & 2 \\ 1 & 1 & -3 \end{bmatrix}$$

and so the three columns of this matrix constitute the new basis.

12. We tabulate the results

A	$R(A)$	A^t	$N(A^t)$	Illustration
$\begin{bmatrix} 1 & 1 \\ -1 & -1 \end{bmatrix}$	$L\{(1,-1)^t\}$	$\begin{bmatrix} 1 & -1 \\ 1 & -1 \end{bmatrix}$	$\{y:y_1 - y_2 = 0\}$ or $L\{(1,1)^t\}$	
$\begin{bmatrix} 5 & 0 \\ 3 & 0 \end{bmatrix}$	$L\{(5,3)^t\}$	$\begin{bmatrix} 5 & 3 \\ 0 & 0 \end{bmatrix}$	$\{y:5y_1 + 3y_2 = 0\}$ or $L\{(3,-5)\}$	
$\begin{bmatrix} 1 & -2 \\ -3 & 6 \end{bmatrix}$	$L\{(1,-3)^t\}$	$\begin{bmatrix} 1 & -3 \\ -2 & 6 \end{bmatrix}$	$\{y:y_1 - 3y_2 = 0\}$ or $L\{(3,1)^t\}$	
$\begin{bmatrix} 1 & 2 \\ 3 & 4 \end{bmatrix}$	$\mathbb{R}^2$	$\begin{bmatrix} 1 & 3 \\ 2 & 4 \end{bmatrix}$	$\{0\}$	

14. We are to prove the complex form of the duality result of Section 3.8, viz. $R(A^*) = N(A)^\perp$.

First we prove $R(A)^\perp \subseteq N(A^*)$. Let $z \in R(A)^\perp$ and $y \in R(A)$. Write $y = Ax$. Then for all x

$$0 = \langle z, y \rangle = \langle z, Ax \rangle = (Ax)^* z = x^* A^* z = \langle A^* z, x \rangle.$$

So $A^* z$ is orthogonal to all vectors x. Hence $A^* z = 0$, i.e. $z \in N(A^*)$.

Now we prove that $N(A^*) \subseteq R(A)^\perp$. Let $z \in N(A^*)$ and $y \in R(A)$. Write $y = Ax$. Then for all x we have:

$$A^* z = 0 \Rightarrow 0 = x^* A^* z = (Ax)^* z = \langle z, Ax \rangle = \langle z, y \rangle.$$

So z is orthogonal to all $y \in R(A)$. Hence $z \in R(A)^\perp$.

15. We note that $Ax = b \Rightarrow PAx = Pb$; similarly, $PAx = Pb \Rightarrow P^{-1}PAx = P^{-1}Pb = b$. Thus the two systems are equivalent. Now suppose

$$A = \begin{bmatrix} a_{11} & \cdots & a_{1n} \\ a_{21} & & a_{2n} \\ \vdots & & \\ a_{m1} & & a_{mn} \end{bmatrix}$$

and assume $a_{11} \neq 0$. [Otherwise exchange the *first* equation with one below where there is a non-zero coefficient at x_1.] To eliminate x_1 from all the other equations subtract from the jth equation (for $j = 2, 3, \ldots, m$) a multiple of the first equation (by a factor of a_{j1}/a_{11}), giving a new jth equation:

$$0x_1 + \left(a_{j2} - \frac{a_{j1}}{a_{11}} a_{12}\right) x_2 + \cdots + \left(a_{jn} - \frac{a_{j1}}{a_{11}} a_{1n}\right) = b_j - \frac{a_{j1}}{a_{11}} b_1.$$

Let us write this as

$$c_{j2} x_2 + \cdots + c_{jn} x_n = b'_j.$$

It is easily seen that the row reduction from $[A:b]$ to

$$\begin{bmatrix} a_{11} & a_{12} & \cdots & a_{1n} & b'_1 \\ 0 & c_{22} & & c_{2n} & b'_2 \\ \vdots & & & & \\ 0 & c_{m2} & & c_{mn} & b'_n \end{bmatrix}$$

corresponds precisely (is equivalent) to the manipulation of the system of equations. Now we observe that these row operations are equivalent to pre-multiplication by elementary matrices. Continuing the elimination process described we arrive at the existence of a sequence of elementary matrices $E_1, E_2, \ldots, E_s$ (corresponding to the row manipulations) such that

$$E_s \cdots E_2 E_1 [A|\mathbf{b}] = [D|\mathbf{d}],$$

where D is in echelon form. Let $P = E_s \cdots E_2 E_1$. Then

$$[D|\mathbf{d}] = P[A|\mathbf{b}] = [PA|P\mathbf{b}].$$

By the first part the equations $A\mathbf{x} = \mathbf{b}$ and $PA\mathbf{x} = P\mathbf{b}$ [i.e. $D\mathbf{x} = \mathbf{d}$] have precisely the same solutions. Thus the first part justifies the fact that to solve $A\mathbf{x} = \mathbf{b}$ we may first reduce $[A|\mathbf{b}]$ to the echelon form $[D|\mathbf{d}]$ by row operations and then solve $D\mathbf{x} = \mathbf{d}$.

16. The equations $a_i x + b_i y = c_i$ have a common solution if and only if

$$\begin{bmatrix} a_1 & b_1 \\ a_2 & b_2 \\ \vdots & \\ a_n & b_n \end{bmatrix} \begin{bmatrix} x \\ y \end{bmatrix} = \begin{bmatrix} c_1 \\ c_2 \\ \vdots \\ c_n \end{bmatrix}$$

has a solution. Writing $\mathbf{u}$ for $(x, y)^t$ and $\mathbf{a}$, $\mathbf{b}$ and $\mathbf{c}$ for the obvious column vectors, the system is soluble if and only if $\mathbf{c} = x\mathbf{a} + y\mathbf{b}$ for some x and y. This is so if $\mathbf{c}$ is linearly dependent on $\mathbf{a}$ and $\mathbf{b}$; that in turn may be restated as $\text{rank}[\mathbf{a}, \mathbf{b}, \mathbf{c}] = \text{rank}[\mathbf{a}, \mathbf{b}]$.

19. Say A is $m \times k$ and B is $k \times k$ with the rank of B being k. Then

$$\text{rank } A = \text{rank } A + \text{rank } B - k \leqslant \text{rank } AB \leqslant \text{rank } A.$$

Alternatively, one may prove $R(AB) = R(A)$ thus: $A\mathbf{x} = AB(B^{-1}\mathbf{x})$ so $R(A) \subseteq R(AB)$; but obviously, $R(AB) \subseteq R(A)$.

Solutions to selected exercises on projections

Exercises 4.8

6. Assume the relationship is $y = mx + c$. Let r_{ij} be the residual for the jth measurement of y corresponding to $x = x_i$, then we have

$$y_{ij} + r_{ij} = mx_i + c.$$

Let $\tilde{\mathbf{b}}^t$ be the partitioned vector $(y_{11}, \ldots, y_{1k} | y_{21}, \ldots, y_{2k} | \cdots)$ and let $\tilde{A}$ be the corresponding partitioned matrix of readings taken on the variable x:

$$\tilde{A} = \begin{bmatrix} x_1 & 1 \\ x_1 & 1 \\ \vdots & \\ x_1 & 1 \\ x_2 & 1 \\ \vdots & \\ x_2 & 1 \\ x_3 & 1 \\ \vdots & \end{bmatrix}.$$

The least squares fit is $(m, c)^t = \tilde{L}\tilde{\mathbf{b}}$ where $\tilde{L} = (\tilde{A}^t \tilde{A})^{-1} \tilde{A}^t$. But

$$\tilde{A}^t \tilde{A} = \begin{bmatrix} x_1 & \cdots & x_2 & \cdots & \cdots & x_n & \cdots \\ 1 & & 1 & & & 1 & \end{bmatrix} \begin{bmatrix} x_1 & 1 \\ & \text{etc} \end{bmatrix}$$

$$= \begin{bmatrix} k\Sigma x_i^2 & k\Sigma x_i \\ k\Sigma x_i & k \end{bmatrix} = k(A^t A),$$

where A (as opposed to $\tilde{A}$) denotes the usual matrix used in least squares fit using each x_i once only. Thus

$$\tilde{L}\tilde{\mathbf{b}} = \frac{1}{k}(A^t A)^{-1} \begin{bmatrix} x_1 & \cdots & x_2 & \cdots & \cdots & x_n & \cdots \\ 1 & & 1 & & & 1 & \end{bmatrix} \begin{bmatrix} y_{11} \\ \cdots \\ y_{21} \\ \cdots \end{bmatrix}$$

$$= \frac{1}{k}(A^t A)^{-1} \begin{bmatrix} x_1 \Sigma y_{1j} + x_2 \Sigma y_{2j} + \cdots \\ \Sigma y_{1j} + \Sigma y_{2j} + \cdots \end{bmatrix}$$

$$= (A^t A)^{-1} \begin{bmatrix} x_1 \bar{y}_1 + x_2 \bar{y}_2 + \cdots \\ \bar{y}_1 + \bar{y}_2 + \cdots \end{bmatrix}$$

$$= (A^t A)^{-1} \begin{bmatrix} x_1 & x_2 & \cdots & x_n \\ 1 & 1 & \cdots & 1 \end{bmatrix} \begin{bmatrix} \bar{y}_1 \\ \bar{y}_2 \\ \cdots \end{bmatrix}$$

$$= (A^t A)^{-1} A^t \mathbf{b} = L\mathbf{b},$$

where $\mathbf{b}$ is the usual column using averaged data and L is the usual matrix derived from A and $\mathbf{b}$. Thus the result for repeated readings is identical to that for the averaged data.

Solutions to selected exercises on spectral theory

Exercises 5.9

1. (i) $p(\lambda) = \lambda^2 + 1$. For $\lambda = i$ we solve the equation $(A - iI)\mathbf{x} = 0$, i.e.

$$\left. \begin{array}{l} -ix_1 - x_2 = 0 \\ x_1 - ix_2 = 0 \end{array} \right\}.$$

We note that the second equation is a multiple of the first (by a factor of $-i$). The solution space consists of vectors $(x_1, -ix_1)^t = x_1(1, -i)^t$, so is spanned by $(1, -i)^t$. Clearly, the solution space for $\lambda = -i$ is spanned by $(1, i)^t$.' [Replace i throughout by $-i$.] Since the matrix is 2×2 and there are two distinct eigenvalues the matrix is non-defective. Indeed, we see that the two eigenvectors just mentioned are linearly independent, they therefore give a basis for $\mathbb{C}^2$ relative to which the original matrix is represented by the digonal matrix with entries $\pm i$.

(iii)   We make use of the hint in calculating the characteristic polynomial:

$$\begin{vmatrix} 53 - \lambda & 4 & 1 \\ 4 & 38 - \lambda & -4 \\ 1 & -4 & 53 - \lambda \end{vmatrix} = \begin{vmatrix} 54 - \lambda & 0 & 54 - \lambda \\ 4 & 38 - \lambda & -4 \\ 1 & -4 & 53 - \lambda \end{vmatrix} \tag{1}$$

$$= (54 - \lambda)\{(38 - \lambda)(53 - \lambda) - 16 - 16 - 38 + \lambda\}$$
$$= (54 - \lambda)\{\lambda^2 - 90\lambda + 1944\}$$
$$= (54 - \lambda)(\lambda - 36)(\lambda - 54).$$

We find the eigenvectors for $\lambda = 54$ by solving the equation
$$-x_1 + 4x_2 + x_3 = 0$$
(we see from the second array in (1) that the matrix system reduces to only one equation). The solutions take the form $(4x_2 + x_3, x_2, x_3)^t$ and are spanned by $(1, 0, 1)^t$ and $(4, 1, 0)^t$. Let us take $\mathbf{u}_1 = (1, 0, 1)^t$ as one of the basis elements. We pick another orthogonal to this. So we select x_2 and x_3 so that $0 = \langle (4x_2 + x_3, x_2, x_3)^t, (1, 0, 1)^t \rangle = 4x_2 + 2x_3$. Thus the eigenvector has to have $x_3 = -2x_2$ and so takes the form $(4x_2 - 2x_2, x_2, -2x_2)^t = x_2(2, 1, -2)^t$. We take $\mathbf{u}_2 = (2, 1, -2)^t$ and leave normalisation till later.

Now we find eigenvectors for $\lambda = 36$. The equations reduce to:
$$\left.\begin{array}{r} 17x_1 + 4x_2 + x_3 = 0 \\ x_1 - 4x_2 + 17x_3 = 0 \end{array}\right\}.$$

Adding the equations gives $18x_1 + 18x_3 = 0$ so that $x_3 = -x_1$ and hence from the first equation $16x_1 + 4x_2 = 0$. Eigenvectors thus take the form $x_1(-1, 4, 1)^t$. We take $\mathbf{u}_3$ to be a rescaled version of $(-1, 4, 1)^t$ and so the orthogonal matrix P which reduces the given matrix to diagonal form is $[\mathbf{u}_1/\sqrt{2}, \mathbf{u}_2/3, \mathbf{u}_3/\sqrt{18}]$. Thus

$$P^t A P = \begin{bmatrix} 54 & & \\ & 54 & \\ & & 36 \end{bmatrix}.$$

2. (iii)   Rewriting $-x^2 - 2\sqrt{3}xz - 4yz + 3z^2 = 25$ as $\mathbf{x}^t A \mathbf{x} = 25$ we have

$$A = \begin{bmatrix} -1 & 0 & -\sqrt{3} \\ 0 & 0 & -2 \\ -\sqrt{3} & -2 & 3 \end{bmatrix}.$$

After some calculation we have $p(\lambda) = -\lambda^3 + 2\lambda^2 + 10\lambda + 4 = -(\lambda + 2)\{\lambda^2 - 4\lambda - 2\}$ so that the eigenvalues are -2 and $2 \pm \sqrt{6}$. We find the eigenvectors to value $\lambda = -2$ by solving

$$\begin{aligned} x_1 \quad\quad - \sqrt{3}x_3 = 0 \\ 2x_2 - 2x_3 = 0 \end{aligned} \Big\},$$

which has solution set spanned by $(\sqrt{3}, 1, 1)$. Next we solve for $\lambda = 2 \pm \sqrt{6}$

$$\begin{aligned} (-1 - 2 \mp \sqrt{6})x_1 \quad\quad\quad - \sqrt{3}x_3 = 0 \\ (-2 \mp \sqrt{6})x_2 - \quad 2x_3 = 0 \end{aligned} \Big\}$$

and this has solutions spanned by the vector:

$$(-\sqrt{3} \pm \sqrt{2}, +2 \mp \sqrt{6}, 1)^t.$$

In new co-ordinates the quadratic form is
$-2X^2 + (2 - \sqrt{6})Y^2 + (2 + \sqrt{6})Z^2$ which is neither positive, nor negative, nor non-positive, nor non-negative definite. The implied change of variable is given by

$$\begin{bmatrix} x \\ y \\ z \end{bmatrix} = \begin{bmatrix} \dfrac{\sqrt{3}}{\sqrt{5}} & \dfrac{-\sqrt{3}+\sqrt{2}}{\sqrt{\{16-6\sqrt{6}\}}} & \dfrac{-\sqrt{3}-\sqrt{2}}{\sqrt{\{16+6\sqrt{6}\}}} \\[3ex] \dfrac{1}{\sqrt{5}} & \dfrac{+2-\sqrt{6}}{\sqrt{\{16-6\sqrt{6}\}}} & \dfrac{+2+\sqrt{6}}{\sqrt{\{16+6\sqrt{6}\}}} \\[3ex] \dfrac{1}{\sqrt{5}} & \dfrac{1}{\sqrt{\{16-6\sqrt{6}\}}} & \dfrac{1}{\sqrt{\{16+6\sqrt{6}\}}} \end{bmatrix} \begin{bmatrix} X \\ Y \\ Z \end{bmatrix}.$$

To find the principal axes *either* remember that the matrix here is orthogonal so that we may read from its transpose the equations of the principal axes *or* just use the eigenvectors selected earlier. In either case we have:

$$\begin{aligned} X = 0 &\Leftrightarrow \quad\quad \sqrt{3}x + \quad\quad\quad y + z = 0, \\ Y = 0 &\Leftrightarrow (-\sqrt{3} + \sqrt{2})x + (+2 - \sqrt{6})y + z = 0, \\ Z = 0 &\Leftrightarrow (-\sqrt{3} - \sqrt{2})x + (+2 + \sqrt{6})y + z = 0. \end{aligned}$$

3. (i) We calculate the principal subdeterminants: 0, $-1/4$, $\{-1/4 + 1/4\}$, $\{-2 \cdot [(1/2) \cdot (-1)] + 1 \ [0]\}$. The last determinant was expanded by its bottom row. The sequence $0, -1/4, 0, 1$ does not obey either test for positive or negative definiteness. The last determinant is non-zero so there are no zero eigenvalues, this rules out non-negative definiteness and non-positive definiteness.

(ii) Evidently, the determinant of the given matrix is zero. The tests are not available to us and we need to calculate the characteristic polynomial. This turns out to be
$-\lambda^3 + 3\lambda^2 = -\lambda^2(\lambda - 3)$ and the quadratic form is non-negative definite.

(iii) The principal determinants are $2, -2, 6$. As in (i) the matrix is indefinite.

4. By Question 1 (ii) we have the decomposition

$$2 \cdot \begin{bmatrix} 1/\sqrt{2} \\ 0 \\ 1/\sqrt{2} \end{bmatrix} (1/\sqrt{2}, 0, 1/\sqrt{2}) + 4 \begin{bmatrix} 0 \\ 1 \\ 0 \end{bmatrix} (0, 1, 0).$$

5. If P is the matrix consisting of the normalised eigenvectors obtained in 1(i) $P^t B^2 P$ is equal to

$$\begin{bmatrix} 0 & 0 & 0 \\ 0 & 2 & 0 \\ 0 & 0 & 4 \end{bmatrix}.$$

Consider the matrix A defined by

$$A = P \begin{bmatrix} 0 & 0 & 0 \\ 0 & \sqrt{2} & 0 \\ 0 & 0 & 2 \end{bmatrix} P^t = \begin{bmatrix} 1/\sqrt{2} & 1/\sqrt{2} & 0 \\ 0 & 0 & 1 \\ -1/\sqrt{2} & 1/\sqrt{2} & 0 \end{bmatrix} \Lambda \begin{bmatrix} 1/\sqrt{2} & 0 & -1/\sqrt{2} \\ 1/\sqrt{2} & 0 & 1/\sqrt{2} \\ 0 & 1 & 0 \end{bmatrix}$$

$$= \begin{bmatrix} 1/\sqrt{2} & 0 & 1/\sqrt{2} \\ 0 & 2 & 0 \\ 1/\sqrt{2} & 0 & 1/2 \end{bmatrix},$$

where Λ is the diagonal matrix in the definition of A. Now we have $A^2 = P\Lambda P^t P \Lambda P^t = P\Lambda^2 P^t = B^2$. hence A solves the problem.

10. It is not true in general that if μ is an eigenvalue of A^2 then $\sqrt{\mu}$ is an eigenvalue of A. All we know is that at least one of $\pm\sqrt{\mu}$ is, since if λ is an eigenvalue of A then λ^2 is an eigenvalue of A^2. For instance, if $A^2 = I$ and A is 2×2 it might be that $A = I$ or that $A = -I$ or A may be the diagonal matrix with entries 1 and -1.

11. Let $Ax = \lambda x$. Then $0 = \|(A - \lambda I)x\|^2 = x^*(A - \lambda I)^*(A - \lambda I)x =$
$x^*\{A^*A - \lambda A^* - \bar{\lambda}A + \lambda\bar{\lambda}I\}x =$
$x^*\{AA^* - \bar{\lambda}A - \lambda A^* + \lambda\bar{\lambda}I\}x = x^*(A^* - \bar{\lambda}I)^*(A^* - \lambda I)x$
$= \|(A^* - \bar{\lambda})x\|^2$. Thus $A^*x = \bar{\lambda}x$.

12. Suppose $Ax = \lambda x$ then, since $x^*A^* = \bar{\lambda}x^*$, we have $x^*A^*Ax = \lambda\bar{\lambda}x^*x$. Thus if $A^*A = I$ we have $\|x\|^2 = \lambda\bar{\lambda}\|x\|^2$. But $x \neq 0$ so $\lambda\bar{\lambda} = 1$ as required.

13. Since A is normal there is an orthogonal P so that $P^*AP =$ diag$(\lambda_1, \ldots, \lambda_n) = \Lambda$ where $\lambda_1, \ldots, \lambda_n$ are the eigenvalues, assumed real, of A. Hence $A^* = (P\Lambda P^*)^* = P\Lambda^* P^* = P\Lambda P^* = A$ and A is Hermitian.

14. We partition the familiar 'eigenvalue equation' to read:

$$\left[\begin{array}{c|c} A & C \\ \hline 0 & B \end{array} \right] \begin{bmatrix} u \\ v \end{bmatrix} = \lambda \begin{bmatrix} u \\ v \end{bmatrix}$$

and this is equivalent to $Au + Cv = \lambda u$ and $Bv = \lambda v$. Thus if $(u, v)^t \neq 0$ we have that either $v \neq 0$ and λ is an eigenvalue of B, or $v = 0$ and then $Au = \lambda u$ so that, since $u \neq 0$, λ is an eigenvalue of A.

Conversely, let λ be an eigenvalue of B which is not eigenvalue of A.

Then we may find $v \neq 0$ so that $Bv = \lambda v$. Further for this v we can choose u so that $(A - \lambda I)u = -Cv$. This is possible since $\det(A - \lambda I) \neq 0$ and so $(A - \lambda I)^{-1}$ exists. If on the other hand λ is an eigenvalue of A then we may choose $v = 0$ and select $u \neq 0$ so that $Au = \lambda u$.

In either case $(u, v)^t \neq 0$ and the 'eigenvalue equation' has been solved.

15. Since A, B are Hermitian and A is positive definite $A^{-1/2}$ exists and we have

$$A^{-1/2}BA^{-1/2} = A^{*-1/2}B^*A^{*-1/2} = (A^{-1/2}BA^{-1/2})^*,$$

so this matrix is Hermitian too. Choose an orthogonal matrix Q so that for some diagonal matrix D we have

$$Q^*A^{-1/2}BA^{-1/2}Q = D.$$

Let $P = A^{-1/2}Q$ then $P^* = Q^*A^{*-1/2} = Q^*A^{-1/2}$. Thus

$$P^*AP = Q^*A^{-1/2}AA^{-1/2}Q = Q^*Q = I$$

and

$$P^*BP = D.$$

Solutions to selected exercises on inverses

Exercises 8.6

3. By analogy with the (numerical) identity

$$\frac{1}{a} + \frac{1}{b} = \frac{b}{ab} + \frac{a}{ab} = \frac{a+b}{ab},$$

we argue:

$$B^{-1}(A + B)A^{-1} = (B^{-1}A + I)A^{-1} = (B^{-1} + A^{-1})$$

so

$$(A^{-1} + B^{-1})^{-1} = A(A + B)^{-1}B.$$

Now substituting A for B and B for A will yield the other result.

5. The equation $Ax = b$ then A is non-singular has solution $x = A^{-1}b$. By Cramer's rule we have $x = A^{-1}b = (\det A)^{-1}\operatorname{adj}(A)b$. Now write $B = \operatorname{adj}(A)$. Then

$$x_j = (B_{j1}b_1 + B_{j2}b_2 + \cdots + B_{jn}b_n)/\det A$$
$$= \{(-1)^{j+1}b_1A_{1j} + (-1)^{j+2}b_2A_{2j} + \cdots + (-1)^{j+n}b_nA_{nj}\}/\det A,$$

where we have substituted for the co-factors. But the contents of the curly { } brackets are precisely the expansion of the following determinant by its jth column:

$$\begin{vmatrix} a_{11}, \ldots, b_1, a_{1j+1}, \ldots, a_{1n} \\ a_{21}, \ldots, b_2, a_{2j+1}, \ldots, a_{2n} \\ \cdot \qquad \cdot \\ a_{n1}, \ldots, b_n, a_{nj+1}, \ldots, a_{nn} \end{vmatrix}.$$

6. Second part. We may reduce B to A by some obvious column operations. Hence we see that

$$A = B \begin{bmatrix} 0 & 0 & 0 & 0 & 1 \\ 1 & 0 & 0 & 0 & 0 \\ 0 & 1 & 0 & 0 & 0 \\ 0 & 0 & 0 & 1 & 0 \\ 0 & 0 & 1 & 0 & 0 \end{bmatrix} = BP, \quad \text{say.}$$

Thus, since $AA^g A = A$, we have $BP(A^g)BP = BP$ or, since P^{-1} exists, $B(PA^g)B = B$. We may therefore obtain a weak generalised inverse for B by applying the row operations implied by P. Thus

$$A^g = \frac{1}{18} \begin{bmatrix} 0 & 0 & 0 & 0 & 0 & 0 \\ -5 & 7 & 1 & 0 & 0 & 0 \\ 1 & -5 & 7 & 0 & 0 & 0 \\ 0 & 0 & 0 & 0 & 0 & 0 \\ 7 & 1 & -5 & 0 & 0 & 0 \end{bmatrix}.$$

8. Observe that rank $A = 2$. We therefore take, say, the first and third column of A for B. Hence we have

$$A = BC = \begin{bmatrix} 1 & 0 \\ 0 & 1 \\ 1 & 0 \end{bmatrix} \begin{bmatrix} 1 & 1 & 0 & -1 \\ 0 & 1 & 1 & 1 \end{bmatrix}.$$

The strong inverse will be $C^t(CC^t)^{-1}(B^tB)^{-1}B^t$. Now

$$CC^t = \begin{bmatrix} 1 & 1 & 0 & -1 \\ 0 & 1 & 1 & 1 \end{bmatrix} \begin{bmatrix} 1 & 0 \\ 1 & 1 \\ 0 & 1 \\ -1 & 1 \end{bmatrix} = \begin{bmatrix} 3 & 0 \\ 0 & 3 \end{bmatrix},$$

$$B^tB = \begin{bmatrix} 1 & 0 & 1 \\ 0 & 1 & 0 \end{bmatrix} \begin{bmatrix} 1 & 0 \\ 0 & 1 \\ 1 & 0 \end{bmatrix} = \begin{bmatrix} 2 & 0 \\ 0 & 1 \end{bmatrix}.$$

Thus the generalised inverse of A is

$$\begin{bmatrix} 1 & 0 \\ 1 & 1 \\ 0 & 1 \\ -1 & 1 \end{bmatrix} \frac{1}{3} I \frac{1}{2} \begin{bmatrix} 1 & 0 \\ 0 & 2 \end{bmatrix} \begin{bmatrix} 1 & 0 & 1 \\ 0 & 1 & 0 \end{bmatrix} = \frac{1}{6} \begin{bmatrix} 1 & 0 \\ 1 & 1 \\ 0 & 1 \\ -1 & 1 \end{bmatrix} \begin{bmatrix} 1 & 0 & 1 \\ 0 & 2 & 0 \end{bmatrix}$$

$$= \frac{1}{6} \begin{bmatrix} 1 & 0 & 1 \\ 1 & 2 & 1 \\ 0 & 2 & 0 \\ -1 & 2 & -1 \end{bmatrix}.$$

The required projection matrix is AA^g, viz.

$$= \frac{1}{6}\begin{bmatrix} 1 & 1 & 0 & -1 \\ 0 & 1 & 1 & 1 \\ 1 & 1 & 0 & -1 \end{bmatrix} \begin{bmatrix} 1 & 0 & 1 \\ 1 & 2 & 1 \\ 0 & 2 & 0 \\ -1 & 2 & -1 \end{bmatrix} = \frac{1}{6}\begin{bmatrix} 3 & 0 & 3 \\ 0 & 6 & 0 \\ 3 & 0 & 3 \end{bmatrix}.$$

9. Multiplying out the right-hand side of the asserted equation we get

$$\begin{bmatrix} B \\ D \end{bmatrix}[I \mid B^{-1}C] = \left[\begin{array}{c|c} B & C \\ \hline D & DB^{-1}C \end{array}\right]$$

and, as in Section 8.5, $DB^{-1}C = E$. This justifies the assertion. Clearly the rank of $[I \mid B^{-1}C]$ is k. The same is true of the leftmost matrix above. We may thus use method II to obtain a generalised inverse. We content ourselves with the computation:

$$[I \mid B^{-1}C]\begin{bmatrix} I \\ C^t(B^{-1})^t \end{bmatrix} = I + B^{-1}CC^t(B^{-1})^t$$

and

$$[B^t \mid D^t]\begin{bmatrix} B \\ D \end{bmatrix} = B^t B + D^t D.$$

10. We have $A = BC$ and $G = C^t(CC^t)^{-1}(B^tB)^{-1}B^t$. Now let us introduce the matrices $\mathbf{b} = C^t(CC^t)^{-1}$ and $\mathbf{c} = (B^tB)^{-1}B^t$. Then certainly $G = \mathbf{bc}$ and moreover both $\mathbf{b}$ and $\mathbf{c}$ have rank k (why?). Next notice that $\mathbf{b}^t\mathbf{b} = (CC^t)^{-1t}C \cdot C^t(CC^t)^{-1} = (CC^t)^{-1}$ and that therefore $(\mathbf{b}^t\mathbf{b})^{-1}\mathbf{b}^t = (CC^t)(CC^t)^{-1}C^{tt} = C$. Similarly, we have $\mathbf{c}^t(\mathbf{c}\cdot\mathbf{c}^t)^{-1} = B$. The formula for a strong generalised inverse for G starting from $G = \mathbf{b}\cdot\mathbf{c}$ now yields BC i.e. A itself. So these are 'mutual inverses'. We leave the verification that $GAG = G$ aside.

(i) We know that AG projects onto $R(A)$, we have to show $N(AG) = N(G)$. But $G\mathbf{x} = 0$ implies $AG\mathbf{x} = 0$ so $N(G)$ is a subset of $N(AG)$. Now if $AG\mathbf{z} = 0$, then $GAG\mathbf{z} = 0$ or $G\mathbf{z} = 0$, hence $N(AG)$ is a subset of $N(G)$.

(ii) We know that GA projects parallel to $N(A)$ into $R(G)$ and since $G\mathbf{x} = GA(G\mathbf{x})$ we see that this is onto.

By definition AG is an orthogonal projection onto $R(A)$, so from $N(AG) = N(G)$ we deduce $N(G)^{\perp} = R(A)$; but $R(A)^{\perp} = N(A^t)$.

It may be verified that GA is an orthogonal projection (it is symmetric and $G \cdot AGA = GA$) and since this is parallel to $N(A)$ we have $R(G) = N(A)^{\perp} = R(A^t)$ as required.

11. The affine subspace defined by $A\mathbf{x} = \mathbf{b}$ is given as non-empty. It is parallel to the space $N(A)$, hence is also orthogonal to $R(G)$, by the last question. Now if $A\mathbf{w} = \mathbf{b}$ then $GA\mathbf{w} = G\mathbf{b}$ lies in $R(G)$. But $\mathbf{w} - GA\mathbf{w}$ is in $N(A)$, hence is orthogonal to $G\mathbf{b}$ and so by Pythagoras' theorem

$$\|\mathbf{w}\|^2 = \|G\mathbf{b}\|^2 + \|\mathbf{w} - G\mathbf{b}\|^2.$$

The expression in $\mathbf{w}$ on the right is clearly at a minimum if $\mathbf{w}$ satisfies $\mathbf{w} = G\mathbf{b}$.

12. Recall that the least squares solution for the system $A\mathbf{x} = \mathbf{b}$ gives a vector $\mathbf{x}$ for which $\|A\mathbf{x} - \mathbf{b}\|$ is a minimum. The point $A\mathbf{x}$ is obtained as the orthogonal projection of $\mathbf{b}$ onto $R(A)$, so that $A\mathbf{x}$ is $AG\mathbf{b}$. Possible values of $\mathbf{x}$ are therefore solutions of the equation $A\mathbf{x} = AG\mathbf{b}$. A particular solution is clearly $\mathbf{x} = G\mathbf{b}$, so the general solution is

$$\mathbf{x} = G(AG\mathbf{b}) + (I - GA)\mathbf{z} = G\mathbf{b} + (I - GA)\mathbf{z}.$$

14. The special case $A = I$ gives the game away. For any matrix C the diagonal elements of $C^t C$ may be rewritten as $\mathbf{c}_1^t \mathbf{c}_1, \ldots, \mathbf{c}_m^t \mathbf{c}_m$ where $\mathbf{c}_1, \ldots$ etc, are the columns of C. Each term is non-negative ($\mathbf{c}_1^t \mathbf{c}_1 = \|\mathbf{c}_1\|^2$) so if $C^t C = 0$, all terms must be zero. Hence each of $\mathbf{c}_1, \ldots$ etc are zero and hence C is zero.

Now if $AB^t B = 0$ then $AB^t BA^t = 0$ i.e. $(BA^t)^t BA^t = 0$. Hence $BA^t = 0$.

15. Let B be a weak generalised inverse of $A^t A$. Then $A^t ABA^t A = A^t A$ so $0 = (A^t AB - I)A^t A$. We deduce, by way of the last question, that

$$A^t ABA^t = A^t. \tag{1}$$

Similarly, by taking transposes of the initial equation, we may deduce that

$$A^t AB^t A^t = A^t. \tag{2}$$

But this says that $A = ABA^t A$.

Now observe that $(ABA^t)(ABA^t) = AB \cdot A^t A \cdot BA^t = ABA^t$. So ABA^t is a projection. Since $A\mathbf{x} = (ABA^t A)\mathbf{x}$ we see that the projection is onto $R(A)$. To prove orthogonality we need to show symmetry of ABA^t. We have, using (1) and (2), $(ABA^t) = AB^t A^t = (ABA^t A)B^t A^t = AB(A^t AB^t A^t) = ABA^t$.

Solutions to selected exercises on convexity

Exercises 9.9

1. These may be decided by reference to diagrams.
   (i)   the halfplane is convex;
   (ii) the line is convex;
   (iii) the interior of the ellipse is convex;
   (iv) the ellipse together with its interior is convex;
   (v)   the exterior of the ellipse is *not* convex;
   (vi) the ellipse and its exterior is *not* convex;
   (vii) the epigraph of the parabola is convex;
   (viii) the region above the sine curve is not convex.

Remark. All these assertions may be proved 'analytically' either by *ad hoc* means or, better, by using the 'test for a convex function' developed in Chapter 15.

3. Let $\mathbf{x} = (x_1, x_2, \ldots, x_n)^t$ and $\mathbf{y} = (y_1, y_2, \ldots, y_n)^t$ be in Θ. Consider α with $0 \leqslant \alpha \leqslant 1$ and let $\mathbf{z} = \alpha \mathbf{x} + (1 - \alpha)\mathbf{y}$. Then since $0 \leqslant 1 - \alpha$ we have for each i, $z_i = \alpha x_i + (1 - \alpha) y_i \geqslant 0$. Hence $\mathbf{z}$ is in Θ.

5. For S convex consider $T = \{\mathbf{y} : \mathbf{y} = A\mathbf{x} \text{ for some } \mathbf{x} \text{ in } S\}$. Let $\mathbf{u} = A\mathbf{x}$ and $\mathbf{v} = A\mathbf{z}$ with $\mathbf{x}$ and $\mathbf{z}$ in S. Consider α with $0 \leqslant \alpha \leqslant 1$ and let $\mathbf{w} = \alpha \mathbf{u} + (1 - \alpha)\mathbf{v}$. Then $\mathbf{w} = \alpha A\mathbf{x} + (1 - \alpha)A\mathbf{z} = A\{\alpha \mathbf{x} + (1 - \alpha)\mathbf{z}\}$. But since S is convex $\alpha \mathbf{x} + (1 - \alpha)\mathbf{z}$ is in S and so $\mathbf{w}$ is in T.

6. (i) The convex set consists of two tetrahedra with a common base. Its affine hull is $\mathbb{R}^3$ so the dimension is 3.

(ii) $(0, 1, 1)^t$ lies in the interior of the triangle defined by the other three points. [It is in fact its centre of gravity.] Note that all four points satisfy $x_1 = 0$ and that $x_1 = 0$ is in the affine hull of the points, so the dimension is 2.

7. Let $K(a_1, a_2, \ldots, a_n, b) = \{x : a_1 x_1 + a_2 x_2 + \cdots + a_n x_n \leqslant b\}$. We check that this is convex. Consider α with $0 \leqslant \alpha \leqslant 1$ and let $\mathbf{x} = (x_1, x_2, \ldots, x_n)^t$ and $\mathbf{y} = (y_1, y_2, \ldots, y_n)^t$ be in $K(a_1, a_2, \ldots, a_n, b)$. Write $\beta = (1 - \alpha)$. Then

$$a_1(\alpha x_1 + \beta y_1) + a_2(\alpha x_2 + \beta y_2) + \cdots + a_n(\alpha x_n + \beta y_n)$$
$$\leqslant \alpha(a_1 x_1 + a_2 x_2 + \cdots + a_n x_n) + \beta(a_1 y_1 + a_2 y_2 + \cdots + a_n y_n)$$
$$\leqslant \alpha b + \beta b \leqslant (\alpha + \beta)b \leqslant b.$$

So $\alpha \mathbf{x} + (1 - \alpha)\mathbf{y}$ is in $K(a_1, a_2, \ldots, a_n, b)$.

The set of points $\mathbf{x}$ satisfying the simultaneous system of equations is equal to the intersection

$$\bigcap_{j=1}^{m} K(a_{j1}, a_{j2}, \ldots, a_{jn}, b_j),$$

but any intersection of convex sets is convex.

8. (i) A is the open disc centered at the origin with radius 1. $(3, 3)^t + A$ is the set of all translates of points $\mathbf{a}$ of A by the vector $(3, 3)^t$. This is therefore a disc of unit radius centered at $(3, 3)^t$.

(ii) Evidently $A + B = B + A$. Now

$$B + A = \bigcup_{\mathbf{b} \in B} \mathbf{b} + A$$

and is thus the union of discs of radius 1 centered at the points of B. The boundary of this set may be sketched out by following round the discs whose centres are placed on the boundary of the square B. Observe, though, that the set $A + B$ does *not* include its boundary. The shape is roughly like a television screen.

(iii) The trick is to notice that $A = -A$ and so $A - B = -A - B = -(A + B)$, so the set is a central reflection of the set obtained in part (ii).

Exercises 9.10

1. Let $\mathbf{x} \in A + \text{conv } B$, thus for some $\mathbf{a} \in A$ and for some scalars $\beta_1, \ldots, \beta_r \geqslant 0$ with $\beta_1 + \cdots + \beta_r = 1$ and for some $\mathbf{b}_1, \ldots, \mathbf{b}_r$ in B we have

$$\mathbf{x} = \mathbf{a} + (\beta_1 \mathbf{b}_1 + \cdots + \beta_r \mathbf{b}_r)$$
$$= (\beta_1(\mathbf{a} + \mathbf{b}_1) + \cdots + \beta_r(\mathbf{a} + \mathbf{b}_r)).$$

Thus $\mathbf{x} \in \text{conv}(A + B)$. It follows that

$$\text{conv}(A + \text{conv } B) \subseteq \text{conv}(A + B).$$

But, since

$$(A + B) \subseteq A + \text{conv } B \subseteq \text{conv}(A + \text{conv } B),$$

and the right-most term is convex we also have

$$\text{conv}(A + B) \subseteq \text{conv}(A + \text{conv } B).$$

This gives us the desired equation

$$\operatorname{conv}(A + B) = \operatorname{conv}(A + \operatorname{conv} B). \qquad (*)$$

Applying this last result with A and B interchanged we have:

$$\operatorname{conv}(A + B) = \operatorname{conv}(\operatorname{conv} A + B).$$

But by the result (*) with conv A for A we have

$$\operatorname{conv}(\operatorname{conv} A + B) = \operatorname{conv}(\operatorname{conv} A + \operatorname{conv} B).$$

But this last term is already equal to conv A + conv B, since that is convex. We thus obtain the final result that

$$\operatorname{conv}(A + B) = \operatorname{conv} A + \operatorname{conv} B.$$

3. We rewrite the expression for $\mathbf{x}$:

$$\mathbf{x} = \alpha \mathbf{a} + \beta \mathbf{b} + \gamma \mathbf{c}$$

$$= (\alpha + \beta) \left\{ \frac{\alpha}{\alpha + \beta} \mathbf{a} + \frac{\beta}{\alpha + \beta} \mathbf{b} \right\} + \gamma \mathbf{c}.$$

So we take for $\mathbf{z}$ the point

$$\left\{ \frac{\alpha}{\alpha + \beta} \mathbf{a} + \frac{\beta}{\alpha + \beta} \mathbf{b} \right\}.$$

4. Since the $n + 1$ vectors $(\mathbf{x}_1 - \mathbf{x}_0), (\mathbf{x}_2 - \mathbf{x}_0), \ldots (\mathbf{x}_{n+1} - \mathbf{x}_0)$ lie in $\mathbb{R}^n$ they cannot be linearly independent, so there are constants $\alpha_1, \ldots, \alpha_{n+1}$ not all zero satisfying

$$\alpha_1(\mathbf{x}_1 - \mathbf{x}_0) + \cdots + \alpha_{n+1}(\mathbf{x}_{n+1} - \mathbf{x}_0) = 0.$$

Let us define the scalar α_0 by

$$\alpha_0 = -(\alpha_1 + \cdots + \alpha_{n+1}).$$

Then

$$\alpha_0 \mathbf{x}_0 + \alpha_1 \mathbf{x}_1 + \cdots + \alpha_{n+1} \mathbf{x}_{n+1} = 0 \quad \alpha_0 + \alpha_1 + \cdots + \alpha_{n+1} = 0$$

as required. Suppose without loss of generality that $\alpha_0, \ldots, \alpha_k$ are positive and $\alpha_{k+1}, \ldots, \alpha_{n+1}$ are non-positive, then

$$\alpha_0 \mathbf{x}_0 + \alpha_1 \mathbf{x}_1 + \cdots + \alpha_k \mathbf{x}_k = (-\alpha_{k+1}) \mathbf{x}_{k+1} + \cdots + (-\alpha_{n+1}) \mathbf{x}_{n+1}.$$

Observe that on both sides of the equation we have non-negative scalars. Note also that

$$\gamma = \alpha_0 + \alpha_1 + \cdots + \alpha_k = (-\alpha_{k+1}) + \cdots + (-\alpha_{n+1}) > 0.$$

Dividing by γ we obtain from our earlier observation

$$\frac{\alpha_0}{\gamma} \mathbf{x}_0 + \frac{\alpha_1}{\gamma} \mathbf{x}_1 + \cdots + \frac{\alpha_k}{\gamma} \mathbf{x}_k = \frac{-\alpha_{k+1}}{\gamma} \mathbf{x}_{k+1} + \cdots + \frac{-\alpha_{n+1}}{\gamma} \mathbf{x}_{n+1}.$$

The common value, call it z, is seen to be on the left-hand side a member of
conv $\{\mathbf{x}_i : i \in I\}$ and on the right-hand side a member of conv $\{\mathbf{x}_j : j \notin I\}$. This solves
the problem.

Solutions to selected exercises on the separating hyperplane

Exercises 10.7

4. Let C be a convex cone. Let H be a supporting hyperplane for C. Say
$H = \{\mathbf{x} : \langle \mathbf{v}, \mathbf{x} \rangle = p\}$ and $C \subseteq \{\mathbf{x} : \langle \mathbf{v}, \mathbf{x} \rangle \leqslant p\}$. Suppose H supports C at
$\mathbf{x}_0$. Thus we have $\langle \mathbf{v}, \mathbf{x}_0 \rangle = p$.
   Now notice that, for any natural number n, $n\mathbf{x}_0$ is in C. Hence

$$\langle \mathbf{v}, n\mathbf{x}_0 \rangle \leqslant p,$$

so

$$\langle \mathbf{v}, \mathbf{x}_0 \rangle \leqslant \frac{p}{n}.$$

Now taking limits as $n \to \infty$ we obtain

$$\langle \mathbf{v}, \mathbf{x}_0 \rangle \leqslant 0.$$

Hence

$$p = \langle \mathbf{v}, \mathbf{x}_0 \rangle \leqslant 0.$$

But $\mathbf{0}$ is in C, so

$$0 = \langle \mathbf{v}, \mathbf{0} \rangle \leqslant p.$$

Thus $p = 0$ and the hyperplane H passes through $\mathbf{0}$.

5. Suppose $\mathbf{a}$ is a point not belonging to the closed convex set K. The
distance from $\mathbf{a}$ to the nearest point of K is say d (the nearest point exists
since K contains its boundary). Thus no point of the ball
$B = \{\mathbf{x} : \|\mathbf{x} - \mathbf{a}\| < \frac{1}{2}d\}$ belongs to K. There is therefore a hyperplane
$H = \{\mathbf{x} : \langle \mathbf{v}, \mathbf{x} \rangle = p\}$ which separates B from K. Say, $K \subseteq \{\mathbf{x} : \langle \mathbf{v}, \mathbf{x} \rangle \leqslant p\} \equiv H^-$
and $B \subseteq \{\mathbf{x} : \langle \mathbf{v}, \mathbf{x} \rangle \geqslant p\}$, then $\mathbf{a}$ necessarily belongs to $\{\mathbf{x} : \langle \mathbf{v}, \mathbf{x} \rangle > p\}$.
Consequently the half-space H^- excludes $\mathbf{a}$. Now the intersection of all the
half-spaces which (like H^-) contain K will exclude all points not belonging
to K. Hence the intersection is identical with K.

6. With the obvious notation, let $K = \{\mathbf{x} : A\mathbf{x} \leqslant \mathbf{b}, \mathbf{x} \geqslant \mathbf{0}\}$. K is a convex set
(see Exercises 9.9, Question 7). It is closed (because each of the half-spaces such
as $\{\mathbf{x} : a_{11}x_1 + a_{12}x_2 + \cdots + \leqslant b_1\}$ is closed). Further, K is contained in the
non-negative orthant $\Theta = \{\mathbf{x} : \mathbf{x} \geqslant \mathbf{0}\}$. If K has no extreme points then either
$K = \varnothing$ or K is a cylinder. In the latter case K contains the whole of a line,
say l, but the line l lies wholly in Θ which is absurd. Thus K does have an
extreme point.
   Intuitively, the extreme points are to be found among the points of
intersection of n hyperplanes taken from among those given by the
equations $x_i = 0$ or $\mathbf{a}_j^t \mathbf{x} = 0$ (where the $\mathbf{a}_j^t$ are the rows of A).

Solutions to selected exercises on linear inequalities
Exercises 11.5

9. We start with the system

$$
\left.\begin{matrix} A'y = 0 \\ y \geqslant 0 \\ b'y < 0 \end{matrix}\right\} \text{ or equivalently } \left.\begin{matrix} A'y \geqslant 0 \\ -A'y \geqslant 0 \\ Iy \geqslant 0 \\ b'y < 0 \end{matrix}\right\} \text{ or } \begin{bmatrix} A^t \\ -A^t \\ I \end{bmatrix} y \geqslant 0,
$$

and

$$ b'y < 0. $$

By Farkas' lemma this is dual to:

$$ [A|-A|I]X = b \text{ and } X \geqslant 0. $$

Let us write X in the partitioned form $(u^t, v^t, w^t)^t$. Then the dual may be rewritten as

$$ Au - Av + w = b \quad \text{and} \quad u \geqslant 0, v \geqslant 0, w \geqslant 0, $$

or

$$ w = -Au + Av + b \geqslant 0 \quad \text{and} \quad u \geqslant 0, v \geqslant 0, $$

or

$$ A(u-v) \leqslant b \quad \text{and} \quad u \geqslant 0, v \geqslant 0. $$

Now notice that *any* vector x can be expressed as a difference $(u - v)$ with $u \geqslant 0$ and $v \geqslant 0$; use the positive co-ordinates of x to form u and its negative co-ordinates to form $-v$. For example $(3, -2, 1)^t = (3, 0, 1)^t - (0, 2, 0)^t$. We see that the last system is equivalent to saying only that

$$ Ax \leqslant b, $$

as required.

10. We rewrite the equation $Ax = b$ as the inequality

$$ \begin{bmatrix} A \\ -A \end{bmatrix} x \leqslant \begin{bmatrix} b \\ -b \end{bmatrix}. $$

By the last question this is dual to

$$ [A^t|-A^t]Y = 0, \ Y \geqslant 0, (b^t, -b^t)Y < 0. $$

We write Y in the partitioned form $(u^t, v^t)^t$ and the system reduces to:

$$ A^t u - A^t v = 0, b^t u - b^t v < 0 \quad \text{and} \quad u \geqslant 0, v \geqslant 0, $$

or

$$ A^t(u - v) = 0, b^t(u - v) < 0 \quad \text{and} \quad u \geqslant 0, v \geqslant 0. $$

The last system is equivalent to $A'y = 0$ with $b'y \neq 0$ for the following reason. If u and v are as satisfying the last system then $b'(u - v) < 0$ so that with $y = (u - v)$ we have $b't \neq 0$ and $A'y = 0$. Conversely, if $b'y \neq 0$ and $A'y = 0$ then we may assume that $b'y < 0$. [Otherwise replace y in the following by $-\mathbf{y}$, and note that $A'(-\mathbf{y}) = \mathbf{0}$.] Now use the same trick as in the last question to rewrite $\mathbf{y}$ as $(\mathbf{u} - \mathbf{v})$ with $\mathbf{u} \geqslant \mathbf{0}$ and $\mathbf{v} \geqslant \mathbf{0}$.

An alternative proof based on $R(A) = N(A')^\perp$ proceeds as follows. $A\mathbf{x} = \mathbf{b}$ is soluble $\Leftrightarrow \mathbf{b} \in R(A) = N(A')^\perp \Leftrightarrow [\mathbf{b}'\mathbf{y} = 0$ for every $\mathbf{y}$ with $A'\mathbf{y} = \mathbf{0}]$.

The last assertion says that

$$\left.\begin{array}{c} A'\mathbf{y} = \mathbf{0} \\ \mathbf{b}'\mathbf{y} \neq 0 \end{array}\right\}$$

is insoluble.

Solutions to selected exercises on linear programming and game theory

Exercise 12.8

4. To manufacture (x, y) the entrepreneur requires (u, v) where

$$\begin{bmatrix} u \\ v \end{bmatrix} = \begin{bmatrix} 1 & 2 \\ 2 & 1 \end{bmatrix}\begin{bmatrix} x \\ y \end{bmatrix} \quad \text{or } A\mathbf{x}, \text{ say.}$$

With selling prices at $\mathbf{c} = (1, 1)'$ and a stock of $\mathbf{b} = (3, 4)'$ the shadow prices $\mathbf{p} = (p_1, p_2)'$ are given by the optimal solution to the problem: minimise $\mathbf{b}'\mathbf{p} = 3p_1 + 4p_2$ subject to $A'\mathbf{p} \geqslant \mathbf{c}$ and $\mathbf{p} \geqslant 0$. The constraints in full are:

$$\left.\begin{array}{c} p_1 + 2p_2 \geqslant 1 \\ 2p_1 + p_2 \geqslant 1 \end{array}\right\}$$
$$p_1, p_2 \geqslant 0.$$

A sketch determines the extreme points to be at $(0, 1)'$, $(1, 0)'$ and at the point where the first two inequalities are equations. The latter point satisfies (by standard elimination) $4p_1 - p_1 = 2 - 1$, i.e. $p_1 = \frac{1}{3}$ and similarly $p_2 = \frac{1}{3}$. The objective function values are respectively $4, 3, \frac{7}{3}$ and the minimum thus occurs at $(\frac{1}{3}, \frac{1}{3})$. These values give the prices below which it pays the entrepreneur to buy a little extra stock.

5. We recast the constraint into the standard form.

$$\begin{bmatrix} A \\ -A \end{bmatrix}\mathbf{x} \geqslant \begin{bmatrix} \mathbf{b} \\ -\mathbf{b} \end{bmatrix},$$
$$\mathbf{x} \geqslant \mathbf{0}.$$

The dual to this is to minimise $(\mathbf{b}' | -\mathbf{b}')Y$ subject to

$$[A' | -A']Y \geqslant \mathbf{c} \quad \text{and} \quad Y \geqslant \mathbf{0}.$$

We partition Y as $(\mathbf{u}', \mathbf{v}')'$. We thus have to minimise $\mathbf{b}'\mathbf{u} - \mathbf{b}'\mathbf{v}$ subject to

$$A'\mathbf{u} - A'\mathbf{v} \geqslant \mathbf{c} \quad \text{and} \quad \mathbf{u} \geqslant 0, \mathbf{v} \geqslant 0.$$

Since any vector may be written as a difference $(\mathbf{u} - \mathbf{v})$ with $\mathbf{u} \geqslant \mathbf{0}$ and $\mathbf{v} \geqslant \mathbf{0}$, we conclude that an equivalent problem is to minimise $\mathbf{b}'\mathbf{z}$ subject to $A'\mathbf{z} \geqslant \mathbf{c}$.

7. Player I maximises λ subject to there being $\mathbf{p} = (p_1, p_2)' \geqslant \mathbf{0}$ with $\mathbf{e}'\mathbf{p} = 1$ and $A'\mathbf{p} \geqslant \lambda\mathbf{e}$. The set of $A'\mathbf{p}$ with $\mathbf{e}'\mathbf{p} = 1$ and $\mathbf{p} \geqslant \mathbf{0}$ is just the convex hull of the

columns of A^t (i.e. of the rows of A). From the diagrams we see that the cone $\{y: y \geqslant \lambda e\}$ intersects that convex hull with λ as large as possible when $\lambda = 1$ and the intersection point is given by $\mathbf{p} = (0, 1)^t$.

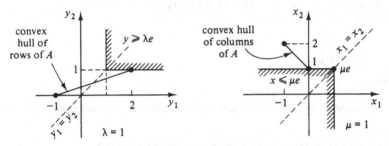

$$\lambda = 1 \qquad\qquad \mu = 1$$

Player II minimises μ subject to there being $\mathbf{q} = (q_1, q_2)^t \geqslant 0$ with $\mathbf{e}^t\mathbf{q} = 1$ and $A\mathbf{q} \leqslant \mu e$. The set of $A\mathbf{q}$ with $\mathbf{e}^t\mathbf{q} = 1$ and $\mathbf{q} \geqslant 0$ is just the convex hull of the columns of A. From the diagrams we see that the smallest value of μ for which the cone $\{\mathbf{x}: \mathbf{x} \leqslant \mu e\}$ intersects that convex hull, occurs when $\mathbf{q} = (0, 1)^t$ and $\mu = 1$.

8. Observe that the entries in the first row of A are greater than the corresponding entries in the second row. Hence Player I is better off playing his first pure strategy, i.e. $(1, 0)^t$, no matter what Player II does. Given this information Player II will minimise his losses by playing his second pure strategy $(0, 1, 0)^t$.

The result can be obtained graphically by inspecting $\mathfrak{A}$ the convex hull of the columns of A. The least μ for which the cone $\{\mathbf{x}: \mathbf{x} \leqslant \mu(1, 1)^t\}$ intersects $\mathfrak{A}$ is seen to be $\mu = 0$, for which value of μ the intersection point is $(0, -1)^t$. So II chooses $\mathbf{q} = (0, 1, 0)^t$.

9. The points awarded for a win are 1 to the winner or $\frac{1}{2}$ each in case of a tie. The pay-off is thus constant-sum. By a shift of origin the constant sum is zero.

10. In general the Lagrangian is $\mathbf{c}^t\mathbf{x} + \mathbf{y}^t(\mathbf{b} - A\mathbf{x})$. In our case

$$L = \frac{\mathbf{e}^t\mathbf{q}}{\lambda} + \frac{\mathbf{e}^t\mathbf{p}}{\mu} - \frac{\mathbf{p}^t A\mathbf{q}}{\lambda\mu}$$

and noting that for our application $\mathbf{e}^t\mathbf{p} = \mathbf{e}^t\mathbf{q} = 1$ we have

$$L = \frac{1}{\lambda} + \frac{1}{\mu} - \frac{\mathbf{p}^t A\mathbf{q}}{\lambda\mu}.$$

The assertion that L has a saddle point at $(\tilde{\mathbf{x}}, \tilde{\mathbf{y}})$ amounts to

$$L(\mathbf{x}, \tilde{\mathbf{y}}) \leqslant L(\tilde{\mathbf{x}}, \tilde{\mathbf{y}}) \leqslant L(\tilde{\mathbf{x}}, \mathbf{y}).$$

To see what this signifies in our case observe that

$$\tilde{\lambda} = \tilde{\mu} = \tilde{\mathbf{p}}^t A\tilde{\mathbf{q}}.$$

Using this in the leftmost inequality gives

$$\frac{1}{\tilde{\lambda}} + \frac{1}{\mu} - \frac{\tilde{\mathbf{p}}^t A\mathbf{q}}{\tilde{\lambda}\mu} \leqslant \frac{1}{\tilde{\lambda}} + \frac{1}{\tilde{\mu}} - \frac{\tilde{\lambda}}{\tilde{\lambda}\cdot\tilde{\mu}},$$

which reduces to

$$\frac{1}{\mu} \leqslant \frac{\tilde{\mathbf{p}}^t A \mathbf{q}}{\tilde{\lambda}\mu}.$$

Assuming $\mu > 0$ we obtain

$$\tilde{\lambda} \leqslant \tilde{\mathbf{p}}^t A \mathbf{q},$$

or

$$\tilde{\mathbf{p}}^t A \tilde{\mathbf{q}} \leqslant \tilde{\mathbf{p}}^t A \mathbf{q}.$$

This says that if Player I plays the strategy selected for him by the linear program, Player II is best off playing the strategy selected for him by his linear program. The other half of the saddle-point inequality will say analogously

$$\mathbf{p}^t A \tilde{\mathbf{q}} \leqslant \tilde{\mathbf{p}}^t A \tilde{\mathbf{q}}.$$

Solutions to selected exercises on simplex method

Exercises 13.6

2. We use the full tableau layout

M	x	y	z	u	v	w	const.	Ratios
	6	3	2	1	0	0	10	$\frac{10}{6}$
	2	1	2	0	1	0	6	$\frac{6}{2}$
	⑥	-6	1	0	0	1	6	$\frac{6}{6}$
1	-3	-2	-1	0	0	0	0	
	↑							
	0	⑨	1	1	0	-1	4	$\frac{4}{9}$
	0	3	$\frac{5}{3}$	0	1	$-\frac{1}{3}$	4	$\frac{4}{3}$
	1	-1	$\frac{1}{6}$	0	0	$\frac{1}{6}$	1	—
	0	-5	$-\frac{1}{2}$	0	0	$\frac{1}{2}$	3	
		↑						
	0	1	$\frac{1}{9}$	$\frac{1}{9}$	0	$-\frac{1}{9}$	$\frac{4}{9}$	—
	0	0	$\frac{4}{3}$	$-\frac{1}{3}$	1	0	$\frac{8}{3}$	—
	1	0	$\frac{5}{18}$	$\frac{1}{9}$	0	⑴⁄₁₈	$\frac{13}{9}$	26
	0	0	$\frac{1}{18}$	$\frac{5}{9}$	0	$-\frac{1}{18}$	$\frac{47}{9}$	
						↑		
	•	•	•	•	•	•	•	
	18	0	5	2	0	1	26	
	$+$	0	$+$	$+$	0	0	$\frac{60}{9}$	

Thus the maximum value is $\frac{60}{9}$.

3. There is a choice of pivot here. We begin by selecting a pivot in the first column.

M	x	y	z	u	v	w	const.	Ratios
	3	3	4	1	0	0	12	$\frac{12}{3}=4$
	4	6	3	0	1	0	12	$\frac{12}{4}=3$
	3	-6	1	0	0	1	3	1
1	-1	-1	-1	0	0	0	0	
	$\uparrow$							
	0	9	3	1	0	-0	9	$\frac{9}{9}=1$
	0	14	$\frac{5}{3}$	0	1	$-\frac{4}{3}$	8	$\frac{8}{14}=\frac{4}{7}$
	1	-2	$\frac{1}{3}$	0	0	$\frac{1}{3}$	1	—
	0	-3	$-\frac{2}{3}$	0	0	$\frac{1}{3}$	1	
		$\uparrow$						
	0	0	$\frac{27}{14}$	1	$-\frac{9}{14}$	$-\frac{1}{7}$	$\frac{27}{7}$	$\frac{14}{7}=2$
	0	1	$\frac{5}{42}$	0	$\frac{1}{14}$	$-\frac{4}{42}$	$\frac{4}{7}$	$\frac{24}{5}$
	1	0	$\frac{12}{21}$	0	$\frac{1}{7}$	$\frac{1}{7}$	$\frac{15}{7}$	$\frac{15}{4}$
	0	0	$-\frac{13}{42}$	0	$\frac{3}{14}$	$\frac{2}{42}$	$\frac{19}{7}$	
			$\uparrow$					
			1				2	
		1					$\frac{1}{3}$	
	1						1	
	0	0	0	+	+	+	$\frac{10}{3}$	

Thus the maximum is $\frac{10}{3}$ and is attained at $(1,\frac{1}{3},2)^t$.

Solutions to selected exercises on convex functions

Exercises 15.5

1. If X is not convex, the set $\{(x,y):x\in X \ \& \ f(x)\leqslant y\}$ cannot be convex. It is possible, however, to alter the clause which defines a function to be convex to read: f is convex on X if for every x_1 and x_2 in X the inequality

$$f(\alpha x_1 + (1-\alpha)x_2)\leqslant \alpha f(x_1)+(1-\alpha)f(x_2)$$

holds *whenever* $\alpha x_1 + (1-\alpha)x_2$ is in X.

3. If $f(x,y)=\exp\{-x^2-y^2\}$, then $Df=f(x,y)(-2x,-2y)$. Thus

$$D^2f = f(x,y)\begin{bmatrix} 4x^2-2 & 4xy \\ 4xy & 4y^2-2 \end{bmatrix}.$$

Hence

$$\det D^2 f = \{f(x, y)\}^2 \{(4x^2 - 2)(4y^2 - 2) - 16x^2 y^2\}$$
$$= \{f(x, y)\}^2 (4 - 8(x^2 + y^2)).$$

Thus the determinant is non-negative when $x^2 + y^2 \leqslant \frac{1}{2}$. Consequently f is concave in the disc centered at the origin and of radius $\frac{1}{2}\sqrt{2}$; indeed, $f_{xx} \leqslant 0$ when $4x^2 - 2 \leqslant 0$ i.e. for $|x| \leqslant \frac{1}{2}\sqrt{2}$. Similarly for f_{yy}. Thus f is concave in X.

Since the point $(1/3, 1/2, \exp\{-13/36\})$ lies on the surface $z = f(x, y)$ and $(1/3, 1/2)$ lies in the interior of X the required supporting hyperplane is the tangent plane at the given point. Its equation is

$$z - \exp(-\tfrac{13}{36}) = -\tfrac{2}{3}\exp\{-\tfrac{13}{36}\}(x - \tfrac{1}{3}) - \exp\{-\tfrac{13}{36}\}(y - \tfrac{1}{2}).$$

4. Put $f(\mathbf{x}) = A\mathbf{x} + \mathbf{b}$. Let $\alpha, \beta \geqslant 0$ be such that $\alpha + \beta = 1$, then

$$f(\alpha\mathbf{x} + \beta\mathbf{y}) = A(\alpha\mathbf{x} + \beta\mathbf{y}) + \mathbf{b}$$
$$= A(\alpha\mathbf{x} + \beta\mathbf{y}) + (\alpha + \beta)\mathbf{b}$$
$$= \alpha(A\mathbf{x} + \mathbf{b}) + \beta(A\mathbf{y} + \mathbf{b})$$
$$= \alpha f(\mathbf{x}) + \beta f(\mathbf{y}).$$

Hence f is both convex and concave.

5. If $f(\mathbf{x}) = \mathbf{x}^t A \mathbf{x}$ and A is symmetric observe that $Df = 2A\mathbf{x}$. This is because

$$f(\mathbf{x}) = \sum_{i,j} a_{ij} x_i x_j,$$

so

$$\frac{\partial f}{\partial x_k} = \sum_{i,j} a_{ij}\left(\frac{\partial x_i}{\partial x_k} x_j + x_i \frac{\partial x_j}{\partial x_k}\right)$$

$$= \sum_j a_{kj} x_j + \sum_i a_{ik} x_i = 2\sum_j a_{kj} x_j.$$

The last equation holds by symmetry of $A(a_{ik} = a_{ki})$. Thus $D^2 f = 2A$. So f is convex if A is non-negative definite; it is concave if A is non-positive definite.

6. By assumption we have, for all $\mathbf{x}$ that

$$f(\mathbf{x}) \leqslant f(\xi) + \nabla f(\xi)(\mathbf{x} - \xi),$$

but $\nabla f(\xi) = 0$ so the inequality reduces to

$$f(\mathbf{x}) \leqslant f(\xi),$$

as required.

7. There are two techniques for this problem.

   Epigraph approach. Since f is convex, the set $C = \{(\mathbf{x}, t): f(\mathbf{x}) \leqslant t\}$ is convex. If $\mathbf{x}_1, \ldots, \mathbf{x}_n$ lie in X and $\alpha_1, \ldots, \alpha_n \geqslant 0$ satisfy $\alpha_1 + \cdots + \alpha_n = 1$ then, since the points $(\mathbf{x}_i, f(\mathbf{x}_i))$ are in C, we have that

$$\alpha_1(\mathbf{x}_1, f(\mathbf{x}_1)) + \cdots + \alpha_n(\mathbf{x}_n, f(\mathbf{x}_n)) \in C,$$

i.e.

$$(\alpha_1 \mathbf{x}_1 + \cdots + \alpha_n \mathbf{x}_n, \alpha_1 f(\mathbf{x}_1) + \cdots + \alpha_n f(\mathbf{x}_n)) \in C,$$

but this says that

$$f(\alpha_1 \mathbf{x}_1 + \cdots + \alpha_n \mathbf{x}_n) \leqslant \alpha_1 f(\mathbf{x}_1) + \cdots + \alpha_n f(\mathbf{x}_n),$$

as required.

Chordal inequality approach. We take our starting point at the inequality

$$f(\alpha \mathbf{x}_1 + \beta \mathbf{x}_2) \leqslant \alpha f(\mathbf{x}_1) + \beta f(\mathbf{x}_2)$$

which is what Jensen's inequality reduces to when $n = 2$. This of course is the defining formula for convexity of f. We consider the case $n = 3$. Assume given $\alpha_1, \alpha_2, \alpha_3 \geqslant 0$ such that $\alpha_1 + \alpha_2 \neq 0$ (otherwise interchange the αs). Then the trick is to note that

$$\frac{\alpha_1}{\alpha_1 + \alpha_2} + \frac{\alpha_2}{\alpha_1 + \alpha_2} = 1,$$

hence

$$f\left(\frac{\alpha_1}{\alpha_1 + \alpha_2}\mathbf{x}_1 + \frac{\alpha_2}{\alpha_1 + \alpha_2}\mathbf{x}_2\right) \leqslant \frac{\alpha_1}{\alpha_1 + \alpha_2}f(\mathbf{x}_1) + \frac{\alpha_2}{\alpha_1 + \alpha_2}f(\mathbf{x}_2).$$

Now we turn to the desired convex combination:

$$f(\alpha_1 \mathbf{x}_1 + \alpha_2 \mathbf{x}_2 + \alpha_3 \mathbf{x}_3) = f\left((\alpha_1 + \alpha_2)\left\{\frac{\alpha_1}{\alpha_1 + \alpha_2}\mathbf{x}_1 + \frac{\alpha_2}{\alpha_1 + \alpha_2}\mathbf{x}_2\right\} + \alpha_3 \mathbf{x}_3\right)$$

$$\leqslant (\alpha_1 + \alpha_2)f\left(\left\{\frac{\alpha_1}{\alpha_1 + \alpha_2}\mathbf{x}_1 + \frac{\alpha_2}{\alpha_1 + \alpha_2}\mathbf{x}_2\right\}\right) + \alpha_3 f(\mathbf{x}_3)$$

$$\leqslant (\alpha_1 + \alpha_2)\left\{\frac{\alpha_1}{\alpha_1 + \alpha_2}f(\mathbf{x}_1) + \frac{\alpha_2}{\alpha_1 + \alpha_2}f(\mathbf{x}_2)\right\} + \alpha_3 f(\mathbf{x}_3)$$

$$= \alpha_1 f(\mathbf{x}_1) + \alpha_2 f(\mathbf{x}_2) + \alpha_3 f(\mathbf{x}_3).$$

This trick and calculation can be repeated for all cases n.

Solutions to selected exercises on non-linear programming

Exercises 16.7

1. We have

$$\begin{aligned}
g_1 &= 1 - x - y, & \nabla g_1 &= (-1, -1), \\
g_2 &= 5 - x^2 - y^2, & \nabla g_2 &= (-2x, -2y), \\
g_3 &= x, & \nabla g_3 &= (1, 0).
\end{aligned}$$

If $\boldsymbol{\xi} = (2, -1)^t$ is a maximum over the constraint region indicated, then there will be constants $\lambda_1, \lambda_2, \lambda_3 \geqslant 0$ such that

$$\lambda_1 g_1(\boldsymbol{\xi}) = \lambda_2 g_2(\boldsymbol{\xi}) = \lambda_3 g_3(\boldsymbol{\xi}) = 0$$

and

$$\nabla g(\xi) + \lambda_1 \nabla g_1(\xi) + \lambda_2 \nabla g_3(\xi) + \lambda_3 \nabla g_3(\xi) = 0.$$

Since $g_3(\xi) = 2 \neq 0$ we have that $\lambda_3 = 0$ so the latter condition becomes

$$\nabla g(\xi) + \lambda_1(-1, -1) + \lambda_2(-4, 2) = 0,$$

or

$$\nabla g(\xi) = \lambda_1(1, 1) + \lambda_2(4, -2).$$

which together with $\lambda_1 \geqslant 0$ and $\lambda_2 \geqslant 0$ says that $\nabla g(\xi)$ points into the normal cone at $(-2, 1)^t$.

Remark
The vector $(-2, 1)^t$ is orthogonal to the circle $g_2 = 0$ at $(-2, 1)^t$, since that vector lies along the radius; $(1, 1)^t$ is orthogonal to the straight line $1 - x - y = 0$.

2. Assume, of course that $A, B \geqslant 0$. If a, b, c are not all positive the constraint region will contain a point (x_0, y_0) with x_0 being the maximum value of x in the constraint region and y_0 being likewise the maximum value of y in the region. In all three cases arising from different sign choices it follows that the objective function is also maximised at the point (x_0, y_0).

The more interesting case occurs when a, b, c are all positive and this we now consider in detail.

Observe that to minimise $\{x^{-p} + y^{-q}\}$ we need the smallest λ such that the curve $x^{-p} + y^{-q} = \lambda$ has points in common with the feasible set. The shape of this curve merits attention. It is *convex* for $x, y \geqslant 0$ since

$$D^2 f = \begin{bmatrix} p(p+1)x^{-p-2} & 0 \\ 0 & q(q+1)y^{-q-2} \end{bmatrix},$$

which is positive definite. The set $\{(x, y): x^{-p} + y^{-q} \geqslant \lambda\}$ is thus convex. Further it has two asymptotes $x = \lambda^{-1/p}$ and $y = \lambda^{-1/q}$.

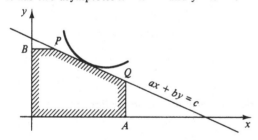

It is again clear that the minimum occurs somewhere along PQ but the exact point will depend on the relative sizes of a, b, c, A, B. One possibility is as drawn above. Since we are dealing with a convex programme the condition for the maximum is given by the Kuhn–Tucker theorem in necessary and sufficient form. The case illustrated above occurs if

$$\nabla f + \lambda \nabla(c - ax - by) = 0$$

for some $\lambda \geqslant 0$ viz.

$$(px^{-p-1}, qx^{-q-1}) + \lambda(-a, -b) = 0,$$

or

$$px^{-p-1} = \lambda a, \qquad qx^{-q-1} = \lambda b,$$

or

$$x = \left(\frac{p}{\lambda a}\right)^{1/(p+1)}, \quad y = \left(\frac{q}{\lambda b}\right)^{1/(q+1)}$$

Now x and y are subject to $ax + by = c$ so this requires

$$a\left(\frac{p}{\lambda a}\right)^{1/(p+1)} + b\left(\frac{q}{\lambda b}\right)^{1/(q+1)} = c \tag{1}$$

and this will surely be satisfied for some value of λ. We need still to check if

$$x \leqslant A, \qquad y \leqslant B. \tag{2}$$

To check (2) we consider the situation when, instead, the optimum occurs at P. The Kuhn–Tucker condition then becomes

$$\nabla f + \lambda \nabla(c - ax - by) + \mu \nabla(B - y) = 0.$$

Similarly at Q this is replaced by

$$\nabla f + \lambda \nabla(c - ax - by) + \mu \nabla(A - x) = 0.$$

In the first case we have

$$px^{-p-1} = \lambda a, \qquad qy^{-q-1} = \lambda b + \mu.$$

Since $y = B$ we have

$$qB^{-q-1} = \lambda b + \mu.$$

We thus need to satisfy

$$a\left(\frac{p}{\lambda a}\right)^{1/(p+1)} + bB = c$$

Since we also need

$$x = \left(\frac{p}{\lambda a}\right)^{1/(q+1)} \leqslant A$$

we obtain the following condition (by substitution from the last equation):

$$c \leqslant aA + bB,$$

which is certainly satisfied (by the hypothesis for this subcase). It remains to check $\mu \geqslant 0$. To do this we note that λ is px^{-p-1}/a and, remembering that $ax + bB = c$, we compute that

$$\mu = qB^{-q-1} - b\frac{p}{a}\left(\frac{a}{c - bB}\right)^{p+1}, \tag{3}$$

and this would have to be non-negative for an optimum at P.

A similar calculation at Q produces the requirement that the following

$$\mu = pA^{-p-1} - a\frac{q}{b}\left(\frac{b}{c-aA}\right)^{q+1} \tag{4}$$

should be non-negative for an optimum at Q.

If neither (3) nor (4) yields a non-negative result the Kuhn–Tucker theorem tells us that (1) and (2) are soluble.

3. We have

$$g_1 = 4 - x^2 - 2y^2, \qquad \nabla g_1 = (-2x, -4y),$$
$$g_2 = 6x - x^2 - 2y^2 - 5, \quad \nabla g_2 = (6 - 2x, -4y),$$
$$g_3 = \tfrac{1}{2}x - y, \qquad\qquad \nabla g_3 = (\tfrac{1}{2}, -1).$$

The Kuhn–Tucker condition requires that at a maximum $\xi = (x, y)^t$

$$\nabla \log(xy) + \lambda_1 \nabla g_1(\xi) + \lambda_2 \nabla g_2(\xi) + \lambda_3 \nabla g_3(\xi) = 0.$$

and

$$\lambda_1 g_1(\xi) = \lambda_2 g_2(\xi) = \lambda_3 g_3(\xi) = 0$$

for some $\lambda_1, \lambda_2, \lambda_3 \geq 0$. Thus we require

$$\left(\frac{1}{x}, \frac{1}{y}\right) + \lambda_1(-2x, -4y) + \lambda_2(6 - 2x, -4y) + \lambda_3(\tfrac{1}{2}, -1) = 0$$

i.e.

$$\left.\begin{array}{l} \dfrac{1}{x} + \lambda_1(-2x) + \lambda_2(6 - 2x) + \lambda_3(\tfrac{1}{2}) = 0 \\[2mm] \dfrac{1}{y} + \lambda_1(-4y) + \lambda_2(-4y) + \lambda_3(-1) = 0 \end{array}\right\},$$

or

$$\left.\begin{array}{l} 1 - 2\lambda_1 x^2 + \lambda_2 x(6 - 2x) + \tfrac{1}{2}\lambda_3 x = 0 \\ 1 - 4\lambda_1 y^2 - 4\lambda_2 y^2 - \lambda_3 y = 0 \end{array}\right\}.$$

Adding the two equations we obtain

$$2 - 2\lambda_1(x^2 + 2y^2) - 2\lambda_2(x^2 + 2y^2) + 6\lambda_2 x + \lambda_3(\tfrac{1}{2}x - y) = 0$$

and the last term disappears since $\lambda_3 g_3 = 0$. Thus

$$2 - 2\lambda_1(x^2 + 2y^2) - 2\lambda_2(x^2 + 2y^2) + 6\lambda_2 x = 0.$$

Suppose $\lambda_1 = 0$, then this reduces still further to

$$\begin{aligned} 2 &= 2\lambda_2(x^2 + 2y^2) - 6\lambda_2 x \\ &= 2\lambda_2(x^2 + 2y^2) - \lambda_2(x^2 + 2y^2 + 5) \\ &= \lambda_2(x^2 + 2y^2 - 5). \end{aligned}$$

Here we have used the fact that $\lambda_2 g_2 = 0$. The expression above on the right-hand side of the equation is negative since $x^2 + 2y^2 \leq 4$ and $\lambda_2 \geq 0$. This is a contradiction. So after all $\lambda_1 > 0$. Hence $g_1 = 0$, i.e. $x^2 + 2y^2 = 4$ and the minimum lies on the first of the two ellipses. This implies that it does not lie on the other ellipse. So at the minimum $g_2 \neq 0$ and $\lambda_2 = 0$. We now have

$$
\begin{aligned}
2 &= 2\lambda_1(x^2 + 2y^2) \\
&= 2\lambda_1(4).
\end{aligned}
$$

Thus $\lambda_1 = \frac{1}{4}$. Suppose $\lambda_3 = 0$, then

$$
\frac{1}{x} + \lambda_1(-2x) = 0,
$$

so $2x^2 = 4$ or $x = \pm\sqrt{2}$. Similarly $y = \pm 1$. But neither point $(\pm\sqrt{2}, \pm 1)$ lies in the feasible set. So $g_3 = 0$ and we have localised the minimum to the intersection of $g_3 = 0$ with $g_1 = 0$. Since $y = \frac{1}{2}x$ we have $3/2x^2 = 4$; so $x = \sqrt{(8/3)} = 2\sqrt{(2/3)}$ and $y = \sqrt{(2/3)}$.

4. The constraint region consists of the one point $(0,0)$. No path can properly enter the constraint set and the constraint qualification is not satisfied.

Exercises 16.8

2. We form the Lagrangian

$$
L = x^2 + y^2 + z^2 + \lambda(x^2 + y^2 + 4z^2 - 1) + \mu(x + 3y + 2z).
$$

We thus obtain

$$
\frac{\partial L}{\partial x} = 2x + 2\lambda x + \mu = 0,
$$

$$
\frac{\partial L}{\partial y} = 2y + 2\lambda y + 3\mu = 0,
$$

$$
\frac{\partial L}{\partial z} = 2z + 8\lambda z + 2\mu = 0.
$$

Multiplying the equations respectively by x, y and z and adding we obtain:

$$
2(x^2 + y^2 + z^2) + 2\lambda(x^2 + y^2 + 4z^2) + \mu(x + 3y + 2z) = 2M + 2\lambda = 0.
$$

Thus $M = -\lambda$ where M is the maximum. Evidently we can also solve the equations for x, y and z in terms of λ and μ. Thus, assuming $\lambda \neq -1$ and $\lambda \neq -\frac{1}{4}$, we have

$$
\left.
\begin{aligned}
x &= \frac{-\mu}{2(1 + \lambda)} \\[2mm]
y &= \frac{-3\mu}{2(1 + \lambda)} \\[2mm]
z &= \frac{-\mu}{(1 + 4\lambda)}
\end{aligned}
\right\}.
$$

Substitution into the linear constraint equation gives

$$
\frac{-\mu}{2(1 + \lambda)} + \frac{-9\mu}{2(1 + \lambda)} + \frac{-2\mu}{(1 + 4\lambda)} = 0.
$$

One solution to this equation is $\mu = 0$. That, however, would give $x = y = z = 0$ which cannot be optimal. Thus, since $\mu \neq 0$, we obtain

$$\frac{5}{(1 + \lambda)} + \frac{2}{(1 + 4\lambda)} = 0.$$

Thus $22\lambda + 7 = 0$, or $\lambda = -7/22$. Consequently, our assumptions on λ lead to $M = 7/22$. The other possibilities are: $\lambda = -1$ (so that $\mu = 0$ and $M = 1$) which fixes z at 0 and x and y are respectively $\pm 3/\sqrt{10}$, $\mp 1/\sqrt{10}$; and, $\lambda = -\frac{1}{4}$ (so that $\mu = 0$ and $M = \frac{1}{4}$) which implies $x = y = 0$ and then the two constraints cannot be simultaneously satisfied.

The above discussion shows that the maximum is 1 and the minimum is 7/22.

3. We form the Lagrangian

$$L = xyz + \lambda(x + y + z - 5) + \mu(xy + yz + zx - 8).$$

We thus obtain

$$\frac{\partial L}{\partial x} = yz + \lambda + \mu(y + z) = 0,$$

$$\frac{\partial L}{\partial y} = zx + \lambda + \mu(z + x) = 0,$$

$$\frac{\partial L}{\partial z} = xy + \lambda + \mu(x + y) = 0.$$

In problems like this it is always worth investigating relationships which follow from symmetry properties of the Lagrange equations and the constraint equations. Two such relations are immediately available.

First we add all three equations and using the constraints obtain

$$xy + yz + zx + 3\lambda + \mu(2x + 2y + 2z) = 8 + 3\lambda + 10\mu = 0.$$

We arrive at:

$$3\lambda + 10\mu = -8.$$

Next multiplying the three equations respectively by x, y and z and adding we also obtain:

$$3xyz + \lambda(x + y + z) + \mu(xy + zx + zy + xy + xz + xy) = 0.$$

Thus letting the maximum be M we deduce that

$$3M + 5\lambda + 16\mu = 0.$$

A further symmetry can be called into play by subtracting the equations pairwise. We then have for instance:

$$yz - xz + \mu(y - x) = 0$$

or

$$(y - x)(z + \mu) = 0.$$

Thus either $x = y$ or $\mu = -z$. This is not the only alternative. By symmetry the full set of alternatives is

$$x = y \quad \text{or} \quad z = -\mu,$$

and

$$y = z \quad \text{or} \quad x = -\mu,$$

and

$$z = x \quad \text{or} \quad y = -\mu.$$

We deduce that the possible solutions are: (i) $x = y = z = -\mu$, (ii) $x = y = -\mu$ with z unspecified; etc.

On substitution into the constraint equations the first of these leads to the equations $3\mu^2 = 8$ and $3\mu = -5$ which contradict each other. So the first solution is excluded.

The second solution leads to $y = 5 + 2\mu$ and $-2\mu y + \mu^2 = 8$. Eliminating y gives

$$3\mu^2 + 10\mu + 8 = 0,$$
$$(3\mu + 4)(\mu + 2) = 0.$$

We note that, by symmetry all of the other solutions leads to the same equation in μ.

Now $\mu = -2$ gives $3\lambda + 10(-2) = -8$, so that $\lambda = 4$ and hence $M = -(5 \cdot 4 + 16 \cdot (-2))/3 = 4$.

On the other hand $\mu = -4/3$ leads to $3\lambda + 10(-4/3) = -8$ so that $\lambda = 16/9$ with the result that $M = -(5 \cdot 16/9 + 16 \cdot (-4/3))/3 = 16.7/27 = 112/27$.

Of the two possible choices for M the smaller is 4. Hence the minimum value here is 4. Note that this value is achieved when $x = z = 2$ and $y = 1$. Evidently any permutation of $2, 2, 1$ will give a minimum point.

Solutions to selected exercises on the integration process

Exercises 17.19

1. Let $P = \{x_0, x_1, \ldots, x_n\}$ be a partition of $[0, 1]$. The R–S sums will be

$$t_1^3\{[x_1] - [x_0]\} + t_2^3\{[x_2] - [x_1]\} + \cdots + t_n^3\{[x_n] - [x_{n-1}]\}.$$

Now $[x_r] = 0$ for $r = 0, 1, \ldots, n - 1$ since $0 \leqslant x_r < 1$ while $[x_n] = [1] = 1$. Thus the sum reduces to

$$t_n^3\{1 - 0\}$$

and this will be close to 1 if t_n is close to unity.

The precise argument looks like this. To control how close t_n^3 is to 1, we consider what happens when t_n satisfies

$$1 - \delta \leqslant t_n \leqslant 1.$$

We suppose, of course, that δ is selected less than 1. Then

$$(1 - \delta)^3 = 1 - 3\delta + 3\delta^2 - \delta^3 \geqslant 1 - 3\delta - \delta^3 \geqslant 1 - 3\delta - \delta = 1 - 4\delta.$$

This is because $\delta^3 = \delta^2 \cdot \delta \leqslant 1 \cdot \delta$. Evidently, $1 - 4\delta \leqslant t_n^3 \leqslant 1$. So we can arrange to have t_n^3 to within ε of 1 if we take $\delta = \varepsilon/4$, for then

$$1 - \varepsilon \leqslant t_n^3 \leqslant 1.$$

Thus, if P is finer than the partition $\{0, 1 - \varepsilon/4, 1\}$, we shall have $1 - \varepsilon/4 \leqslant x_{n-1}$ (and so $1 - \varepsilon/4 \leqslant t_n$) and the R–S sum will be to within ε of 1.

4.

$$\int_0^2 x d([2^x]) = \int_1^4 \log_2 t d[t] = \sum_2^4 \log_2 n = \log_2 2 \cdot 3 \cdot 4 = 3 + \log_2 3,$$

since

$$t = 2^x \Rightarrow x = \log_2 t.$$

$$\int_0^3 x d(x[x^2]) = [x^2[x^2]]_0^3 - \int_0^3 [x^2] x dx = 81 - \frac{1}{2} \int_0^3 [x^2] d(x^2)$$

$$= 81 - \frac{1}{2}\left\{[[x^2]x^2]_0^3 - \int_0^3 x^2 d[x^2]\right\}$$

$$= \frac{81}{2} + \frac{1}{2}\int_0^9 t d[t] = \frac{81}{2} + \frac{1}{2}\sum_{n=1}^9 n = \frac{81}{2} + \frac{1}{4} \cdot 9 \cdot 10 = \frac{126}{2} = 63.$$

$$\int_2^5 [5 - x] d(\log[x]) = \sum_{n=3}^5 (5 - n)\{\log n - \log(n - 1)\}$$

$$= 2 \cdot \log \tfrac{3}{2} + \log \tfrac{4}{3} = \log \frac{3^2 4}{2^2 3} = \log 3.$$

5. Let

$$L = \int_0^n x^3 d[x]$$

$$= [x \tfrac{1}{6}[x]([x] + 1)(2[x] + 1)]_0^n - \int_0^n \frac{1}{6}[x]([x] + 1)(2[x] + 1) dx$$

$$6L = n^2(n + 1)(2n + 1) - \int_0^n [x]\{2[x]^2 + 3[x] + 1\} dx$$

$$= n^2(n + 1)(2n + 1) - 2(L - n^3) - \int_0^n \{3[x]^2 + [x]\} dx$$

$$8L = n^2(n + 1)(2n + 1) - \tfrac{3}{6}n(n - 1)(2n - 1) - \tfrac{1}{2}n(n - 1) + 2n^3$$

$$= n^2(n + 1)(2n + 1) - \tfrac{1}{2}n(n - 1)(2n) + 2n^3$$

$$= n^2\{(n + 1)(2n + 1) - (n - 1) + 2n\}$$

$$= n^2\{2n^2 + 4n + 2\},$$

so

$$L = \tfrac{1}{4}n^2(n + 1)^2.$$

6.

$$\int_0^n 2xd(x[x]) = [2x^2[x]]_0^n - \int_0^n [x]2xdx$$

$$= 2n^3 - \int_0^n [x]d(x^2) = 2n^3 - [x^2[x]]_0^n + \int_0^n x^2d[x]$$

$$= n^3 + \sum_{r=1}^n r^2.$$

Exercises 17.20

1. (c) Two methods come to mind.
   Integrating by parts:

$$\int_0^x f(t)dt = [f(t)t]_0^x - \int_0^x tdf(t)$$

$$= xf(x) - \sum_N^\infty 2^{-n} \cdot 2^{-n},$$

where N is the smallest integer with $2^{-N} < x \ (\leqslant 2^{-N+1})$. The formula is based on the observation that for each $n \geqslant N$, f is constant in the interval $(2^{-n}, 2^{-n+1}]$ and the jump in f at 2^{-n} is $2^{-n+1} - 2^{-n} = 2^{-n}$.
This explains the contribution

$$\int_{2^{-n}}^{2^{-n+1}} tdf(t) = 2^{-n} \times \text{jump in } f \text{ at } 2^{-n}.$$

Now the geometric series sums to

$$\frac{4^{-N}}{1 - \frac{1}{4}} = \frac{4}{3}4^{-N} = \frac{1}{3}(2^{-N+1})^2 = \frac{1}{3}\{f(x)\}^2.$$

Alternatively, we can compute the area under the curve directly as a sum of contributions like

$$\tfrac{1}{2}2^{-n}(=\text{base}) \times 2^{-n}(=\text{height}),$$

arising from full sized rectangles (this is so for $n \geqslant N$) plus the incomplete rectangle whose base has length $(x - 2^{-N})$ and height 2^{-N+1}. Thus the area comes to

$$\frac{1}{2}\sum_N^\infty 2^{-n}\cdot 2^{-n} + (x - 2^{-N})2^{-N+1}$$

$$= \frac{1}{2}\frac{4^{-N}}{1 - \frac{1}{4}} + x2^{-N+1} - \tfrac{1}{2}(2^{-N+1})^2$$

$$= \tfrac{1}{6}(2^{-N+1})^2 + x2^{-N+1} - \tfrac{1}{2}(2^{-N+1})^2$$

$$= -\tfrac{1}{3}(2^{-N+1})^2 + x2^{-N+1}$$

$$= -\tfrac{1}{3}\{f(x)\}^2 + xf(x).$$

2.

$$\int_0^n f(x)d(2x(1+[x])) = [f(x)(2x(1+[x]))]_0^n$$

$$+ \int_0^n 2x(1+[x])\exp(-x^2)dx$$

$$= 2n(1+n)f(n) - \int_0^n (\exp(-x^2))(-2x)dx$$

$$- \int_0^n [x](\exp(-x^2))(-2x)dx$$

$$= 2n(1+n)f(n) - [\exp(-x^2)]_0^n$$

$$- \int_0^n [x]d(\exp(-x^2))$$

$$= 2n(1+n)f(n) - \exp(-n^2)$$

$$+ 1 - [[x]\exp(-x^2)]_0^n + \int_0^n \exp(-x^2)d[x]$$

$$= 2n(1+n)f(n) - \exp(-n^2) + 1 - n\exp(-n^2)$$

$$+ \sum_1^n \exp(-r^2)$$

$$= 2n(1+n)\int_n^\infty \exp(-t^2)dt - (1+n)\exp(-n^2)$$

$$+ \sum_0^n \exp(-r^2)$$

$$\to \sum_0^\infty \exp(-r^2).$$

3. (a)

$$\int_1^n [x]sx^{-s-1}dx = \int_1^n [x]d(-x^{-s}) = [-[x]x^{-s}]_1^n - \int_1^n -x^{-s}d[x]$$

$$= -\frac{1}{n^{s-1}} + 1 + \sum_2^n \frac{1}{r^s}.$$

Thus

$$\sum_1^n \frac{1}{r^s} = \frac{1}{n^{s-1}} + s\int_1^n \frac{[x]dx}{x^{s+1}}.$$

(b) Let

$$I = \int_1^{2n} (2[x/2] - [x]) \cdot sx^{-s-1} dx$$

$$= [-x^{-s}(2[x/2] - [x])]_1^{2n} + \int_1^{2n} x^{-s} d(2[x/2] - [x]).$$

At this stage it is important to plot the graph of the function $(2[x/2] - [x])$ to examine the size and location of jump discontinuities. This done we see that

$$I = 0 - 1 + \sum_2^{2n} \frac{1}{r^s}(-1)^r = \sum_1^{2n}(-1)^r \frac{1}{r^s}.$$

4. We assume that f is continuous (in fact left-sided continuity at the integer points is all we require). Note that $A(x) = 0$ for $x < 1$.

$$\sum_{n \leqslant a} a_n f(n) = \sum_1^{[a]} a_n f(n) = \int_0^{[a]} f(x) dA(x)$$

$$= [f(x)A(x)]_0^{[a]} - \int_0^{[a]} A(x) df(x)$$

$$= f([a])A([a]) - \sum_0^{[a]-1} \int_n^{n+1} A(x) df(x) \qquad (+)$$

$$= f([a])A([a]) - \sum_0^{[a]-1} A(n)\{f(n+1) - f(n)\}$$

$$= \sum_0^{[a]} A(n)\{f(n) - f(n+1)\} - (A([a])\{f([a]) - f([a]+1)\})$$

$$+ f([a])A([a])$$

$$= \sum_{n \leqslant a} A(n)\{f(n) - f(n+1)\} + A([a])f([a]+1).$$

We justify step $(+)$ by noting that $A(x)$ is constant most of the time; more precisely, note that for any small δ

$$\int_n^{n+1} A(x) df(x) = \int_n^{n+1-\delta} A(x) df(x) + \int_{n+1-\delta}^{n+1} A(x) df(x)$$

and the first of the summands is

$$A(n)\{f(n+1-\delta) - f(n)\}$$

whilst the second is nearly zero for small δ. (To see this refer to the R–S sums.)
5. We have

$$\int_a^b \langle x \rangle f'(x) dx = [\langle x \rangle f(x)]_a^b - \int_a^b f(x) d(x - [x])$$

$$= [\langle x \rangle f(x)]_a^b - \int_a^b f(x) dx + \int_a^b f(x) d[x]$$

$$= \langle b \rangle f(b) - \langle a \rangle f(a) - \int_a^b f(x)dx + \sum_{a < n \leqslant b} f(n).$$

Rearrangement of the terms gives the desired result.

Exercises 17.21

2.

$$\int_0^y e^x d[x] = \sum_{n=1}^{[y]} e^n = \frac{e^{[y]+1} - e}{e - 1}$$

$$\int_0^y e^{[x]} dx = \sum_{n=0}^{[y]-1} e^n + (y - [y])e^{[y]} = \frac{e^{[y]} - 1}{e - 1} + (y - [y])e^{[y]}$$

$$\int_0^n x e^x d[x] = \left[x \frac{e^{[x]+1} - e}{e - 1} \right]_0^n - \int_0^n \frac{e^{[x]+1} - e}{e - 1} dx$$

$$= \frac{ne^{n+1} - ne}{e - 1} + \frac{ne}{e - 1} - \frac{e}{e - 1} \left\{ \frac{e^n - 1}{e - 1} \right\}$$

$$= \frac{ne^{n+1}}{e - 1} - \frac{e^{n+1} - e}{(e - 1)^2}$$

$$= \frac{ne^{n+2} - (n + 1)e^{n+1} + e}{(e - 1)^2}.$$

Exercises 17.22

2. For each n, we consider the particular partition

$$P_n = \left\{ 0, \frac{1}{2n}, \ldots, \frac{1}{6}, \frac{1}{4}, \frac{1}{2}, 1 \right\}.$$

In each of the subintervals $I_r = [1/(2r), 1/(2r - 2)]$ the function is either increasing or decreasing. Thus for $r \geqslant 2$

$$v_f \left[\frac{1}{(2r - 2)}, \frac{1}{(2r)} \right] = \frac{1}{(2r - 2)^2} - \frac{1}{(2r)^2} \leqslant \frac{1}{2r^2}.$$

Now, given any partition $P = \{x_0, x_1, \ldots, x_m\}$, choose k so large that $1/(2k) < x_1$. Let Q be the common refinement of P and P_k. Then $v(P) \leqslant v(Q)$ and since on each interval I_r the function f is monotonic we have

$$v(Q) \leqslant \frac{1}{(2k)^2} + \left(\sum_2^k \frac{1}{(2r - 2)^2} - \frac{1}{(2r)^2} \right) + 1$$

$$\leqslant \frac{1}{4k^2} + \frac{1}{2} \left(\frac{1}{2^2} + \frac{1}{3^2} + \cdots + \frac{1}{k^2} \right) + 1$$

$$< 1 + \frac{1}{2} \sum \frac{1}{r^2} + \frac{1}{4} < \infty.$$

3. Let P be a partition of $[a,c]$. Then $P_c = P \cup \{b\}$ is a partition of $[a,b]$. Thus for all P

$$v_{[a,c]}(P) \leqslant v_{[a,b]}(P_c) \leqslant V_f[a,b] < \infty.$$

Similar considerations apply to $[c,b]$.

To prove the equation we will find it convenient to prove inequalities both ways round. Let $P = \{x_0, x_1, \ldots, x_n\}$ be a partition of $[a,b]$. Choose r so that $x_r \leqslant c < x_{r+1}$. Then $P_1 = \{x_0, x_1, \ldots, x_r, c\}$ is a partition of $[a,c]$ and $P_2 = \{c, x_{r+1}, \ldots, x_n\}$ is a partition of $[c,b]$. Thus

$$v_{[a,b]}(P) \leqslant v_{[a,c]}(P_1) + v_{[c,b]}(P_2) \leqslant V_f[a,c] + V_f[c,b].$$

It now follows (taking suprema over all P) that

$$V_f[a,b] \leqslant V_f[a,c] + V_f[c,b].$$

Next we prove the reverse inequality. Since $V_f[a,c]$ is the supremum of all $v_{[a,c]}(P)$ let us take a small number δ. Then there will be a partition P_δ such that

$$V_f[a,c] - \delta \leqslant v_{[a,c]}(P_\delta).$$

Similarly there is Q_δ, a partition of $[c,b]$, such that

$$V_f[c,b] - \delta \leqslant v_{[c,b]}(Q_\delta).$$

Hence, adding we obtain

$$V_f[a,c] + V_f[c,b] - 2\delta \leqslant v_{[a,c]}(P_\delta) + v_{[c,b]}(Q_\delta)$$
$$= v_{[a,b]}(P_\delta \cup Q_\delta) \leqslant V_f[a,b].$$

This is because $P_\delta \cup Q_\delta$ is a partition of $[a,b]$.

Thus for every δ we have

$$V_f[a,c] + V_f[c,b] - 2\delta \leqslant V_f[a,b],$$

and taking the limit as δ tends to zero we obtain

$$V_f[a,c] + V_f[c,b] \leqslant V_f[a,b],$$

and the two inequalities taken together imply the required equation.

If $a \leqslant x \leqslant y \leqslant b$ we have, by the same reasoning as for the last equation,

$$V_f[a,x] + V_f[x,y] = V_f[a,y],$$

but $V_f[x,y] \geqslant 0$ so

$$V_f[a,x] \leqslant V_f[a,y],$$

i.e. $\alpha(x)$ is increasing.

Since $\{x,y\}$ is a partition of $[x,y]$ we have

$$|f(x) - f(y)| \leqslant V_f[x,y],$$

hence

$$-V_f[a,b] \leqslant -V_f[a,x] \leqslant f(x) - f(a) \leqslant V_f[a,x] \leqslant V_f[a,b],$$

so

$$- V_f[a, b] + f(a) \leqslant f(x) \leqslant V_f[a, b] + f(a),$$

i.e. $f(x)$ is bounded in $[a, b]$. We consider $\beta(x) = f(x) - \alpha(x)$. We have for $a \leqslant x \leqslant y \leqslant b$

$$\begin{aligned}
\beta(y) - \beta(x) &= f(y) - f(x) - \{\alpha(y) - \alpha(x)\} \\
&= f(y) - f(x) - \{V_f[a, y] - V_f[a, x]\} \\
&= f(y) - f(x) - \{V_f[x, y]\} \\
&\leqslant 0.
\end{aligned}$$

so β is decreasing.

4. For α of bounded variation the inequality

$$|\alpha(x) - \alpha(y)| \leqslant V_\alpha[x, y],$$

enables one to prove

$$\left| \int_a^b f(x) d\alpha(x) \right| \leqslant \int_a^b |f(x)| dV_\alpha[a, x].$$

5. We consider the case of $\alpha\beta$. Let $P = \{x_0, x_1, \ldots, x_n\}$ be a partition of $[a, b]$, then

$$\begin{aligned}
|\alpha\beta(x_{r+1}) - \alpha\beta(x_r)| &= |\alpha(x_{r+1})\beta(x_{r+1}) - \alpha(x_r)\beta(x_{r+1}) \\
&\quad + \alpha(x_r)\beta(x_{r+1}) - \alpha(x_r)\beta(x_r)| \\
&\leqslant |\alpha(x_{r+1}) - \alpha(x_r)| |\beta(x_{r+1})| \\
&\quad + |\alpha(x_r)| |\beta(x_{r+1}) - \beta(x_r)| \\
&\leqslant |\alpha(x_{r+1}) - \alpha(x_r)| B + A |\beta(x_{r+1}) - \beta(x_r)|,
\end{aligned}$$

where A and B are bounds for α and β (see Question 3). It follows, on summation, etc., that

$$V_{\alpha\beta}[a, b] \leqslant B V_\alpha[a, b] + A V_\beta[a, b].$$

If β is bounded away from zero, suppose $K \leqslant |\beta(x)|$ in $[a, b]$, where $K > 0$. We have

$$\left| \frac{1}{\beta(x_{r+1})} - \frac{1}{\beta(x_r)} \right| = \left| \frac{\beta(x_{r+1}) - \beta(x_r)}{\beta(x_{r+1})\beta(x_r)} \right| \leqslant \frac{|\beta(x_{r+1}) - \beta(x_r)|}{K^2}$$

Thus we see that the variation of β^{-1} is in this case bounded by

$$K^{-2} V_\beta[a, b].$$

Solutions to selected exercises on manipulation of integrals

Exercises 18.14

1. We differentiate with respect to x the expression

$$\phi(x) = \arctan\left(\frac{x + y}{1 - xy}\right) - \arctan(x)$$

using the defining integral. Thus we obtain

$$\frac{1}{1+\left(\dfrac{x+y}{1-xy}\right)^2}\cdot\frac{1\cdot(1-xy)-(x+y)(-y)}{(1-xy)^2}-\frac{1}{1+x^2},$$

but the first term is

$$\frac{1-xy+xy+y^2}{1-2xy+x^2y^2+x^2+2xy+y^2}=\frac{1+y^2}{(1+y^2)(1+x^2)}.$$

Thus the derivative of the expression $\phi(x)$ is zero. Hence $\phi(x)$ is independent of x. To obtain its value put $x=0$. The expression reduces to $\arctan(y)$. Hence:

$$\arctan\left(\frac{x+y}{1-xy}\right)-\arctan(x)=\arctan(y).$$

2. Let us denote the value of the integral by $I(t)$. We notice that the integral is minus the derivative of the following integral, which we denote by $\phi(t)$:

$$\phi(t)=\int_0^{\pi/2}\frac{dx}{(1+t\sin^2 x)}.$$

We obtain the value of $\phi(t)$. Since the integrand is even with respect to $\sin x$ we use the standard substitution

$$\tan x=u,$$

so that

$$\sin x=\frac{\sin x}{\sqrt{\{\sin^2 x+\cos^2 x\}}}=\frac{u}{\sqrt{(1+u^2)}},$$

$$\frac{dx}{du}=\frac{1}{1+u^2}.$$

Thus

$$\phi(t)=\int_0^\infty\frac{du}{(1+u^2+tu^2)}$$

$$=\int_0^\infty\frac{du}{\{1+(1+t)u^2\}}=\left[\frac{1}{\sqrt{(1+t)}}\arctan u\sqrt{(1+t)}\right]_0^\infty$$

$$=\frac{\pi}{2\sqrt{(1+t)}}.$$

Now we differentiate with respect to t and obtain

$$\phi'(t)=-\int_0^{\pi/2}\frac{\sin^2 x\,dx}{(1+t\sin^2 x)^2}=-\frac{\pi}{4\sqrt{(1+t)^3}}.$$

The manipulation is justified provided $t\neq-1$.

3. (iii) The integrand decays a little slower than x^{-p}. We therefore compare the integrand against something manageable which is less than x^{-p}; we try x^{-q} where $1 < q < p$, the idea being that the difference between p and q counterbalances $\log x$. We keep $q > 1$ for convergence [the integral of x^{-q} is convergent in the range $[1, \infty)$]. Applying the ratio test (and l'Hôpital's Rule en route), we have, taking limits as x tends to infinity:

$$\lim \frac{\log x}{x^p} \div \frac{1}{x^q} = \lim \frac{\log x}{x^{p-q}} = \lim \frac{1/x}{(p-q)x^{(p-q)-1}}$$

$$= \lim \frac{1}{(p-q)x^{(p-q)}} = 0, \text{ since } (p-q) > 0.$$

In this case the ratio is zero but the integrand of x^{-q} converges hence so does the integral in question. [Reassure yourself, by noting that for large values of x the test ratio is less than 1, so for such x, we have

$$0 < \frac{\log x}{x^p} < \frac{1}{x^q}$$

and the comparison method clinches the convergence. See also Question 8.]

(iv) It is instructive (though we omit this) to test by comparing with powers of x, as in the last question. The results are usually inconclusive. It is best therefore to make the substitution $y = \log x$ in order to bring into focus the effects of the factor $(\log x)^p$. The integral transforms very nicely to

$$\int_{\log 2}^{\infty} \frac{dy}{y^p}$$

and this converges, since $p > 1$.

(v) We begin by rationalizing the denominator. The integrand becomes

$$\frac{\sqrt{(x+1)} + \sqrt{x}}{x},$$

so this is approximately $2x^{-1/2}$. The integral of the latter converges in the range $[1, \infty)$.

(vi) Here again it is best to use the identity $a^3 - b^3 = (a - b)(a^2 + ab + b^2)$, with $b = x$ and $a = (x^3 + 4)^{1/3}$. The integrand is then

$$\frac{a^3 - b^3}{a^2 + ab + b^2}$$

which for large x is approximately

$$\frac{4}{3x^2}.$$

Now the integral of this converges in the interval $[1, \infty)$. It remains to check that the ratio of the given integrand to this approximation is in limit 1.

(vii) The integrand only apparently has a blow-up at $x = 0$. Actually of course,

$$\lim_{x \to 0} \frac{\sin x}{x} = 1,$$

so the integrand may be regarded as continuous at 0 provided we read the value of the integrand at 0 to be 1.

(viii) We follow the hint, integrating by parts from 1 to X:

$$\int_1^X \frac{\sin x}{x} dx = \left[\frac{-\cos x}{x} \right]_1^X - \int_1^X \frac{\cos x}{x^2} dx$$

$$\to \frac{\cos 1}{1} - \int_1^\infty \frac{\cos x}{x^2} dx.$$

This last integral exists by comparison with

$$\int_1^\infty \frac{1}{x^2} dx.$$

4. (i) Close to 0 the integrand is approximately x^p. But

$$\int_\delta^1 x^p dx = \left[\frac{x^{p+1}}{p+1} \right]_\delta^1 = \frac{1}{p+1}(1 - \delta^{p+1})$$

and this exists in the limit as δ tends to zero if $p + 1 > 0$. Thus we require $p > -1$.

Similarly close to $x = 1$ the integrand is approximately $2^q(1 - x)^q$. The integral of the latter function is convergent in $[0, 1]$ provided $q + 1 > 0$. [Put $y = 1 - x$ and apply the previous argument.] Thus we also require $q > -1$.

(ii) We examine the factor x^x first. This is the same as $e^{x \log x}$. But $x \log x$ tends to 0 as x tends to 0. So the integral converges in the range $(0,1]$ provided that the other factor does not blow up. Now consider that

$$-x^p + x \log x = -x^p(1 - x^{1-p} \log x).$$

As x tends to infinity the expression $x^{1-p} \log x$ tends to zero if $1 - p < 0$ and to infinity if $1 - p \geqslant 0$. Hence we require $1 < p$ (otherwise our integrand tends to infinity). In this case, as x tends to infinity

$$1 - x^{1-p} \log x$$

is approximately 1 and so the integrand is approximately $\exp\{-x^p\}$. The latter function has convergent integral in the range $(1, \infty)$ provided $p > 0$. Thus we require just $p > 1$ to secure convergence. To finish off the argument one should use the limit ratio test. If we compute the limiting ratio of $\exp\{-x^p\}$ to $\exp\{x \log x - x^p\}$, this, however, turns out to be ∞ which would be inconclusive. The trick is therefore to take any q with $1 < q < p$ and consider for comparison the function $\exp\{-x^q\}$. This time the ratio as x tends to infinity turns out to be

$$\exp\{x \log x - x^p\}/\exp\{-x^q\} = \exp\{x^q(1 + x^{1-q} \log x) - x^p\}$$

$$= \exp\{-x^p[1 - x^{q-p}(1 + x^{1-q} \log x)]\}$$

$$\to \exp\{-\infty \cdot [1 - 0 \cdot 1]\} = 0$$

and so since the function $\exp\{-x^q\}$ has a convergent integral so does the given function.

(iii) In the range $[0,1]$ we require $p < 1$; in $[1, \infty)$ we need $p > 1$. No convergence.

5. (i). Let us write out in full the sum to n terms, call it s, and use a similar trick to that for deriving the sum of a geometric progression.

$$s = x + 2x^2 + 3x^3 + 4x^4 + \cdots + nx^n$$
$$xs = x^2 + 2x^3 + 3x^4 + \cdots + (n-1)x^n + nx^{n+1}.$$

Thus

$$(1-x)s = x + x^2 + x^3 + x^4 + \cdots + x^n - nx^{n+1}$$
$$= \frac{x - x^{n+1}}{1-x} - nx^{n+1},$$

consequently

$$s = \frac{x - x^{n+1}}{(1-x)^2} - \frac{nx^{n+1}(1-x)}{(1-x)^2} = \frac{x - (n+1)x^{n+1} + nx^{n+2}}{(1-x)^2}.$$

The infinite sum exists provided $|x| < 1$ and equals

$$\frac{x}{(1-x)^2}.$$

(ii) Denote the nth term by a_n then

$$\tfrac{3}{4} \leqslant a_n \leqslant \tfrac{3}{4}\{1+1\}^{1/n}$$

and since the nth term does not converge to zero (it converges to $\tfrac{3}{4}$) the sum is divergent.

7. We first note that the discontinuities in $P(x)$ occur at the integer points and that $P(x)$ is piece-wise constant (just like the staircase function). We thus have

$$\int_{-1}^{\infty} x \, dP(x) = \sum_{n=0}^{\infty} n \frac{\lambda^n e^{-\lambda}}{n!} = e^{-\lambda}\lambda \sum_{n=1}^{\infty} \frac{\lambda^{n-1}}{(n-1)!} = e^{-\lambda}\lambda \sum_{n=0}^{\infty} \frac{\lambda^n}{n!}$$
$$= e^{-\lambda}\lambda e^{\lambda} = \lambda.$$

Similarly we have

$$\int_{-1}^{\infty} x^2 \, dP(x) = \sum_{n=0}^{\infty} n^2 \frac{\lambda^n e^{-\lambda}}{n!} = e^{-\lambda}\lambda \sum_{n=1}^{\infty} n \frac{\lambda^{n-1}}{(n-1)!}.$$

To calculate the above sum we are naturally led to consider differentiating the identity:

$$\lambda e^{\lambda} = \sum_{0}^{\infty} \frac{\lambda^{n+1}}{n!}$$

since we expect a similar series. In fact we obtain

$$(\lambda + 1)e^{\lambda} = \sum_{n=0}^{\infty} (n+1)\frac{\lambda^n}{n!} = \sum_{n=1}^{\infty} n \frac{\lambda^{n-1}}{(n-1)!}.$$

This, fortunately, is precisely the series occurring in our earlier calculation, so we may now deduce

$$\int_{-1}^{\infty} x^2 \, dP(x) = e^{-\lambda} \lambda(\lambda + 1) e^{\lambda} = \lambda(\lambda + 1).$$

8. We expect the logarithm to arise after differentiation of an exponentiated parameter, so we start from

$$\int_{1}^{\infty} x^t \, dx = \left[\frac{x^{t+1}}{t+1} \right]_{1}^{\infty} = -\frac{1}{t+1}$$

where we assume, of course, $t + 1 < 0$. Now a straight differentiation with respect to t on both sides, assuming the manipulation is justified, yields

$$\int_{1}^{\infty} \log x \cdot x^t \, dx = \frac{1}{(t+1)^2}.$$

Putting $t = -(n+1) < 0$ gives the result

$$\int_{1}^{\infty} \frac{\log x}{x^{n+1}} \, dx = \frac{1}{n^2}.$$

We note that by Question 3(iii) the integral

$$\int_{1}^{\infty} \frac{\log x}{x^p} \, dx$$

converges for $p > 1$. Now if $n > 0$ choose any p with $1 < p < n + 1$ (e.g. $p = \frac{3}{2}$), then we have the domination

$$\left| \frac{\log x}{x^t} \right| \leqslant \frac{\log x}{x^p} \quad \text{for} \quad 1 \leqslant x < \infty \quad \text{and} \quad p \leqslant t \leqslant n + 2$$

and since this includes the case $t = n + 1$ we are done. Note that the result breaks down for $n = 0$ and indeed the integral in question diverges. This is easiest seen by the integral test in alliance with the condensation test, since

$$\sum 2^n \cdot \frac{\log 2^n}{2^n} = \sum n \log 2 = \infty.$$

9. Clearly the integral of $K(x, t)$ is 1. (Area of the triangle.) On the other hand the limit of $K(x, t)$ as t tends to zero is 0 for every x.

11. Assuming the manipulation is justified we have

$$f'(t) = \int_{0}^{\infty} -e^{-xt} \sin x \, dx = -\frac{1}{t^2 + 1}.$$

Clearly the manipulation is justified for any $t > 0$. Given such a t choose a with $0 < a < t$ and let $b = t + 1$. We then have the domination for $a \leqslant s \leqslant b$:

$$|e^{-xs} \sin x| \leqslant e^{-xa}.$$

Hence for $t > 0$ we conclude

$$f(t) = C - \arctan t.$$

Let us divide the range into three parts $[0, \delta], [\delta, 1], [1, \infty)$ and estimate in each case the behaviour of the integrals when t is large. In $[1, \infty)$ the integral is no bigger than

$$\int_1^\infty e^{-xt} dx = \frac{1}{t}.$$

In $[\delta, 1]$ we have seen by the mean value theorem (cf. Section 17.18) that the integral is no bigger than

$$2 \frac{e^{-t\delta}}{\delta}.$$

In $[0, \delta]$ we shall have

$$\frac{\sin x}{x} \leqslant 1$$

provided δ is small enough. [Actually $\sin x \leqslant x$ for all $x \geqslant 0$.] So in this range the integral is no bigger than

$$\int_0^\delta 1 \, dx = \delta.$$

Altogether, we obtain the total estimate for the integral amounting to

$$\delta + \frac{e^{-t\delta}}{\delta} + \frac{1}{t}.$$

This can be made as small as we wish by taking first δ small enough and then t large enough to make the second and third terms small. Thus

$$\lim_{t \to \infty} f(t) = 0.$$

Hence $C = \pi/2$.

We follow the hint regarding integration by parts; for $t > 0$ we have:

$$\int_1^\infty \frac{\exp\{x(i - t)\}}{x} dx = \left[\frac{\exp\{(i - t)x\}}{(i - t)x} \right]_1^\infty + \int_1^\infty \frac{\exp\{(i - t)x\}}{(i - t)x^2} dx$$

$$= -\frac{\exp\{1(i - t)\}}{(i - t)} + \int_1^\infty \frac{\exp\{(i - t)x\}}{(i - t)x^2} dx.$$

Now the integral appearing in the second line is dominatedly convergent since

$$\int_1^\infty \frac{dx}{x^2} < \infty.$$

Taking limits as t tends to zero we have

$$-\frac{\exp\{1(i-t)\}}{(i-t)} + \int_1^\infty \frac{\exp\{(i-t)x\}}{(i-t)x^2}dx.$$

$$\rightarrow -\frac{e^i}{i} + \int_1^\infty \frac{e^{ix}}{ix^2}dx.$$

But, (by integration by parts again) this is equal to

$$\int_1^\infty \frac{e^{ix}}{x}dx.$$

In the finite range $[0,1]$ we may take limits as t tends to zero to obtain

$$f(0) = \lim_{t\to 0} f(t) = C.$$

Solutions to selected exercises on multiple integrals exercises 1

Exercises 19.17

1. We have:

$$u = \tfrac{1}{2}\{x^2 - y^2\},$$
$$v = xy.$$

So it is easiest to invert the transformation by noting that $v = xy$ implies $-v^2 = x^2.(-y^2)$ so that x^2 and $-y^2$ are roots of

$$t^2 - 2ut - v^2 = 0.$$

Hence x^2 and $-y^2$ are in some order

$$u \pm \sqrt{\{u^2 + v^2\}}.$$

To determine the order notice that only one of the above roots is negative and only one is positive. We thus have, depending on the quadrant, that

$$x = \pm\sqrt{[u + \sqrt{\{u^2 + v^2\}}]},$$
$$y = \pm\sqrt{[\sqrt{\{u^2 + v^2\}} - u]}.$$

The transform of the positive quadrant is therefore the set of all u, v with $v > 0$ (since $x, y > 0$ imply $v > 0$). Notice that the parenthetical remark is needed since we threw some information away when squaring v.

We note the Jacobian is equal to $x^2 + y^2$.

2. The formulas given define a transformation from $\mathbb{R}^2 \backslash \{(0,0)^t\}$. We note that $u = $ constant ($\neq 0$) describes a circle centred at $(1/(2u), 0)^t$ of radius $1/(2u)$ since

$$x^2 + y^2 - x/u = 0.$$

The circle passes through $(0,0)^t$. Similarly we note that $v = \text{constant}$ ($\neq 0$) describes a circle centred at $(0, 1/(2v))^t$ of radius $1/(2v)$ since

$$x^2 + y^2 - y/v = 0$$

and this circle also passes through the origin. It follows that in the domain of definition of the transformation, viz. the punctured plane $\mathbb{R}^2 \backslash \{(0,0)^t\}$, any circle $v = \text{constant}$ will cut any circle $u = \text{constant}$ at most once. This explains away the apparent ambiguity in the specification of the transformation.

To obtain the inverse transformation observe that

$$u^2 + v^2 = \frac{x^2 + y^2}{(x^2 + y^2)^2} = \frac{1}{x^2 + y^2}.$$

Consequently,

$$u = x(u^2 + v^2).$$

Hence

$$x = \frac{u}{u^2 + v^2}.$$

Similarly

$$y = \frac{v}{u^2 + v^2}.$$

Finally we note that the Jacobian of the transformation is $-(x^2 + y^2)^{-4}$.
3. Recall from Section 19.7 that as $v = x + y$ and $u = xy$ the roots of $t^2 - vt + u = 0$ are x and y. To have an inverse transformation we divide the region of interest (here the disc is centered at the origin of radius $\sqrt{2}$) into two regions D_1 where $y \geq x$ and D_2 where $y \leq x$. In D_2 we have

$$x = \frac{v + \sqrt{\{v^2 - 4u\}}}{2}, \quad y = \frac{v - \sqrt{\{v^2 - 4u\}}}{2}.$$

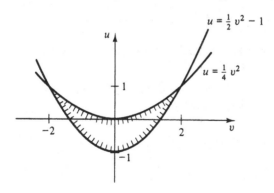

We determine the transformed region analytically. We have to deal with two conditions: (i) x and y are real so $v^2 - 4u \geqslant 0$, and (ii) $x^2 + y^2 \leqslant 2$. To translate the latter we observe that

$$x^2 + y^2 = (x + y)^2 - 2xy = v^2 - 2u$$

so the transformed region Δ is bounded by the two parabolas $v^2 = 4u$ and $v^2 = 2u + 2$. From the illustration we see that $-2 \leqslant v \leqslant 2$ throughout the region Δ.

4. If $x, y > 0$ we certainly have $u, v > 0$. Moreover

$$\frac{u}{v} = \frac{x}{y}$$

so

$$y = x + y - x$$

$$= v - y\frac{u}{v}$$

$$y\left(1 + \frac{u}{v}\right) = v$$

$$y = \frac{v^2}{u + v}$$

and so

$$x = \frac{uv}{u + v}.$$

It follows that the transformed region is likewise the positive quadrant. We note that the Jacobian is

$$\frac{\partial(u, v)}{\partial(x, y)} = \begin{vmatrix} \dfrac{2x + y}{y} & -\dfrac{x^2}{y^2} \\ 1 & 1 \end{vmatrix} = \frac{2xy + y^2}{y^2} + \frac{x^2}{y^2}$$

$$= \frac{(x + y)^2}{y^2} = v^2\frac{(u + v)^2}{v^4} = \frac{(u + v)^2}{v^2}.$$

5. We observe that $u \geqslant 0$. To find x and y in terms of u and v we note that $v^2 = x^2 y^2$ so x^2, y^2 are in some order the zeros of

$$t^2 - ut + v^2 = 0.$$

So in some order they are

$$\frac{u \pm \sqrt{\{u^2 - 4v^2\}}}{2}.$$

We are concerned with the region where $0 < y < x$ so in fact we see that

$$x = \sqrt{\left(\frac{u + \sqrt{\{u^2 - 4v^2\}}}{2}\right)}$$
$$y = \sqrt{\left(\frac{u - \sqrt{\{u^2 - 4v^2\}}}{2}\right)}.$$

Evidently these formulas require $u^2 - 4v^2 \geq 0$ and, since in the sector stated $x, y > 0$ (so that $v > 0$), we have the requirements: $u \geq 2v$. Thus the transformed region is given by $u > 0$, $v > 0$, $u \geq 2v$.

We note the the Jacobian is $2x^2 - 2y^2$.

6. We begin by computing the Jacobian.

$$\frac{\partial(u, v)}{\partial(x, y)} = \begin{vmatrix} -2x & 1 \\ 2x & -2y \end{vmatrix} = 4xy - 2x = 2x(2y - 1).$$

The Jacobian is thus of constant sign in the four quadrants defined by the vertical and horizontal passing through the point $x = 0$, $y = \frac{1}{2}$.

To obtain inverse tranformations we first note that

$$u + v = y - y^2,$$

so we obtain

$$y = \frac{1 \pm \sqrt{\{1 - 4(u + v)\}}}{2}.$$

Thus in the quadrant $y < \frac{1}{2}$ we have

$$y = \frac{1 - \sqrt{\{1 - 4(u + v)\}}}{2}, \tag{1}$$

the other sign being appropriate to the other half-plane. Also we have

$$x^2 = -u + y$$
$$= \tfrac{1}{2}\{(1 - 2u) - \sqrt{[1 - 4(u + v)]}\} \quad (\text{if } y < \tfrac{1}{2}).$$

Thus depending on quadrant we have

$$x = \pm \sqrt{(\tfrac{1}{2}\{(1 - 2u) \pm \sqrt{[1 - 4(u + v)]}\})}.$$

To find the image of the half-strip $x > 0$, $0 < y < \frac{1}{2}$ we argue from the appropriate formulas as follows. Note that to use the formula as at (1) is to imply already that $y < \frac{1}{2}$.

$$x > 0 \Rightarrow (1 - 2u) - \sqrt{[1 - 4(u + v)]} \geq 0$$
$$\Rightarrow (1 - 2u) \geq \sqrt{[1 - 4(u + v)]}.$$

In particular $1 - 2u \geq 0$, i.e. $u \leq \frac{1}{2}$. Moreover, squaring we obtain

$$1 - 4u + 4u^2 \geq 1 - 4(u + v),$$

whence $u^2 \geq -v$ and $u \leq \frac{1}{2}$.

Now y real implies

$$[1 - 4(u + v)] \geq 0$$

so that $u + v \leq \frac{1}{4}$. Moreover $y > 0$ implies

$$\{1 - \sqrt{[1 - 4(u + v)]}\} > 0,$$

or

$$1 > \sqrt{[1 - 4(u + v)]},$$

or

$$0 > -4(u + v),$$

so that also $u + v > 0$. These steps are reversable. We are thus led to the region as illustrated.

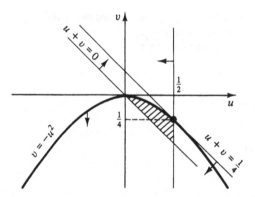

At $u = \frac{1}{2}$ the parabola has slope $v' = -2u = -1$

7. Observe that the formula is undefined for $x = 0$.

It is instructive to go through both the analytic argument as well as the geometric. First, we deal with the analytic argument.

$$u = x + vx,$$
$$= x(1 + v),$$
$$x = \frac{u}{1 + v},$$
$$y = \frac{uv}{1 + v}.$$

We now see how $y \geq 0$ translates. If $u \geq 0$, we must have $v(1 + v) \geq 0$, which requires that $v \leq -1$ or $v \geq 0$. On the other hand, if $u \leq 0$ then $v(1 + v) \leq 0$ so that we then require $-1 \leq v \leq 0$. We illustrate these three regions below.

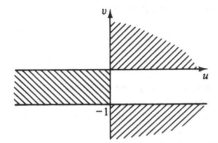

Next we consider the condition $y \leqslant 1$, that is

$$\frac{uv}{1+v} \leqslant 1.$$

Thus if $v > -1$ (so that v is in one of two intervals mentioned earlier viz. $(-1, 0)$ or $(0, \infty)$), we have

$$uv \leqslant 1 + v,$$

or

$$v(u - 1) \leqslant 1.$$

The transformed points lie therefore under the hyperbola $v(u - 1) = 1$ the axes of which are $v = 0$, $u = 1$.

On the other hand, if $v \leqslant -1$, so that v lies in the third interval, we have the reverse inequality

$$v(u - 1) \geqslant 1$$

and the transformed points are bounded by $u = 0$ and the same hyperbola as before.

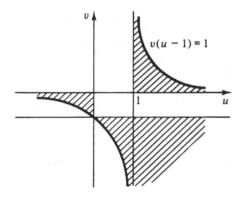

Now for the geometric argument. The lines $v = $ constant intersect the lines $u = $ constant in one point only unless $v = -1$. For a fixed v the line $y = vx$ meets a line $x + y = u$ in the region $0 \leqslant y \leqslant 1$ only in a specified range. To determine the range note that when $y = 1, x = u - 1$ and so

$$v = \frac{1}{u-1},$$

i.e. the hyperbola $v(u - 1) = 1$ gives one end of the range. the other end of the range is inevitably on $u = 0$. The range is degenerate when $v = 0$, in that there are no endpoints (since they are at $u = \pm \infty$). A second degeneracy occurs when the two endpoints coincide. This happens when $u = 0$ and $v = -1$.

For each v the horizontal line segments for that value of v and with one endpoint on $u = 0$ and the other on the hyperbola may be sketched as above. Evidently, we obtain the same result as by the analytic method. The three regions arise from the 'degeneracy' positions.

Exercises 19.18

1. Put $u = x + y \quad v = x - y$. Then

$$\frac{\partial(u, v)}{\partial(x, y)} = \begin{vmatrix} 1 & 1 \\ 1 & -1 \end{vmatrix} = -2.$$

It is convenient to divide the region of integration into two regions D_1 and D_2 as indicated. We observe that, for given u, D_1 is scanned by v

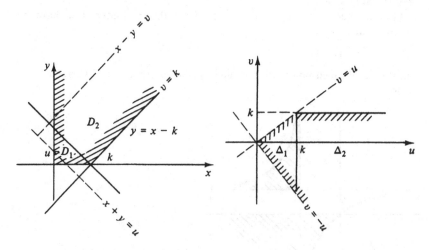

running from $-u$ to u, whereas D_2 is scanned by v running from $-u$ to k. Thus in the obvious notation

$$I = I_1(\Delta_1) + I_2(\Delta_2).$$

$$I_1(\Delta_1) = \int_0^k du \int_{-u}^u e^{-u} \tfrac{1}{2} dv = \frac{1}{2} \int_0^k e^{-u} 2u \, du$$

$$= [-ue^{-u}]_0^k + \int_0^k e^{-u}\,du$$

$$= 1 - ke^{-k} - e^{-k}$$

and

$$I_2(\Delta_2) = \int_k^\infty du \int_{-u}^k e^{-u}\tfrac{1}{2}dv = \frac{1}{2}\int_k^\infty (k+u)e^{-u}\,du$$

$$= \frac{1}{2}\left\{ ke^{-k} + [-ue^{-u}]_k^\infty + \int_k^\infty e^{-u}\,du \right\} = \tfrac{1}{2}\{2ke^{-k} + e^{-k}\}.$$

So

$$I = 1 - \tfrac{1}{2}e^{-k}.$$

3. The natural choice of new variables is $u = xy$ and $v = x^2 - y^2$. Thus

$$\frac{\partial(u,v)}{\partial(x,y)} = \begin{vmatrix} y & x \\ 2x & -2y \end{vmatrix} = -2(x^2 + y^2).$$

We also note the identity

$$(x^2 - y^2)^2 = x^4 - 2x^2y^2 + y^4 = (x^2 + y^2)^2 - 4x^2y^2.$$

The integration proceeds as follows. (See below for more details.)

$$I = \frac{1}{2}\int_1^2 du \int_1^2 \frac{dv}{\sqrt{\{v^2 + 4u^2\}}} = \frac{1}{2}\int_1^2 \left[sh^{-1}\frac{v}{2u} \right]_1^2 du$$

$$= \frac{1}{2}\int_1^2 \left\{ sh^{-1}\frac{1}{u} - sh^{-1}\frac{1}{2u} \right\} du$$

$$= \frac{1}{2}\int_1^2 \log\frac{1 + \sqrt{(u^2 + 1)}}{\tfrac{1}{2}\{1 + \sqrt{(4u^2 + 1)}\}} du$$

$$= \frac{1}{2}\int_1^2 \{\log(1 + \sqrt{(u^2 + 1)}) - \log(1 + \sqrt{(4u^2 + 1)})\} du + \tfrac{1}{2}\log 2.$$

Before we go on to use integration by parts, we stop to explain the first step above. What we have made use of are the two formulas:

$$\int \frac{dv}{\sqrt{\{v^2 + a^2\}}} = sh^{-1}\frac{v}{a}$$

and

$$sh^{-1}u = \log|u + \sqrt{(u^2 + 1)}|.$$

Now consider that

$$\int \log(1 + \sqrt{(u^2 + 1)})\,du$$

$$= u\log(1 + \sqrt{(u^2 + 1)}) - \int \frac{1}{(1 + \sqrt{(u^2 + 1)})}\cdot\frac{u}{\sqrt{(u^2 + 1)}} u\,du$$

$$= u \log(1 + \sqrt{(u^2 + 1)}) - \int \frac{\{1 - \sqrt{(u^2 + 1)}\}u^2 \, du}{(1 - (u^2 + 1))\sqrt{(u^2 + 1)}}$$

$$= u \log(1 + \sqrt{(u^2 + 1)}) + \int \frac{\{1 - \sqrt{(u^2 + 1)}\}}{\sqrt{(u^2 + 1)}} \, du$$

$$= u \log(1 + \sqrt{(u^2 + 1)}) - u + \log(u + \sqrt{(u^2 + 1)}).$$

Similarly

$$\int \log(1 + \sqrt{(4u^2 + 1)}) \, du = u \log(1 + \sqrt{(4u^2 + 1)})$$
$$- u + \tfrac{1}{2}\log(2u + \sqrt{(4u^2 + 1)})$$

Thus the required answer is

$$[u \log(1 + \sqrt{(u^2 + 1)}) + \log(u + \sqrt{(u^2 + 1)})$$
$$- u \log(1 + \sqrt{(4u^2 + 1)}) - \tfrac{1}{2}\log(u + \sqrt{(4u^2 + 1)})]_1^2 + \tfrac{1}{2}\log 2$$
$$= 2\log(1 + \sqrt{5}) + \log(2 + \sqrt{5}) - 2\log(1 + \sqrt{17}) - \tfrac{1}{2}\log(4 + \sqrt{17})$$
$$- \{\log(1 + \sqrt{2}) + \log(2 + \sqrt{2}) - \log(1 + \sqrt{5})$$
$$- \tfrac{1}{2}\log(4 + \sqrt{5})\} + \tfrac{1}{2}\log 2$$

and we refrain from simplifying this.

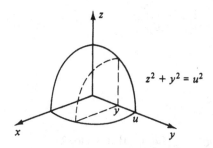

4. Working in one octant only (see figure) we see that u runs from 0 to R. For any u the variable v ($= y$) runs from 0 to u and once v is fixed the segment is described by the variable w ($= z$) running from 0 to $\sqrt{(u^2 - v^2)}$. We calculate the Jacobian

$$\frac{\partial(u, v, w)}{\partial(x, y, z)} = \begin{vmatrix} \dfrac{x}{\sqrt{(x^2 + y^2 + z^2)}} & \dfrac{y}{\sqrt{(x^2 + y^2 + z^2)}} & \dfrac{z}{\sqrt{(x^2 + y^2 + z^2)}} \\ 0 & 1 & 0 \\ 0 & 0 & 1 \end{vmatrix}$$

$$= \frac{x}{\sqrt{(x^2 + y^2 + z^2)}} = \frac{\sqrt{(u^2 - v^2 - w^2)}}{u}.$$

Hence

$$I = 8 \int_0^R du \int_0^u dv \int_0^{\sqrt{(u^2 - w^2)}} \frac{u}{\sqrt{(u^2 - v^2 - w^2)}} dw$$

$$= 8 \int_0^R du \int_0^u u \, dv \int_0^{\pi/2} d\theta$$

$$= 4\pi \int_0^R u^2 \, du = \tfrac{4}{3}\pi R^3.$$

Note that we have substituted $w = \sqrt{(u^2 - v^2)} \sin \theta$.

5. Referring to Question 3 we see that

$$\frac{\partial(u, v)}{\partial(x, y)} = -4(x^2 + y^2).$$

So

$$I = \frac{1}{4} \iint_A \frac{e^{-v}}{1 + u^2} \, du \, dv = \frac{1}{4} \int_{-\infty}^{+\infty} \frac{du}{1 + u^2} \int_0^\infty e^{-v} \, dv$$

$$= \tfrac{1}{4} [\arctan u]_{-\infty}^{+\infty} \cdot 1 = \frac{\pi}{4}.$$

6. In the region of interest we have $x \leqslant y$ and, also since $x, y \geqslant 0$, we have, unless $x = y = 0$, that

$$-x + y \leqslant x + y \Rightarrow 0 \leqslant \frac{y - x}{x + y} \leqslant 1.$$

The integrand is thus bounded in the region and undefined at the origin (different approaches towards the origin will yield various limiting values for the integrand ranging from 0 to 1). We must therefore interpret the integral

$$\iint_D \left(\frac{x - y}{x + y} \right)^2 dx \, dy,$$

where D is the region shown in the figure, as an improper integral and need to justify its existence.

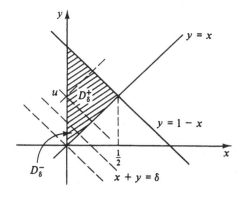

So denote by D_δ^- that part of the original region including $(0,0)$ which is bounded by $x + y = \delta$ and let the remaining region the denoted by D_δ^+. We consider the limit

$$\lim_{\delta \to 0} \iint_{D_\delta^+} \left(\frac{x-y}{x+y} \right)^2 dx\, dy.$$

Since the integrand is positive the limiting integral must exist either as $+\infty$ or a finite number. But the integrand is bounded and the region of integration has finite area. So the limit here is not ∞. Thus the improper integral does converge.

We go on to compute this limit. Reference to Question 1 shows that the Jacobian for the required transformation is 2. Thus

$$I(D_\delta) = \int_\delta^1 du \int_0^u \frac{v^2}{u^2} \frac{1}{2} dv = \frac{1}{2} \int_\delta^1 \frac{u^3}{3u^2} du = \frac{1}{12}[1 - \delta^2]_\delta^1 \to \frac{1}{12}.$$

7. The right-hand side is the negative of the left (transpose x and y). The two answers are different $(\pm\frac{1}{2})$. This comes as no surprise since the integrand is unbounded (consider its value when $y = 2x$).

Exercise 19.19

1. The region of integration is $\{(x, y, z): x < y < z\}$. Thus for a fixed z with $0 \leqslant z \leqslant 1$ the variable y runs from 0 to z. Finally, given values for z and y the variable x runs from 0 to y. Thus the required probability is

$$\int_0^1 dz \int_0^z dy \int_0^y 8xyz\, dx = \int_0^1 dz \int_0^z 4y^3 z\, dy = \int_0^1 z^5\, dz = \frac{1}{6}.$$

2. We require to know the value for any k of Prob$\{Z \leqslant k\}$. So we need to

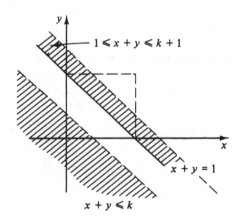

integrate over the region (shown in the figure).

$$D(k) = \{(x, y):(x + y) \leqslant k \text{ if } (x + y) < 1, \text{ or } (x + y) - 1 \leqslant k \text{ if } (x + y) \geqslant 1\}.$$

The alternative calls for (x, y) such that $x + y \geqslant 1$ and $x + y \leqslant k + 1$. Such points exist if and only if $1 \leqslant k + 1$, i.e. $0 \leqslant k$. Thus the second alternative is vacuous if $k < 0$. Thus for $k < 0$ we have

$$\text{Prob}\,\{Z \leqslant k\} = \iint_{x+y \leqslant k} f(x, y)\, dx\, dy = 0.$$

For $k = 0$ the same result holds. Now for $0 < k < 1$ we have

$$\text{Prob}\,\{Z \leqslant k\} = \iint_{x+y \leqslant k} f(x, y)\, dx\, dy + \iint_{1 \leqslant x+y \leqslant k+1} f(x, y)\, dx\, dy.$$

Evidently if $k \geqslant 1$ the probability that Z is less than k is 1. So consider $0 < k < 1$. We have

$$\text{Prob}\,\{Z \leqslant k\} = \tfrac{1}{2}k^2 + \tfrac{1}{2}(\sqrt{2} + (1-k)\sqrt{2})\frac{k}{\sqrt{2}} = k.$$

Here we have used the formula for the area of a trapezoid $\tfrac{1}{2}h(k + l)$ where k, l are the lengths of the parallel edges and h is their distance apart.

3. (i) The region we are concerned with is clearly the disc of radius $\sqrt{(2k)}\,(k \geqslant 0)$. Thus the required probability is

$$\frac{1}{2\pi}\iint \exp\{-(x^2 + y^2)/2\}\, dx\, dy = \frac{1}{2\pi}\int_0^{2\pi} d\theta \int_0^{\sqrt{(2k)}} \exp(-r^2/2) r\, dr$$

$$= [-\exp(-r^2/2)]_0^{\sqrt{(2k)}} = 1 - e^{-k}.$$

(ii) The relation $kz_1 \leqslant \sqrt{(z_1^2 + z_2^2)}$, for $k > 0$, is equivalent to: either $z_1 \leqslant 0$ or $k^2 z_1^2 \leqslant (z_1^2 + z_2^2)$. The later inequality reads $(k^2 - 1)z_1^2 \leqslant z_2^2$. Thus the relation is equivalent to: either $z_1 \leqslant 0$ or $[z_1 \geqslant 0$ and either $\sqrt{(k^2 - 1)}z_1 \leqslant z_2$ or

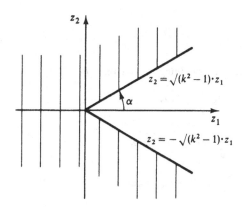

$z_2 \leqslant -\sqrt{(k^2 - 1)z_1}]$. The region is illustrated in the figure. To find the required probability we prefer to consider the complementary probability. It is

$$\frac{1}{2\pi}\int\int \exp\{-(x^2 + y^2)/2\} \, dx \, dy = \frac{1}{2\pi}\int_{-\alpha}^{+\alpha} d\theta \int_0^\infty \exp(-r^2/2) r \, dr$$

$$= \frac{2\alpha}{2\pi} = \frac{1}{\pi}\arctan\sqrt{\{k^2 - 1\}}.$$

The tangent of α is seen to be the slope of one of the bounding lines.

(iii) The relation $|z_1| \leqslant k\sqrt{(z_2^2 + z_3^2)}$ is equivalent to $z_1^2 \leqslant k^2(z_2^2 + z_3^2)$ and this region is complementary to the conical body illustrated below. Passing

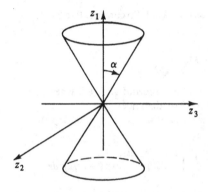

to cylindrical polar co-ordinates we have

$$z_1 = r\cos\phi$$
$$z_2 = r\sin\phi\cos\theta$$
$$z_3 = r\sin\phi\sin\theta$$

and

$$\frac{\partial(z_1, z_2, z_3)}{\partial(r, \theta, \phi)} = r^2 \sin\phi.$$

This is positive for $0 \leqslant \phi \leqslant \alpha$. We calculate the integral over the cone. It is

$$\frac{1}{(2\pi)^{3/2}}\int\int\int \exp\{-(x^2 + y^2 + z^2)/2\} \, dx \, dy \, dz$$

$$= \frac{1}{(2\pi)^{3/2}}\int_0^{2\pi} d\theta \int_0^\alpha \sin\phi \, d\phi \int_0^\infty \exp(-r^2/2) r^2 \, dr$$

$$= \frac{2\pi}{(2\pi)^{3/2}}(1 - \cos\alpha)\int_0^\infty \exp(-r^2/2) \, dr = (1 - \cos\alpha),$$

since

$$\int_0^\infty \exp(-r^2/2)dr = \sqrt{(\pi/2)}.$$

Note that if $\cot \alpha = k$ then $\cos \alpha = k/\sqrt{(1+k^2)}$, hence the required answer.

We are required to integrate over the region $D(k)$ illustrated below in the expression

$$\frac{\exp\{-(z_1^2 + z_2^2)/2\}}{2\pi}.$$

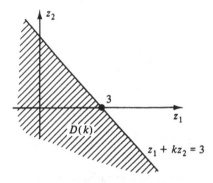

Clearly a change of variable to

$$u = z_1 + kz_2$$

is called for. Thus

$$z_2^2 = \left(\frac{u - z_1}{k}\right)^2 = \frac{u^2 - 2uz_1 + z_1^2}{k^2}.$$

Unfortunately a mixed term arises. To avoid this we take a new variable v so that lines $v =$ constant are orthogonal to the lines $u =$ constant. So take

$$v = kz_1 - z_2,$$

and then

$$z_1^2 + z_2^2 = \frac{(kz_1 - z_2)^2 + (z_1 + kz_2)^2}{1 + k^2}.$$

Noting that

$$\frac{\partial(u, v)}{\partial(z_1, z_2)} = \begin{vmatrix} 1 & k \\ k & -1 \end{vmatrix} = -(k^2 + 1)$$

the required integral now reads

$$\iint_{D(k)} \frac{\exp\{-(z_1^2 + z_2^2)/2\}}{2\pi} dz_2 \, dz_1$$

$$= \iint \frac{1}{2\pi(1 + k^2)} \exp\left\{-\frac{1}{2}\left(\frac{u^2 + v^2}{1 + k^2}\right)\right\} du \, dv$$

$$= \frac{1}{2\pi(1 + k^2)} \int_{-\infty}^{+\infty} \exp\left(\frac{-v^2}{2 + 2k^2}\right) dv \int_{-\infty}^{3} \exp\left(\frac{-u}{2 + 2k^2}\right) du.$$

We now put

$$s = \frac{u}{\sqrt{(1 + k^2)}}$$

$$t = \frac{v}{\sqrt{(1 + k^2)}}$$

$$h = \frac{3}{\sqrt{(1 + k^2)}}$$

and the integral reduces to

$$\frac{1}{2\pi} \int_{-\infty}^{+\infty} \exp(-t^2/2) dt \int_{-\infty}^{h} \exp(-s^2/2) ds = \frac{1}{\sqrt{(2\pi)}} \int_{-\infty}^{h} \exp(-s^2/2) ds.$$

It is well known to statisticians that if X and Y are random variables which are independent and have identical normal distribution with zero mean and unit variance then

$$\frac{aX + bY}{\sqrt{(a^2 + b^2)}}$$

has a normal distribution with zero mean and unit variance. This is proved by exactly the argument above.

Solutions to selected exercises on differential equations.

Exercises 20.4

6. There is a problem here in that our characteristic equation reads

$$\frac{dx}{x + y} = \frac{dy}{x - y} = \frac{dz}{0}.$$

Recall that $(\dot{x}, \dot{y}, \dot{z})$ is to be proportional to $(P, Q, R) = (x + y, x - y, 0)$. Thus in fact we have

$$\frac{dx}{x + y} = \frac{dy}{x - y} \quad \text{and} \quad dz = 0.$$

Thus $z = c_1$. Also

$$(x - y) - (x + y)\frac{dy}{dx} = 0. \tag{*}$$

This equation is exact, since if we seek a solution of (*) of form $g(x, y) = 0$ we want

$$\frac{\partial g}{\partial x} = x - y \quad \text{and} \quad \frac{\partial g}{\partial y} = -(x + y). \tag{+}$$

But exactness requires

$$\frac{\partial}{\partial y}(x - y) = \frac{\partial^2 g}{\partial x\,\partial y} = \frac{\partial}{\partial x}(-(x + y))$$

which is satisfied. Solving (+) we get

$$g(x, y) = \tfrac{1}{2}x^2 - xy + A(y),$$
$$g(x, y) = -xy - \tfrac{1}{2}y^2 + B(x).$$

So

$$\tfrac{1}{2}x^2 - xy + A(y) = -xy - \tfrac{1}{2}y^2 + B(x),$$

or

$$\tfrac{1}{2}x^2 - B(x) = -\tfrac{1}{2}y^2 - A(y) = \text{const.} = k.$$

So

$$0 = g(x, y) \equiv \tfrac{1}{2}x^2 - xy - \tfrac{1}{2}y^2 - k,$$

or

$$y^2 + 2xy - x^2 = c_2,$$

giving

$$z = F(y^2 + 2xy - x^2),$$

where F is an arbitrary function.

7. (a) We suppose that $r \neq 0$ and then we are to solve for some λ

$$\dot{x} = \lambda x \quad \dot{y} = \lambda y \quad \dot{z} = \lambda r z.$$

Thus

$$\frac{dy}{dx} = \frac{y}{x} \quad \text{and} \quad \frac{dz}{dx} = \frac{rz}{x}$$

or

$$\int \frac{1}{y}dy = \int \frac{1}{x}dx \quad \text{and} \quad \int \frac{1}{z}dz = \int \frac{r}{x}dx.$$

Hence

$$\log y = \log x + c \quad \text{and} \quad \log z = r\log x + d$$

or, renaming constants

$$y/x = c_1 \quad \text{and} \quad z/x^r = c_2.$$

This yields the solution $z/x^r = F(y/x)$ for some function F. Thus

$$z = x^r F(y/x),$$

and this is homogeneous of degree r.

(b) We have

$$\frac{dx}{1/x} = \frac{dy}{1/y} = \frac{dz}{x^2 - y^2}.$$

Hence

$$\int x\,dx = \int y\,dy \quad \text{or} \quad \tfrac{1}{2}x^2 = \tfrac{1}{2}y^2 + c_0.$$

Thus

$$x^2 - y^2 = c_1. \tag{1}$$

Also

$$\frac{dz}{dy} = y(x^2 - y^2)$$

$$= c_1 y.$$

So

$$z = \tfrac{1}{2}c_1 y^2 + c_2.$$

Our second solution is thus

$$z - \tfrac{1}{2}(x^2 - y^2)y^2 = c_2. \tag{2}$$

The general solution is thus of the form

$$z - \tfrac{1}{2}(x^2 - y^2)y^2 = F(x^2 - y^2),$$

$$z = \tfrac{1}{2}y^2(x^2 - y^2) + F(x^2 - y^2). \tag{3}$$

Remark. Since F is an arbitrary function equation (3) can be recast in various forms. An example follows.
   Since

$$2y^2 = (x^2 + y^2) - (x^2 - y^2)$$

$$z = \tfrac{1}{4}\{(x^2 + y^2) - (x^2 - y^2)\}(x^2 - y^2) + F(x^2 - y^2)$$

$$= \tfrac{1}{4}(x^2 + y^2)(x^2 - y^2) + \{F(x^2 - y^2) - \tfrac{1}{4}(x^2 - y^2)^2\}$$

and the curly bracket is a function of $(x^2 - y^2)$.

7. (d) This is nasty in detail. We have:

$$\frac{dx}{x + y} = \frac{dy}{x - y} = \frac{dz}{1};$$

as before (see Question 6)

$$(x - y) - (x + y)\frac{dy}{dx} = 0$$

gives

$$y^2 - x^2 + 2xy = c_1.$$

The equations with z as parameter, viz.,

$$\frac{dx}{dz} = x + y$$

$$\frac{dy}{dz} = x - y \qquad (*)$$

are awkward. We need to eliminate variables:

$$\frac{d^2x}{dz^2} = \frac{dx}{dz} + \frac{dy}{dz} = (x+y) + (x-y) = 2x.$$

We could solve $d^2x/dz^2 = 2x$ outright. But that turns out to be a long way round. Note that

$$2\frac{dx}{dz}\frac{d^2x}{dz^2} = 4x\frac{dx}{dz}.$$

So

$$\left[\frac{dx}{dz}\right]^2 = 2x^2 + c_0.$$

But if $x = 0$ and $y = 0$ then by (*) $dx/dz = 0$; so here $c_0 = 0$. Thus

$$\frac{dx}{dz} = \sqrt{2x}$$

and so

$$\int \frac{1}{x}\, dx = \sqrt{2} \int dz \ .$$

giving

$$\log x = \sqrt{2z} - c_2.$$

Hence

$$\sqrt{2z} - \log x = F(y^2 - x^2 + 2xy).$$

8. Again '$R = 0$' so we have

$$\frac{dx}{1} = \frac{dy}{-(ax+by)} \quad \text{and} \quad dz = 0.$$

Thus

$$z = c_1.$$

Moreover

$$\frac{dy}{dx} = -(ax+by),$$

or

$$\frac{dy}{dx} + by = -ax.$$

Now the integrating factor is e^{bx}, so

$$\frac{d}{dx}(e^{bx}y) = -axe^{bx}.$$

Thus

$$e^{bx}y = \left[-ax\frac{e^{bx}}{b}\right] - \int\frac{e^{bx}}{b}(-a)\,dx.$$

$$= -\frac{a}{b}xe^{bx} + \frac{a}{b^2}e^{bx} + c_0,$$

i.e.

$$e^{bx}(abx - a + b^2y) = b^2c_0.$$

Renaming the constant to c_2, we get

$$e^{bx}(abx - a + b^2y) = c_2.$$

Hence

$$z = F(e^{bx}(abx - a + b^2y)),$$

where F is an arbitrary function.

9. Let the surface be $z = f(x, y)$. Say (X, Y, Z) is on the surface. The normal vector to this surface is thus

$$\left[\frac{\partial f}{\partial x}, \frac{\partial f}{\partial y}, 1\right]^t.$$

Hence the line through (X, Y, Z) in this direction has the parametric equations

$$x = X + t\frac{\partial f}{\partial y}, \quad y = Y + t\frac{\partial f}{\partial x}, \quad z = Z + t.$$

If this line meets the z − axis at $x = 0, y = 0, z = b$ we have

$$0 = X + t\frac{\partial f}{\partial x} \quad 0 = Y + t\frac{\partial f}{\partial y} \quad b = Z + t.$$

Eliminating t gives

$$Y\frac{\partial f}{\partial x}(X, Y) = X\frac{\partial f}{\partial y}(X, Y).$$

This holds at all X, Y. We solve

$$y\frac{\partial f}{\partial x} - x\frac{\partial f}{\partial y} = 0$$

by the method of characteristic curves by setting, for some λ,

$$\dot{x} = \lambda y \quad \dot{y} = -\lambda x \quad \dot{z} = 0.$$

Thus $z = c_1$. Also

$$\frac{dy}{dx} = -\frac{x}{y}$$

with solution $y^2 + x^2 = c_2$. Thus we have

$$z = F(x^2 + y^2)$$

for some function F. But, if this surface is to pass through the line $x = a, y = z$, we have that

$$y = F(a^2 + y^2).$$

Taking $t = a^2 + y^2$ we obtain $F(t) = \sqrt{\{t - a^2\}}$ so that in general

$$z = F(x^2 + y^2)$$
$$= \sqrt{\{x^2 + y^2 - a^2\}}.$$

Solutions to selected exercises on Laplace transforms

Exercises 21.15

1.
$$L\{2e^{4t}\} = \frac{2}{s-4}, \quad L\{3e^{-2t}\} = \frac{3}{s+2}, \quad L\{5t - 3\} = \frac{5}{s^2} - \frac{3}{s},$$

$$L\{t^2 - 2e^{-t}\} = \frac{2}{s^3} - \frac{2}{s+1}, \quad L\{3\cos 5t\} = \frac{3s}{s^2 + 25},$$

$$L\{(t^2 + 1)^2\} = L\{t^4 + 2t^2 + 1\} = \frac{24}{s^5} + \frac{4}{s^3} + \frac{1}{s},$$

$$L\{t^3 e^{-3t}\} = \frac{6}{(s+3)^4},$$

$$L\{(t + 2)^2 e^t\} = L\{(t^2 + 4t + 4)e^t\} = \frac{2}{(s-1)^3} + \frac{4}{(s-1)^2} + \frac{4}{(s-1)}.$$

2.
$$L^{-1}\left\{\frac{1}{s-a}\right\} = e^{at}, \quad L^{-1}\left\{\frac{1}{s^2 + a^2}\right\} = \frac{1}{a}\sin at, \quad L^{-1}\left\{\frac{1}{s^n}\right\} = \frac{t^{n-1}}{(n-1)!},$$

$$L^{-1}\left\{\frac{5s + 4}{s^3}\right\} = L^{-1}\left\{\frac{5}{s^2} + \frac{4}{s^3}\right\} = 5t + \frac{4t^2}{2},$$

$$L^{-1}\left\{\frac{6}{2s - 3}\right\} = L^{-1}\left\{\frac{3}{s - 3/2}\right\} = 3e^{3t/2},$$

$$L^{-1}\left\{\frac{2s - 18}{s^2 + 9} + \frac{24 - 30\sqrt{s}}{s^4}\right\} = L^{-1}\left\{\frac{2s}{s^2 + 3^2} - \frac{6\cdot3}{s^2 + 3^2} + \frac{24}{s^4} - \frac{30}{s^{7/2}}\right\}$$

$$= 2\cos 3t - 6\sin 3t + 4t^3 - \frac{30t^{5/2}}{\Gamma(7/2)},$$

$$L^{-1}\left\{\frac{3s+7}{s^2-2s-3}\right\}=L^{-1}\left\{\frac{3s+7}{(s-3)(s+1)}\right\}$$

$$=L^{-1}\left\{\frac{-3+7}{-4(s+1)}+\frac{9+7}{4(s-3)}\right\}$$

$$=-e^{-t}+4e^{3t}.$$

3. (i)   We shift the origin so that the side conditions are at the origin instead of at $t=2$. We do this by introducing a function $g(t)$ with

$$g(t)=f(t+2).$$

Thus $g(0)=f(2)$; we also have $g'(t)=f'(t+2)$ (so that $g'(0)=f'(2)$) and $g''(t)=f''(t+2)$. Next we substitute $t+2$ for t in the given differential equation:

$$f''(t+2)+2f'(t+2)+5f(t+2)=(t+2)-2,$$

so we now wish to solve

$$g''(t)+2g'(t)+5g(t)=t,$$

with $g(0)=g'(0)=1$. We take transforms

$$s^2\tilde{g}-s-1+2(s\tilde{g}-1)+5\tilde{g}=\frac{1}{s^2}$$

i.e.

$$(s^2+2s+5)\tilde{g}=\frac{1}{s^2}+s+3.$$

Thus

$$\tilde{g}=\frac{s+3}{(s+1)^2+2^2}+\frac{1}{s^2\{(s+1)^2+4\}}$$

$$=\frac{s+1}{(s+1)^2+2^2}+\frac{2}{(s+1)^2+2^2}+\frac{1}{s^2\{(s+1)^2+4\}}.$$

Now

$$\frac{1}{s^2\{(s+1)^2+4\}}=\frac{1}{5s^2}+Q_1,$$

$$Q_1=\frac{5-\{(s+1)^2+4\}}{5s^2\{(s+1)^2+4\}}=-\frac{s^2+2s}{5s^2\{(s+1)^2+4\}}$$

$$=-\frac{s+2}{5s\{(s+1)^2+4\}}=-\frac{2}{5s(5)}+Q_2$$

$$Q_2=\frac{-5s-10+(2s^2+4s+10)}{25s\{(s+1)^2+4\}}=\frac{2s^2-s}{25s\{(s+1)^2+4\}}$$

$$=\frac{2s-1}{25\{(s+1)^2+4\}}.$$

Thus

$$\tilde{g} = \frac{s+1}{\{(s+1)^2+4\}} + \frac{2}{25} \cdot \frac{s+1}{\{(s+1)^2+4\}} - \frac{3}{25\{(s+1)^2+4\}}$$

$$- \frac{2}{25s} + \frac{1}{5s^2}.$$

We now see that

$$g(t) = e^{-t}\cos 2t \cdot \left(1 + \frac{2}{25}\right) - e^{-t}\frac{3}{2 \cdot 25}\sin 2t - \frac{2}{25} + \frac{t}{5}.$$

So, since $f(t) = g(t-2)$ we have

$$f(t) = e^{-t+2}\cos(2t-4) \cdot \left(1 + \frac{2}{25}\right) - e^{-t+2}\frac{3}{2 \cdot 25}\sin(2t-4) - \frac{2}{25} + \frac{(t-2)}{5}.$$

(ii)   Since $f(0) = 0 = g(0)$, we have on taking transforms:

$$\left.\begin{array}{c} s\tilde{g} + \tilde{g} + 2\tilde{f} = \dfrac{1}{s-2} \\[2mm] 2s\tilde{g} - \tilde{g} + s\tilde{f} = 0 \end{array}\right\}$$

Solving, we obtain

$$(s^2 + s - 2[2s-1])\tilde{g} = \frac{s}{s-2}.$$

But

$$(s^2 - 3s + 2) = (s-1)(s-2),$$

so

$$\tilde{g} = \frac{s}{(s-1)(s-2)^2} = \frac{2}{(s-2)^2} + Q_1$$

$$Q_1 = \frac{s - 2(s-1)}{(s-1)(s-2)^2} = -\frac{s-2}{(s-1)(s-2)^2} = \frac{-1}{(s-1)(s-2)}$$

$$= \frac{-1}{(s-1)(-1)} + \frac{-1}{1 \cdot (s-2)}.$$

Hence

$$g = 2te^{2t} + e^t - e^{2t}$$

and we now deduce that

$$\begin{aligned} f &= \tfrac{1}{2}\{e^{2t} - g' - g\} \\ &= \tfrac{1}{2}\{e^{2t} - [2 \cdot 2te^{2t} + 2e^{2t} + e^t - 2e^{2t}] - [2te^{2t} + e^t - e^{2t}]\} \\ &= \tfrac{1}{2}\{2e^{2t} - 6te^{2t} - 2e^t\} \\ &= e^{2t} - 3te^{2t} - e^t. \end{aligned}$$

4. $$L^{-1}\left\{\frac{1}{(s+3)(s-1)}\right\} = \int_0^t e^{-3u}e^{(t-u)}\,du = \int_0^t e^t e^{-4u}\,du$$

$$= e^t[-\tfrac{1}{4}e^{-4u}]_0^t = \tfrac{1}{4}(e^t - e^{-3t})$$

$$L^{-1}\left\{\frac{1}{(s+1)(s^2+1)}\right\} = \int_0^t \sin u\, e^{-(t-u)}\,du = e^{-t}\int_0^t e^u \sin u\,du.$$

The most direct way of solving this integral is to see $\sin u$ as the imaginary part of e^{iu}. [Alternatively, one can use integration by parts.]

$$\int_0^t e^u e^{iu}\,du = \left[\frac{\exp\{(i+1)u\}}{i+1}\right]_0^t = \left[\frac{\exp\{(i+1)t\}(1-i)}{2} - \frac{1-i}{2}\right].$$

The imaginary part here is

$$\tfrac{1}{2}e^t\{-\cos t + \sin t\} + \tfrac{1}{2}.$$

Hence the required function is

$$\tfrac{1}{2}\{-\cos t + \sin t\} + \tfrac{1}{2}e^{-t}.$$

5. We use the substitution $t = \sin^2\theta$.

$$\int_0^{\pi/2} \sin^{2p-1}\theta\cos^{2q-1}\theta\,d\theta = \frac{1}{2}\int_0^{\pi/2}\sin^{2p-2}\theta\cos^{2q-2}\theta\cdot 2\sin\theta\cos\theta\,d\theta$$

$$= \frac{1}{2}\int_0^1 t^{p-1}(1-t)^{q-1}\,dt = \tfrac{1}{2}B(p,q).$$

In particular

$$\int_0^{\pi/2}\sin^4\theta\cos^6\theta\,d\theta = \int_0^{\pi/2}\sin^{2(5/2)-1}\theta\cos^{2(7/2)-1}\theta\,d\theta$$

10. $$\frac{d}{ds}\int_0^\infty e^{-st}f(t)\,dt = \int_0^\infty \frac{\partial}{\partial s}\{e^{-st}f(t)\}\,dt$$

$$= \int_0^\infty e^{-st}(-t)f(t)\,dt$$

$$= -L\{tf(t)\}$$

and the result follows by repeated application of the above formula. (Evidently the manipulation is justified for s large enough, provided f is of exponential growth at most.)

If a linear ordinary differential equation of order n with independent variable t has coefficients that are polynomials in t of degree at most m, its transform will be a differential equation of order m with polynomial coefficients of degree at most n. The transformed equation may therefore represent a lower order equation if $m < n$. Otherwise the transform does not reduce the problem; at best in some cases the algebra may prove to be easier.

12. By Section 21.11 we observe

$$L\left\{\int_0^t F(u)du\right\} = \frac{1}{s}\tilde{F}.$$

Hence, if we put

$$F(u) = \frac{f(u)}{u},$$

we have $uF(u) = f(u)$ and using the last question

$$\tilde{f} = L\{uF(u)\} = -\frac{d}{ds}L\{F(u)\},$$

so that, integrating from s to infinity,

$$\int_s^\infty \tilde{f}(v)\,dv = [-\tilde{F}(v)]_s^\infty = \tilde{F}(s).$$

Here we have used the fact that the transform vanishes at infinity. Finally, using the observation made at the beginning of the question we have

$$L\left\{\int_0^t \frac{f(u)}{u}\,du\right\} = \frac{1}{s}\tilde{F} = \frac{1}{s}\int_s^\infty \tilde{f}(v)\,dv.$$

Applying this result when $f(u) = \sin u$, we have

$$L\left\{\int_0^t \frac{\sin u}{u}\,du\right\} = \frac{1}{s}\int_s^\infty \frac{1}{v^2+1}\,dv = \frac{1}{s}[\tan^{-1}v]_s^\infty = \frac{1}{s}\left[\frac{\pi}{2} - \tan^{-1}s\right].$$

Hence in particular, by the final value theorem,

$$\int_0^\infty \frac{\sin u}{u}\,du = \lim_{s\to 0} s\cdot\frac{1}{s}\left[\frac{\pi}{2} - \tan^{-1}s\right] = \frac{\pi}{2}.$$

We recognise that

$$\frac{d}{ds}\{\log(s+a) - \log(s+b)\} = \frac{1}{s+a} - \frac{1}{s+b}$$

But

$$= L\{e^{-at} - e^{-bt}\}.$$

$$L\{tf(t)\} = -\frac{d}{ds}\tilde{f}$$

so

$$-\frac{1}{t}\{e^{-at} - e^{-bt}\} = L^{-1}\left\{\log\frac{s+a}{s+b}\right\}.$$

13. We follow the hint

$$\frac{d}{da}\int_0^\infty e^{-st}\frac{\sin at}{t}\,dt = \int_0^\infty e^{-st}\cos at\,dt = \frac{s}{s^2+a^2},$$

so, integrating with respect to a between 0 and a, we obtain

$$\Phi(a) \equiv \int_0^\infty e^{-st}\frac{\sin at}{t}\,dt = \tan^{-1}(a/s).$$

[Note that $\int \Phi'\,da = \Phi(a) - \Phi(0) = \Phi(a)$.] Finally putting $a = 1$, we obtain the desired result

$$L\left\{\frac{\sin t}{t}\right\} = \int_0^\infty e^{-st}\frac{\sin t}{t}\,dt = \tan^{-1}(1/s).$$

Exercises 21.16

3. Following the hint we put $\psi = 2\theta$ and obtain

$$\int_0^{\pi/2}\sin^{2p}2\theta\,d\theta = \int_0^\pi \tfrac{1}{2}\sin^{2p}\psi\,d\psi$$

$$= 2\int_0^{\pi/2}\tfrac{1}{2}\sin^{2p}\psi\,d\psi.$$

The latter equation holds by symmetry (reflection in the vertical $\psi = \pi/2$).
Thus we have using the result of Question 5 in the last set of exercises:

$$B(p + \tfrac{1}{2}, \tfrac{1}{2}) = 2\int_0^{\pi/2}\sin^{2p}2\theta\,d\theta = 2\int_0^{\pi/2}2^{2p}\sin^{2p}\theta\cos^{2p}\theta\,d\theta$$

$$= 2^{2p}B(p + \tfrac{1}{2}, p + \tfrac{1}{2}).$$

We make use of the gamma function and obtain:

$$\frac{\Gamma(p + \tfrac{1}{2})\Gamma(\tfrac{1}{2})}{\Gamma(p + 1)} = 2^{2p}\frac{\Gamma(p + \tfrac{1}{2})\Gamma(p + \tfrac{1}{2})}{\Gamma(2p + 1)}.$$

Hence, using $\Gamma(\alpha + 1) = \alpha\Gamma(\alpha)$, we conclude that

$$\Gamma(2p)2p\,\Gamma(\tfrac{1}{2}) = 2^{2p}p\,\Gamma(p)\Gamma(p + \tfrac{1}{2})$$

or, finally,

$$\Gamma(2p)\Gamma(\tfrac{1}{2}) = 2^{2p-1}\Gamma(p)\Gamma(p + \tfrac{1}{2}).$$

5. We have

$$\frac{d}{ds}\tilde{f}(s) = -e^{-\sqrt{s}}\cdot\frac{1}{2\sqrt{s}}$$

$$\frac{d^2}{ds^2}\tilde{f}(s) = e^{-\sqrt{s}}\cdot\frac{1}{4s} + \frac{e^{-\sqrt{s}}}{4s^{3/2}}.$$

Hence

$$4s\frac{d^2\tilde{f}}{ds^2} + 2\frac{d\tilde{f}}{ds} - \tilde{f} = 0.$$

Now recall from an earlier question (Section 21.15 Question 10) that

$$-\frac{d\tilde{f}}{ds} = L\{tf(t)\},$$

$$\frac{d^2\tilde{f}}{ds^2} = L\{t^2 f(t)\}.$$

We are almost ready to take the inverse transform of the differential equation. We need, though, to compute the effect of the multiplication by s. Using $L\{g'\} = sL\{g\} - g(0)$, we have here

$$s\frac{d^2\tilde{f}}{ds^2} = sL\{t^2 f(t)\} = L\left\{\frac{d}{dt}[t^2 f(t)]\right\} + 0 \cdot f(0).$$

We therefore conclude that the inverse transform of the equation is

$$4\frac{d}{dt}[t^2 f(t)] - 2tf(t) - f(t) = 0$$

or

so

$$8tf(t) + 4t^2 f'(t) - 2tf(t) - f(t) = 0,$$

$$4t^2 f'(t) + (6t - 1)f(t) = 0,$$

or

$$\frac{f'(t)}{f(t)} = -\frac{(6t-1)}{4t^2} = \frac{1}{4t^2} - \frac{3}{2t}.$$

Thus

$$\log f(t) = -\frac{1}{4t} - \tfrac{3}{2}\log t + C$$

and so

$$f(t) = \frac{C'}{t^{3/2}}\exp\{-1/(4t)\},$$

which is continuous at $t = 0$. We can determine the constant C' by some cavalier approximation. We note that

$$L\{tf(t)\} = -\frac{d}{ds}\tilde{f}(s) = \frac{e^{-\sqrt{s}}}{2\sqrt{s}}.$$

On the other hand, for large t,

$$tf(t) \simeq \frac{C'}{t^{1/2}}$$

and

$$L\left\{\frac{C'}{t^{1/2}}\right\} = \frac{C'\sqrt{\pi}}{\sqrt{s}}$$

so we deduce, by way of the final values theorem, that

$$C' = \frac{1}{2\sqrt{\pi}}.$$

6. We take Laplace transforms

$$L\{t^{-\alpha}\}L\{f\} = L\{g\}$$

or

$$\frac{\Gamma(1-\alpha)}{s^{1-\alpha}}L\{f\} = L\{g\}$$

so

$$L\{f\} = \frac{1}{\Gamma(1-\alpha)}s^{1-\alpha}L\{g\}.$$

We consider the transform of $\phi'(t)$ where

$$\phi(t) = \int_0^t (t-u)^{\alpha-1}g(u)\,du.$$

Now $L\{\phi'\}$ equals $sL\{\phi\} - \phi(0)$. But $\phi(0) = 0$, so, using the convolution theorem,

$$L\{\phi'\} = sL\{t^{\alpha-1}\}L\{g\}$$

$$= s\frac{\Gamma(\alpha)}{s^{\alpha}}L\{g\} = \Gamma(\alpha)s^{1-\alpha}L\{g\}.$$

Thus

$$L\{f\} = \frac{L\{\phi'\}}{\Gamma(\alpha)\Gamma(1-\alpha)}$$

and this verifies the assertion of the exercise. Note that, of course, $L\{t^{\alpha-1}\}$ exists provided $\alpha > 0$ and $L\{t^{-\alpha}\}$ exists provided $-\alpha > -1$, i.e. $\alpha < 1$; this explains the assumption $0 < \alpha < 1$. Observe that by Question 10.

$$f = \frac{\sin \pi\alpha}{\pi}\phi'.$$

7. We use the substitution $u = t - n$ in the following formula

$$L\{f(t-n)\} = \int_0^\infty e^{-st}f(t-n)\,dt$$

$$= \int_{-n}^\infty \exp\{-s(u+n)\}f(u)\,du$$

so, since $f(u) = 0$ for $u \leqslant 0$, we obtain

$$L\{f(t-n)\} = e^{-sn}L\{f\}. \tag{$*$}$$

We take transforms of the lag equation and obtain

$$s\bar{f} + e^{-s}\bar{f} = \frac{1}{s^2}.$$

whence, for large enough s,

$$\tilde{f} = \frac{1}{s^2(s + e^{-s})} = \frac{1}{s^3\{1 + (e^{-s}/s)\}}$$

$$= \frac{1}{s^3}\{1 - e^{-s}/s + e^{-2s}/s^2 - e^{-3s}/s^3 + \cdots\}.$$

Now it would be *WRONG* to conclude:

$$f(t) = \tfrac{1}{2}t^2 - \frac{1}{3!}(t-1)^3 + \frac{1}{4!}(t-2)^4 - \cdots + \frac{(-1)^n}{n!}(t-n+2)^n + \cdots.$$

Recall that the formula (*) was derived on the assumption that the function to be transformed is zero to the left of the origin. To rectify the formula we would need to *interpret* each of the functions $t^2, t^3, t^4, \ldots$ as being zero to the left of the origin. Hence for any t the summation stops as soon as n is such that $t - n + 2 \leqslant 0$. i.e. when $n \geqslant t + 2$. Thus when $n = [t] + 3$ (and beyond), the appropriate term is interpreted as zero. The summation thus terminates at $n = [t] + 2$.

8. We follow the usual argument:

$$L_\delta\{f'\} = \int_\delta^\infty e^{-st} f'(t)\,dt$$

$$= [e^{-st} f(t)]_\delta^\infty - \int_\delta^\infty -se^{-st} f(t)\,dt$$

$$= sL_\delta\{f\} - e^{-s\delta} f(\delta).$$

We perform the required differentiation:

$$g(t) = \int_0^{\sqrt{t}} \exp(-u^2)\,du,$$

$$g'(t) = \tfrac{1}{2} e^{-t} \cdot t^{-1/2},$$

$$g''(t) = -\tfrac{1}{2} e^{-t} \cdot t^{-1/2} - \tfrac{1}{4} t^{-3/2} e^{-t}.$$

Thus

$$g''(t) = -g'(t) - \frac{1}{2t} g'(t),$$

or

$$2tg''(t) + (2t + 1)g'(t) = 0. \qquad (*)$$

It is not possible to take a Laplace transform in the usual way, since $g(0)$ is undefined and we cannot use the formula $L\{g'\} = sL\{g\} - g(0)$. Nor can we consider working with $h(t) = g'(t)$ and taking ordinary transforms of the equation (*) since $h(0)$ is undefined. Instead we therefore use the L_δ transform. We observe first that

$$L_\delta\{f''\} = sL_\delta\{f'\} - e^{-s\delta}f'(\delta)$$
$$= s\{sL_\delta\{f\} - e^{-s\delta}f(\delta)\} - e^{-s\delta}f'(\delta)$$
$$= s^2L_\delta\{f\} - se^{-s\delta}f(\delta) - e^{-s\delta}f'(\delta)$$

and that

$$L_\delta\{tf(t)\} = -\frac{d}{ds}L_\delta\{f(t)\}.$$

If we use ° to denote this new transform, our equation transforms to

$$-2\frac{d}{ds}\{s^2\mathring{g} - se^{-s\delta}g(\delta) - e^{-s\delta}g'(\delta)\} - 2\frac{d}{ds}\{s\mathring{g} - e^{-s\delta}g(\delta)\} + s\mathring{g} - e^{-s\delta}g(\delta) = 0.$$

Thus

$$-2\{2s\mathring{g} + s^2\mathring{g}' - e^{-s\delta}g(\delta) + s\delta e^{-s\delta}g(\delta) + \delta e^{-s\delta}g'(\delta)\}$$
$$-2\{\mathring{g} + s\mathring{g}' + \delta e^{-s\delta}g(\delta)\} + s\mathring{g} - e^{-s\delta}g(\delta) = 0,$$

where $\mathring{g}'$ denotes the s derivative of $\mathring{g}$. We have

$$\mathring{g}'[-2s(s+1)] + \mathring{g}[-3s-2] = e^{-s\delta}[-g(\delta) + 2\delta g'(\delta) + 2\delta g(\delta) + 2s\delta g(\delta)]$$

i.e.

$$\mathring{g}' + \frac{3s+2}{2s(s+1)}\mathring{g} = \frac{e^{-s\delta}(K(\delta) + sL(\delta))}{s(s+1)},$$

where K and L tend to zero with δ. We rewrite

$$\frac{3s+2}{2s(s+1)} = \frac{1}{s} + \frac{-1}{(-2)(s+1)}.$$

The integral of this last expression is $\log s + \frac{1}{2}\log(s+1)$, the integrating factor is thus the exponential of this, i.e. $s\sqrt{(s+1)}$. Thus

$$\frac{d}{ds}\{\mathring{g}s\sqrt{(s+1)}\} = \frac{e^{-s\delta}(K(\delta) + sL(\delta))}{\sqrt{(s+1)}}.$$

Integrating up from 1 to s, and letting the constant on the left-hand side be $-C(\delta)$, we obtain

$$\mathring{g}s\sqrt{(s+1)} = C(\delta) + K(\delta)I_1(s) + L(\delta)I_2(s),$$

where I_1 and I_2 are the appropriate integrals. Now since $e^{-s\delta} \leqslant e^{-\delta}$ on $[1, s]$, note that, for example

$$|I_1(s)| \leqslant e^{-\delta}\int_1^s \frac{ds}{\sqrt{(s+1)}},$$

which for a fixed s remains bounded as δ tends to zero. Letting δ tend to zero we obtain:

$$\tilde{g} = \frac{C}{s\sqrt{(s+1)}}.$$

Here C is the limiting value of $C(\delta)$ as δ tends to zero. We can now invoke the final value theorem. The limit of $g(t)$ as t tends to infinity is the well known infinite integral with value $\sqrt{\pi}/2$. On the other hand

$$\lim_{s \to 0} s\tilde{g}(s) = \lim_{s \to 0} \frac{C}{\sqrt{(s+1)}} = C.$$

Thus finally,

$$\tilde{g} = \frac{\sqrt{\pi}}{2s\sqrt{(s+1)}}.$$

9. Recall that

$$\lim_{n \to \infty} \left(1 + \frac{x}{n}\right)^n = e^x.$$

Hence, for fixed t we have

$$\lim_{n \to \infty} f(n, t) = e^{-t}t^{x-1}.$$

Hence for $x > 0$

$$\Gamma(x) = \int_0^\infty e^{-t}t^{x-1}\,dt$$

$$= \int_0^\infty \lim f(n, t)\,dt$$

$$= \lim \int_0^\infty f(n, t)\,dt$$

$$= \lim \int_0^n \left(1 - \frac{t}{n}\right)^n t^{x-1}\,dt,$$

where it remains to justify the limit manipulations. Before we do this we put $v = t/n$ and obtain:

$$\int_0^n \left(1 - \frac{t}{n}\right)^n t^{x-1}\,dt$$

$$= \int_0^1 (1 - v)^n n^{x-1} v^{x-1} n\,dv$$

$$= n^x \int_0^1 (1 - v)^n v^{x-1}\,dv$$

$$= n^x \left\{ \left[(1 - v)^n \frac{v^x}{x} \right]_0^1 + \int_0^1 n(1 - v)^{n-1} \frac{v^x}{x} \, dv \right\}$$

$$= n^x \cdot \frac{n}{x} \cdot \int_0^1 (1 - v)^{n-1} v^x \, dv$$

$$= n^x \cdot \frac{n}{x} \cdot \frac{n-1}{x+1} \cdot \frac{n-2}{x+2} \cdots \cdot \frac{1}{n+x-1} \int_0^1 v^{n+x-1} \, dv$$

$$= n^x \frac{n!}{x \cdot (x+1) \cdots (x+n)}.$$

We have assumed implicitly that x is not an integer. As regards the manipulation we note that

$$\log \left(1 - \frac{t}{n} \right) \leqslant -\frac{t}{n},$$

so

$$n \log \left(1 - \frac{t}{n} \right) \leqslant -t,$$

hence

$$\left(1 - \frac{t}{n} \right)^n \leqslant e^{-t}.$$

Consequently,

$$|f(n, t)| \leqslant e^{-t} t^{x-1} \equiv g(t).$$

But $\int g(t) \, dt = \Gamma(x)$ exists for $x > 0$, so g provides the required domination.
10. As x and $1 - x$ are positive and are not integers we may use the Gauss formula.

$$\Gamma(x)\Gamma(1-x) = \lim \frac{n^x n!}{x(x+1)\cdots(x+n)} \frac{n^{1-x} n!}{(1-x)(1-x+1)\cdots(1-x+n)}$$

$$= \lim \frac{n!}{x(1+x)(1-x)\cdots(n+x)(n-x)\cdot(n+1-x)}$$

$$= \lim \frac{1}{x} \frac{n! \quad n!}{(1-x^2)(4-x^2)\cdots(n^2-x^2)} \cdot \frac{1}{\{1+(1-x)/n\}}$$

$$= \lim \frac{1}{x} \frac{1}{(1-x^2)(1-x^2/4)\cdots(1-x^2/n^2)}$$

$$= \frac{1}{x} \{(\pi x)^{-1} \sin(\pi x)\}^{-1} = \frac{\pi}{\sin \pi x}.$$

Solutions to selected exercises on series solutions of differential equations

Exercises 22.10

1. Substituting into the equation we have

$$\sum_0^\infty a_n(n+\gamma)(n+\gamma-1)x^{n+\gamma-2} - \sum_0^\infty a_n x^{n+\gamma+1} = 0.$$

The first few lowest powers are contributed by the first series; these are $x^{\gamma-2}$, $x^{\gamma-1}$, x^γ. Thereafter (i.e. from $x^{\gamma+1}$ onwards) contributions arise from both series. We thus have

$$a_0\gamma(\gamma+1) = 0, \text{ and so } \gamma = 0 \text{ or } \gamma = -1,$$
$$a_1(\gamma+1)(\gamma+2) = 0,$$
$$a_2(\gamma+2)(\gamma+3) = 0 \text{ so } a_2 = 0.$$

The general recurrence relation is

$$(n+3+\gamma)(n+2+\gamma)a_{n+3} = a_n.$$

Taking $\gamma = 0$ we obtain $a_1 = a_2 = 0$ and hence $a_4 = a_5 = 0$, etc. The nonzero coefficients are of the form a_{3k} and we have

$$a_{3(k+1)} = \frac{a_{3k}}{(3k+3)(3k+2)}.$$

Thus the series may be regarded as being in the variable x^3. To obtain the radius of convergence we calculate the ratio of consecutive coefficients:

$$\frac{a_{3(k+1)}}{a_{3k}} = \frac{1}{(3k+3)(3k+2)} \to 0$$

and so the radius of convergence is ∞.

Now we consider the case $\gamma = -1$. This time a_1 is arbitrary. It may be checked, however, that the part of the series with coefficients $a_1, a_4, a_7, \ldots$ is the same as the series obtained for $\gamma = 0$, apart, of course, from a constant multiplier. We therefore lose nothing by putting $a_1 = 0$ (so that $a_4 = a_7 = \cdots = 0$). The recurrence relation for the coefficients $a_0, a_3, a_6, \ldots$ is

$$a_{3(k+1)} = \frac{a_{3k}}{(3k+2)(3k+1)}.$$

Here again the radius of convergence is ∞.

2. We make the change of variable suggested: $t = mx$. Let D denote differentiation with respect to t, then $D_x = m^{-1}D$. The equation thus transforms to

$$m^2 t^2 m^{-2} D^2 y + m t m^{-1} D y + (t^2 - n^2)y = 0,$$

or
$$t^2 D^2 y + t Dy + (t^2 - n^2)y = 0.$$

But this is the same example as in Section 22.3. Thus $y = J_n(mx)$ is one of the solutions.

3. This is known as Legendre's Equation. Carrying out the usual substitution we have on the left-hand side:

$$(1 - x^2) \sum_0^\infty a_n(n + \gamma)(n + \gamma - 1)x^{n+\gamma-2} - 2x \sum_0^\infty a_n(n + \gamma)x^{n+\gamma-1}$$

$$+ k(k + 1) \sum_0^\infty a_n x^{n+\gamma},$$

so the equation reads

$$\sum_0^\infty a_n(n + \gamma)(n + \gamma - 1)x^{n+\gamma-2} + \sum_0^\infty a_n\{k(k + 1) - (n + \gamma)(n + \gamma - 1)$$

$$- 2(n + \gamma)\}x^{n+\gamma} = 0,$$

or

$$\sum_0^\infty a_n(n + \gamma)(n + \gamma - 1)x^{n+\gamma-2} + \sum_0^\infty a_n\{k(k + 1) - (n + \gamma)(n + \gamma + 1)\}x^{n+\gamma} = 0.$$

We first inspect those lowest powers of x which occur in only one summand (viz. $x^{\gamma-2}, x^{\gamma-1}$) and obtain the equations

$$a_0\gamma(\gamma - 1) = 0, \quad \text{so } \gamma = 0 \text{ or } \gamma = 1,$$

$$a_1(\gamma + 1)\gamma = 0.$$

The general recurrence relation is

$$a_{n+2}(n + 2 + \gamma)(n + 1 + \gamma) = -a_n\{k(k + 1) - (n + \gamma)(n + \gamma + 1)\}.$$

First consider $\gamma = 0$. The recurrence relation now reads

$$a_{n+2}(n + 2)(n + 1) = -a_n\{k(k + 1) - n(n + 1)\}.$$

Thus we have

$$a_{n+2} = -\frac{\{k(k + 1) - n(n + 1)\}}{(n + 2)(n + 1)}a_n = -\frac{(k - n)(k + n + 1)}{(n + 2)(n + 1)}a_n, \qquad (1)$$

so

$$a_{2m} = (-1)^m \frac{\{(k - 2m + 2)(k + 2m - 1)\} \cdots}{(2m)!}a_0$$

and

$$a_{2m+1} = (-1)^m \frac{\{(k - 2m + 1)(k + 2m)\} \cdots}{(2m)!}a_1.$$

Let us put $b_m = a_{2m}/a_0$ and $c_m = a_{2m+1}/a_1$. Then we obtain the solution

$$y = a_0 \sum_0^\infty (-1)^m b_m x^{2m} + a_1 \sum_0^\infty (-1)^m c_m x^{2m+1}$$

and there are already before us two linearly independent solutions corresponding to the two summation signs (so we need not consider $\gamma = 1$). Observe that

$$\frac{a_{n+2}}{a_n} = -\frac{(k-n)(k+n+1)}{(n+2)(n+1)} \to 1,$$

as n tends to infinity. Thus the radius of convergence is 1.

If k is an integer, we see from (1) that $a_{k+2} = 0$, so there is a polynomial solution to the equation.

4. Since $f(-x) = -f(x)$ the cosine coefficients are zero. Thus

$$\frac{\pi}{2} b_n = \int_0^\pi x \sin x \, dx = [-x \cos x]_0^\pi + \int_0^\pi \cos x \, dx$$

$$= \pi + [\sin x]_0^\pi = \pi,$$

i.e. $b_n = 2$. Thus $f(x) \sim \Sigma 2 \sin nx$.

6. One way to deal with this is to use Riemann–Stieltjes integration. We let $v(x)$ be any continuous function with $v(n) = v_n$ (e.g. the function whose graph consists of straight edges passing through the points (n, v_n)). Let $u(x)$ be defined similarly and let $U(x)$ be defined by

$$U(k) = \int_{-1}^k u(x) d[x] = \sum_{0 \le n \le k} u_n.$$

Then we have

$$\sum_n^m u_r v_r = \int_{n-1}^m v(x)u(x) \, d[x] = [v(x)U(x)]_{n-1}^m - \int_{n-1}^m U(x) \, dv(x)$$

$$= v_m U_m - v_{n-1} U_{n-1} - \sum_{n-1}^{m-1} \int_r^{r+1} U(x) \, dv(x).$$

But in $[r, r+1)$ $U(x)$ is constantly equal to U_r. The right-hand side is thus equal to

$$v_m U_m - v_{n-1} U_{n-1} - \sum_{n-1}^{m-1} U_r \{v(r+1) - v(r)\}$$

$$= v_m U_m - v_{n-1} U_{n-1} - \sum_n^{m-1} U_r \{v_{r+1} - v_r\} - U_{n-1}\{v_n - v_{n-1}\}$$

$$= v_m U_m - v_n U_{n-1} - \sum_n^{m-1} U_r \{v_{r+1} - v_r\}. \tag{1}$$

Let $S(x)$ denote the sum to infinity $u_0(x) + u_1(x) + u_2(x) + \cdots$. We assume that the functions $u_i(x)$ are bounded (in view of the application). Let a small number δ be given. By uniform convergence there is N so that whenever n and m are beyond N we have

$$\left| \sum_n^m u_r(x) \right| < \delta$$

independently of the choice of x. It follows that $S(x)$ is bounded as a function of x. (The function is no bigger than $u_0(x) + u_1(x) + \cdots + u_N(x) + \delta$ and that is bounded.) Evidently, if

$$\sum_1^\infty |v_i - v_{i-1}| < \infty,$$

then for any small number δ there is K so that, for n and $m \geqslant K$,

$$\sum_n^m |v_i - v_{i-1}| < \delta,$$

i.e. the tail end of the series has to be small and most of the sum resides in the first K terms.

Thus, if M is a bound on $S(x)$ and m and n are large enough, we have by (1):

$$\left| \sum_n^m v_r u_r(x) \right| \leqslant |S(x)| \sum_n^m |v_r - v_{r+1}| + |S(x)| \{|v_n| + |v_m|\}$$

$$\leqslant M \sum_n^m |v_r - v_{r+1}| + M\{|v_n| + |v_m|\}.$$

So assuming $v_n \to 0$ as $n \to \infty$ we see that the tail-end sum may be made small independently of x. Actually of course, we can make the tail-end sum small without $v_n \to 0$. The point is that

$$v_m = v_m - v_{m-1} + v_{m-1} - v_{m-2} + \cdots + v_{n+1} - v_n + v_n$$

so

$$|v_n - v_m| \leqslant \sum_{n+1}^m |v_i - v_{i-1}| \leqslant \delta.$$

In particular the sequence v_m is bounded, say by V. [Actually, this calculation tells us that v_m converges.] Thus the term

$$|U_m(x)v_m - U_{n-1}(x)v_n| = |(v_n - v_m)U_{n-1}(x) + v_m(U_{n-1}(x) - U_m(x))| \quad (2)$$

may be estimated as follows. Suppose $S(x) > 0$ (so that for large m and n the same is true of $U_n(x)$ and $U_m(x)$). Say $v_m > v_n$, then the term in (2) is no greater than

$$\delta U_{n-1}(x) + |v_m| \, |U_{n-1}(x) - U_m(x)| \leqslant \delta M + V|U_{n-1}(x) - U_m(x)|$$

which can be made small for n, and m large, provided also δ was selected small enough.

The three applications are now obvious.

7. We follow the hint

$$0 \leqslant \int_{-\pi}^\pi \{f(x) - g(x)\}^2 \, dx$$

$$= \int_{-\pi}^\pi \{f(x)^2 - 2f(x)g(x) + g(x)^2\} \, dx$$

$$= \int_{-\pi}^{\pi} f^2 \, dx - 2 \sum_{r=1}^{n} \left\{ \int_{-\pi}^{\pi} a_r f(x) \cos rx \, dx + \int_{-\pi}^{\pi} b_r f(x) \sin rx \, dx \right\}$$

$$- 2 \int_{-\pi}^{\pi} f(x) \tfrac{1}{2} a_0 \, dx + \int_{-\pi}^{\pi} g^2 \, dx$$

$$= \int_{-\pi}^{\pi} f^2 \, dx - 2 \tfrac{1}{2} a_0 \pi a_0 - 2 \sum_{r=1}^{n} \{ \pi a_r{}^2 + \pi b_r{}^2 \}$$

$$+ \left\{ \int_{-\pi}^{\pi} \sum_{r,t=1}^{n} \{ a_r a_t \cos rx \cos tx + 2 a_r b_t \cos rx \sin tx + b_r b_t \sin rx \sin tx \} \right.$$

$$\left. + \sum_{r=1}^{n} a_0 a_r \cos rx + \sum_{t=1}^{n} a_0 b_t \sin tx \right\} dx + \tfrac{1}{4} a_0^2 2\pi$$

$$= \int_{-\pi}^{\pi} f^2 \, dx - a_0 \pi a_0 - 2 \sum \{ \pi a_r^2 + \pi b_r^2 \} + \sum \{ a_r^2 \pi + b_r^2 \pi \} + \tfrac{1}{2} \pi a_0^2.$$

Hence

$$\frac{1}{\pi} \int_{-\pi}^{\pi} f^2 \, dx \geqslant \tfrac{1}{2} a_0^2 + \sum \{ a_r^2 + a_r^2 \},$$

as required.

A Miscellany

1. Mass of constant density 2 is distributed uniformly along $[1,3]$. Additionally unit point masses are placed at $x = 1, 2, 3$. Compute $\alpha(x) =$ the total mass lying in $[1, x]$ and sketch its graph. Find the following 'moment of inertia':

$$\int_1^3 x^2 \, d\alpha(x).$$

2. Compute

$$\iiint_V \frac{dx \, dy \, dz}{(1 + x + y + z)^3}$$

where V is the tetrahedron $x + y + z \leqslant 1$, $x, y, z \geqslant 0$.

3. Prove that

$$\left\{ \int_a^b f(x) dx \right\} \times \left\{ \int_a^b \frac{dx}{f(x)} \right\} \geqslant (b - a)^2.$$

It may be helpful to consider

$$\iint_V \left\{ \frac{f(x)}{f(y)} + \frac{f(y)}{f(x)} \right\} dx \, dy,$$

where R is the rectangle $a \leqslant x \leqslant b$, $a \leqslant y \leqslant b$.

4. Find the volume common to the sphere $x^2 + y^2 = a^2$ and the cylinder $x^2 + y^2 = ax$ (which is centred at $(a/2, 0)$).

5. Show that $(A^T)^G = (A^G)^T$, where A^G is the strong generalized inverse of A.

6. Show that the vector x nearest the origin which satisfies $Ax = b$ (assumed consistent) is $x = A^G b$. [Hint: Consider $x = A^G b + (I - A^G A)z$ and use Pythagoras' theorem.]

7. Find the minimum and maximum of $x^2 + y^2 + z^2$ subject to

$$\left. \begin{array}{r} x^2 + 2y^2 + z^2 + 2yz = 1, \\ y + z = 0. \end{array} \right\}$$

8. Show that

$$\int_0^n \frac{d(x[x])}{x} = n(1 + \log n) - \log(n!).$$

9. Show that

$$\int_0^{\pi/2} \frac{d\theta}{\sqrt{1 - \frac{1}{2}\sin^2\theta}} = \frac{\Gamma(\frac{1}{4})^2}{4\sqrt{\pi}}.$$

[Hint: $B(p, p) = 2 \int_0^{1/2} t^{p-1}(1-t)^{p-1} dt = 4 \int_0^{\pi/4} \sin^{2p-1}\theta \cos^{2p-1}\theta \, d\theta.$]

10. Find the volume of the wedge of the cylinder $x^2 + y^2 \leqslant 2ax$ contained between the planes $z = mx$ and $z = nx$ where $m > n$.

11. Evaluate $\iint_R x \, dx \, dy$ where R is the interior of the curve $r = 2a(1 + \cos\theta)$.

12. Evaluate $\int_0^1 x^{t-1}(t \log x + 1) \, dx$.

13. Solve the equation

$$\int_0^t u^2 f(u - t) = 2f'(t) - 4$$

with $f(0) = 0$.

14. Show that for any point $(x, y) \neq (0, 0)$ there are two real values for λ^2 satisfying

$$\frac{x^2}{\lambda^2} + \frac{y^2}{\lambda^2 - c^2} = 1$$

with one value, call it s^2, lying in $(0, c)$ and the other, call it t^2, lying in (c, ∞). [Hint: If $f(\lambda) = \lambda^2(\lambda^2 - c^2) - y^2\lambda^2 - x^2(\lambda^2 - c^2)$, consider $f(0)$ and $f(c)$.] Show that

$$x = \pm\frac{st}{c}, \quad y = \pm\frac{\sqrt{(t^2 - c^2)(c^2 - s^2)}}{c}.$$

Describe the contours $s = $ constant and $t = $ constant. What contours are obtained in the limiting cases when (i) $s \to 0$, (ii) $s \to c$, (iii) $t \to c$. What is the transform of (a) $x = 1$, and (b) $y = 1$.

Compute the Jacobian $\partial(x, y)/\partial(s, t)$.

15. Evaluate

$$\int_0^\infty \frac{dx}{(x^2 + 1)^2}.$$

[Hint: For $a \geqslant 1$ evaluate

$$\int_0^\infty \frac{dx}{x^2 + a}$$

and differentiate your result with respect to *a*.] How does your result generalize for powers higher than 2 outside the bracket?

16. A transformation of the positive quadrant is given by

$$u = \sqrt{x} + \sqrt{y}, \\ v = xy$$

Show that the *u*-contours are parabolas tangential to the *x*- and *y*-axes. [Hint: Square *u* and consider $X = x + y$ and $Y = x - y$.) Determine the vanishing points of the Jacobian of the transformation and interpret their geometric significance in terms of the *u*- and *v*-contours. By considering an appropriate local inverse transformation, find the (u, v)-image of the set $\{(x, y): 1 \leqslant x \& y \leqslant x\}$ and sketch it on a diagram.

17. Repeat the problem when

$$u = x^2 + y \\ v = x^2 y$$

and the region to be transformed is $\{(x, y): 0 \leqslant x \leqslant 1 \text{ and } y \geqslant x^2\}$.

18. A function $f_\delta(t)$ is defined for $\delta > 0$ by the integral

$$f_\delta(t) = \int_\delta^1 \frac{x^t - 1}{\log x} dx.$$

Find $f_\delta(t)$ for $t > -1$. (You should justify your manipulations.) By integrating back your answer with respect to *t* from 0 to α show that for $\alpha > -1$:

$$f_\delta(\alpha) \to \log(1 + \alpha) \quad \text{as} \quad \delta \to 0.$$

Deduce that

$$\int_0^1 \frac{x - 1}{\log x} dx = \log 2.$$

19. The incomplete Laplace transform L_δ for $\delta \geqslant 0$ is defined by

$$L_\delta(f) = \int_\delta^\infty f(t) e^{-st} dt.$$

Verify that

$$L_\delta(f') = sL_\delta(f) - e^{-s\delta} f(\delta), \tag{1}$$

and

$$L_\delta(t f(t)) = -\frac{d}{ds} L_\delta(f). \tag{2}$$

A function $g(t)$ is defined for $t > 0$ by the integral

$$g(t) = \int_t^\infty \frac{1}{u} e^{-u} du.$$

By splitting the range into $[t, 1]$ and $[1, \infty)$ show that

$$0 \leqslant g(t) \leqslant \log t + e^{-1}.$$

Deduce that $\lim_{\delta \to 0} \delta g(\delta) = 0.$ (3)

Show by changing the order of integration that $\lim_{s \to \infty} s L_\delta(g) = 0.$ (4)

Starting from $tg'(t) = -e^{-t}$ derive the result for $\delta > 0$ that:

$$-vL_\delta(g) = \int_{v+1}^{1} e^{-\delta w} \frac{1}{w} \, dw + (1 - e^{-v\delta}) \int_{\delta}^{\infty} \frac{1}{u} e^{-u} \, du.$$

[Hint: Apply (1) and (2) then using (4) integrate with respect to s from v to infinity. You will need to make the change of variable $u = (v + 1)\delta$ and split the range: $[\delta, \infty)$ into $[\delta, \delta(v + 1)]$ and $[\delta(v + 1), \infty)$.]
Use (3) to deduce that $sL_0(g) = \log(1 + s)$.
Why could we not apply the Laplace transform L_0 directly to $tg'(t)$?

20. (i) Assuming that the series $\sum a_n$ converges show by considering $a_n = s_n - s_{n-1}$ (where s_n is the sum to n terms) that $a_n \to 0$ as $n \to \infty$.
   (ii) Give an example of a series where $a_n \to 0$, but $\sum a_n$ diverges.
   (iii) Use the graph of $x \sin x$, to explain why

$$\int_0^\infty x \sin x \, dx$$

     does not exist. (Put $a_n = \int_{n\pi}^{(n+1)\pi} x \sin x \, dx$.)
   (iv) Show also that

$$\int_0^\infty \frac{x^3 \sin x}{1 + x^2} \, dx$$

     does not exist. Note that $1/\sqrt{2} \leqslant |\sin x|$ for
     $n\pi + \pi/4 \leqslant x \leqslant n\pi + 3\pi/4$ and that for $x \geqslant 1$, $\frac{1}{2}x \leqslant x^3/(1 + x^2)$.
   (v) Does $\int_0^\infty [x^2 \sin x/(1 + x^2)] dx$ diverge? Justify your answer.
   (vi) Show that $\int_0^\infty [x \sin x/(1 + x^2)] dx$ converges (conditionally).
   (vii) Show that $x^p/(1 + x^2)$ is eventually decreasing when $p < 2$. Deduce that for $p < 2$,

$$\int_\pi^\infty \frac{x^p \sin x}{1 + x^2} \, dx.$$

     converges.

21. By dissecting the four-dimensional simplex S

$$x + y + z + t \leqslant a$$
$$x, y, z, t \geqslant 0$$

into tetrahedra (obtained by holding t constant) show that

$$\iint_{S}\iint dx\,dy\,dz\,dt = \frac{1}{4!}a^4.$$

[Hint: The volume of the corresponding three dimensional simplex is $\frac{1}{6}a^3$.]

22. Show that AB and BA for square matrices A, B of size $n \times n$ have the same eigenvalues. Deduce that $AB - BA = I$ is impossible.

23. Show that the non-zero eigenvalues of a real antisymmetric matrix are all purely imaginary.

24. Show that for the matrix

$$C = \begin{bmatrix} A & B \\ B & A \end{bmatrix},$$

where A and B are square, Spectrum (C) = Spectrum $(A + B) \cup$ Spectrum $(A - B)$. (Spectrum (M) denotes the set of eigenvalues of the matrix M.)

25. If A is a real symmetric matrix and λ is an eigenvalue of multiplicity m, show that $m = \dim N(A - \lambda I)$.

[Hint: Diagonalize A.]

26. Using row reductions find the rank of the following $(m + 1) \times m$ matrix.

$$\begin{bmatrix} 1 & 1 & \cdots & 1 \\ -m & 1 & \cdots & 1 \\ 1 & -m & \cdots & 1 \\ & & \cdots & \\ 1 & 1 & \cdots & -m \end{bmatrix}.$$

27. Find the eigenvalue of the matrix

$$\begin{bmatrix} a & b & b & \cdots & b \\ b & a & b & \cdots & b \\ b & b & a & \cdots & b \\ & & & \cdots & \\ b & b & b & \cdots & a \end{bmatrix}.$$

corresponding to the eigenvector $e = (1, 1, \ldots, 1)'$? What is its multiplicity? [Hint: Use Questions 25 and 26.]

28. By writing $A = aI + bU$ find A^2 as a combination of I and U and hence determine constants c and d such that $A^2 - cA + dI = 0$. Hence find Spectrum (A).

29. If A is positive definite prove that

(i) $a_{ii} > 0$ for each i;

(ii) $a_{ii}a_{jj} > |a_{ij}|^2$ for $i \neq j$.

30. Suppose λ is an eigenvalue of $A = \{a_{ij}\}$ and $Ax = \lambda x$. By considering r such that $|x_r| = \max \{|x_1|, \ldots, |x_n|\}$, show that $|a_{rr} - \lambda| \leqslant \sum_{j \neq r}|a_{rj}|$. Deduce that the tridiagonal matrix

$$\begin{bmatrix} a & -1 & & & & \\ -1 & a & -1 & & & \\ & -1 & a & & & \\ & & \cdot & \cdot & & -1 \\ & & & & -1 & a \end{bmatrix}$$

is positive definite for $a > 2$. Solve the standard recurrence relation for the Sturm sequence to show that for $a = 2$ the matrix is likewise positive definite.

31. Show, given any $\varepsilon > 0$ that if $A = \{a_{ij}\}$ is any square $n \times n$ matrix there is an $n \times n$ matrix $B = \{b_{ij}\}$ with all its eigenvalues distinct such that $|b_{ij} - a_{ij}| < \varepsilon$ for $1 \leqslant i, j \leqslant n$; i.e. B approximately equals A. [Hint: Reduce A to upper triangular form.]

32. If $f : \mathbb{R} \to \mathbb{R}$ is an increasing convex function and $T = f[\mathbb{R}]$ is convex, show that $f^{-1} : T \to \mathbb{R}$ is a concave function.
Remark. If f is continuous then T will necessarily be convex.

33. A point $\mathbf{x}$ is a convex combination of $n + 2$ points of $S \subseteq \mathbb{R}^n$, say of $\mathbf{x}_0, \mathbf{x}_1, \ldots, \mathbf{x}_{n+1}$. Show by considering the $n + 1$ linearly dependent vectors $\mathbf{x}_1 - \mathbf{x}_0, \ldots, \mathbf{x}_{n+1} - \mathbf{x}_0$ that there are constants $\alpha_0, \alpha_1, \ldots, \alpha_{n+1}$ not all zero such that:

$$\alpha_0 \mathbf{x}_0 + \alpha_1 \mathbf{x}_1 + \cdots + \alpha_{n+1} \mathbf{x}_{n+1} = 0,$$
$$\alpha_0 + \alpha_1 + \cdots + \alpha_{n+1} = 0.$$

If $\mathbf{x} = \gamma_0 \mathbf{x}_0 + \cdots + \gamma_{n+1} \mathbf{x}_{n+1}$ show by selecting an appropriate λ that

$$\mathbf{x} = (\gamma_0 - \lambda\alpha_0)\mathbf{x}_0 + \cdots + (\gamma_{n+1} - \lambda\alpha_{n+1})\mathbf{x}_{n+1}$$

is in fact a convex combination of fewer than $n + 2$ vectors.
[Hint: $\lambda = \gamma_i/\alpha_i$ for an appropriate positive α_i.]

Deduce Caratheodory's Theorem that if $S \subseteq \mathbb{R}^n$, then every vector of conv(S) is a convex combination of $n + 1$ (or fewer) vectors in S.

34. If C is a convex set and $z \in C$ define the 'face determined by z' to be the largest convex subset F of C such that z is in the relative interior of F. The dimension of F is then the 'facial dimension of z', denoted f-dim(z). If C is the triangle conv$\{x_1, x_2, x_3\}$ what is the face determined by z when (i) z does not lie on any edge of the triangle; (ii) z lies on an edge but is not a vertex; (iii) z is a vertex? In each case what is the facial dimension of z?

35. If $C = \text{conv}(S)$ where S is a *finite* subset of $\mathbb{R}^2$ and $z \in C$ show that there is always a point of S in the face F determined by z. [Hint: By Caratheodory's theorem $z \in F \cap \text{conv}\{s_1, s_2, s_3\}$ for some three points of S.] Let $z(t) = z + t(s - z)$, so that $z(0) = z$ and $z(1) = s$. Show that there is a largest value t_0 of t such that $z(t) \in F$. Deduce that $z(t_0)$ has lower facial dimension than dim(F). [Hint: $z(t_0)$ is not in the relative interior of F.]

36. Let S_1, S_2, S_3 be three finite subsets of $\mathbb{R}^2$ and suppose $z_i \in \text{conv}(S_i)$ for $i = 1, 2, 3$. Pick s_i in S_i to lie in the face determined by z_i (as in the last question). Show that for some scalars $\alpha_1, \alpha_2, \alpha_3$ not all zero.

$$\alpha_1(s_1 - z_1) + \alpha_2(s_2 - z_2) + \alpha_3(s_3 - z_3) = 0.$$

Define $z_i(t) = z_i + t\alpha_i(s_i - z_i)$. Show that

- (i) $z_1 + z_2 + z_3 = z_1(t) + z_2(t) + z_3(t)$, and
- (ii) for some value of t one of the vectors $z_i(t)$ has lower facial dimension than z_i.

37. Conclude from the last question that if $x \in \text{conv}(S_1 + S_2 + S_3)$ then for some y_1, y_2, y_3 we have $x = y_1 + y_2 + y_3$, where each $y_i \in \text{conv}(S_i)$ and for at most two exceptions $y_i \in S_i$. [Hint: Suppose that $f\text{-dim}(z_1) + f\text{-dim}(z_2) + f\text{-dim}(z_3)$ is at a minimum.]

Remark. The Shapley–Folkman theorem asserts that if $S_1, S_2, \ldots, S_m$ are compact (e.g. finite) subsets of $\mathbb{R}^n$ and $m > n$ then any $x \in \text{conv}(S_1 + \cdots + S_m)$ may be expressed in the form

$$x = y_1 + \cdots + y_m,$$

where, for each i, $y_i \in \text{conv}(S_i)$ and with at most n exceptions $y_i \in S_i$. (The proof is a simple generalization of the argument leading up to the last question.)

38. (Homogeneous programming) For the problem: maximize $f(x, y, z)$ subject to $g_1(x, y, z) \geq c$ and $g_2(x, y, z) \geq d$, where f, g_1, g_2 are homogeneous with respective degrees l, m, n, and c, d are constants, show that the Kuhn–Tucker conditions imply the following equation.

$$lf_{\max} + \lambda_1 mc + \lambda_2 nd = 0.$$

Show that the equation is also valid for maximization with equational constraints. [Hint: Euler's equation].

Appendix A

Existence of the Riemann integral and of the Riemann–Stieltjes integral

In Section 17.3 it was asserted that both the over-estimate and the under-estimate approach to the definition of the Riemann integral agreed when the integrand is continuous. We give a proof of this fact; the same argument easily extends to a proof that the Riemann–Stieltjes integral of a continuous function with respect to an increasing function exists (Section 17.6).

Let $f(x)$ be continuous on $[a, b]$. Our objective is to show that for appropriate partitions P the lower estimate $L(P)$ and the upper estimate $U(P)$ are approximately equal. We begin by setting ourselves a target discrepancy of at most $\varepsilon > 0$ between the lower and upper estimates. Since f is continuous (Section 17.4) at $x = a$, there is a $\delta_1 > 0$ such that $|f(a) - f(x)| \leqslant \varepsilon$ whenever $a \leqslant x \leqslant a + \delta_1$. Hence if $a \leqslant x, x' \leqslant a + \delta_1$ we have

$$|f(x) - f(x')| \leqslant |f(x) - f(a)| + |f(a) - f(x')| \leqslant 2\varepsilon.$$

So if P is a partition with $x_1 = a + \delta_1$ we have $0 \leqslant M_1 - m_1 \leqslant 2\varepsilon$. Now we can repeat the continuity argument at $a_1 = a + \delta_1$ and obtain δ_2 so that $|f(a_1) - f(x)| \leqslant \varepsilon$ whenever $a_1 \leqslant x \leqslant a_1 + \delta_2$. Thus $|f(x) - f(x')| \leqslant 2\varepsilon$ whenever $a_1 \leqslant x, x' \leqslant a_1 + \delta_2$. So if P is a partition with $x_1 = a_1$ and $x_2 = a_1 + \delta_2$ we have

$$0 \leqslant M_1 - m_1 \leqslant 2\varepsilon \quad \text{and} \quad 0 \leqslant M_2 - m_2 \leqslant 2\varepsilon.$$

It would seem natural to define $a_2 = a_1 + \delta_2$ and to continue the same argument until for some n we reach $a_n = b$. Some care, however, is needed: the choice of δ_1, δ_2 etc must not be arbitrary. Suppose we always choose the δ_n to be as large as possible, applying the rule that if $a_n < b$, then

$$\delta_{n+1} = \max \{\delta > 0 : a_n + \delta \leqslant b \quad \text{and} \quad (\forall x, x' \text{ in}$$
$$[a_n, a_n + \delta]) | f(x) - f(x')| \leqslant 2\varepsilon\}.$$

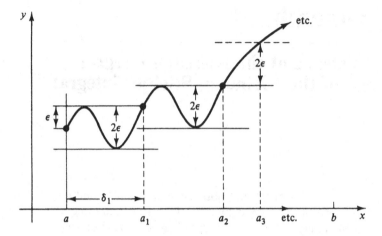

(See the remark at the end.) Suppose that the defined points
$a_1 < a_2 < a_3 < \cdots$ satisfy $a_n < b$ for *all* $n = 1, 2, 3, \ldots$. Let
$a' = \sup\{a_n : n = 1, 2, 3, \ldots\}$ so that $a_n < a' \leqslant b$ for all natural numbers
n. We appeal to continuity at a'. We have for some $\delta' > 0$ that
$|f(a') - f(x)| \leqslant \varepsilon$ whenever $a' - \delta' < x \leqslant a'$. But for some n,
$a' - \delta' < a_n < a'$. Consequently, if x, x' satisfy: $a_n \leqslant x, x' \leqslant a'$, we have

$$|f(x) - f(x')| \leqslant |f(x) - f(a')| + |f(a') - f(x)| \leqslant 2\varepsilon.$$

Thus $a_{n+1} = a_n + \delta_{n+1} \geqslant a'$ and we have a contradiction.

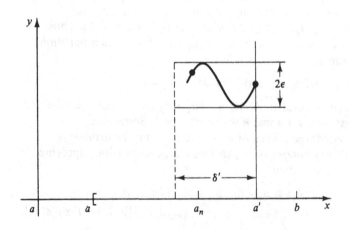

So there is, for some n, a partition $P^* = \{a, a_1, \ldots, a_n\}$ of $[a, b]$ such that for each $r, 0 \leqslant M_r - m_r \leqslant 2\varepsilon$. It now follows that

$$U(P^*) - L(P^*) = \sum (M_r - m_r)(a_r - a_{r-1})$$
$$\leqslant 2\varepsilon \sum (a_r - a_{r-1})$$
$$= 2\varepsilon(b - a).$$

Thus the discrepancy was out by the factor $2(b-a)$, never mind! Evidently if P is any partition finer than P^* we shall still have

$$U(P) - L(P) \leqslant 2\varepsilon(b - a).$$

Our conclusion is that with fine enough partitions we can control the discrepancy between upper and lower estimate to be as small as we wish.

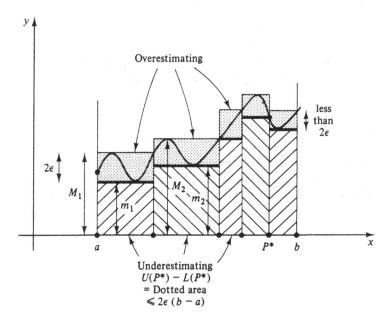

Remark In the above proof we have tacitly used the fact that for any $c < b$ the supremum of the set

$$\{\delta > 0 : c + \delta \leqslant b \,\&\, (\forall x, x' \text{ in } [c, c + \delta])(|f(x) - f(x')| \leqslant 2\varepsilon)\}$$

is in the set, i.e. that its maximum exists. This is easily checked and relies on the continuity of f.

How does this proof extend to the Riemann–Stieltjes case?

Consider the partition P^* above. Suppose α is increasing. Given any choice of t_r with $a_{r-1} \leqslant t_r \leqslant a_r$ we have that $m_r \leqslant f(t_r) \leqslant M_r$. Hence since α is increasing

$$m_r(\alpha(a_r) - \alpha(a_{r-1})) \leqslant f(t_r)(\alpha(a_r) - \alpha(a_{r-1})) \leqslant M_r(\alpha(a_r) - \alpha(a_{r-1})).$$

With the obvious notation for lower and upper estimates we therefore have:

$$\begin{aligned}
L_\alpha(P^*) \equiv \sum m_r(\alpha(a_r) - \alpha(a_{r-1})) &\leqslant \sum f(t_r)(\alpha(a_r) - \alpha(a_{r-1})) \\
&\leqslant \Sigma M_r(\alpha(a_r) - \alpha(a_{r-1})) \\
&\equiv U_\alpha(P^*).
\end{aligned} \tag{1}$$

Thus the general Riemann–Stieltjes sum lies between $L_\alpha(P^*)$ and $U_\alpha(P^*)$. Again we have

$$\begin{aligned}
U_\alpha(P^*) - L_\alpha(P^*) &= \sum (M_r - m_r)(\alpha(a_r) - \alpha(a_{r-1})) \\
&\leqslant 2\varepsilon \sum (\alpha(a_r) - \alpha(a_{r-1})) \\
&= 2\varepsilon(\alpha(b) - \alpha(a)).
\end{aligned} \tag{2}$$

The results (1) and (2) remain true for any P finer than P^*. Thus the lower and upper estimates are close provided the partition is fine enough. Hence

$$\sup_P L_\alpha(P) = \inf_P U_\alpha(P). \tag{3}$$

It follows that the limit of the Riemann–Stieltjes sum also exists – being caught between the upper and lower estimates – and that the Riemann–Stieltjes integral equals the common value in (3).

Appendix B

Lagrange multipliers

The argument presented in Chapter 16 justifying the Kuhn–Tucker theorem does not apply to the non-linear programming problem with equality constraints. This section is devoted to a sketchy justification in the special case of two variables and the more general case of several variables (where the argument is similar in spirit).

The simple case of two variables

Consider the problem

maximize $f(x, y)$

subject to $g(x, y) = 0$.

Let us suppose that the maximum occurs at $(\hat{x}, \hat{y})$ and that, in principle, we can 'solve' the equation $g(x, y) = 0$, i.e. express y explicitly as a function of x (at least near $x = \hat{x}$), say as $y = h(x)$. We use the existence of $h(x)$ to turn the constrained problem into the unconstrained problem of finding the (local) maximum of $f(x, h(x))$ which occurs at $x = \hat{x}$. The Lagrange multiplier then allows us to eliminate all reference to $h(x)$.

We begin by examining the stationarity condition for

$$z = f(x, h(x)).$$

Applying the chain rule we have:

$$\frac{dz}{dx} = \frac{\partial f}{\partial x} + \frac{\partial f}{\partial y}\frac{dh}{dx} = 0$$

so

$$\frac{dh}{dx} = -\frac{\partial f}{\partial x} \bigg/ \frac{\partial f}{\partial y}. \quad \left(\text{assuming } \frac{\partial f}{\partial y} \neq 0\right). \tag{1}$$

But we have the identity

$$g(x, h(x)) \equiv 0,$$

which we may also differentiate. Thus

$$\frac{\partial g}{\partial x} + \frac{\partial g}{\partial y}\frac{dh}{dx} = 0,$$

hence

$$\frac{dh}{dx} = -\frac{\partial g}{\partial x}\bigg/\frac{\partial g}{\partial y} \quad \left(\text{assuming } \frac{\partial g}{\partial y} \neq 0\right). \tag{2}$$

We may now eliminate $h'(x)$ between and (1) and (2) to obtain

$$\frac{\partial f}{\partial x}\bigg/\frac{\partial f}{\partial y} = \frac{\partial g}{\partial x}\bigg/\frac{\partial g}{\partial y}$$

or

$$\frac{\partial f}{\partial x}\bigg/\frac{\partial g}{\partial x} = \frac{\partial f}{\partial y}\bigg/\frac{\partial g}{\partial y}. \tag{3}$$

Calling the common value in (3) λ we have

$$\left.\begin{aligned}
\frac{\partial f}{\partial x} &= \lambda\frac{\partial g}{\partial x} \\
\frac{\partial f}{\partial y} &= \lambda\frac{\partial g}{\partial y}
\end{aligned}\right\}, \tag{4}$$

or

$$\left.\begin{aligned}
\frac{\partial}{\partial x}(f - \lambda g) &= 0 \\
\frac{\partial}{\partial y}(f - \lambda g) &= 0
\end{aligned}\right\}. \tag{5}$$

The conditions (5) assert that the expression in x, y, λ

$$L(x, y, \lambda) \equiv f(x, y) - \lambda g(x, y)$$

known as the *Lagrangian* of the problem has a *stationary* point. We have thus converted a constrained optimization problem to an unconstrained one. The term 'λ' is called a *Lagrange multiplier*. Notice, that the problem of identifying the optimum reduces now to solving the three equations

$$\frac{\partial L}{\partial x} \equiv \frac{\partial f}{\partial x} - \lambda \frac{\partial g}{\partial x} = 0$$

$$\frac{\partial L}{\partial y} \equiv \frac{\partial f}{\partial y} - \lambda \frac{\partial g}{\partial y} = 0$$

$$\frac{\partial L}{\partial \lambda} \equiv g(x, y) = 0$$

for the three unknowns. This may, in principle at least, be carried out and might not require the calculation of $h(x)$. The beauty of this method lies in the fact that it easily generalizes to problems involving *more* variables and *more* constraints. For instance, to solve the problem

> maximize $f(x, y, z)$
>
> subject to $g_1(x, y, z) = 0$
> and $g_2(x, y, z) = 0$

we introduce *two* Lagrange multipliers, one for each constraint, say λ_1 and λ_2, and look for stationary points of the Lagrangian

$$L(x, y, z, \lambda_1, \lambda_2) = f(x, y, z) - \lambda_1 g_1(x, y, z) - \lambda_2 g_2(x, y, z).$$

Geometric interpretation

The stationarity of the Lagrangian may also be written in the form

$$\nabla f = \lambda \nabla g.$$

It is instructive to interpret this condition geometrically.

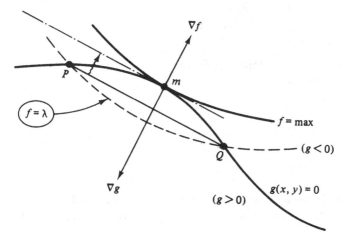

In general the contour $f(x, y) = \mu$, when μ is near the optimal value, intersects the contour $g(x, y) = 0$ in two points, say P and Q. As λ approaches the optimal value the chord PQ approaches the position of the tangent to the g-contour at M. Thus the normal to PQ becomes parallel, in the limit, to $\nabla g(\hat{x}, \hat{y})$. But PQ is also a chord to the contour $f(x, y) = \mu$ and, surely, therefore tends to the tangent at M of the contour $f(x, y) = \max$. Hence the normal to PQ also becomes parallel, in the limit, to $\nabla f(\hat{x}, \hat{y})$. Consequently, as both gradient vectors are parallel, we have

$$\nabla f = \lambda \nabla g.$$

The general case

We wish to maximise $f(x_1, \ldots, x_n)$ subject to the m constraints

$$\left.\begin{aligned}
g_1(x_1, \ldots, x_n) &= 0, \\
g_2(x_1, \ldots, x_n) &= 0, \\
&\cdots \\
g_m(x_1, \ldots, x_n) &= 0.
\end{aligned}\right\} \tag{6}$$

We shall suppose that the optimal point is $\hat{x} = (\hat{x}_1, \ldots, \hat{x}_n)$. Following the special case, we attempt to solve the m simultaneous equations explicity in such a way that some of the variables are expressed as functions perhaps of the first few, say of $x_1, \ldots, x_k$. What should k be? Leaving this question aside, let us formulate our objective properly. Apart from k, we also seek functions $h_1(x_1, \ldots, x_k), \ldots, h_m(x_1, \ldots, x_k)$ so that

$$\left.\begin{aligned}
g_1(x_1, \ldots, x_k, h_1(x_1, \ldots, x_k)) &= 0, \\
g_2(x_1, \ldots, x_k, h_2(x_1, \ldots, x_k)) &= 0,
\end{aligned}\right\} \tag{7}$$

$$\cdots$$

Let us think about the constraint set Ω which is given by the simultaneous equations

$$\left.\begin{aligned}
g_1(x_1, \ldots, x_n) &= 0, \\
g_2(x_1, \ldots, x_n) &= 0, \\
&\cdots \\
g_m(x_1, \ldots, x_n) &= 0.
\end{aligned}\right\}$$

We have m non-linear equations in n variables, so it is natural to regard Ω as the intersection of m many n-dimensional 'hypersurfaces' (non-linear, or, distorted versions of hyperplanes), since their linear approximations near $x = \hat{x}$ will be of the form:

$$\frac{\partial g_1}{\partial x_1}\cdot(x_1 - \hat{x}_1) + \frac{\partial g_1}{\partial x_2}\cdot(x - \hat{x}_2) + \cdots + \frac{\partial g_1}{\partial x_n}\cdot(x - \hat{x}_n) = 0,$$

etc. When $n = 3$ and $m = 2$ this is clear to visualize. Ω is the intersection of two surfaces and so is, presumably, a (one-dimensional) curve in

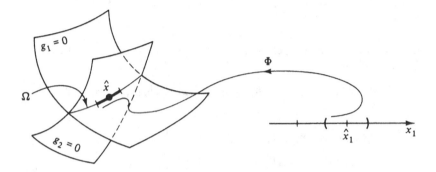

$\mathbb{R}^3$. In the general case, where $m < n$, one expects Ω to be an $n - m$ dimensional 'hypersurface' passing through the optimal point $\hat{x}$. This is indeed true locally at $\hat{x}$ provided the gradient vectors

$$\nabla g_1(\hat{x}), \nabla g_2(\hat{x}), \ldots, \nabla g_m(\hat{x}) \tag{8}$$

are linearly independent, but a proof of this fact belongs to more advanced courses in calculus. More is true. The 'implicit function theorem', which gives conditions for the solubility of an implicitly given relationship $g(\mathbf{u}, \mathbf{v}) = 0$ between $\mathbf{v}$ and $\mathbf{u}$, tells us that near $\hat{x}$ we can indeed solve (6) and, in fact, also 'parametrize' a piece of Ω. By 'parametrize' we mean here that there exists a differentiable function Φ defined on some open ball in $\mathbb{R}^{n-m}$ with values taken on Ω which include the point $\hat{x}$. We make this a little more precise. The assumption of linear independence made on the gradient vectors (8) implies that the Jacobian matrix of the vector-valued function

$$\mathbf{y} = g(\mathbf{x}) = (g_1(x_1, \ldots, x_n), \ldots, g_m(x_1, \ldots, x_n))^t$$

etc. namely

$$\frac{d\mathbf{y}}{d\mathbf{x}} = \begin{bmatrix} \dfrac{\partial g_1}{\partial x_1}, \ldots, \dfrac{\partial g_1}{\partial x_n} \\ \cdots \\ \dfrac{\partial g_m}{\partial x_1}, \ldots, \dfrac{\partial g_m}{\partial x_n} \end{bmatrix}$$

has rank m. Re-labelling $x_1, \ldots, x_n$, if necessary, we may suppose that the last m columns of this Jacobian are linearly independent. The implicit function theorem then says that taking $\mathbf{u} = (x_1, \ldots, x_{n-m})$ and $\mathbf{v} = (\mathbf{x}_{n-m+1}, \ldots, x_n)$ there is a function $\mathbf{v} = h(\mathbf{u}) = h(x_1, \ldots, x_{n-m})$ defined on some small open ball centred at $\hat{\mathbf{u}} = (\hat{x}_1, \ldots, \hat{x}_{n-m})$ with values in $\mathbb{R}^m$ so that

$$\mathbf{v} = h(x_1, \ldots, x_{n-m}) = (h_1(x_1, \ldots, x_{n-m}), \ldots, h_m(x_1, \ldots, x_{n-m}))^t$$

and

$$\left.\begin{aligned} g_1(x_1, x_2, \ldots, x_{n-m}, h(x_1, \ldots, x_{n-m})) &= 0 \\ g_2(x_1, x_2, \ldots, x_{n-m}, h(x_1, \ldots, x_{n-m})) &= 0 \\ \cdots \\ g_m(x_1, x_2, \ldots, x_{n-m}, h(x_1, \ldots, x_{n-m})) &= 0 \end{aligned}\right\} \tag{9}$$

i.e. $g(x_1, \ldots, x_{n-m}, h(x_1, \ldots, x_{n-m})) \equiv 0$. Compare this with the two variable cases above.

We have thus solved the m equations in terms of the variables $x_1, \ldots, x_{n-m}$. Define Φ by

$$\Phi(\mathbf{u}) = (\mathbf{u}, h(\mathbf{u}))^t \tag{10}$$

i.e.

$$\begin{aligned} \Phi(x_1, \ldots, x_{n-m}) \\ = (x_1, \ldots, x_{n-m}, h_1(x_1, \ldots, x_{n-m}), \ldots, h_m(x_1, \ldots, x_{n-m}))^t. \end{aligned}$$

Then Φ transforms a small ball of $\mathbb{R}^{n-m}$ into a piece of Ω. We thus know that $(\hat{x}_1, \ldots, \hat{x}_{n-m})$ is where $f(\Phi(x_1, \ldots, x_{n-m}))$ is maximized, but now $f\Phi$ is maximized while $u = (x_1, \ldots, x_{n-m})$ is unconstrained near $\hat{\mathbf{u}} = (\hat{x}_1, \ldots, \hat{x}_{n-m})$. Let us make some calculations. First of all, since the maximum is unconstrained and is at $(\hat{x}_1, \ldots, \hat{x}_{n-m})$ we have that, at $(\hat{x}_1, \ldots, \hat{x}_{n-m})$:

$$Df\Phi = 0 \quad \text{i.e.} \quad \frac{d}{d\mathbf{u}}\{f(\Phi(x_1, \ldots, x_{n-m}))\} = 0,$$

or, by the chain rule,

$$\frac{df}{d\mathbf{x}}\cdot\frac{d\Phi}{d\mathbf{u}}=0.$$

Taking transposes, we have

$$A\cdot\nabla f=0,$$

where A is the matrix satisfying

$$A^t=\frac{d\Phi}{d(x_1,\dots,x_{n-m})}.$$

Thus $\nabla f\in\mathcal{N}(A)$. Note that A^T is $n\times(n-m)$, so A is $(n-m)\times n$.

Secondly, for $i=1,2,\dots,m$, we have by (6) near $(\hat{x}_1,\dots,\hat{x}_{n-m})$ the identity:

$$g_i(\Phi(\mathbf{u}))=g_i(\Phi(x_1,\dots,x_{n-m}))\equiv 0.$$

Again, by the chain rule, we may differentiate this identity near $(\hat{x}_1,\dots,\hat{x}_{n-m})$ to obtain

$$\frac{dg_i}{d\mathbf{x}}\cdot\frac{d\Phi}{d\mathbf{u}}=0.$$

Once again taking transposes tells us that, in particular, at $(\hat{x}_1,\dots,\hat{x}_{n-m})$:

$$A\cdot\nabla g_i=0.\quad(i=1,2,\dots,m.)$$

Thus $\nabla g_1(\hat{x})_1,\dots,\nabla g_k(\hat{x})\in\mathcal{N}(A)$. We wish to show that at $\hat{x}$

$$\nabla f=\lambda_1\nabla g_1+\cdots+\lambda_m\nabla g_m.$$

This will follow if we can show that the m linearly independent gradients (8) span $\mathcal{N}(A)$, i.e. we wish to know if the nullity of A is m. But,

$$\text{rank}(A)+\text{nullity}(A)=n.$$

Also, differentiating (10), we have

$$\frac{d\Phi}{d(x_1,\dots,x_{n-m})}=\begin{bmatrix}1,0,\dots,0,\dfrac{\partial h_1}{\partial x_1},\dfrac{\partial h_2}{\partial x_1},\dots\\0,1,\dots,0,\dfrac{\partial h_1}{\partial x_2},\dots\\\dots\dots\dots\dots\dots\\0,0,\dots,1,\dfrac{\partial h_1}{\partial x_{n-m}},\dots\end{bmatrix}^t.$$

So the rank of this matrix and, hence also of A, is $n - m$ (on account of the identity sub-matrix). But now we have

$$(n - m) + \text{nullity}(A) = n,$$

i.e.

$$\text{nullity}(A) = m,$$

as required.

Fig. 1. Beginning the construction.

Fig. 2. The construction must terminate else a contradiction arises from $a' = \lim a_n$.

Fig. 3

Fig. 1

Fig. 2. Ω is the curve of intersection of the two surfaces $g_1 = 0$ and $g_2 = 0$. On a piece of Ω we can solve the two equations simultaneously $g_1(x_1, h(x_1)) = 0 = g_2(x_1, h(x_1))$. Then $(x_1, x_2, x_3) = \Phi(x_1) = (x_1, h_1(x_1), h_2(x_2))$ parametrizes a piece of Ω.

Index

Abel's integral equation 404
absolute convergence 282
admissible variation 426
affine
  combination 116
  functions 327
  set 17
Alice in Wonderland 223
alternating sign test 279
anti-derivative 207
area 207, 224
artificial variable 173
auditor's problem 148

basic variable 167
basis 5, 33
basis change 33
Bessel's equation 408, 420
beta function 398
block diagonal form 80
bordered matrix 82
brachistochrone problem 423, 432

Canonical form
  Jordan 62
  similarity 61
Cauchy principal value 268, 421
Cauchy–Schwartz inequality 24
chain rule 179
change of variable 238, 308, 349
characteristic curves 357, 359
characteristic polynomial 59
chordal inequality 185
closed set xii
co-factor 102
co-ordinates 6
column cone 141
combination
  affine 116
  convex 115
  linear 4, 115

common roots 373
comparison tests 269, 271, 276
concave function 184
concave programming 200
condensation test 278
conditional convergence 282
cone 118
  column-cone 141
connected 313
constraint
  active 197
  qualification 197
continuity
  definition 212
  criteria 217
  joint 259
  recognition 260
  separate 260
contour plotting 323
convergence
  absolute 281
  conditional 282
  dominated 287
  uniform 287
convex
  combination 115
  function 184
  hull 115
  set 113
convolution integral 342
convolution theorem 391
cover-up rule 386
Cramer rule 102
curve length 243
cycling in the simplex tableau 172
cycloid 433
cylinder 131

decomposition
  singular value 76
  spectral 67

defective matrix 62
density function 340
derivative 180
differential equation
  common roots 373
  general 374
  homogeneous 362
  simultaneous 89
dimension 6, 14, 116
direct sum 48
directional derivative 178, 187
Dirichlet's theorem 413, 416
dissection 305, 307, 335
distribution function 400
dual program 143
dual system – inequality 138
duality 137
  result 41
  theorem 145

economics 147
eigenvalue 59, 96
eigenvector 59
elementary matrix 37
entrepreneur 147
epigraph 184
equation – indicial 407
equilibrium point 152
error function 209
Euler's constant 385
Euler's equation 183
Euler–Lagrange equation 428, 431
even function 415, 299
exactness 352
expected reward 153
exponential growth 380
extremal ray 133
extreme point 130

Farkas lemma 138, 140
feasible point 198
feasible set 143
final value theorem 389
finer partition 227
Form
  positive definite 71
  quadratic 69
  upper triangular 80
Fourier series 410
Fubini's theorem 305
function
  affine 327
  beta 398
  concave 184
  continuous 212, 260, 259
  convex 184

distribution 400
  error 209
  exponential growth 380
  gamma 398
  homogeneous 183
  jump 225, 230
  linear 30
  objective 144
  one-to-one 313, xii
  probability density 340
  staircase 236
  transfer 396
  weighting 225
  of bounded variation 241, 245, 246
functional 425

game theory 151 ff
gamma function 398
Gauss' formula 405
Gauss–Jordan elimination 169
generalised inverse
  strong 106
  weak 104
geodesic problem 423, 436
Gram–Schmidt process 27
grid 321

half-space 122
Hermitian matrix 62, 63
Hölder's inequality 193
homogeneous equation 361
homogeneous differential equation 350
homogeneous function 183
Householder's method 92
hull
  affine hull 116
  convex hull 115
hyperplane 18, 22, 122

idempotent 50
improper integral
  multiple integral 333, 334
  single variable 266
impulse response 396
indicial equation 407
inequalities 135
inequality
  Cauchy-Schwarz 24
  chordal 185
  Hölder 193
  Jensen 192, 193
  Minkowski 193
  primal system 138
  solution set 136
infimum 127, 211
initial value theorem 389

inner product 21
input–output systems 395
integrable xiii, 302
integral
    multiple vs repeated 300
    test 277
integrating factor 348, 353
integrating by parts 239
integrator 227
inverse
    generalised strong 106
    generalised weak 104
    left 103
    local 319
    right 103
isoperimetric problem 423, 438

Jacobian 310
Jensen's inequality 192, 193
Joint continuity 258
Jordan canonical form 62

kernel 8, 30
Kuhn–Tucker conditions 199

l'Hôpital's rule 275
Lagrange multipliers 201, Appendix B
Lagrangian 150
Laplace transform 380
    table 386
least squares 53
line 18, 113
line segment 113
linear combination 4, 115
linear dependence 4
linear equations – inconsistent 105
linear part 428
linear transformation 30
local inverse 319

manipulations – finite range 261
matrix
    bordered 82
    defective 62
    elementary 37
    hermitian 62, 63
    non-square 75
    normal 63
    orthogonal 62
    reflection 92
    skew-symmetric 63
    symmetric 63, 92 ff
    tridiagonal 71 ff
    unitary 63
matrix game 152

mean value theorem
    differential calculus 247
    first 249
    second 250
method
    characteristic curves 357, 359
    small parameter 418
minimax theorem 152
Minkowski functional 193
Minkowski inequality 193
mixed strategy 152
moment 400
moment (of k'th order) 234
Mr Tomkins in Wonderland 224
multiple integral 300

non-basic variable 167
non-linear programming 195
non-negative definite 71
non-square matrix 75
norm 20
normal
    cone 130, 196
    matrix 63
null space 30
nullity 9
numerical integration 217

objective function 144
one-to-one 313
open set xii
optimal point 198
optimal solution 144
optimum point 195
orthant 136
orthogonal complement 39
orthogonal matrix 62
orthogonal projection 51, 68
orthogonality 25
orthogonality relations 414
orthonormal set 63
orthonormality 26

parameter 263
parameter range infinite 291
partial differential equations 355
partial fractions 387
particular solution 10, 361
partition 211, 223, 245
    refinement 227
payoff function 151
periodic solutions 416
pivot 169
planes 18
point evaluation 230
polytope 117

positive definite 71
power series 289
primal program 143
primal system – inequality 138
principal axes 71
principal minors 73
probability distribution 340
process
  renewal 397
  stochastic 396
production vector 147
program
  dual 143
  primal 143
programming linear 162
  non-linear 195
projection 48 ff, 105 ff
pseudo-convex 194

quadratic form 69, 71
quantum jump 225
quasi-convex 193

radius of convergence 290
random variables 339
range 8
rank of AB 42
ray 118
reflection matrix 92
renewal process 397
repeated integral 303
resource vector 147
revenue 147
Riemann integral 212
Riemann–Stieltjes integral 223, 229
rule
  cover-up 388
  Cramer's 102
  L'Hôpital's 275
  Simpson's 219
  Trapezoidal 217

saddle point 150, 152
sample
  distribution function 233
  mean 233
  standard deviation 234
scalars 2
separation 122
separation of variables 347
series
  Fourier 410
  power 289
set
  affine 17
  closed xii

connected 313
convex 113
open xii
solution-set 136
Spanning set 5
strategy set 151
shadow price 148, 201
shift operator 368
shift property 386
similarity 61
simplex 117
simplex tableau 168
Simpson's rule 219
simultaneous differential equations 89
simultaneous equations 387
singular values decomposition 76
skew-symmetric matrix 63
slack variables 167
  generalised 167
small parameter method 418
smooth 235
solubility of equations 45
solution set 10
  inequalities 136
span 5
spectral decomposition 67, 75
spectrum 60
spherical polar co-ordinates 329
staircase function 236
statistics 339, 400
stochastic process 396 – Miscellany
  Question 30
strategy 152
strategy set 151
Sturm sequence 96
subdeterminants 43
sum of squares 250
sum theorem 365
supporting hyperplane 129, 187
supremum 127, 211
symmetric matrix 63, 92 ff

tangential hyperplane 129
tearing 103
test
  alternating sign 279
  comparison 269, 271, 276
  condensation 278
  integral 277
theorem
  Carathedory – Miscellany, see Question
    33
  Fubini 305
  Helly 121
  Kuhn–Tucker 199
  mean value 247

mean value – first 249
mean value – second 250
Radon 121
separating hyperplane 122
Shapley–Folkman–Miscellany
   Question
sum 365
Von-Neumann 152
trace 60
transfer function 396
trapezoidal rule 217
triangle inequality 25, 193
tridiagonal matrix 92 ff
triple integral 307

Unconditional convergence 281

uniform convergence 287
unitary matrix 63
upper triangular form 80

value of a game 152
Van Dantzig 162
Van der Monde determinant 14
variation 425
vector space 2
Von-Neumann's theorem 152

Weighting function 225
Winnie the Pooh 223
Wronskian 11

zero-sum game 151